Lecture Notes of the Institute
for Computer Sciences, Social-Informatics
and Telecommunications Engineering 19

Yan Chen   Tassos D. Dimitriou
Jianying Zhou (Eds.)

Security and Privacy in Communication Networks

5th International ICST Conference
SecureComm 2009
Athens, Greece, September 14-18, 2009
Revised Selected Papers

 Springer

Volume Editors

Yan Chen
Northwestern University, Department of Electrical Engineering
and Computer Science, Robert R. McCormick School
of Engineering and Application Science, 2145 Sheridian Road
Evanston, IL 60208-3118, USA
E-mail: ychen@northwestern.edu

Tassos D. Dimitriou
Athens Information Technology
Markopoulo Ave.
GR-19002, Peania, Greece
E-mail: tdim@ait.edu.gr

Jianying Zhou
Institute for Infocomm Research
1 Fusionopolis Way, 21-01
Connexis, South Tower, 138632 Singapore
E-mail: jyzhou@i2r.a-star.edu.sg

Library of Congress Control Number: 2009938038

CR Subject Classification (1998): C.2, D.4.6, K.6.5, K.4.4, J.1

ISSN   1867-8211
ISBN-10   3-642-05283-5 Springer Berlin Heidelberg New York
ISBN-13   978-3-642-05283-5 Springer Berlin Heidelberg New York

springer.com

© ICST Institute for Computer Science, Social Informatics and Telecommunications Engineering 2009
Printed in Germany

Typesetting: Camera-ready by author, data conversion by Scientific Publishing Services, Chennai, India
Printed on acid-free paper   SPIN: 12773008   06/3180   5 4 3 2 1 0

Preface

SecureComm 2009, the 5th International ICST Conference on Security and Privacy in Communication Networks, was held in Athens, Greece, September 14–17, 2009. SecureComm was organized by the Institute for Computer Sciences, Social-Informatics and Telecommunications Engineering (ICST). The General Chair of the conference was Peng Liu from the Pennsylvania State University, USA.

The conference received 76 submissions (one withdrawn) and each submission received at least three reviews, with some papers receiving as many as five reviews. The first phase of the review process (April 7 to May 27) generated 290 reviews overall or about 3.8 reviews per submission. In the second phase (May 28 to June 12), a two-week online discussion was initiated ensuring consensus for each accepted paper. The review process was challenging and we are deeply grateful to the committee members and the external reviewers for their outstanding work. The Program Committee consisted of 64 academics and professionals well known in their corresponding area of expertise.

After meticulous deliberation, the Program Committee, which was chaired by Yan Chen from Northwestern University, USA and Tassos Dimitriou from Athens Information Technology, Greece, selected 19 full papers and 7 short ones for presentation in the academic track and inclusion in this volume. This gives the acceptance rates of 25.3% for the full papers and 34.7% for all papers.

The program also included two invited talks in addition to the academic and industrial tracks. The invited talks were given by Sal Stolfo from Columbia University on "Polymorphic Shellcode: The Demise of Signature-based Detection," and Bart Preneel from Katholieke Universiteit Leuven, Belgium on "Upgrading Cryptographic Algorithms for Network Security." We would like to genuinely thank them for accepting our invitation and for contributing to the success of SecureComm 2009.

Finally, we would like to thank the technical sponsors CreateNet, ICST, and the Institute of Informatics and Telecommunications (IIT) of the Greek National Centre for Scientific Research "Demokritos" for their support. We would like to thank all the people involved in the organization of this conference. In particular, we would like to thank the Publication Chair Jianying Zhou, the Conference Coordinators Gergely Nagy and Eszter Hajdu, the Website Coordinator Kun Bai, and the Steering Committee members, namely, Imrich Chlamtac and Krishna Sivalingam.

September 2009

Yan Chen
Tassos Dimitriou
Peng Liu

Preface

SecureComm 2009

5th International Conference on
Security and Privacy in Communication Networks

Athens, Greece
September 14–18, 2009

Organized and Sponsored by

Institute for Computer Sciences, Social-Informatics
and Telecommunications Engineering (ICST)

General Chair

Peng Liu · Penn State University, USA

Technical Program Chairs

Yan Chen · Northwestern University, USA
Tassos Dimitriou · Athens Information Technology, Greece

Publicity Chair

Morley Mao · University of Michigan, USA

Publication Chair

Jianying Zhou · Institute for Infocomm Research, Singapore

Sponsorship Chair

Effie Makri · Institute of Informatics and
Telecommunications, Greece

Workshop Chair

Reza Curtmola · NJIT, USA

Conference Coordinator

Gergely Nagy   ICST

Website Coordinator

Kun Bai   Penn State University, USA

Technical Program Committee

Ehab Al-Shaer	DePaul University
Feng Bao	Institute for Infocomm Research
Paul Barford	University of Wisconsin-Madison
Nikita Borisov	University of Illinois at Urbana-Champaign
Hao Chen	University of California, Davis
Shuo Chen	Microsoft Research
Songqing Chen	George Mason University
Yingying Chen	Stevens Institute of Technology
Mauro Conti	University of Rome
Bruno Crispo	University of Trento
Reza Cutmola	New Jersey Institute of Technology
George Danezis	Microsoft Research
Sven Dietrich	Stevens Institute of Technology
Xuhua Ding	Singapore Management University
Yingfei Dong	University of Hawaii
Roberto DiPietro	University of Rome
Cristian Estan	University of Wisconsin-Madison
Felix C. Freiling	University of Mannheim
David Galindo	University of Luxembourg
Guofei Gu	Texas A&M University
Yong Guan	Iowa State University
Peter Gutmann	University of Auckland
Markus Jakobsson	Palo Alto Research Center
Brent Hoon Kang	University of North Carolina at Charlotte
Nikos Komninos	Athens Information Technology
Ioannis Krontiris	University of Mannheim
Brian LaMacchia	Microsoft
Loukas Lazos	University of Arizona
Javier Lopez	University of Malaga
Zhichun Li	Northwestern University
Donggang Liu	University of Texas at Arlington
Peng Liu	Penn State University
Kostas Markantonakis	University of London
Jelena Mirkovic	USC Information Sciences Institute

John Mitchell	Stanford University
David Molnar	University of California, Berkeley
Panos Papadimitratos	EPFL
Kenny Paterson	Royal Holloway, University of London
Adrian Perrig	Carnegie Mellon University
Radha Poovendran	University of Washington
Neeli R. Prasad	Center for TeleInFrastruktur
Kui Ren	Illinois Institute of Technology
Pierangela Samarati	University degli Studi di Milano
Sanjeev Setia	George Mason University
Jessica Staddon	Palo Alto Research Center
Yannis Stamatiou	University of Ioannina
Angelos Stavrou	George Mason University
Paul Syverson	Naval Research Laboratory
Patrick Traynor	Georgia Tech
Haining Wang	College of William and Mary
XiaoFeng Wang	Indiana University
Dirk Westhoff	NEC Europe
Avishai Wool	Tel Aviv University
Felix Wu	University of California, Davis
Yinglian Xie	Microsoft Research
Dongyan Xu	Purdue University
Yanjiang Yang	Institute for Infocomm Research
Vinod Yegneswaran	SRI International
Yanchao Zhang	New Jersey Institute of Technology
Ben Y. Zhao	University of California, Santa Barbara
Jianying Zhou	Institute for Infocomm Research
Bo Zhu	Concordia University
Sencun Zhu	Penn State University
Cliff Zou	University of Central Florida

Steering Committee

Imrich Chlamtac (Chair)	Create-Net, Italy
Krishna Sivalingam (Co-chair)	University of Maryland Baltimore County, USA
Andreas Schmid	Novalyst IT, Germany
Peng Liu	Penn State University, USA

Table of Contents

Wireless Network Security II, Sensor Networks

Key Management, Credentials, Authentications

Wireless Network Security III

Secure Multicast, Emerging Technologies

Mitigating DoS Attacks on the Paging Channel by Efficient Encoding in Page Messages

Liang Cai[1], Gabriel Maganis[1], Hui Zang[2], and Hao Chen[1]

[1] Computer Science Department, University of California, Davis
{lngcai,gymaganis}@ucdavis.edu,
hchen@cs.ucdavis.edu
[2] Sprint Advanced Technology Labs
hui.zang@sprint.com

Abstract. Paging is an important mechanism for network bandwidth efficiency and mobile terminal battery life. It has been widely adopted by mobile networks, such as cellular networks, WiMax, and Mobile IP. Due to certain mechanisms for achieving paging efficiency and the convergence of wireless voice and data networks, the paging channel is vulnerable to inexpensive DoS attacks. To mitigate these attacks, we propose to leverage the knowledge of the user population size, the slotted nature of the paging operation, and the quick paging mechanism to reduce the length of terminal identifiers. In the case of a CDMA2000 system, we can reduce each identifier from 34 bits down to 7 bits, effectively doubling the paging channel capacity. Moreover, our scheme incurs no paging latency, missed pages, or false pages. Using a simulator and data collected from a commercial cellular network, we demonstrate that our scheme doubles the cost for DoS attackers.

Keywords: Paging, DoS Attacks, General Page Message, Quick Paging.

1 Introduction

The biggest advantage of mobile networks over wired networks is mobility, which allows users to access the network from different locations. Mobile networks achieve mobility through *Macro-mobility management* and *Micro-mobility management*. The former ensures that mobile terminals (e.g., cell phones and laptops with wireless cards) are addressable when they roam between different domains, while the latter manages the mobile terminal's movement between access points or base stations within the same domain. While implementations vary across mobile networks, macro-mobility management always requires roaming users to notify the network each time they arrive at a new domain. By contrast, a similar scheme, which would require terminals to update their locations every time they move to a new access point or base station, is impractical for micro-mobility management for several reasons. First, location updates would be much more frequent than in macro-mobility management, which would consume significant

Y. Chen et al. (Eds.): SecureComm 2009, LNICST 19, pp. 1–20, 2009.

wireless bandwidth and mobile terminal power. Second, to trace their own locations, terminals must continuously monitor the beacon or pilot channel of base stations, which would drain their batteries even faster.

Paging is a critical mechanism to improve the bandwidth and terminal energy efficiency in micro-mobility management. The designer divides the network into *paging areas* and requires terminals to notify the network about their location only when they enter a new paging area. Each paging area is usually large enough so that location updates are infrequent even for highly mobile terminals. Meanwhile, terminals monitor the network at longer intervals and enter the idle mode in between. When an incoming call arrives, the network controller broadcasts a page message to the entire paging area. If the terminal is located in the paging area, it responds to acquire a traffic channel. As an efficient location management scheme, paging has been widely adopted in mobile networks, including cellular communication systems (GSM[1], W-CDMA and CDMA2000[2]), WiMax[3] and Mobile IP systems[4].

There are a pair of low-bandwidth channels in a cellular network used for location management. The downlink channel, often referred to as the *paging channel*, is used for paging while the uplink channel (the *access channel*) is used for location updates. To lower the bandwidth requirement on the access channel, we desire larger paging areas; however, larger paging areas would increase the load on the paging channel since all the users in the same paging area share the same paging channel. Concentrated flash crowds could even lead to temporary paging channel overload or saturation. This tradeoff between bandwidth requirement and paging area size is known as the *paging efficiency problem*.

The recent convergence of wireless voice and data networks exacerbates this problem. Besides incoming voice calls, incoming short messages (SMSs) and data packets may also increase the load on the paging channel. This provides attackers with an opportunity to launch a DoS attack on wireless networks from the Internet, possibly with very low cost. For example, Serror et al. showed how to saturate the paging channel of a cellular network by sending data packets from the Internet at a very low cost [5], and Enck et al. showed how to disrupt a cellular network in a major city by sending SMS messages of a sufficient rate [6].

If we improve paging efficiency, we could not only accommodate flash crowds but also mitigate DoS attacks. Previous approaches focused on reducing the number of paging requests (e.g., by predicting terminals' locations [7]). In this paper, we take a different approach by increasing the number of paging requests that the paging channel can carry. A page message contains the identifiers of all the paged mobile terminals. Our key insight is that the shorter the lengths of these identifiers are, the more terminals a single page message can page. Towards this goal, we propose a series of methods for shortening terminal IDs by leveraging the knowledge of the population size in a paging area, by grouping terminals based on paging channel slots, and by using special Bloom filters in quick paging. When applying these methods to a CDMA2000 system, we are able to reduce the length of each terminal ID from 34 bits to 7 bits, which we shall show doubles the number of terminals that one page message can contain.

Since our scheme only shortens terminal IDs, it has no adverse effect on paging performance (e.g., paging latency, missed page rate, false paging) and requires little change to the paging protocols.

The rest of the paper is organized as follows: We describe the paging channel operation and page message format in Section 2, and show the importance of improving paging efficiency. We present the optimization schemes for reducing the terminal ID length in Section 3. We then evaluate the scheme using an experiment on a real cellular system and a simulation tool and illustrate the results in Section 4. After comparing our scheme with several related works in Section 5, we conclude in Section 6.

2   Paging Channel Operation

In this section, we first describe the paging channel operation and the page message format in the context of a cellular network using CDMA2000 technology. To show that our scheme is not limited to cellular systems, we explore the page operations in other mobile systems and discuss their differences from cellular systems.

2.1   Paging Channel Operation

A mobile network needs to track the location of the mobile terminals so that it can deliver data and voice calls to their intended recipients. A simple solution would be to require mobile terminals to report their locations through *location updates* whenever their locations change. However, since mobile devices are typically resource (e.g., battery power) constrained, requiring them to remain in the "active state" just to report their locations would be inefficient. Thus, mobile operators typically divide their networks into *location areas*, and mobile terminals report their locations only when they enter a new location area. When a new call or data packet arrives, the network *pages* the recipient mobile terminal in the location area. Therefore, *paging* and location updates are key components in mobility management in a mobile network.

CDMA2000 networks have a dedicated channel, the *paging channel*, that delivers page messages to mobile terminals[1]. Mobile terminals monitor this channel through a Time Division Multiple Access (TDMA) scheme. The network divides a paging cycle (either 2.56 or 5.12 seconds) into *slots* (either 32 or 64 slots, respectively). Thus, a mobile terminal stays in the *idle mode*, when the power consumption is minimal, most of the time and wakes up only during its assigned slot (whose duration is 80 ms) to determine whether it has been paged. The network assigns a mobile terminal to a slot based on the terminal's International Mobile Station Identifier (IMSI). Figure 1 illustrates the structure of 32-slot paging channel.

When an incoming call (or downlink data packet) arrives, the mobile switching center (MSC) broadcasts a General Page Message (GPM) to the location area

---

[1] A CDMA system can be configured to have at most seven paging channels.

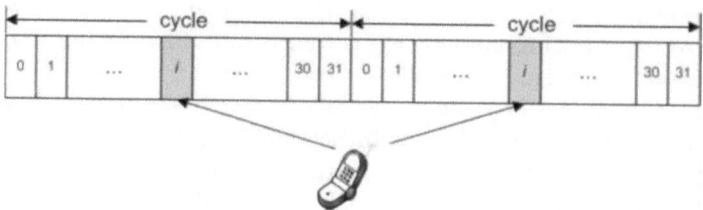

Fig. 1. An example structure of a paging channel [5]. A mobile terminal is in idle mode except during the ith slot, when it wakes up and monitors the paging channel.

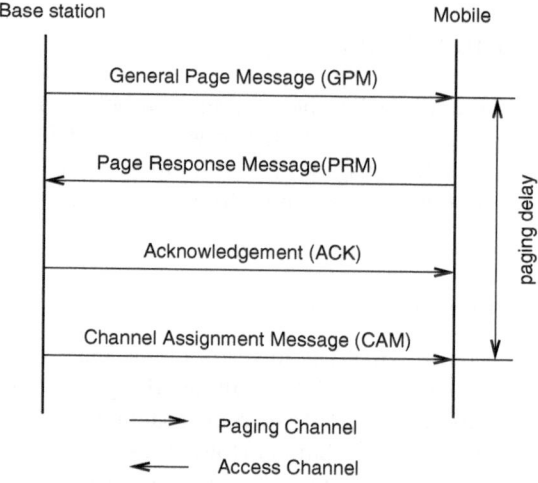

Fig. 2. Messages exchanged between a base station and a mobile terminal during a paging operation

(also known as the *paging area*) during the recipient mobile terminal's assigned slot. When the mobile terminal finds that it has been paged, it will respond through the associated base station with a Page Response Message (PRM). Then, the MSC sends an acknowledgment message (ACK) and a Channel Assignment Message (CAM) to the mobile terminal. The mobile terminal responds to the CAM with an ACK to establish a connection with the base station over the assigned *traffic channel*. Both the ACKs and the CAMs are sometime referred as *non-slotted* messages, since they can be sent during any slot (After sending the PRM, the terminal leaves the slotted mode and monitors all the slots). On the other hand, GPMs are *slotted* messages, which must be sent during the recipient mobile terminal's assigned slot. Figure 2 shows the messages exchanged between the base station and a mobile terminal during the paging process when an incoming call arrives. After this step, the terminal communicates with the base station on the assigned traffic channel only.

The number of slotted messages in the paging channel greatly exceeds that of non-slotted messages. When the MSC sends a slotted message, it does not know the location of the recipient, so it has to broadcast the page message to all the base stations in the paging area. By contrast, it needs to send the subsequent non-slotted messages to only the base station as determined from the PRM. Besides slotted messages and non-slotted messages, the paging channel is also used for transmitting system parameters, which are sometimes regarded as overhead traffic in the paging channel. These messages occupy the paging channel based on the paging load. The more GPMs there are, the fewer overhead messages will be sent. Typically, overhead traffic takes at least 25% of the capacity.

2.2   Paging Message Format

In the CDMA2000 standard, the specification for the GPM format is highly flexible, so GPMs in different systems may vary in length and pattern. Figure 3 shows an actual GPM that we captured from a mobile device in a live commercial CDMA2000 network. We can see that a GPM is composed of a header, several page records, padding, and the CRC.

Page records comprise the main body of a page message. Among all the fields in a page record, the terminal identifier field plays a critical role in the paging

Field	value	length(bit)
General Paging Message		
Message Header		
MSG_LENGTH	0000xxxx	8
MSG_ID	010001	6
CONFIG_MSG_SEQ	000011	6
ACC_MSG_SEQ	011101	6
CLASS_0_DONE	1	1
CLASS_1_DONE	1	1
TMSI_DONE	1	1
ORDERED_TMSIS	0	1
BROADCAST_DONE	1	1
RESERVED	0000	4
ADD_LENGTH	000	3
Mobile Station 1		
PAGE_CLASS	00	2
PAGE_SUBCLASS	00	2
MSG_SEQ	100	3
IMSI_S	(xxx) xxx-xxxx	34
SDU_INCLUDED	1	1
SERVICE_OPTION	xx	16
Mobile Station 2		
...		
Message end Padding		
PDU_PADDING	0000	4
CRC		30

Fig. 3. The format of an actual General Paging Message. A 34-bit IMSI_S is used as the terminal identifier, and the length of each page record is 58 bits, while the header and the tail are 38 and 30 bits long, respectively.

operation. This field may use many identifier types, indicated by the *PAGE_CLASS* and *PAGE_SUBCLASS* fields, listed in the CDMA2000 specification [8]. In our example, the terminal's IMSI_S (a 10-digit number derived from the terminal's International Mobile Subscriber Identity (IMSI)) is used as the identifier. In a GPM, the IMSI_S is encoded into a 34-bit string. Other optional identifier types have similar length (e.g., a TMSI is 32 bits) or longer.

Another important field in page records is the *SERVICE_OPTION*. It informs the terminal of the type of the incoming call. This is important since the message after the GPM could be of different service types. For example, the GPM is always followed by Channel Assignment Message(CAM) in the case of a voice call, but the CAM can be replaced by a Data Burst Message (DBM) in the case of text service.

2.3   Paging Operation in Other Mobile Networks

IEEE802.16 systems such as WiMax [9] or Hpi [10] do not use separated slotted channels in their paging operations. When a WiMax terminal is not engaged in communication, it enters the *idle mode*, which works in four stages: *idle mode initialization, idle mode entry, idle mode operation* and *idle mode exit*. Idle mode can be initiated either by the mobile terminal or the base station. When the mobile terminal initiates the idle mode, it sends a deregistration request message (DREG-REQ); when the base station initiates the idle mode, it sends a DREG-CMD message to the terminal, which responds with a DREG-REQ message. The base station then notifies the paging controller of the terminal's service information. The paging controller decides the PAGING_CYCLE, PAGING_OFFSET, and *Paging Listen Interval* (PLI) parameters and sends them to the terminal in a DREG-CMD message via the base station. Then, the terminal wakes up during the Paging Listen Interval periodically to check for a MOB-PAG-ADV message (the GPM's counterpart in a WiMax system). The message consists of a 48-bit MAC header, several page group IDs, several page records, and padding. The length of each page record is 32 bits. A paged terminal is identified with a hash value of its 24-bit MAC address.

Mobile IP systems also propose a paging operation [11,12,13,4,14]. In most of these schemes, a terminal's home IP address is used as its identifier in the page message, so the identifier is 32 bits in IPv4 systems and 128 bits in IPv6 systems.

2.4   Paging Channel Overload Problem

Recall that the size of a paging area determines how often mobile terminals send location updates. The smaller a paging area is (i.e., containing fewer cells), the more frequently a terminal with high mobility needs to send location updates, which consumes more power and generates more traffic on the access (uplink signaling) channel, which is also a low-bandwidth channel like the paging channel. To avoid this adverse effect, in current cellular networks, a paging area usually consists of hundreds of cells.

Equation 1 calculates the maximum number of terminals that can be paged in each slot per paging area, where we assume that the bandwidth of the paging channel is 9600bps, the duration of a paging slot is 0.08s, the overhead traffic occupies 25% of the channel capacity.[2], the length of the page message header is 38, the length of the CRC value is 30, and the length of each page record is 58.

$$N_{max} = \lfloor \frac{9600 \times 0.08 \times (1 - 0.25) - 38 - 30}{58} \rfloor = 8 \qquad (1)$$

Equation 1 shows that the call arrival rate to a paging area is limited to 100 per second (8 calls / 0.08 second). Given the size of a typical paging area, this maximum call arrival rate is acceptable when only voice calls are paged. However, when the network provides more and more text and data services, this upper bound makes the paging operation an essential bottleneck.

Worse yet, the paging channel has become an ideal target of Denial of Service (DoS) attacks on the cellular network. [5] described such an attack by flooding the network with UDP packets. When a mobile terminal establishes a wireless data connection with the network, it acquires an IP address. The network reserves the address for the terminal until the terminal disconnects from the network, even when it is in the idle mode. The DoS vulnerability lies within the fact that when a data packet arrives at the mobile network, the recipient mobile terminal needs paged. Since it is relatively easy to find the IP subnets assigned to a mobile service provider, an attacker on the Internet can flood these IPs with UDP packets to trigger a flood of page messages within the mobile network. The authors explored the feasibility of this attack by conducting experiments on a live commercial CDMA2000 network. Due to legal and ethical constraints, the goal of the experiments was only to increase the paging channel load by 10%. The authors predicated that the performance of the network would degrade further if they had increased the attack load or if the attacks had been carried out in a busy area.

3   Efficient Encoding in Page Records

To mitigate paging channel overload, we wish that a page record can carry more terminal IDs. However, the length of a page record is determined by its slot duration and the paging channel bandwidth, both of which are constrained by system configurations and physical limitations. Instead, we investigate how to fit more terminal IDs into existing page records.

The CDMA2000 specification, for example, supports different types of terminal IDs [2], but most of them are longer than 30 bits. IMSI_S, one of the most commonly used terminal ID, is 34 bits, and a TMSI is 32 bits. Typically, terminal IDs account for more than half of a page record's size. Therefore, they are a good target for optimization. Moreover, terminals IDs are universal in all paging systems, while other fields in page records are system specific.

For convenience, we describe our scheme for efficiently encoding terminals IDs in the context of a CDMA2000 system, although the principle applies to other

---

[2] 25% is a common overhead load. When the overhead traffic load is less than 25%, we occasionally observe GPMs with 9 records in real paging data.

mobile networks, such as WiMax and Mobile IP. Using a series of techniques, we are able to reduce terminal IDs from 34 bits down to only 7 bits, as described in detail below.

3.1 Approaches

Optimization using Knowledge about Population Size in a Paging Area. One reason why the IMSI_S is long is that it is globally unique. However, the paging operation only needs to differentiate between terminals in the same paging area. Therefore, as the first step, we replace the globally unique IMSI_S with a locally unique identifier. As we observed from a commercial CDMA2000 system, the number of terminals in a single paging area, including the most populated areas such as Manhattan, does not exceeded one million. This indicates that 20 bits suffice for locally unique IDs.

Optimization using the Slotted Nature of the Paging Channel. Section 2.4 showed that a cellular network divides the paging channel into 32 or 64 slots. Each terminal wakes up in only one slot (calculated based on its IMSI) in the paging cycle to listen to the page message. In other words, terminals in a paging area are divided into distinctive *slot groups* by their slot numbers. Since a terminal only listens to one slot, their local IDs need to be unique only within each slot group. A typical CDMA 2000 system has 64 slots. If all the terminals in a paging area are evenly divided into slot groups, no slot group should contain more than $2^{20}/64 = 2^{14}$ terminals. Therefore, we can reduce the length of local IDs further to 14 bits.

Optimization using the Quick Paging Mechanism. Finally, we decrease the length of the local IDs even further by using the Quick Paging channel. Quick Paging is a standardized operation adopted by most mobile networks to reduce terminals' wakeup time to improve their power efficiency. Similar to the Paging Channel operation, the Quick Paging channel is also divided into slots. In fact, a terminal's quick paging channel slot occurs exactly 100ms earlier than its paging channel slot. The purpose of Quick Paging is to convey *"paging indicator bits"* to help terminals pre-determine whether they are paged. Towards this goal, each quick paging slot is divided into four frames, and each frame carries a sequence of indicator bits. Each terminal has two indicator bits. The system calculates the positions of these two bits in the quick paging frames by feeding the terminal's IMSI into two hash functions. The standard requires that these two indicator bits occur in either the first and third frames, or the second and fourth frames (so that a terminal needs to wake up in only half of the frames). If a terminal detects that either one of its indicator bits is not set, it is not paged and therefore will stay idle in the coming paging slot; otherwise, it might be paged, so it will wake up in the coming paging slot. Quick paging increases the wakeup duration of the paged terminals by half, but decreases the wakeup duration of unpaged terminals by at least half (because the terminal only wakes up in two of the four frames of the quick paging slot). Since typically only a small fraction of terminals are paged, quick paging reduces the overall wakeup time of all terminals.

The Quick Paging operation uses a special Bloom filter. Due to the false positives inherent in Bloom filters, quick paging cannot replace the paging operation. However, we can take advantage of quick paging to reduce the length of local IDs further. Since quick paging instructs only a very small fraction of terminals to wake up and listen to their paging slots, the local IDs need to differentiate only between the terminals that are truly paged and those that are not paged but whose paging indicator bits are set due to the inaccuracy of the Bloom filter.

As mentioned earlier, the first indicator bit of a terminal must be in either the first or second frame. In our reference CDMA system, the quick paging channel operates at full speed (9600 bps) and each frame is 20ms, so there are $9600 \times 0.02 \times 2 = 384$ bits in the first two frames. The CDMA2000 specification uses several bits in these frames as broadcast bits so the total number of bits used as paging indicators in the first two frames is 368. Since we only need to differentiate between the terminals whose first indicator bits are at the same location in the first two quick paging frames, we can reduce the local ID space further. Assuming that the locations of the first indicator bits of all terminals are evenly distributed, we can reduce the local ID space by $368 \approx 2^8$. As a result, we would need only $14 - 8 = 6$ bits to represent each local ID. We discuss our scheme below.

If no first indicator bits of the paged terminals share the same location, we can order the local IDs in the page record by the order of their corresponding first indicator bits in the quick page frames. For example, if a terminal's first indicator bit is the i_{th} set bit in the quick page frames, the terminal will check the i_{th} local ID in the page record (to see if it is really paged or if its first indicator bits are set merely due to Bloom filter inaccuracy).[3]

However, the above solution would not work when multiple terminals are paged but their first indicator bits share the same location in the quick paging frames. To solve this problem, in the page record, we group terminals by the locations of their first indicator bits, and prepend a *group bit* to each local ID. We set the group bit of the first terminal in a group to 1, and the group bits of all the other terminals in the same group to 0. For example, in Figure 4 the first indicator bits of both Terminal 1 and 4 are at the same position in the first quick paging frame. Therefore, in the page record, the group bit of Terminal 1 is 1 since it is the first terminal in this group, and the group bit of Terminal 4 is 0 since it is not the first terminal in this group.

We summarize the paging operation from a terminal's perspective. When a terminal joins a paging area, the network assigns a 6-bit ID to the terminal.

---

[3] A subtle complication occurs when the first indicator bit of a mobile terminal is in the second quick page frame. In this case, since the mobile terminal does not listen to the first quick page frame, it does not know how many bits are set there, so it does not know the position of its page record in the page frame. We can solve this problem by a simply trick: rather than calculating its position from the beginning of the frame, the above terminal should calculate its position from the end of the frame. For example, if a mobile terminal's first indicator bit is the i_{th} set bit from the end of the second quick page frame, it should check the i_{th} page record from the end of the page frame for its local ID.

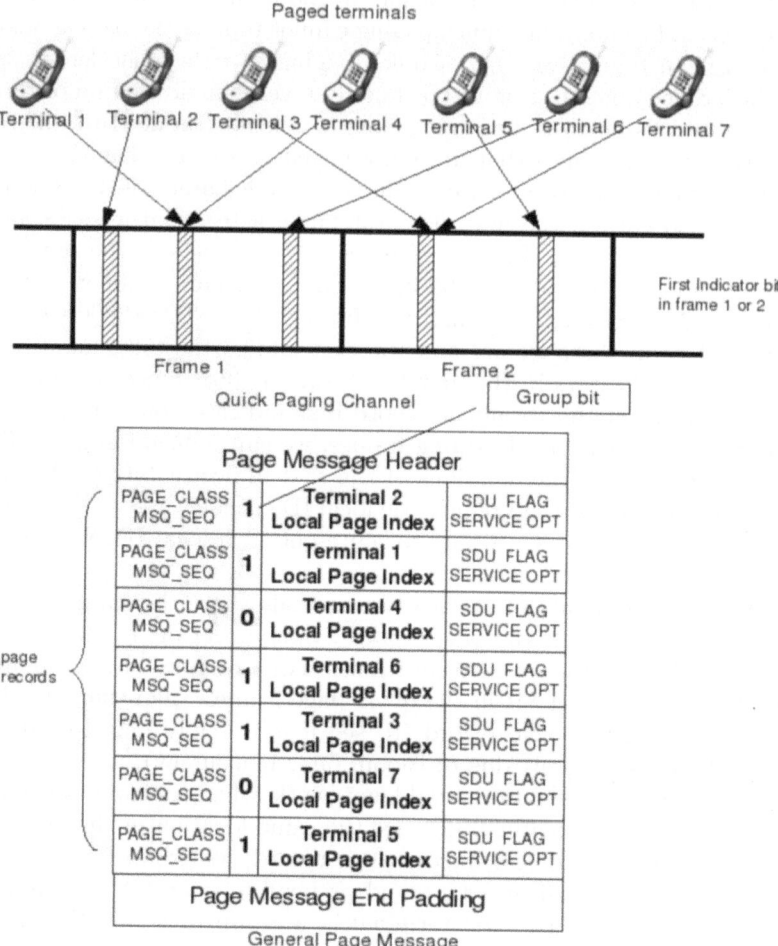

Fig. 4. An example of an optimized GPM based on quick paging. Seven paged terminals are ordered by the position of their corresponding first indicator bits. Those with the same first indicator bits are grouped by a group bit.

The terminal then calculates the position of its slot in the page message and the positions of its first and second indicator bits in the quick page message. In each paging cycle, the terminal wakes up to listen to two of the four frames in its slot in the quick page message. If both the first and second indicator bits are set, the terminal wakes up to listen to its paging channel slot to receive the page record. Using the method described in Section 3.1, the terminal compares its local ID with the corresponding one in the page record. If they match, the terminal is paged.

3.2 Bandwidth Gain

For our reference CDMA2000 system, our scheme reduces the length of local IDs from 34 bits down to 7 bits, and the length of each pag/doubing records from 58 bits to 31 bits. After applying our scheme, the maximum number of page records per slot increases from 8 to 16(Figure 5).

	Without our scheme	With our scheme
Terminal identifier length	34	7
Page record length	58	31
Maximum page records per each slot	**8**	**16**

Fig. 5. Our scheme doubles the maximum page records per slot

3.3 Implementation Requirements

Implementing our scheme is straightforward. It requires only the following modifications to the existing paging operation.

- **Local ID management by Paging Controller.** The paging controller (PC) maintains all the local IDs of terminals in the paging area. When a terminal arrives, the PC searches for an unused local ID and assigns it to the terminal. When an incoming call for the terminal arrives, the PC constructs a GPM using the local ID. When the terminal leaves, the PC reclaims the local ID. Given the high computational power of the paging controller, such management overhead is negligible.
- **Local ID transfer.** Our scheme requires that the paging controller sends the local ID to the terminal. The controller can do this during user registration. Since the local ID is only several bits, the overhead is negligible.
  To determine the length of local IDs, the paging controller must estimate the maximum number of terminals in the same paging area. If the controller finds this estimate insufficient, it may increase the length of local IDs and broadcast the new length to all the terminals in a configuration message.
- **Terminal modification.** Our scheme requires slight modification to the paging module in terminals. Note that our scheme does not change the protocol messages; rather it merely changes the algorithm by which a terminal searches for its local ID in the page record.

3.4 Advantages

Simplicity. Our scheme does not cause any adverse effects, such as paging latency, false paging, and missed paging, that other schemes often suffer from. Our scheme is compatible with and complementary to many other schemes, such as the ones based on location prediction [7].

Versatility. Besides cellular networks, our scheme applies to many other mobile networks. WiMax, for example, can also benefit from this page message optimization, although it uses a very different paging operation. Instead of using slots to determine terminals' wakeup time, the base station and the mobile station in WiMax negotiate the numerical values of the *PAGING_CYCLE*, *PAGING_OFFSET* and *PAGING_LISTEN_INTERVAL* parameters through the *DREG_CMD* and *DREG_REQ* message pair. Such mechanism invalidates our scheme where it groups terminals by their wake up slots. However, [9] proposes to group WiMax terminals by aligning their *PAGING_OFFSET* so that their page messages can be merged into one message. The essence of the scheme is to borrow the concept of slots from cellular networks. This proposal makes our idea of using a shorter identifier local to each slot feasible again. Furthermore, [15] and [16] propose a quick paging channel for IEEE802.16, making our entire scheme applicable to WiMax systems. In WiMax's page message, *MOB-PAG-ADV*, the mobile identifier, is a 24-bit hash value of the MAC address, so false paging is inevitable. Our scheme, by contrast, can eliminate the unnecessary false paging in WiMax.

4   Evaluation

We evaluate the effectiveness of our scheme on mitigating DoS attacks and on increasing the capacity of the paging channel. Since modifying a commercial cellular system and launching a full-fledged DoS attack are prohibited, we demonstrate the performance of our scheme using real paging data collected from a live cellular network as well as using a simulation tool.

4.1   Evaluation Based on Partial DoS Attack on Live Cellular Network

One advantage of our scheme is that it does not change existing paging protocols, as our scheme merely changes the terminal IDs inside the GPM. Therefore, we can use paging data measured on a real paging system to infer the performance of our scheme (such as its impact on reducing channel utilization) with one exception: During high paging load, the paging controller without applying our scheme may not be able to page all the requested terminals in a slot, so it will page some of these terminals in the next paging cycle instead. Since our scheme allows the paging controller to fit more terminal IDs into one page message, it will eliminate some or all of these delays. In this case, the terminals paged in each paging cycle would be different if the paging system had adopted our scheme.

Based on the above observation, we recreated the partial DoS attack experiment described in [5]. We captured GPMs over an CDMA2000 interface. We then launched a partial DoS attack by injecting UDP packets from the Internet to data users of the cellular network. Using the captured GPMs, we calculated the utilization of the paging channel by GPMs. To infer the channel load when

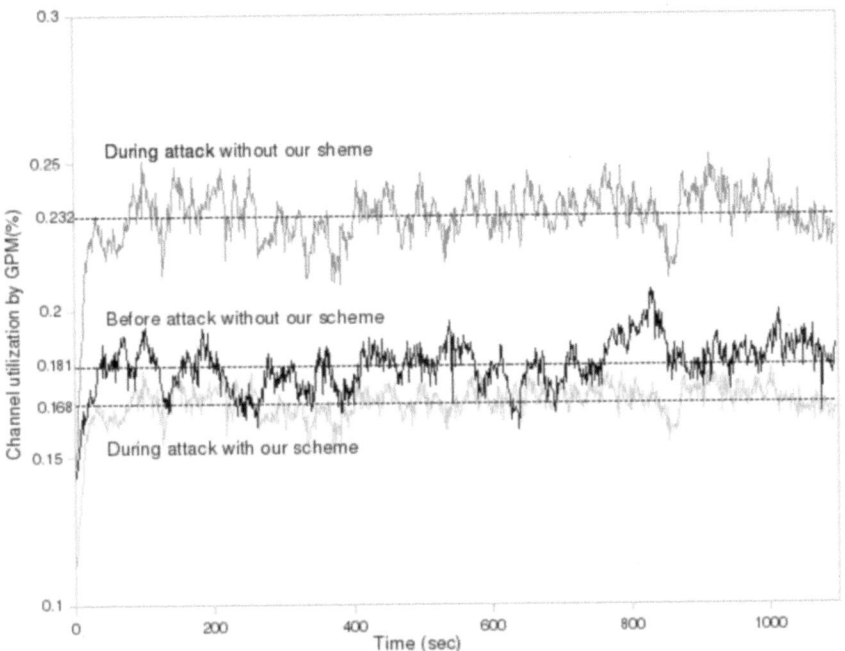

Fig. 6. Paging channel utilization during the attack without our scheme (top), before the attack without our scheme (middle), and during attack with our scheme (bottom)

our scheme is applied, we only need to calculate the length of the GPMs under our scheme, if there were no or negligible paging delays indicated by our captured GPMs. To verify this assumption, we examined the captured GPMs and found only three GPMs (out of more than 20,000 GPMs) that contained the maximum number of paging records (which indicates *potential* paging delays). This validates our assumption that paging delays occurred rarely in the captured GPMs.

Figure 6 depicts the utilization of paging channel by GPM under three different situations. For legibility, we have smoothed the curves using the Exponential Moving Weighted Average (EMWA) algorithm. Before the attack, the average utilization by GPMs in the measured system was 18.1%[4]. The utilization went up to 23.2% during the attack. If the system deployed our scheme, the average utilization would be 14.2% before the attack (not shown in the figure for legibility), and 16.8% after the attack.

As another measurement of the effectiveness of our scheme, we quantified the resources that an attacker must acquire to saturate the paging channel. Since overhead messages occupy at least 25% of the paging channel capacity, an

---

[4] The paging channel utilization is calculated as the total bits of GPM during a certain time period, divided by the product of the length of the time period and the channel capacity (e.g., 9600bps).

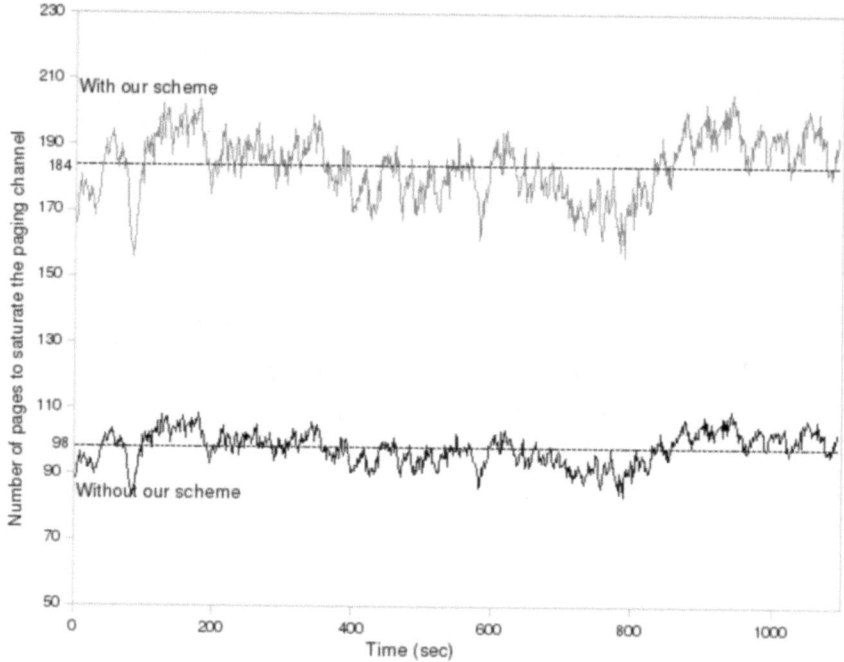

Fig. 7. Number of page records required to completely saturate the channel with and without our scheme

attacker only needs to saturate the remaining 75% of the paging channel. We calculated how many page records the attacker must trigger. We assume that all the messages other than GPMs remain unchanged after the attack begins. Figure 7 shows that our scheme almost doubles the efforts of the attacker to completely saturate the paging channel.

4.2   Simulating a Paging System

In Section 4.1, we examined the effect of our scheme using page messages measured on a live cellular network. In this section, we use simulation to study our scheme under different conditions of the paging system. We simulate the paging channel *at a base station* as a queueing system. There are two main types of messages in a paging system, slotted messages and non-slotted messages. Slotted messages need to be sent during their assigned slots in the paging channel while non-slotted messages can be sent at any time. Non-slotted messages arrive only after the paged terminal moves into the cell (by contrast, a page message is used to locate a terminal in the paging area and hence is not necessarily associated with *this* base station). Therefore, the arrival process of the non-slotted messages is equivalent to the service process of the slotted messages multiplied by a factor of p, where p, the paging success factor, is the probability that a mobile

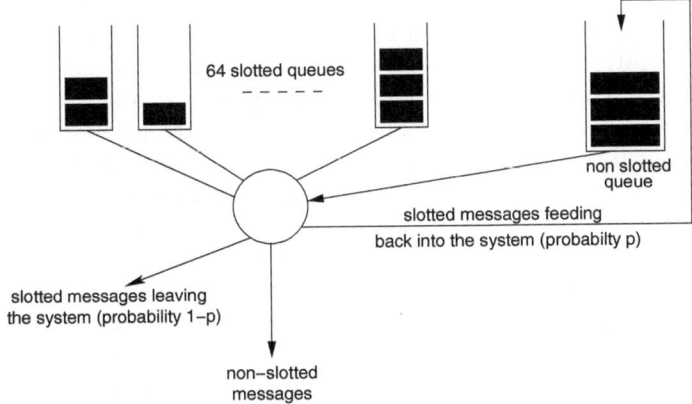

Fig. 8. A queueing system representation of a paging system. There are 64 slotted queues and 1 non-slotted queue. Slotted messages initially arrive according to a Poisson process. For each slotted message, the system generates a non-slotted message with probability p.

terminal is located with a given base station and is inversely proportional to the size of the paging area. We assume that there is no delay between the time when a slotted message is served and the time it triggers a non-slotted message. We also assume that slotted messages initially arrive according to a Poisson process. This is a common assumption for modelling the arrival of events such as calls in phone systems. Figure 8 illustrates this queueing system.

We simulate such a paging system with a paging cycle divided into 64 slots. Hence we have 64 *slotted* queues and 1 *non-slotted* queue, as shown in Figure 8. The simulation program has three main modules, namely, the arrival, slot, and server modules, which we describe in detail below.

Arrival. The arrival module generates slotted messages according to a Poisson process. Then it randomly assigns them to one of the 64 slotted queues.

Slot. The slot module implements the schedule in which the slotted queues are served. Each slotted queue is served in a time division multiplexing manner. Specifically, the slot module calls the server module on each slotted queue in a round-robin schedule. The slot module allows the service of each slotted queue for a fixed duration of $D = 0.08$ seconds (i.e., the slot duration).

Server. The server module dequeues messages (i.e., sending messages) as well as generates non-slotted messages. When invoked by the slot module, it builds a GPM by dequeuing messages up to the maximum capacity (N_{max}). In more detail, it computes the total capacity of the paging channel, subtracts the length of the GPM header, and subtracts up to ($N_{max} \times r_s$) bits where r_s is the length of each page record. Before applying our scheme, N_{max} is only 8 and r_s is 58 bits. After our scheme is applied, N_{max} becomes 16 and r_s becomes 31 bits. For each new page record, the module generates, with a probability p, a subsequent non-slotted message and inserts it into the non-slotted queue.

If the slotted queue has spare bandwidth (i.e., the slotted queue contains fewer than N_{max} messages), the non-slotted queue is serviced for the remainder of the slot duration. To do this, the simulator builds a non-slotted message by subtracting the length of a non-slotted message header followed by the length of a non-slotted message record multiplied by as many non-slotted messages as can be dequeued (sent) during the remainder of the slot duration. If there are insufficient messages in the non-slotted queue, the server sits idle during the rest of the slot duration.

A sent non-slotted message indicates a successfully established call so we use the number of serviced non-slotted messages to calculate the system throughput. We plot the throughput of non-slotted messages with increasing arrival rate (λ) in Figure 9. We find that in both cases, applying our scheme allows the throughput to be sustained for up to about twice the peak arrival rate that the current scheme can sustain. This can be attributed to the increase in N_{max} after applying our scheme.

We simulated the paging system for $p = 0.01$ and $p = 0.05$. Since p represents the success rate of the paging algorithm, (e.g., location management scheme), it is inversely proportional to the size of the paging area. A small value of p (e.g., 0.01) represents a relatively large paging area, while a large value of p (e.g., 0.05) indicates a small paging area. From the data we captured in our experiments (Section 4.1), we observed that p was less than 5%.

Paging delay is the amount of time that it takes to establish a connection between the initiating terminal and the target terminal. It is mainly caused by paging channel overload. Figure 10 shows the average paging delay before and after applying our scheme. Again, we find that our scheme can sustain up to twice the slotted message arrival intensity that the current scheme can before paging delay grows exponentially. Note that the paging delay is roughly the same regardless of p since the number of non-slotted messages in the paging channel only affects the paging delay when λ is small. However, the paging delay is also small when λ is small.

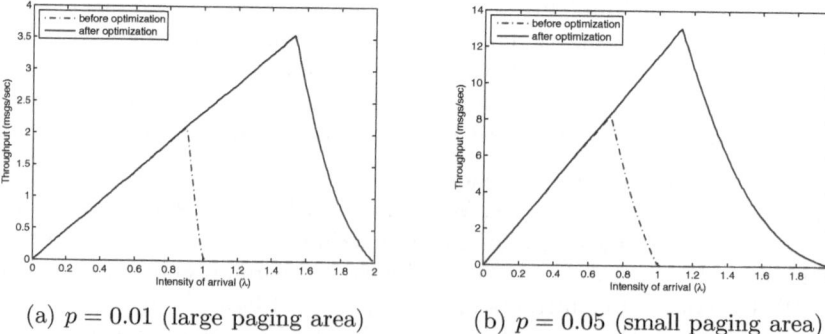

(a) $p = 0.01$ (large paging area)   (b) $p = 0.05$ (small paging area)

Fig. 9. Throughput of non-slotted messages under different arrival rates λ. A small value of p (e.g., 0.01) represents a large paging area while a large value of p (e.g., 0.05) represents a small paging area.

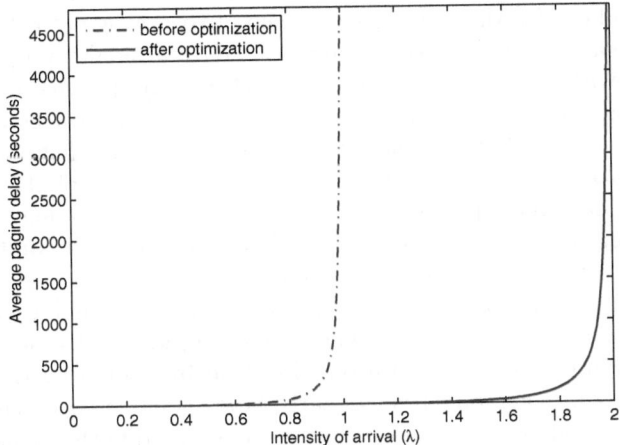

Fig. 10. The average paging delay on different slotted message arrival rates, before and after applying our scheme

Section 4.1 demonstrated that our scheme would force the attacker to spend more resources before he could overload the paging channel (i.e., the attacker would need to generate more calls). Figure 11 shows that our scheme doubles the number of slotted messages required for saturating the paging channel.

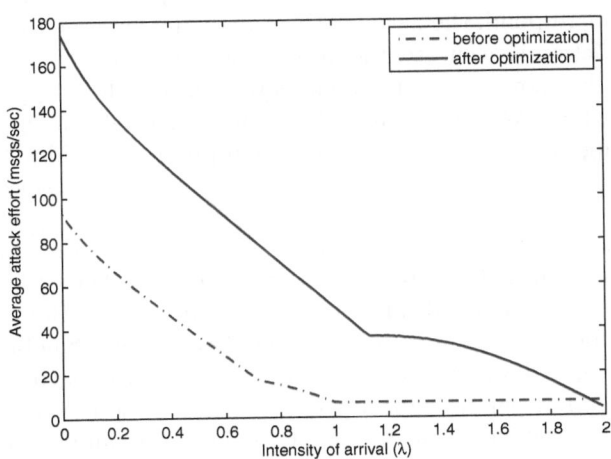

Fig. 11. The average number of slotted messages needed to saturate the paging channel (i.e., the attacker's effort)

5   Related Work

With the ongoing convergence of wireless voice and data networks, denial of service (DoS) attacks on the paging channel of wireless networks have attracted a lot of attention. Enck et al. presented a denial-of-service attack by sending a sufficient number of SMS messages per second to a range of cellular phones in the same area [6]. An attacker would need only a single computer with a broadband network access to disrupt a network in a major city by saturating control channels shared between voice calls and SMSs. Serror et al. provided experimental evidence of the impact of an attack on the paging channel. They injected UDP packets from the Internet to data users in a cellular network to cause increased load on the paging channel [5]. By improving paging efficiency, we aim at mitigating these attacks. Our approach is complementary and orthogonal to other attack detection [17] and mitigation techniques.

Researchers have proposed many solutions to enhance paging efficiency by improving location update schemes. The underlying idea is to increase the estimation precision of the mobile stations' location by exploiting their mobility patterns. [18] provides a good survey of early work. Some of them, such as time-based, distance-based and zone-based, have been adopted in the standard. Other proposed schemes include *movement based*, *cost based* and *velocity based*. One popular approach is profile based location management: each mobile station has a profile that helps the paging controller to predict the paging area. This idea is based on the observation that each individual user has her own mobility pattern. Researches in this area mainly focus on how to establish user's profile. E.g., [7] mined the call history of the users to build their mobility profiles.

To improve paging efficiency, location management schemes break the paging process into two or more stages. In the first stage, the network sends the paging message to a predicted small subset of cells in the paging area. If the mobile terminal does not respond, the network then sends the message to a larger set of cells. If the prediction is accurate, the average number of paged cells is expected to be much smaller; however, if the prediction is wrong, these approaches cause paging latency. The average paging latency in a normal paging operation is half of the paging cycle (2.56s). Each additional phase will add 5.12s to it. Our scheme, by comparison, does not increase the paging latency, since it does not break paging into stages.

Some researchers have explored paging message optimization. [19] proposed a Bloom filter to map multiple page records to one fixed length bitmap. Quick paging, described in Section 3, also uses a special form of Bloom filter. The main difference between them is that the number of hash functions in [19] is dynamically calculated and is transferred as a parameter of each page message. While achieving high paging capacity in certain situations, this Bloom filter based paging system suffers from excessive false page rates and hence low battery efficiency. By contrast, our scheme causes no false page. Another problem with this approach is that it removed single paging records from paging messages; therefore, useful information previously piggybacked with the paging records, such as the Service Option field, was no longer available to terminals.

[9] aims at improving the paging efficiency of WiMax networks. Based on the observation that two individual MOB-PAG-ADV messages use more bandwidth than one MOB-PAG-ADV message with two records, it grouped multiple mobile

station records into one MOB-PAG-ADV message to reduce the overhead and improve the paging efficiency. This solution is specific to WiMax. By contrast, our scheme applies to almost all mobile networks that require paging.

6   Conclusion

We propose a novel approach to improve paging efficiency and to mitigate DoS attacks on the paging channel. We describe a series of mechanisms for efficiently encoding terminal identifiers in page messages to increase the paging channel capacity. For instance, we can shorten the terminal identifier in a CDMA2000 General Page Message from its current length of 34 bits down to 7 bits. We evaluated our scheme using data measured on a live cellular network and using simulation. The results indicate that our scheme can significantly increase the paging throughput and the cost to the attackers, thereby mitigating DoS attacks on the paging channel. Our scheme is simple and is straightforward to implement. It does not incur any adverse effect, such as paging delay, false paging, and higher missed paging rate, that other schemes often suffer from. Furthermore, it is compatible with location-based paging efficiency improving schemes. Although we describe our scheme in the context of cellular networks, the scheme applies to other mobile networks such as WiMax.

Acknowledgment

This paper is based upon work supported by the National Science Foundation under Grant Nos. 0644450 and 0520320 and by a generous gift from Sprint. We thank Jean Bolot, Prasant Mohapatra and Sridhar Machiraju for their valuable comments.

References

1. Mobile radio interface layer 3 specification, 3GPP TS24008 (December 2008)
2. Upper layer (layer 3) signaling standard for cdma2000 spread spectrum systems, 3GPP2 C.S0005-D (September 2005)
3. Air interface for fixed and mobile broadband wireless access systems. IEEE Std 802.16eTM-2005 (February 2006)
4. Haverinen, H., Malinen, J.: Mobile ip regional paging (June 2000), http://draft-haverinen-mobileip-reg-paging-00.txt
5. Serror, J., Zang, H., Bolot, J.C.: Impact of paging channel overloads or attacks on a cellular network. In: WiSe 2006: Proceedings of the 5th ACM workshop on Wireless security, pp. 75–84. ACM, New York (2006)
6. Enck, W., Traynor, P., McDaniel, P., Porta, T.L.: Exploiting open functionality in sms-capable cellular networks. In: CCS 2005: Proceedings of the 12th ACM conference on Computer and communications security, pp. 393–404. ACM, New York (2005)

7. Zang, H., Bolot, J.C.: Mining call and mobility data to improve paging efficiency in cellular networks. In: MobiCom 2007: Proceedings of the 13th annual ACM international conference on Mobile computing and networking, pp. 123–134. ACM, New York (2007)
8. Signaling link access control (lac) standard for cdma1998 spread spectrum systems, 3GPP2 C.S0004-D (September 2005)
9. Mohanty, S., Venkatachalam, M., Yang, X.: A novel algorithm for efficient paging in mobile wimax. In: Mobile WiMAX Symposium, March 2007, pp. 48–53. IEEE, Los Alamitos (2007)
10. Roh, H.S., Lee, S.h.: Paging scheme for high-speed portable internet (hpi) system. In: Advanced Communication Technology, 2006. ICACT 2006, The 8th International Conference, February 2006, vol. 3(4), p. 1732 (2006)
11. Kempf, J.: Dormant mode host alerting ("ip paging") problem statement, rfc3132 (June 2001)
12. Ramjee, R., Varadhan, K., Salgarelli, L., Thuel, S.R., Wang, S.-Y., La Porta, T.: Hawaii: a domain-based approach for supporting mobility in widearea wireless networks. IEEE/ACM Transactions on Networking 10(3), 396–410 (2002)
13. Campbell, A.T., Gomez, J., Valko, A.G.: An overview of cellular ip. In: Wireless Communications and Networking Conference, 1999. WCNC 1999, vol. 2, pp. 606–610. IEEE, Los Alamitos (1999)
14. Zhang, X., Castellanos, J.G., Campbell, A.T.: P-mip: paging extensions for mobile ip. Mob. Netw. Appl. 7(2), 127–141 (2002)
15. Mohanty, S., Venkatachalam, M., Timiri, S., Ahmadi, S.: Proposal for ieee 802.16m quick paging channel design (July 2008)
16. Koorapaty, H., Ernstrm, P.: Quick paging signal for ieee 802.16e (May 2008)
17. Traynor, P., McDaniel, P., La Porta, T.: On attack causality in internet-connected cellular networks. In: Proceedings of 16th USENIX Security Symposium on USENIX Security Symposium (2007)
18. Akyildiz, I.F., Ho, S.M.: On location management for personal communications networks. Communications Magazine, IEEE 34(9), 138–145 (1996)
19. Mutaf, P., Castelluccia, C.: Hash-based paging and location update using bloom filters: a paging algorithm that is best suitable for ipv6. Mob. Netw. Appl. 9(6), 627–631 (2004)

FIJI: Fighting Implicit Jamming in 802.11 WLANs*

Ioannis Broustis[1], Konstantinos Pelechrinis[1], Dimitris Syrivelis[2],
Srikanth V. Krishnamurthy[1], and Leandros Tassiulas[2]

[1] University of California, Riverside
{broustis,kpele,krish}@cs.ucr.edu
[2] University of Thessaly
{jsyr,leandros}@inf.uth.gr

Abstract. The IEEE 802.11 protocol inherently provides the same long-term throughput to all the clients associated with a given access point (AP). In this paper, we first identify a clever, low-power jamming attack that can take advantage of this behavioral trait: *the placement of a low-power jammer in a way that it affects a single legitimate client can cause starvation to all the other clients.* In other words, the *total throughput* provided by the corresponding AP is drastically degraded. To fight against this attack, we design FIJI, a cross-layer anti-jamming system that detects such intelligent jammers and mitigates their impact on network performance. FIJI looks for anomalies in the AP load distribution to efficiently perform jammer detection. It then makes decisions with regards to *optimally* shaping the traffic such that: (a) the clients that are not explicitly jammed are shielded from experiencing starvation and, (b) the jammed clients receive the maximum possible throughput under the given conditions. We implement FIJI in real hardware; we evaluate its efficacy through experiments on a large-scale indoor testbed, under different traffic scenarios, network densities and jammer locations. Our measurements suggest that FIJI detects such jammers in real-time and alleviates their impact by allocating the available bandwidth in a fair and efficient way.

Keywords: IEEE 802.11 WLANs, Fairness, Jamming, Measurement.

1 Introduction

The proliferation of IEEE 802.11 WLANs makes them an attractive target for malicious attackers with jamming devices [1,2]. A jammer typically emits electromagnetic energy thereby causing: *(a)* prolonged packet collisions at collocated devices, and *(b)* packet transmission deferrals due to legitimate nodes detecting continuous medium activity. Hence, jamming attacks can lead to significant throughput degradation, especially when they intelligently exploit the properties of the MAC protocol in use.

In this paper, we first identify a clever jamming attack where the jammer can not only hurt its intended victim, but cause starvation to other clients that are associated with the

* This work was done partially with support from the US Army Research Office under the Multi-University Research Initiative (MURI) grants W911NF-07-1-0318 and the NSF NeTS:WN / Cyber trust grant 0721941.

Y. Chen et al. (Eds.): SecureComm 2009, LNICST 19, pp. 21–40, 2009.

same AP as the victim. We call this attack the *Implicit-Jamming* attack. We design and implement FIJI, a cross-layer anti-jamming system to effectively detect such jammers and mitigate the impact of their attack.

The implicit-jamming attack. An inherent characteristic of the IEEE 802.11 MAC protocol is that under saturated traffic demands, an AP (access point) will provide the *same* long-term throughput to all of its affiliated clients [3]. If a client cannot receive high throughput from its AP for any reason (e.g. long-distance AP→client link or high levels of interference at the client side), the AP will spend a large amount of time serving this client at a low transmission bit-rate; this rate is determined by the rate adaptation algorithm in use. This will compel the AP to serve each of its other "healthier" clients (to which it can support higher transmission rates) for smaller periods. In other words, the AP does not distinguish between clients with low-SINR links and clients with high-SINR links; the long times taken to serve the former class of clients hurts the time available to serve the latter class of clients. This behavior is referred to as *the performance anomaly* of 802.11 [4] and is caused by the inherent design principles of the IEEE 802.11 MAC protocol (described in more detail in section 2).

The implicit jammer exploits this anomaly. To illustrate, consider the scenario depicted in Fig. 1. In this scenario: *(a)* all clients have high-SINR links with their AP in benign conditions, and *(b)* a low power jammer is placed next to a particular client (say client C) such that it does not *directly* affect any other client of the AP. The jammer causes high levels of interference at client C and thus, most of the packets sent by the AP to C are not successfully received. This in turn causes the AP to reduce the transmission rate used to serve C (an inherent property of rate adaptation). As a result, the AP spends more time attempting to serve C, and this reduces the fraction of time that it provides to its other clients. Thus, the throughput of all the clients drops significantly due to the jamming of only client C. In other words, jamming a small subset of clients (even only a single client) implicitly affects all the clients that are affiliated with the same AP.

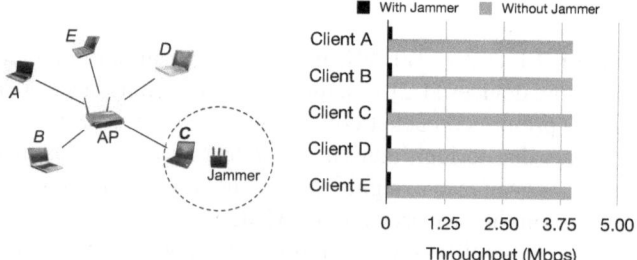

Fig. 1. Implicit Jamming. The jammer takes advantage of the 802.11 performance anomaly. Using very low transmission power, it simply attacks client C. This is sufficient to tremendously degrade the throughput of all clients.

The impact of the implicit-jamming attack. In order to demonstrate the potential impact of this attack on the performance of the network, we conduct a set of preliminary experiments on our wireless testbed (described later in section 4). In particular, we construct the scenario in Fig. 1, where an AP maintains ongoing sessions with 5 clients and transmits saturated unicast traffic to all of these clients. We place a jammer 7 ft. away from one client (C). The jammer emits energy continuously at 0 dBm (1 mW), such that it causes interference to client C only. Fig. 1 depicts our throughput measurements, with and without the jammer. We observe that in the absence of jamming each client receives 4.1 Mbits/sec, on average. When the jammer is enabled, however, the long-term throughput of *all* clients drops to 90 Kbits/sec.

FIJI: An anti-jamming system to mitigate the implicit-jamming attack. In order to alleviate the effects of this intelligent attack, we design and implement FIJI, a distributed software system that is executed locally at the APs. With FIJI, the AP is able to quickly detect an implicit jamming attack and identify the clients that are under the direct influence of the jammer(s). Furthermore, via a minimal set of online calibrating measurements that characterize the impact of the attack, the AP shapes the downlink traffic such that: *(a)* the jammed clients receive the maximum possible throughput given the circumstances, and *(b)* the rest of the clients are unaffected, i.e., shielded from the influence of the jammer(s). Some parts of FIJI are implemented on the *Click* software framework [5] and the rest are implemented on the driver/firmware of our wireless cards. Via extensive experiments, we observe that FIJI effectively mitigates the implicit-jamming attack on an 802.11a/g wireless testbed.

Our work in perspective. FIJI can be potentially applied in scenarios wherein jammers attack APs directly. However, in this work, we focus on addressing intelligent jammers that exploit the performance anomaly at the client side. Moreover, note that the impact of implicit jamming is exacerbated in downlink traffic scenarios; with uplink traffic, jammed clients will simply defer accessing the medium and will thereby allow the other clients to obtain higher levels of access.

The remainder of the paper is structured as follows. In section 2, we provide a brief background on the performance anomaly in 802.11 as well as jamming attacks, and discuss related studies. In section 3, we describe the implicit jamming detection and mitigation with FIJI, our anti-jamming system. We describe the implementation of FIJI and evaluate its effectiveness in section 4. Section 5 provides the scope of our study. We conclude in section 6.

2   Background and Previous Work

In this section, we first describe the so-called performance anomaly with IEEE 802.11 and efforts related to addressing the anomaly. We then discuss jamming attacks in brief as well as prior work related to anti-jamming.

2.1   Performance Anomaly in 802.11 WLANs

Heusse et al. [4] were the first to observe that the long term throughput of all the clients associated with an AP in a WLAN is limited by the client with the poorest link. This effect eventually provides the same long-term throughput to all clients. Although [4] considers uplink traffic, this "anomaly" arises with downlink traffic as well [6,7]. With either uplink or downlink saturated traffic, 802.11 provides equal medium access probability to all links. Let us consider the downlink scenario. An AP→client link with low SINR will coerce the rate adaptation mechanism at the AP to use a low transmission rate for this client. Thus, when attempting to serve this client, the AP will spend large amounts of time. Given that the AP will access the channel with equal probability for low-SINR clients and high-SINR clients (higher bit rate, shorter transmission durations), the latter will be served for smaller proportions of time.

Let us assume that AP α is sending saturated unicast traffic to each of its κ clients. The *theoretical* instantaneous transmission rate from AP α towards client c_i, where $i \in \{1, ..., \kappa\}$, is a step function of the SINR for this client [8]. In this work, we consider f_{c_i} to be the instantaneous *deliverable* rate towards client c_i, which in practice may not always be equal to the transmission rate (especially at high rates). Each client c_i of AP α will receive the ***same*** throughput T_i in the long term; this throughput is given by:

$$T_i = M_\alpha \cdot \frac{B}{\sum_{i=1}^{\kappa} \frac{B}{f_{c_i}}} = M_\alpha \cdot \frac{1}{\sum_{i=1}^{\kappa} \frac{1}{f_{c_i}}} \ . \tag{1}$$

In the above equation, M_α is the fraction of the time that AP α is able to access the medium, given the contention with its co-channel neighbor devices. We assume that AP α transmits data packets of the same length B to all clients. From the above equation it is evident that if a client c_i receives low throughput, *all* clients will also receive equally low throughput under saturated conditions. Note that this phenomenon has been taken into account during the design of previous performance improvement algorithms for WLANs; examples can be found in [3], [6], [7], [8]. All these studies take the anomaly as a given and try to improve the network performance through other intelligent strategies, such as AP load balancing and power control. In other words, such studies are inherently based on the fact that the 802.11 MAC protocol provides long-term fairness. Clearly, when this property of 802.11 is exploited by a malicious attacker, the performance of the schemes that are based on this property is also compromised. Hence, the existence of a mechanism that detects and mitigates such jammers becomes very vital.

Studies on mitigating the performance anomaly in 802.11. There have been numerous efforts on addressing the anomaly in 802.11. Most of them either require significant modifications on the 802.11 protocol functionality or they are very difficult to implement in practice.

Packet aggregation. Razafindralambo et al., [9] propose *PAS*, a technique that involves packet aggregation with dynamic time intervals. With PAS, nodes transmit consecutive packets back-to-back, separated by a SIFS period [10]. As a result, high-rate clients are able to transmit/receive many packets during an allocated time interval. However, packet aggregation requires modifications on the 802.11 protocol, in order to allow back-to-back data frame transmissions.

Contention window manipulation. Kim et al., [11] show that the anomaly can be addressed by tuning the 802.11 contention window size. They compute the minimum value of the window for the elimination of the anomaly. This technique, however, requires modification to the algorithm that selects the value of the contention window in 802.11. In contrast, our proposed scheme (described in the following section) does not require any changes to the 802.11 protocol semantics.

Data traffic manipulation. Bellavista et al., in [12] propose *MUM*, an application-level middleware for facilitating multimedia streaming services. MUM tries to detect the anomaly by monitoring the RSSI of received packets and estimating the goodness of links. It employs the Linux tc/iptables to implement a hierarchical token buffer scheduler [13] that "differentiates" data transmissions towards low-rate nodes. The RSSI, however, cannot accurately capture the levels of contention and interference [14]. In addition, [12] uses a limited set of 4 static rate classes for traffic differentiation; *this setting is not adequate in jamming scenarios, as we show in section 4.* Along the same lines, Dunn et al., [15] propose a heuristic for allocating a packet size to every client, which is proportional to the transmission rate. *We show in section 4 that the use of this heuristic during an implicit-jamming attack leads to some undesirable effects that in turn lead to poorer throughput than what is possible with FIJI.* Similar approaches are followed in [16,17] and [18]. Finally, Yang et al. [19] analytically model a WLAN with stations that support multiple transmission rates in order to demonstrate the performance anomaly. In contrast with these studies, our anti-jamming solution addresses the fact that the maximum transmission rate achieved by a single client can bound the total AP throughput. From the above discussion, as well as our measurements in section 4, it becomes evident that prior efforts on overcoming the performance anomaly problem in 802.11 cannot efficiently mitigate implicit jammers. We approach the 802.11 anomaly from the security point of view; in particular we examine a case where a malicious adversary can remotely exploit this feature as a vulnerability to cause complete starvation to the associated clients. *FIJI is effective against the implicit jamming attack, provides the best trade-offs between throughput and fairness and does not require any modifications on the 802.11 protocol.*

2.2   Jamming in Wireless Networks

Jammers are classified into two main categories based on their behaviors.

- *Constant jammers:* They emit electromagnetic energy all the time. This jamming technique is not usually adopted, since it depletes the battery of mobile jammers rather quickly. This category includes *deceptive* jammers [20], which transmit seemingly legitimate back-to-back data packets. With this, deceptive jammers can mislead other nodes and monitoring systems into believing that legitimate traffic is being sent over the medium.
- *Intermittent jammers.* They conserve battery life by emitting energy intermittently. As examples: (i) *Random* jammers alternate between random jamming and sleeping periods. (ii) *Reactive* jammers emit energy right after the detection of traffic on the medium, and remain inactive as long as the medium is idle. The implementation of reactive jammers is difficult; the detection and alleviation of such attacks is very challenging.

Previously proposed anti-jamming techniques. Prior work has focused on the impact of jamming on the performance of isolated wireless links. To the best of our knowledge,

FIJI is the first system to examine the effects of implicit jamming on the *overall* performance of WLANs. Some previous studies employ frequency hopping techniques to avoid jammers [21,22,23]. We do not adopt such techniques in FIJI, since frequency hopping cannot overcome wide-band jammers [2], which are capable of jamming a plurality of the available bands simultaneously. Moreover, frequency hopping has limited effectiveness when multiple collocated jammers operate on different frequencies. FIJI, however, can be complementary to frequency hopping.

Gummadi *et al.* [21] show that even ultra-low power jammers can corrupt the reception of packets; towards coping with these jammers they propose a rapid frequency hopping strategy. Navda *et al.* [22] implement a proactive frequency hopping protocol with pseudo-random channel switching. They compute the optimal frequency hopping parameters, assuming that the jammer is aware of the frequency hopping procedure that is followed. Xu *et al.* [23] propose two anti jamming techniques: reactive channel surfing and spatial retreats. However, they do not consider 802.11 networks. In [20], efficient mechanisms for jammer detection at the PHY layer are developed. However, the authors do not propose any anti-jamming mechanisms. The work in [24] suggests that the proper adjustment of transmission power and error correction codes could alleviate jamming effects. However, it neither proposes an anti-jamming protocol nor performs evaluations of these strategies. Along the same lines, Lin and Noubir [25] present an analytical evaluation of the use of cryptographic interleavers with various coding schemes to improve the robustness of wireless LANs. In subsequent work, Noubir and Lin [26] investigate the power efficiency of a jammer. They show that in the absence of error-correction codes a jammer can conserve battery power by simply destroying only a portion of a legitimate packet. Finally, Noubir [27] proposes a combination of directional antennae and node-mobility in order to alleviate jammers.

None of these efforts consider the implicit jamming attack; FIJI is the first system to address this attack.

3   FIJI to Combat the Implicit Jamming Attack

In this section, we describe the design of our anti-jamming software system, FIJI. The goal of FIJI is twofold:

1. To detect the attack and restore the throughput on clients that are not explicitly jammed (we call these clients **"healthy"**).
2. To maintain connectivity and provide the highest possible throughput to clients that are explicitly jammed (we call these clients **"jammed"**).

FIJI involves the co-design of two individual modules, executed at the AP: a *detection* module and a *traffic shaping* module. We have implemented the two modules in the kernel space (we provide implementation details in section 4).

Attack model. In this work, we focus on low-power deceptive jammers. In particular, we assume that the jamming device has the following properties:

- It is placed next to legitimate clients. With this, the jammer is able to distort packets destined to the jammed client(s). In addition, the jammer is constantly transmitting packets back-to-back, thereby prohibiting the jammed clients from accessing the medium.

- It operates at very low power. As discussed earlier, the jammer simply needs to explicitly affect one of the clients of the AP. By transmitting at low power the jammer can conserve energy and make the detection of the attack a challenging task.
- It is able to operate on a wide band (covering all the available channels); this makes frequency hopping techniques inappropriate.

We describe the operation of the detection and the traffic shaping modules in what follows.

3.1 Detecting the Implicit-Jamming Attack

The purpose of this module is to make the AP capable of detecting the jammed clients. *Previous jamming detection schemes assume that the jammed node is always the one that performs the detection. However with the implicit-jamming attack, the AP needs to detect the jammed client(s) in order to prevent the throughput starvation of the healthy clients.* As an example, in [20] the jammed node performs a consistency check between the instantaneous PDR (Packet Delivery Ratio), and the RSSI (Received Signal Strength Indicator) that it measures on its antenna. If the PDR is extremely low (i.e., almost zero), while the RSSI is much higher than the CCA threshold[1], the node is considered to be jammed. With the implicit jamming attack, however, the AP does not know the RSSI value that is observed by each of its clients. Thus, the approach in [20] does not allow the AP to detect the implicit jamming attack.

Measuring the transmission delay per client. FIJI relies on measuring the data unit transmission delay $d_{c_i} = B/f_{c_i}$ of every client c_i at the AP. More specifically, the denominator of Eq. (1) is the aggregate transmission delay D_α incurred by AP α in order to serve all of its associated clients once; it is the sum of the individual d_{c_i} values, $i \in \{1, ..., \kappa\}$, of the κ clients that are associated with AP α [3]. In other words, if we assume saturated downlink traffic, D_α corresponds to the average time that AP α needs in order to send one data unit to every client. The value of D_α is the same for all clients, and the transmission delay d_{c_i} of client c_i contributes to the value of D_α. Hence, a sudden, very large increment in D_α indicates that one or more of the d_{c_i} values has suddenly increased; *this would imply that one or more clients are under attack.* Towards calculating D_α, AP α needs to measure the d_{c_i} value for every client c_i (this includes possible retransmission delays and the rate-scaling overhead[2]). Measuring d_{c_i} will directly reveal the jammed clients: the value of $d_{c_i^j}$ for a jammed client c_i^j is likely to be much higher than the delays of the other clients. We adopt this detection strategy in FIJI.

3.2 Shaping the Traffic at the AP to Alleviate Jammers

A trivial solution to the problem of mitigating the attack would be for the AP to simply stop serving the jammed clients. However, this would be unfair, since in many cases

---

[1] The CCA (Clear Channel Assessment) threshold specifies the RSSI value below which, receptions are ignored with regards to carrier sensing [8].

[2] The rate scaling overhead accounts for the higher delays incurred due to transient lower rates that the rate adaptation algorithm invokes.

the jammed clients might still be able to receive data, albeit at lower rates. We opt to provide a **fair** bandwidth allocation solution; our twofold objective is to simultaneously achieve the following:

- **Objective 1.** For each of the healthy clients we seek to provide the same throughput that they would have enjoyed in the absence of the jammer, i.e., prior to the attack.
- **Objective 2.** A jammed client typically cannot receive much throughput as long as the jammer is active. Hence we want to provide to every jammed client the maximum possible throughput that it can receive, given that objective 1 is satisfied.

We refer to the state where these objectives are met as the ***optimal state***.

We propose a real-time, cross-layer software system to mitigate the effects of the implicit-jamming attack. The system is implemented partly in the Click module [5] and partly in the wireless driver/firmware. Click receives information from the MAC Layer with regards to the properties of the jammed clients. The AP→client traffic is then appropriately shaped and forwarded down to the MAC layer at the AP.

i) DPT: Controlling the data packet size. With this strategy, the AP fragments the packets destined to jammed clients; each such smaller fragment is now an independent packet. We call this approach DPT for *Data Packet Tuning*. With DPT, the rate at which these smaller packets are sent to the MAC layer is equal to the rate at which normal packets were forwarded to the MAC layer, prior to jamming. DPT is expected to have the following effects: *(a)* The transmission of small data packets is more robust to interference due to jamming; hence these small packets are more likely to be correctly deciphered by the jammed clients. *(b)* The rate at which the AP accesses the medium for the jammed clients remains unchanged; however, the channel occupancy time that is spent for them is reduced, due to transmitting smaller packets to jammed clients. Hence, the AP will allocate a larger fraction of time for healthy clients.

Deriving the optimal data packet sizes. Our target is to determine the right packet size such that the optimal state is reached. The problem of achieving this state is formulated as follows.

Let us suppose that AP α has κ associated clients, and that n clients are being jammed, with $n \leq \kappa$. Our objective is to *minimize the aggregate transmission delay D_α^J of all the jammed clients c_i^J, $i \in \{1,..,n\}$ of AP α*. In other words, we seek to minimize

$$D_\alpha^J = \sum_{i=1}^n d_{c_i^J} = \sum_{i=1}^n \frac{J_i}{f_{c_i^J}} \, ,$$

where J_i is the data unit length for jammed client c_i^J, while $f_{c_i^J}$ is the deliverable rate at c_i^J.

Constraint. The $d_{c_i^J}$ value of each jammed client c_i^J must be at least equal (and as close as possible) to its data unit transmission delay d_{c_i} in benign conditions:

$$X1: \ d_{c_i^J} \geq d_{c_i} \Rightarrow \frac{J_i}{f_{c_i^J}} \geq \frac{B}{f_{c_i}}, \ \forall i \in n \, ,$$

where B is the default data unit length that the AP is using for all clients, and f_{c_i} is the deliverable rate to c_i^J in benign conditions. As explained earlier, the value of D_α

is the same for all clients that are associated with AP α. If we sum constraint $X1$ over all jammed clients, the left hand side of the inequality is our objective function. With this we make sure that the healthy $\kappa - n$ clients will indeed experience an aggregate transmission delay very close to $D_\alpha = \sum_{i=1}^{\kappa}(B/f_{c_i})$; note that this is the aggregate transmission delay that was experienced by these clients prior to the jamming attack. Hence, by choosing the packet size J_i that results in a transmission delay that is as close to d_{c_i} as possible, we ensure that the throughput of the healthy clients remains unaffected (we elaborate on this later with an example).

Based on the above constraint, our optimization problem can be formulated as follows:

$$minimize : D_\alpha^J = \sum_{i=1}^{n} d_{c_i^J} = \sum_{i=1}^{n} \frac{J_i}{f_{c_i^J}} \tag{2}$$

$$subject\ to : \quad 1 \le J_i \le B,\ \forall i \in \{1, 2, ..., n\}, \tag{3}$$

$$and\ X1. \tag{4}$$

The solution to the above problem provides the values of J_i that minimize (2). Although the problem is an integer programming problem, it is easy to see that its special form ensures that it always has a solution, which can be found in polynomial time w.r.t. the number of variables.

How does DPT operate? Let us consider a case study with AP α, $\kappa = 3$, $n = 1$ and default packet size B. The transmission delays for the healthy clients c_1 and c_2 are d_1 and d_2, respectively; for the jammed client c_3, it is d_3. The long-term throughput of every client in benign conditions will be: $T_b = \frac{B}{d_1+d_2+d_3}$. If c_3 is now being jammed, its transmission delay will be $d_3^J > d_3$ and the new throughput will be: $T_J = \frac{B}{d_1+d_2+d_3^J}$. By applying DPT, the packet size towards c_3 will be J_3^{dpt} and its new transmission delay will be d_3^{dpt}. Since the rest of the clients are to maintain their old transmission delays (they are not explicitly jammed), the throughput with DPT will be: $T_{dpt} = \frac{B}{d_1+d_2+d_3^{dpt}}$. Our minimization problem ensures that $d_3^{dpt} \approx d_3$. Thus, for clients c_1 and c_2: $T_{dpt_1} = T_{dpt_2} \approx T_b$. In other words, DPT restores the throughput at the healthy clients.

Next, we show that the jammed client cannot receive a higher throughput if we further decrease the packet size[3] to a value $J_3^l < J_3^{dpt}$. With packet size J_3^{dpt} the throughput at c_3 will be: $T_{dpt_3} = \frac{J_3^{dpt}}{d_1+d_2+d_3^{dpt}}$. Let us assume that with packet size $J_3^l < J_3^{dpt}$ the transmission delay of c_3 is d_3^l. The throughput at c_3 will then be $T_{l_3} = \frac{J_3^l}{d_1+d_2+d_3^l}$. The required condition $T_{l_3} < T_{dpt_3}$ can be simplified as:

$$T_{l_3} < T_{dpt_3} \Leftrightarrow d_3^l > \frac{J_3^l}{J_3^{dpt}} \cdot (d_1 + d_2 + d_3^{dpt}) - d_1 - d_2.$$

---

[3] For larger packet sizes, objective 1 cannot be satisfied; hence we do not need to consider such a case.

Since the packet delivery rate f_{c_3} is the same, we have:

$$\frac{J_3^l}{J_3^{dpt}} = \frac{d_3^l}{d_3^{dpt}} \Leftrightarrow d_3^l = d_3^{dpt} \cdot \frac{J_3^l}{J_3^{dpt}}$$

Thus: $\frac{J_3^l}{J_3^{dpt}} \cdot d_3^{dpt} > \frac{J_3^l}{J_3^{dpt}} \cdot (d_1 + d_2 + d_3^{dpt}) - d_1 - d_2 \Leftrightarrow$

$$0 > (\frac{J_3^l}{J_3^{dpt}} - 1)(d_1 + d_2).$$

The last inequality is always true; hence, $T_{l_3} < T_{dpt_3}$.

Similar steps can be followed in order to show that DPT operates in the same manner in scenarios with multiple jammed clients. We adopt DPT in FIJI.

ii) DRT: An alternate approach. An alternative strategy would be to *explicitly tune the rate at which the packets are delivered* at the MAC layer (the packet size is now kept unchanged), destined to jammed clients. Fewer packets would arrive at the MAC layer for transmission towards the jammed clients, thereby allowing the AP to send traffic to healthy clients more frequently. Let us call this approach DRT for *Data Rate Tuning*. DRT operates as follows. Based on the measured d_{c_i} for each client c_i, the deliverable rate to every jammed client would be:

$$f_{c_i^J} = B/d_{c_i^J}. \tag{5}$$

DRT would bound the packet generation rate such that the data rate to the jammed client c_i^J is at most $f_{c_i^J}$. As a result, the rest of the (healthy) clients would share the remaining bandwidth. Thus, they would enjoy a share that is in fact higher than what they had prior to the attack. However, the packets destined to the jammed clients could be potentially lost due to channel or interference effects. Hence with DRT, the jammed clients will eventually receive *lower long-term throughput* than the specified (by DRT) rate of $f_{c_i^J}$. Clearly, while both DPT and DRT shape the traffic in order to overcome the implicit jamming effects, they essentially differ in the way they allocate the bandwidth. With DPT the healthy clients receive the *same throughput as before the attack*, while the jammed clients achieve the *maximum possible* throughput under the circumstances. On the other hand, with DRT the healthy clients have a higher share of the bandwidth than in benign settings and receive *more throughput than before the attack*; the APs will spend more time serving the healthy clients, since most of the traffic is now destined to them. However, since the jammed clients do not reach their capacity, they are treated rather "unfairly". We evaluate this fairness versus throughput trade-off in section 4.3.

4   Implementation and Evaluation

In this section, we first describe our implementation of FIJI. Next we apply FIJI on a WLAN testbed and evaluate its efficacy in overcoming the implicit jamming attack.

4.1   The Implementation of FIJI

FIJI is implemented entirely at the AP; no client software modifications are needed. In addition, FIJI does not require any special functionalities at the APs or at the clients; the only requirement is for the AP to be able to measure the d_{c_i} value for each affiliated client. Hence, FIJI can be applied on commercial APs through a driver/firmware update. In order to implement the two modules of FIJI we perform modifications on the driver and firmware of the AP, and we develop specific traffic shaping functionalities on the Click framework [5].

Implementing the implicit-jamming detection module. As explained in section 3.1, the AP needs to measure d_{c_i} for every client c_i. This will reveal, with high probability, the set of jammed clients. However, the value of d_{c_i} cannot be directly obtained from the driver of the wireless card; modifications in the firmware are required in order to compute this value. We use a prototype version of the *Intel ipw2200* AP driver/firmware; for every client we measure the time duration between the placement of the packet at the head of the MAC queue until an 802.11 ACK frame is received for this packet. The value is then passed up to the driver. The AP maintains a table in the driver space with the d_{c_i} value for every client c_i. It also computes D_α^J (when jammers are active) and D_α (when jammers are inactive), by summing up the corresponding client delays. Temporary variations of the d_{c_i} values are handled by FIJI by using weighted moving average filtering; the previously maintained average is assigned a weight of 0.9 while the new sample has an associated weight of 0.1 (similar values are used in [3,6]). Using these values, the AP constructs a table with the appropriate data packet sizes for the jammed clients. If the weighted $d_{c_i(new)}/d_{c_i(old)}$ value (for one or more clients) exceeds a pre-specified threshold δ, the AP computes the new packet sizes, updates the table and subsequently feeds it into the traffic shaping module, described below.

Implementation of the traffic shaping module. We implement the traffic shaper in Click. The module receives the table from the driver with suggested parameter settings for every client and shapes the traffic accordingly. We implement both DPT and DRT for comparison purposes. For DPT we have also developed an application-level script, which reads the table with the suggested packet sizes and inputs these values to the rude/crude measurement tool [28]. For DRT one may use two different Click elements, namely either the BandwidthShaper(bandwidth) or the LinkUnqueue (latency, bandwidth) element; we utilize the latter. Finally, we configure the AP to periodically flush the stored transmission delay values for every client and perform fresh delay measurements, using the default packet size. With this, we address scenarios of mobile jammers, which may move to the proximity of different clients, jammers with variable transmission power as well as jammers that stop operating.

4.2   Experimental Set-Up and Methodology

Testbed description. Our testbed consists of 28 Soekris net4826 nodes [29], which mount a Debian Linux distribution with kernel v2.6 over NFS. The testbed is deployed in the 3rd floor of our campus building; the node layout is depicted in Fig. 2. Each node is equipped with an *Intel-2915* mini PCI WiFi card, connected to two 5-dBi gain external omnidirectional antennae. We use both the *main* and *aux* antenna connectors of

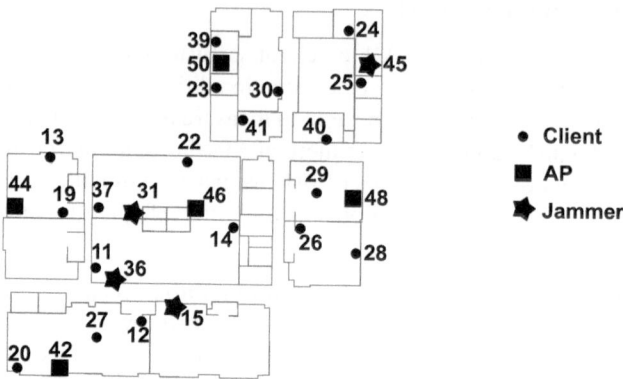

Fig. 2. The deployment of our indoor 802.11a/g WLAN testbed in the 3rd floor of a campus building

the card for diversity. As mentioned earlier, we use a proprietary version of the *ipw2200* AP driver/firmware of the *Intel-2915* card. With this version we are able to (a) measure the D_α and D_α^J values at the AP, and (b) experiment with both 802.11a and 802.11g.

Constant jammer implementation. We have implemented our own deceptive jammer (instead of purchasing a commercial one [2]) since this gives us the freedom of tuning various jamming parameters. Our implementation of a constant jammer is based on a specific card configuration and a user space utility that sends broadcast packets as fast as possible. Our jammers are also equipped with the Intel-2915 cards; our ipw2200 prototype firmware for these cards allows the tuning of the CCA threshold parameter. By setting the CCA threshold to 0 dBm, we force the WiFi card to ignore all 802.11 signals during carrier sensing (packets arrive at the jammer's circuitry with powers much less than 0 dBm, even if the distances between the jammer and the legitimate transceivers are very small). The jammer transmits broadcast UDP traffic. This ensures that its packets are transmitted back-to-back and that the jammer does not wait for any ACK messages (the back-off functionality is disabled in 802.11 for broadcast traffic). We have developed an application-layer utility that employs *raw sockets*, allowing the construction of UDP packets and the forwarding of each packet directly down to the hardware.

Experimental methodology. For each experiment we first enable traffic from the AP to its clients and subsequently we activate the jammer(s). The duration of each experiment is 10 minutes; during each minute, the jammer is inactive for the first k sec, where $k \in [5, 20]$, and active for the other $60 - k$ sec. We use a subset of 4 nodes as the jamming devices (nodes 15, 31, 36 and 45 in Fig. 2). We collect throughput and transmission delay (d_{c_i}) measurements once every 500 msec, for each client. We experiment with many different topological settings, with different numbers of APs and clients. By default all legitimate nodes set their transmission powers to the maximum value of 20 dBm and their CCA thresholds to -80 dBm. We examine both 802.11a and 802.11g links (unless otherwise stated, we observe the same behavior for 802.11a and 802.11g).

The experiments are performed late at night in order to avoid interference from col-located WLANs, as well as not to cause interference to them. We use saturated UDP traffic with a default data packet size $B = 1500$ bytes. We also experiment with TCP traffic[4]. We use the *iperf* measurement tool to generate data traffic among legitimate nodes. We also use the *rude* tool to test DPT.

4.3 Does FIJI Deliver?

Next, we apply our anti-jamming framework on the testbed and evaluate its efficiency in alleviating the effects of implicit-jamming on the WLAN performance.

i) The efficacy of the detection module. We seek to observe two properties of this module:

1. *Efficiency of Detection*: How quickly can FIJI detect the presence of implicit jammers?
2. *Accuracy of Detection*: How accurately can FIJI determine if there is an ongoing jamming attack?

We conduct experiments with 5 APs and different numbers of clients with various link qualities. We configure the jammers to transmit at 0 dBm (1 mW) with CCA = 0 dBm, such that they affect one or more clients without affecting the APs.

a) On the speed of detection. Our measurements indicate that the transmission delay $d_{c_i^J}$ of a client increases sharply upon experiencing the implicit jamming attack. This increase is seen in less than 700 msecs; this time includes the transient periods before the weighted average $d_{c_i^J}$ converges to a stable value. Fig. 3 depicts a delay snapshot with one AP and four clients with moderate-quality links. We observe that the $d_{c_i^J}$ value increases significantly (by 26 times in this experiment). Other experiments provided similar results. In summary, these results show that FIJI can quickly detect implicit jamming attacks.

b) On the accuracy of detection. We seek to evaluate FIJI in terms of its ability to detect an implicit jamming attack in the presence of interference. Note that the d_{c_i} value for a client c_i is affected by the levels of interference on the AP $\rightarrow c_i$ link. The higher the level of interference, the higher the d_{c_i} value. In order to evaluate this ability of FIJI, we perform experiments with multiple overlapping cells (each with its own AP), so that some clients suffer interference from one or more APs; in this setting, we activate our low-power jammers.

Detecting jamming on good quality links. We first consider links that have a high SINR. Fig. 4 depicts sample experimental results. In the snapshot of Fig. 4, a jammer is placed such that it affects 2 out of the 4 clients of an AP. We observe that FIJI is able to perform a successful detection. In general, our empirical observations suggest that

---

[4] The anomaly exists with TCP traffic as well [4]. Even though we do not present our TCP measurements, we observe that FIJI is similarly efficient with TCP traffic; we discuss this briefly in section 5.

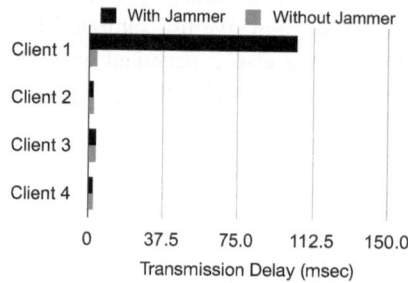

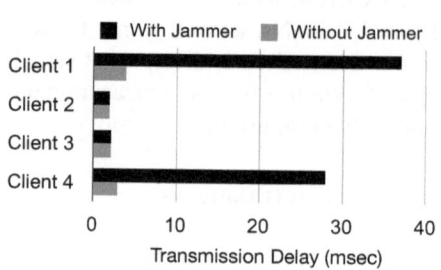

Fig. 3. FIJI detects jammed clients by measuring their data unit transmission delays

Fig. 4. The jammer detection functionality of FIJI is accurate in most cases

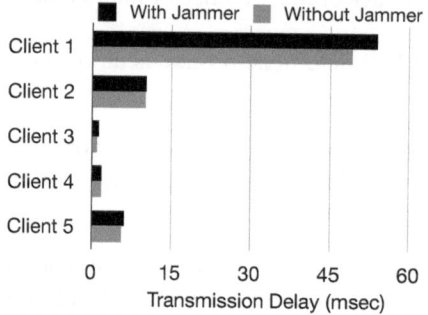

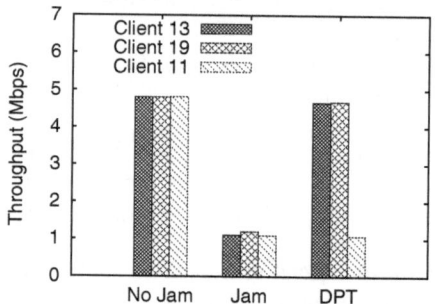

Fig. 5. The jammer detection with FIJI is less accurate in scenarios with very poor links

Fig. 6. DPT restores the performance of healthy clients to that in benign settings

when threshold $\delta \geq 9$, FIJI can effectively detect the attack (Fig. 4). In the experiment described above, the value of δ was 9.

FIJI and poor quality links. With poor quality links (SINR is low), FIJI cannot easily decide if a client is under attack or not. This effect is captured in Fig. 5, where the jammer affects a very poor link. In particular, the link $46 \rightarrow 25$ is considered with the node 45 acting as a jammer (Fig. 2). The link achieves 190 Kbits/sec in the absence of jamming and 164 Kbits/sec under jamming. Since the jammer does not significantly increase the delay experienced on such poor links, FIJI cannot decipher whether the increased $d_{node-25}$ value is due to jamming or legitimate interference. However, in such conditions, the overall change in the network performance due to the jammer is unlikely to be significant; the presence of the poor link already hurts the network performance. Furthermore note that a jammer is unlikely to attack such poor quality links if it aims to harm the network to the extent possible.

In some extreme cases, a poor quality link (exposed perhaps to other interfering APs that are hidden from its own AP) might cause a client to experience large delays. In such scenarios with healthy but poor-quality links, FIJI may incorrectly classify such links as being *jammed*. Classifying such cases as attacks, though, is perhaps appealing in terms of improving performance for the rest of the network.

FIJI and high power jammers. An implicit-jamming attacker is likely to place its jammer(s) very close to one or more clients so as to:

- degrade the client's observed SINR value to the extent possible, and
- use a very low transmission power, in order to conserve energy and avoid detection.

As our experiments indicate, under these conditions, FIJI can identify the jammed clients in real time since all measured d_{c_j} values are usually extremely high for those clients. In contrast, a jammer could use high transmission power (although this could increase the chance of its detection and result in high energy consumption). Such a high power jammer is likely to affect multiple clients and even the AP itself, directly. The delays of all these clients may go up and in this case, given its design principles, FIJI may not be able to detect the jammer. However, there are other jammer detection techniques that can be used in conjunction with FIJI to detect such jammers [20].

ii) The traffic shaping module in action. Next we evaluate the efficacy of DPT and compare it against DRT.

DPT is the most fair solution. In a nutshell we observe that as long as the jammer is successfully detected, DPT restores the throughput at the healthy clients. A sample case is depicted in Fig. 6. Here, AP 44 transmits unicast traffic to clients 11, 13 and 19; node 36 is jamming client 11. In the absence of jamming each client receives 4.8 Mbits/sec on average. When the jammer is active, without enabling DPT, all clients receive 1.1 Mbits/sec on average. The solution to the problem formulated in (2) suggests that J_{11} should be set to 345 bytes. When DPT is enabled and this packet size is chosen for the jammed client, we observe that the throughput of the healthy clients 13 and 19 is restored to 4.66 Mbits/sec, while the jammed client 11 achieves about 1.1 Mbits/sec. Note that the healthy clients do not achieve their jamming-free throughput of 4.8 Mbits/sec. This is because in our solution the equality in the constraint $X1$ is achieved for a non-integral value of J_{11}; we round the value of J_{11} up to the nearest integer. With this, the transmission delay for the jammed client is a bit higher as compared to the delay under benign conditions and this slightly degrades the throughput at the healthy clients.

In order to validate that DPT provides the most fair bandwidth allocation, we experiment with many different J_{11} values. Fig. 7 depicts the results that correspond to the settings with two J_{11} values: 166 and 700 bytes. We observe that:

- With packet sizes smaller than J_{11}^{dpt} (case with 166 bytes), the jammed client does not reach its capacity (receives 360 Kbits/sec) and the AP spends more time serving the healthy clients (as discussed in section 3): each healthy client now receives 5.1 Mbits/sec. Note that the value $J_{11} = 166$ bytes is computed using the approach proposed in [15] for the considered scenario and *it clearly does not provide the desirable fairness in terms of throughput.*
- When the packet size is higher than J_{11}^{dpt} (case with 700 bytes), the throughput at the jammed client is lower than 1.1 Mbits/sec; the healthy clients also underperform.

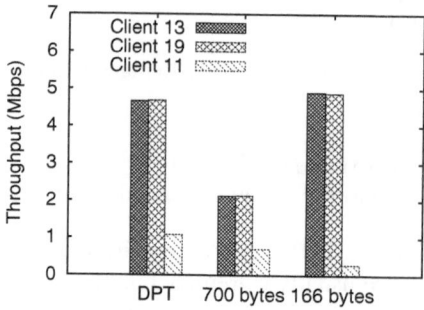

Fig. 7. DPT always manages to provide a fair allocation of throughput among clients

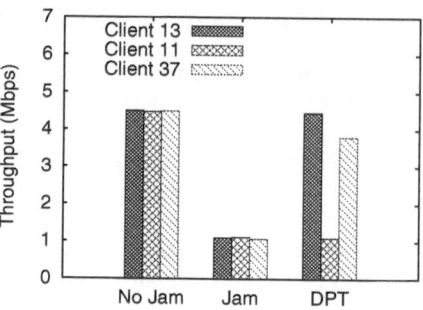

Fig. 8. DPT can easily handle scenarios with multiple clients that are simultaneously jammed

This is again conformant with our analytical assessments in section 3 with regards to the maximum achievable throughput.

Multiple jammed clients. We have so far considered scenarios wherein a single client was jammed. Next, we examine scenarios with multiple jammed clients per AP. Our experiments reveal that DPT is also able to effectively mitigate the implicit jamming attack in such scenarios. Fig. 8 presents a sample case with AP 46 and clients 11, 37 and 14; the jammer-node 36 explicitly affects both clients 11 and 37. Under benign conditions all clients receive approximately 4.5 Mbits/sec on average. As soon as the jammer is activated, without enabling DPT, all clients receive about 1.1 Mbits/sec. DPT sets the value of J_{11} to be 367 bytes and J_{37} to be 1266 bytes. With this, *DPT is able to restore the throughput at the healthy clients.*

DPT vs. DRT. Using the same methodology, we examine the effectiveness of the DRT solution. Our measurements demonstrate that DRT provides much higher throughput to healthy clients. On the other hand, DRT results in an additional unfair degradation at the jammed client. Fig. 9 represents the behaviors in an example scenario, with the same topological configuration as before (AP 44, clients 11, 13 and 19, jammer 36); the figure depicts the throughput prior to the attack (benign settings), with the jammer without DRT, and after the application of DRT. We observe that DRT overcomes the implicit impacts of the attack. Upon enabling DRT, clients 13 and 19 are no longer affected by the jammer and they receive 5.12 Mbits/sec each. Although DRT sets the maximum allowable data rate towards client 11 to be 1.1 Mbits/sec, the observed throughput at this client is significantly lower i.e., 680 Kbits/sec on average. This behavior of DRT conforms with our discussion in section 3.2; we observe similar trends in all our measurements with one or more jammed clients. *To summarize, with DRT the healthy clients receive more throughput than before the attack; however the jammed clients are penalized further.*

The choice between DPT and DRT depends on the performance objectives; one has to decide between fairness (with DPT) and bandwidth utilization (with DRT). DPT is fair: the healthy clients receive the same throughput as before the attack, while the

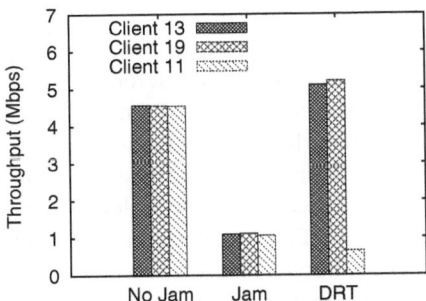

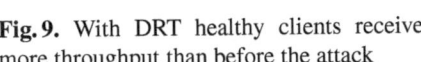

Fig. 9. With DRT healthy clients receive more throughput than before the attack

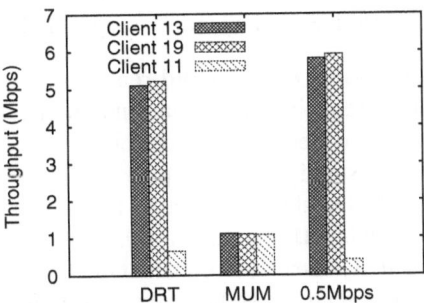

Fig. 10. DRT satisfies our objectives better than other data rate allocation approaches

jammed clients achieve the maximum possible throughput under the circumstances. On the other hand, DRT increases the throughput at the healthy clients and potentially, the total network throughput. However, the jammed clients cannot receive the maximum throughput that they can achieve in the presence of the jammer.

Note that DRT also relies on the online measurement and use of d_{c_i}. With this, DRT seeks to eliminate the effects of implicit jamming at healthy clients, while at the same time not degrade the throughput at jammed clients. Fig. 10 depicts a case with 802.11a where DRT sets the data rate at 1.1 Mbits/sec, while MUM [12] (recall our discussion in section 2) sets 6 Mbits/sec. We observe that by using data rates higher than the one chosen by DRT, the healthy clients are still affected by the attack, since in this case the downlink traffic for the jammed client is still saturated. Moreover, if we use lower data rates than the one chosen by DRT, the healthy clients get more service time, however the jammed clients receive much lower throughput than with DRT.

5   The Scope of Our Study

FIJI and previous studies on traffic shaping. Our work is the first to analytically derive the *optimal* settings for traffic shaping at the AP to mitigate the implicit-jamming attack. Traffic shapers have also been previously proposed in [12,16,17,15]. Clearly, FIJI could also be considered as another traffic shaper, simply to overcome the performance degradation due to the 802.11 anomaly. Unlike FIJI however, previous traffic shaping schemes cannot overcome the effects of an implicit-jamming attack, as explained in sections 2 and 4. Other schemes that provide fair access to the WLAN resources [31,3] would also be inadequate in combating an implicit-jamming attack since they are not designed for this purpose.

FIJI versus power control. Power control has been suggested as a means of mitigating legitimate interference [7,31]. Typically with power control, nodes tune their transmission power and CCA settings in order to reduce the amount of interference from/to their neighbors. However, if the jammer is very close to one or more clients, its signal cannot be ignored through CCA adaptation. If a client increases its CCA threshold to a high level (to ignore the jammer's signal), the connectivity to the AP will be lost.

Addressing random and reactive jammers. FIJI can mitigate the interference due to any type of jammer, even random or reactive jammers. With prolonged random jamming and sleeping periods (order of seconds), FIJI can perform a rapid detection and then customize the data packet size, as per the observed data unit transmission delay $d_{c_i^J}$. If the sleep and active periods of the random jammer are of the order of milliseconds, FIJI can monitor the *average* $d_{c_i^J}$ value instead. FIJI is expected to alleviate reactive jammers, too, since it only needs to monitor the impact of reactive jamming by measuring $d_{c_i^J}$. We have not experimented with reactive jammers, since implementing such a jammer is a very difficult task.

FIJI against other attacks. The two modules of FIJI can arguably be effective against any attempt to exploit the 802.11 performance anomaly in order to degrade the client throughput. As examples, a *compromised* device x could *deliberately* decide to *(a)* associate to a very distant AP α, or *(b)* accept traffic at a very low reception rate only (e.g. by discarding a large volume of correctly received packets). In both cases, x would receive a few Kbits/sec. Note here that, legitimate, non-compromised devices would follow such an approach only if they cannot associate with a better APs. However, given that (a) dense deployments of WLANs make the presence of an AP with a good quality link likely [7], and (b) distant poor quality APs are likely to be beyond the administrative domain of the client (the client will not be able to associate with such APs), the possibility of this is small in practice. FIJI can arguably be effective against such attacks. In particular, FIJI considers such clients to be jammed clients and ensures that the other clients remain unaffected.

FIJI and TCP. FIJI is implemented above the 802.11 MAC and below the transport layer at the AP. We have done measurements with TCP, which have demonstrated that: *(a)* Without FIJI, the performance anomaly also exists with downlink TCP traffic. The TCP packets that are destined to the jammed clients require a significant amount of time for successful delivery. As a consequence, the healthy clients are affected; they do not achieve the same throughput as before the attack. *(b)* Our experiments also demonstrated that the application of FIJI in TCP traffic scenarios is beneficial. By reducing the rate at which packets are delivered to the MAC for the jammed clients, DPT shapes the TCP traffic in a way that the healthy clients are unfettered. Note that the packet fragmentation with FIJI is executed after any TCP layer fragmentation; hence, FIJI does not intervene with TCP operations.

6   Conclusion

In this paper we identify a low-power jamming attack that we call the *implicit jamming attack*. With this attack, a jammer exploits a performance trait of the IEEE 802.11 MAC protocol to cause starvation to not only an explicitly jammed client, but all the clients associated with the same AP as that client. Since the 802.11 MAC provides long term fairness (under saturation conditions) to the associated clients in terms of equal throughput, the attacker can nullify the AP throughput by affecting only one or at most a few clients.

We design, implement and evaluate FIJI, a cross layer software system for mitigating the implicit-jamming attack. FIJI is comprised of two modules, for detecting such an attack and shaping the traffic appropriately in order to alleviate the jamming effects. We evaluate FIJI on an 802.11a/g testbed, and under many different jamming scenarios. We show that FIJI can quickly detect the attack and effectively restore the throughput at the implicitly affected clients. FIJI also ensures that the jammed clients get as much throughput as they can under the circumstances.

Acknowledgment. We thank Intel Research for providing the wireless driver.

References

1. SESP jammers, `http://www.sesp.com/`
2. ISM wideband jammers. http://69.6.206.229/e-commerce-solutions-catalog1.0.4.html
3. Sundaresan, K., Papagiannaki, K.: The Need for Cross-Layer Information in Access Point Selection Algorithms. In: ACM IMC (2006)
4. Heusse, M., Rousseau, F., Berger-Sabbatel, G., Duda, A.: Performance Anomaly of 802.11b. In: IEEE INFOCOM (2003)
5. Click web page, `http://read.cs.ucla.edu/click/`
6. Kauffmann, B., et al.: Measurement-Based Self Organization of Interfering 802.11 Wireless Access Networks. In: IEEE INFOCOM (2007)
7. Broustis, I., Papagiannaki, K., Krishnamurthy, S.V., Faloutsos, M., Mhatre, V.: MDG: Measurement-Driven Guidelines for 802.11 WLAN Design. In: ACM MOBICOM (2007)
8. Mhatre, V., Papagiannaki, K., Baccelli, F.: Interference Mitigation through Power Control in High Density 802.11 WLANs. In: IEEE INFOCOM (2007)
9. Razafindralambo, T., Lassous, I.G., Iannone, L., Fdida, S.: Dynamic Packet Aggregation to Solve Performance Anomaly in 802.11 Wireless Networks. In: ACM MSWiM (October 2006)
10. ANSI/IEEE 802.11-Standard. 1999 edn.
11. Kim, H., Yun, S., Kang, I., Bahk, S.: Resolving 802.11 Performance Anomalies through QoS Differentiation. IEEE Comm. Letters 9(7) (July 2005)
12. Bellavista, P., Corradi, A., Foschini, L.: The MUM Middleware to Counteract IEEE 802.11 Performance Anomaly in Context-aware Multimedia Provisioning. International Journal of Multimedia and Ubiquitous Engineering 2(2) (July 2007)
13. Hierarchical Token Bucket, `http://luxik.cdi.cz/~devik/qos/htb/`
14. Vlavianos, A., Law, E., Broustis, I., Krishnamurthy, S.V., Faloutsos, M.: Assessing Link Quality in IEEE 802.11 Wireless Networks: Which is the Right Metric? In: IEEE PIMRC (2008)
15. Dunn, J., Neufeld, M., Sheth, A., Grunwald, D., Bennett, J.: A Practical Cross-Layer Mechanism For Fairness in 802.11 Networks. In: IEEE BROADNETS (2004)
16. Portoles, M., Zhong, Z., Choi, S.: IEEE 802.11 Downlink Traffic Shaping Scheme for Multi-User Service. In: IEEE PIMRC (2003)
17. Iannone, L., Fdida, S.: Sdt. 11b: Un Schema a Division de Temps Pour Eviter l'anomalie de la Couche MAC 802.11b. In: CFIP (April 2005)
18. Yoo, S., Choi, J., Hwang, J.-H., Yoo, C.: Eliminating the Performance Anomaly of 802.11b. In: ICN (2005)

19. Yang, D., et al.: Performance Enhancement of Multi-Rate IEEE 802.11 WLANs with Geographically-Scattered Stations. IEEE Trans. Mob. Comp. 5(7) (July 2006)
20. Xu, W., Trappe, W., Zhang, Y., Wood, T.: The Feasibility of Launching and Detecting Jamming Attacks in Wireless Networks. In: ACM MOBIHOC (2005)
21. Gummadi, R., et al.: Understanding and Mitigating the Impact of RF Interference on 802.11 Networks. In: ACM SIGCOMM (2007)
22. Navda, V., et al.: Using Channel Hopping to Increase 802.11 Resilience to Jamming Attacks. In: IEEE INFOCOM mini-conference (2007)
23. Hu, W., Wood, T., Trappe, W., Zhang, Y.: Channel Surfing and Spatial Retreats: Defenses Against Wireless Denial of Service. In: WISE (2004)
24. Xu, W., Ma, K., Trappe, W., Zhang, Y.: Jamming Sensor Networks: Attacks and Defense Strategies. In: IEEE Network (May/June 2006)
25. Lin, G., Noubir, G.: On Link Layer Denial of Service in Data Wireless LANs. In: Wireless Communications and Mobile Computing (May 2003)
26. Noubir, G., Lin, G.: Low-power DoS Attacks in Data Wireless LANs and Countermeasures. In: ACM MOBIHOC (2003) (poster)
27. Noubir, G.: On Connectivity in Ad Hoc Network under Jamming Using Directional Antennas and Mobility. In: Langendoerfer, P., Liu, M., Matta, I., Tsaoussidis, V. (eds.) WWIC 2004. LNCS, vol. 2957, pp. 186–200. Springer, Heidelberg (2004)
28. Rude/Crude measurement tool, http://rude.sourceforge.net/
29. Soekris-net4826, http://www.soekris.com/net4826.htm
30. Jardosh, A., et al.: IQU: Practical Queue-Based User Association. In: ACM MOBICOM (2006)
31. Akella, A., Judd, G., Seshan, S., Steenkiste, P.: Self-Management in Chaotic Wireless Deployments. In: ACM MOBICOM (2005)

Deny-by-Default Distributed Security Policy Enforcement in Mobile Ad Hoc Networks

Mansoor Alicherry[1], Angelos D. Keromytis[1], and Angelos Stavrou[2]

[1] Department of Computer Science, Columbia University
[2] Department of Computer Science, George Mason University

Abstract. Mobile Ad-hoc Networks (MANETs) are increasingly employed in tactical military and civil rapid-deployment networks, including emergency rescue operations and *ad hoc* disaster-relief networks. However, this flexibility of MANETs comes at a price, when compared to wired and base station-based wireless networks: MANETs are susceptible to both insider and outsider attacks. This is mainly because of the lack of a well-defined defense perimeter preventing the effective use of wired defenses including firewalls and intrusion detection systems.

We introduce a novel distributed security policy enforcement architecture that is designed specifically for MANETs. Our approach harnesses and extends the concept of *network capabilities* and is especially suited for mobile and heterogeneous communication environments. Our model imposes communication restrictions between MANET nodes by enforcing hop-by-hop policies in a distributed manner. We use a *deny-by-default* principle, allowing compromised nodes to access only authorized services. This significantly limits their ability disrupt or even interfere with end-to-end connectivity and nodes beyond their local communication radius. In this short paper, we only present the overall architecture of the system.

Keywords: MANETs, Capabilities, Distributed Firewall.

1 Introduction

Recent advances in low-power computing and communications have led to the proliferation of handheld and portable devices equipped with wireless connectivity. These mobile wireless devices appear to be ideal for situations where fixed infrastructure is too costly or dangerous to deploy, or has been rendered inoperable. However, because of radio power consumption, physical obstacles, and channel capacity, a mobile node may not be able to reach all other nodes within a single broadcast. Therefore, to achieve end-to-end connectivity, nodes have to form mobile *ad hoc* wireless networks (MANETs), which allow data to be routed through intermediate nodes. MANETs are fundamentally different from the Internet because all peers act as both sources and routers using the other participants to relay packets to their final destination. Due to their flexibility, MANETs are currently employed in both military and commercial applications.

Unfortunately, not all MANET nodes are equally capable, nor can all users be equally trusted. Worse yet, mobile nodes in tactical environments run the

Y. Chen et al. (Eds.): SecureComm 2009, LNICST 19, pp. 41–50, 2009.

danger of being captured or malfunction. Even a small number of misbehaving nodes can successfully render the entire MANET inoperable: malicious peers can abuse the network exhausting all network and power resources.

In traditional networks, malicious nodes and traffic are kept away from a set of nodes belonging to an organization or a group using *firewalls*. This is feasible because of the existence of a well defined network perimeter. All incoming and outgoing traffic needs to transit through these firewall nodes, which enforce the policies at the perimeter. Within the perimeter, smaller sub-groups can have more stringent policies by deploying their own firewalls. Unfortunately, the concept of a network perimeter does not exist in MANETs, and policies need to be enforced in a distributed manner while taking into consideration node mobility.

To address this, we propose an architecture that enforces trust relationships and traffic accountability between mobile nodes through a novel policy enforcement scheme designed specifically for MANETs. We extend the network capability framework [8,2] and we tailor it to the resource-constrained MANET environment. A capability is a token of authority that has associated rights. In our model, capabilities propagate both access control rules and traffic-shaping parameters that should govern a node's traffic. To that end, we define a protocol for communicating capabilities, which are treated as soft state, across the MANET.

Our architecture enables the enforcement of adaptive bandwidth constraints inside the network, denying by default unauthorized traffic. Nodes can only access the services and hosts they are authorized for by the capabilities given to them. Compromised or malicious nodes cannot exceed their authority and expose the whole network to an adversary. Upon detection, we can prevent a compromised node from further attacking the network simply by revoking its capabilities. Moreover, our architecture helps mitigate the impact of denial of service (DoS) attacks because excess or unauthorized packets are dropped closer to the attack source. Thus, we avoid unnecessary data processing and forwarding at the target node and the network itself.

Even though we focus on MANETs, our system can also be used in wired networks. However, MANETs provide our architecture both advantages and challenges. Specifically, the ratio of CPU cycles to available bandwidths (Hz/kbit) is normally higher in MANET nodes compared to their wired counterparts. This enables us to do more intelligent processing (and use cryptography) on most or all of the packets transiting through a MANET node. The number of traffic flows handled by a MANET node is also small due to the small network size. However, frequent route changes between a source and a destination node due to node mobility represents a difficult challenge in an distributed enforcement environment such as ours.

The rest of the paper is organized as follows. We begin by describing the threat model in Section 2. We then present the system architecture and a high-level overview of our scheme, including the security analysis, in Section 3. Related work is discussed in Section 4.

2   Threat Model

Our goal is to protect network resources and end-node services from denial of service attacks, and to enforce access control rules in the absence of a fixed topology. Thus, we want a node to be able to access only the services it is entitled to, and to limit the amount of traffic that can be sent to any such service. To preserve bandwidth and power, we need to filter any unauthorized traffic early on.

We assume MANET environments where an adversary may be an existing node that has been compromised (insider) or a malicious external node that might want to participate in the MANET. In addition, there may be multiple cooperating adversaries; and compromised nodes may not be detected as such immediately, or ever (depending on their actions).

The resources needed to access a service are allocated by the *group controller(s)* (GCs) of the MANET. Group controllers are nodes responsible for maintaining the group membership for a set of MANET nodes, and *a priori* authorize communications within the group. This means that GCs do not participate in the actual communications, nor do they need to be consulted by nodes in real time; in fact, if they distribute the appropriate policies ahead of time, they need not even be members of the MANET. In most cases, the GC may be reachable through a high-energy-consumption, high-latency, low-bandwidth long-range link (*e.g.*, a satellite connection); interactions in such an environment should be kept to a minimum, and only for exceptional circumstances (*e.g.*, for revoking access for compromised nodes).

Without compromising a GC, an external node can participate in a MANET only by stealing the authorization credentials that are bound to the identity of a legitimate node. Because we envision GCs as being primarily offline or, at best, intermittently reachable (with respect to the MANET), we are not addressing the issue of compromised controllers in this paper.

If a node is compromised, an adversary can only access the services and bandwidth that node is authorized to access. If other MANET nodes are adhering to our architecture, a compromised node does not have the ability to disrupt or interfere with end-to-end service connectivity and other nodes beyond its local radio communication radius. The nodes providing services will receive only the traffic that the compromised node is authorized to transmit, unless the adversary is in the local communication radius.

3   System Architecture

In our architecture, there is one or more pre-defined nodes that act as a *group controller* (GC). These nodes are trusted by all the group nodes. For simplicity and without loss of generality, we will assume that all the MANET nodes are part of a single group. A group controller has authority to assign resources to the nodes in MANET. These resources are expressed in terms of limits on the number of packets or on bandwidth rates that a MANET participant is permitted to transmit toward another node. The resource allocation by the GC to a node is represented using a credential called *policy token* that all the nodes can verify. The policy tokens are typically provisioned ahead of time, and represent the projections of centralized policy, even though an on-demand allocation from the GC is possible. The GC may be offline after it distributes the policy tokens, and may be reachable sporadically at best after that (as external connectivity permits). The presence of the GC is not required, after the initial policy token distribution, for the normal working of the protocol.

When a node (initiator) requests a service from another MANET node (responder) using the policy token assigned to the initiator, the responder can provide a capability back to the initiator. This is called a *network capability*, and it is generated based on the resource policy assigned to the responder and its dynamic conditions (*e.g.*, level of utilization).

All the nodes in the path from an initiator to a responder (*i.e.*, nodes relaying the packets) are required to enforce and abide by the resource allocation encoded by the GC in the policy token and the responder in the network capability. The enforcement involves both accessibility and bandwidth allocation. A responder accepts packets (except for the first one) from an initiator only if the initiator has authorization to send, in the form of a valid network capability. An intermediate node will forward the packets from a node only if the packets have an associated policy token and network capability, and if they do not violate the conditions contained therein. Note that the possession of a network capability does not imply resource reservation; they are the maximum limit a node can use. Available resources are allocated by the intermediate nodes in a fair manner, in proportion to the allocations defined in the policy token and network capability. Intermediate nodes cache policy tokens and network capabilities in a *capability database*, treating them soft state.

Figure 1 gives an overview of the protocol exchanges when an initiator wants to communicate with a responder. The initiator has a policy token previously issued by the GC that authorizes the communication with the responder (step 1). The initiator sends a communication request (and, optionally, initial data), along with its policy token toward the responder (step 2). This packet also contains a *transaction id* that the initiator will use in subsequent packets to the same responder. The packet may also contain a network capability that the initiator generates; this can be used by the responder to communicate back to the initiator. Here, we assume that the initiator has a routing table entry for the responder. Otherwise the underlying routing protocol will be invoked to get the route. An intermediate node will forward the packet only after validating it (step 3). The validation involves cryptographic verification of the capability, and verification of the constraints (*e.g.*, bandwidth usage, service and destination address) specified in the policy token. If the validation is successful, the intermediate node also records the policy token in its capability database, along with other attributes of the packet, such as source and destination node address and the transaction id.

The responder, on receiving the packet verifies the policy token and creates a network capability for the initiator (step 4). The responder sends the response

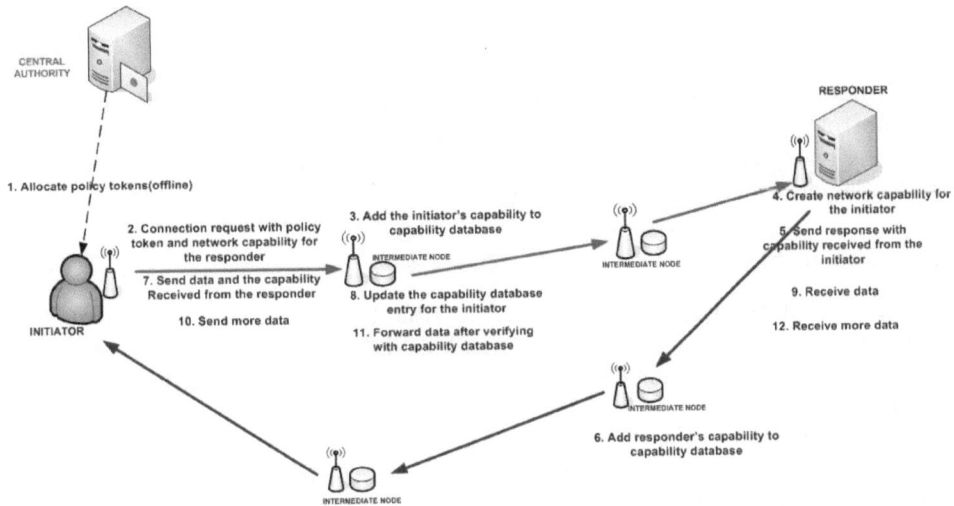

Fig. 1. System overview

to the request as well as the newly created network capability for the initiator (step 5). The responder also creates a transaction id for the communication, and includes it in the response. The responder also needs to include the network capability it received from the initiator in the first message, which authorizes it to communicate back; alternatively (or in addition), it may use a policy token issued by the GC to responder that is authorizing the communication with the initiator. Intermediate nodes, on receiving this packet from the responder, validate the packet and adds the responder's policy token and network capability to its capability database (step 6). In the diagram, the reverse path is shown to be different from the forward path; the paths can also be the same. The initiator will then have to include the responder-issued network capability in subsequent packets it transmits (step 7); intermediate nodes will add this credential to their capability database (steps 8, 9).

Any further data traffic between the initiator and the responder does not contain the policy token and network capability; instead, it contains only the transaction id that was included in the initial handshake (steps 10-12). The packets are signed by the sender, and can be verified by the intermediate nodes. If the cost of the cryptographic operations is too high (in terms of latency or power consumption), cryptographic validation may be done probabilistically. The intermediate nodes can validate the packets by looking at the policy token and network capability contained in the capability database corresponding to the transaction id in the packet. This process ensures that the packet does not exceed the resource limit allowed in the policy token and the network capability, and is authorized to reach the destination by both the GC and the destination itself. For this validation, the intermediate node also maintains the resource usage against each capability in its capability database. The only time the initiator or responder need to re-send the capability is when the path between them changes due to node mobility, or when the network capability expires and is reissued by the peer.

We note that our solution can be used to protect multicast traffic and routing control packets. Furthermore, we can bound the probability of an adversary injecting traffic that remains undetected, when probabilistic cryptographic validation is performed. We omit the details due to lack of space.

3.1 Feasibility

We argue that the proposed solution is feasible for MANETs, even though the memory and processing power are lower in MANET nodes compared to routers in wired networks. Our scheme requires memory to store the information about the traffic sessions, and CPU cycles for the cryptographic operations. The feasibility comes from the fact that the bandwidth in MANETs is significantly lower than that of wired networks, while the nodes are relatively powerful (*e.g.,* normal laptops, or high-end cellphone devices). As a result, the available memory and processing power per packet is higher in MANETs than in wired networks. The processing power per packet for MANET nodes are increasing everyday with the advent of faster but less power-hungry processors for portable devices.

Furthermore, the per-packet cryptographic operations, which involve a public key signature verification, can be achieved with very small key sizes. This is because, unlike traditional uses of public keys, these keys are useful only for the short duration of the session. For longer sessions, new keys can be generated and old ones discarded.

3.2 Capability Definition

Each node has authority to send traffic to its peers at certain rates. This authority is encoded in the policy token and network capability. Both of these are represented by KeyNote-style credentials [3]. Each credential contains

1. Identity of the node (principal)
2. (Optional) Identity of the destination node; if left unspecified, it applies to all destinations
3. Type of service and amount of data the principal is allowed to send
4. An expiration time
5. Signature of the GC (for policy tokens) or peer (for network capabilities)

All nodes in the MANET know the public key of the GCs, so that they can verify policy tokens issued by them. Identities are expressed in term of the long-term public key of the node to which a credential is assigned. The destination node can be a host, subnet, or public key. Type of service refers to the transport protocol identifiers (*e.g.,* TCP ports) a credential authorizes.

Typically, the bandwidth available to a node on a network capability is higher than that of its policy token. Policy tokens are assigned by the GC, which has no knowledge of network load at the time the communication takes place. Hence, the central authority will consider the worst case scenario while assigning the policy token and permit only enough communication to take place for a handshake to occur. It is up to the responder to provide a network capability with enough bandwidth allocation to enable the communication to proceed. Note, also, that it is in the interest of a node to issue short-lived network capabilities to its communicating peers, so that it can quickly respond to changing network dynamics or (more importantly) to peer misbehavior (*e.g.*, a flood-based DoS).

Policy tokens and network capabilities have the same syntactic representation. Following is an example:

```
serial: 130745
owner: unit01.nj.army.mil (public key)
destination: *.nj.army.mil
service: https
bandwidth: 50kbps
expiration: 2010-12-31 23:59:59
issuer: captain.nj.army.mil
signature: sig-rsa 23455656767543566678
```

The above represents a policy token assigned by node captain.nj.army.mil to unit01. The unit can use this policy token to send the traffic to any node in the domain *nj.army.mil*. The peak data rate using this credential cannot exceed 50kbps.

If unit01 wants to communicate with unit02, it will send a message to unit02 using this policy token. Unit02 will issue a network capability for unit01, if the communication needs more bandwidth than available in the policy token.

```
serial: 1567
owner: unit01.nj.army.mil (public key)
destination: unit02.nj.army.mil
bandwidth: 150kbps
expiration: 2007:10:21 13:05:35
issuer: unit02.nj.army.mil
comment: Policy allowing the receiver
    to issue this capability.
signature: sig-rsa 238769789789898
```

This capability is restricted to be used only by unit01 for communication with unit02. It specifies a higher bandwidth, but a shorter expiration date. The issuer of the capability is the same as the destination of the capability.

After receiving this capability, unit01 will use this capability for communication with unit02. The more general policy token can be used by unit01 for communicating with other nodes.

If the communication from unit01 to unit02 was short and required low bandwidth, unit01 could have used its policy token for the entire duration of the communication, without requesting for a network capability from unit02. This will be faster for short communication as there is no capability request/reply, and unit02 does not have to issue any capabilities. If unit01 expects some messages from unit 2 that require more capabilities than the one that is available to unit02 in the form of its corresponding policy token, then unit01 could issue a network capability to unit02.

3.3 Security Analysis

We now discuss how our architecture relates to the threat model described in Section 2.

Since the capabilities are signed by a GC and are verifiable by all nodes, adversaries cannot generate their own valid capabilities. Adversaries can create valid capabilities only if the GC is compromised. Since the individual packets are signed, an adversary cannot use a transaction id that does not belong to it to transmit packets.

A compromised or malicious node that does not enforce the capability protocol can only have impact within its communication radius. Packets generated without the capability or with a snooped transaction id by a malicious node will be dropped by the neighboring nodes due to invalid signatures. A compromised node can only access the services it is authorized to. Packets of nodes trying to use more bandwidth than is allocated to them will be rejected. A malicious node frequently doing this can be detected and isolated.

A receiver can protect against DoS attacks by controlling the issuance of network capabilities to its peers. A malicious node can use its policy tokens or network capabilities to send duplicate packets in multiple disjoint paths; we do not currently protect against this attack, which allows a node to transmit more traffic that it is authorized to. We note, however, that local nodes in the radio perimeter of the misbehaving node can detect this scenario. Since the network capability can be created only based on the policy allowed by the GC, it is not possible for two compromised nodes to collaborate and create arbitrarily large network capabilities.

4 Related Work

Security for mobile *ad hoc* network is an active area of research. Most of the prior work on MANET security focused on solving specific problems or retrofitting security into an existing IP-based network architecture; we are trying to introduce a new architecture where security is built into the network. Surveys of research in MANETs can be found elsewhere [11,13,9].

The concept of capabilities was used in operating system for securing resources [10]. There was work on allowing controlled exposure of resources at the network layer using the concept of "visas" for packets [4], which is similar to network capabilities. More recently, network capabilities were proposed to prevent DoS in wired networks [2]. We extend the concept to MANET and use it for both access control rules and traffic shaping parameters. In the original approach, the capabilities were assigned only by the receivers, and there is no limit on the amount of capability that a receiver can assign. Though it achieves the goal of preventing the DoS attack at the receiver, it does not prevent two nodes from taking up all the available network resources. Their solution also assumes that the links in the path between a sender and receiver cannot be snooped, and the path is fixed. These assumptions are valid for the wire line system that their solution is designed for, but does not work for MANETs. Previous work on distributed firewalls [5] focused on wired fixed-network environments, and attempts to protect only the end hosts using a host-based solution. Our solution is for a mobile network, using a combination of network and host-based solutions that attempt to protect both the network and end-host resources.

Signing and verification of packets between a sender and a receiver were commercially available in early 1990s. Novell's Netware 3.11 and 4.x supported *NCP Packet Signature Option*, where a unique signature was appended to each packet sent between the client and the server [7]. The keys for the signatures were negotiated at login time. Intermediate nodes were not involved in packet verification.

Mitigating the denial of service attacks by including a message authentication code and the certificate of the sender for each packet has been previously proposed [12]. That work does not study the high overhead associated with sending a large signature or a large certificate on each packet. The authors use game theory to study the problem of dealing with selfish nodes that do not verify the packet signatures, using incentives and punishments. This mechanism or any other reputation based mechanism [6] can also be used in our scheme to deal with selfish nodes.

HEAP [1] mitigates various MANET attacks from outsider nodes by doing a hop-by-hop packet authentication using HMAC. MACs (end-to-end or hop-by-hop) cannot deal with insider attacks. They also cannot provide access control unless different MAC keys are used for different policies. Even with different keys, MACs allow rogue nodes to "hide" since MACs are repudiable as all the intermediate nodes in the path between a sender and a receiver need to know the key. Only asymmetric key mechanisms can allow validation by all the intermediate nodes that the packets indeed sent by the source node of the packet.

5   Conclusions and Future Work

We presented a novel architecture for enforcing security policies in MANETs. Our scheme, based on the concept of network capabilities and following a deny-by-default paradigm, can protect both end-host resources and network bandwidth from denial of service attacks, as well as limit the exposure of the MANET to compromised and malicious nodes. We discussed the details of the architecture and protocol used for propagating policy tokens and receivers, and discussed the various scenarios of use. For our future work, we plan to study the impact of our scheme on throughput and latency for different topologies and classes of traffic. In addition, we intend to quantify the performance of multicast traffic on mobility scenarios, and to implement and deploy on MANET testbeds with real traffic.

Acknowledgements

This work was supported in part by the National Science Foundation through Grant CNS-07-14277. Any opinions, findings, conclusions, and recommendations expressed in this paper are those of the authors and do not necessarily reflect the views of the NSF or the US Goverment.

Mansoor Alicherry was supported by Alcatel-Lucent, Murray Hill, New Jersey.

References

1. Akbania, R., Korkmaz, T., Raju, G.: HEAP: A packet authentication scheme for mobile ad hoc networks. In: Communications and Networking Simulation Symposium (2007)
2. Anderson, T., Roscoe, T., Wetherall, D.: Preventing internet denial-of-service with capabilities. In: Proc. of Hotnets-II (2003)
3. Blaze, M., Ioannidis, J., Keromytis, A.: Trust management for ipsec. In: Symposium on Network and Distributed Systems Security, SNDSS (2001)
4. Estrin, D., Mogul, J.C., Tsudik, G.: Visa protocls for controlling interorganizational datagram flow. IEEE Journal on Selected Areas in Communications (May 1989)
5. Ioannidis, S., Keromytis, A.D., Bellovin, S.M., Smith, J.M.: Implementing a distributed firewall, pp. 190–199 (2000)
6. Jaramillo, J., Srikant, R.: Darwin: Distributed and adaptive reputation mechanism for wireless ad-hoc networks. In: MOBICOM (2007)
7. Lee, R.: Netware 4.x performance tuning and optimization: Part 3 (October 1993), http://support.novell.com/techcenter/articles/ana19931001.html
8. Parno, B., Wendlandt, D., Shi, E., Perrig, A., Maggs, B., Hu, Y.-C.: Portcullis: protecting connection setup from denial-of-capability attacks. SIGCOMM Comput. Commun. Rev. 37(4), 289–300 (2007)
9. Shi, E., Perrig, A.: Designing secure sensor networks. IEEE Wireless Communications (2004)
10. Wobber, E., Abadi, M., Burrows, M., Lampson, B.: Authentication in the taos operating system. ACM Transactions on Computer Systems 12 (February 1994)
11. Wu, B., Chen, J., Wu, J., Cardei, M.: A survey on attacks and countermeasures in manets. In: Wireless/Mobile Network Security, ch. 12. Springer, Heidelberg (2006)
12. Wu, X., Yau, D.K.Y.: Mitigating Denial-of-Service Attacks in MANET by Distributed Packet Filtering: A Game-theoretic Approach. In: ASIACCS (March 2007)
13. Yang, H., Luo, H., Ye, F., Lu, S., Zhang, L.: Security in mobile ad hoc networks: Challenges and solutions. IEEE Wireless Communications (2004)

Baiting Inside Attackers Using Decoy Documents

Brian M. Bowen, Shlomo Hershkop, Angelos D. Keromytis,
and Salvatore J. Stolfo

Department of Computer Science Columbia University

Abstract. The insider threat remains one of the most vexing problems in computer security. A number of approaches have been proposed to detect nefarious insider actions including user modeling and profiling techniques, policy and access enforcement techniques, and misuse detection. In this work we propose trap-based defense mechanisms and a deployment platform for addressing the problem of insiders attempting to exfiltrate and use sensitive information. The goal is to confuse and confound an adversary requiring more effort to identify real information from bogus information and provide a means of detecting when an attempt to exploit sensitive information has occurred. "Decoy Documents" are automatically generated and stored on a file system by the D^3 System with the aim of enticing a malicious user. We introduce and formalize a number of properties of decoys as a guide to design trap-based defenses to increase the likelihood of detecting an insider attack. The decoy documents contain several different types of bogus credentials that when used, trigger an alert. We also embed "stealthy beacons" inside the documents that cause a signal to be emitted to a server indicating when and where the particular decoy was opened. We evaluate decoy documents on honeypots penetrated by attackers demonstrating the feasibility of the method.

1 Introduction

Much research in computer security has focused on the means of preventing unauthorized and illegitimate access to systems and information. Unfortunately, the most damaging malicious activity is the result of internal misuse within an organization, perhaps since far less attention has been focused inward. Despite classic internal operating system security mechanisms and the body of work on formal specification of security and access control policies, including Bell-LaPadula [1] and the Clark-Wilson models [4], we still have an extensive insider attack problem. Indeed in many cases, formal security policies are incomplete and implicit or they are purposely ignored in order to get business goals accomplished. There seems to be little technology available to address the insider threat problem.

Insider attack has overtaken viruses and worm attacks as the most reported security incident according to a report from the US Computer Security Institute

Y. Chen et al. (Eds.): SecureComm 2009, LNICST 19, pp. 51–70, 2009.
© Institute for Computer Science, Social-Informatics and Telecommunications Engineering 2009

(CSI) [19]. The annual Computer Crime and Security Survey for 2007 surveyed 494 security personnel members from US corporations and government agencies, finding that insider incidents were cited by 59 percent of respondents, while only 52 percent said they had encountered a conventional virus in the previous year. The state-of-the-art seems to be still driven by forensics analysis after an attack, rather than technologies that prevent, detect, and deter insider attack.

We define insider threats by differentiating between Masqueraders (attackers who impersonate another inside user) and Traitors (an inside attacker using their own legitimate credentials). One possible solution for masquerade detection involves anomaly detection [27]. In this approach, users actions are profiled to form a baseline of normal behavior. Subsequent monitoring for abnormal behaviors that exhibit large deviations from this baseline [16] signal a potential insider attack. The common strategy to prevent inside attacks involves policy-based access control techniques to limit the scope of systems and information an insider is authorized to use, and hence, limit the damage the organization may incur when an insider goes awry. Prevention techniques may not always succeed, and thus, monitoring and detection techniques are needed when prevention fails. In this paper, we are focused on different techniques aimed at detecting masqueraders and traitors.

We note that some external attackers can become insiders when an outsider attains internal network access. Many attacks use spyware and rootkits [3], which give outsiders internal access. Such software can easily be installed on systems from physical or digital media (e.g., email, downloads) and allow an attacker administrator or "root" access on a machine along with a means to gather sensitive data. Rootkits have the ability to conceal themselves and elude detection, especially when the rootkit is previously unknown, as is true in zero-day attacks [8]. An external attacker that manages to install rootkits internally in effect becomes an insider, thereby multiplying the ability to inflict harm. Although the techniques described in this paper may have utility for these cases, in this paper our primary focus is on human insiders attempting to exfiltrate sensitive information. By exfiltration we mean unauthorized copying and transmission of information by any means.

The insider attack defense system described in this paper is of an offensive nature, intended to confuse and deceive a traitor by leveraging uncertainty, to reduce the knowledge they ordinarily have of the systems and data they might be authorized to use. This work considers methods to detect insider actions against enterprise systems as well as individual hosts and laptops. We introduce a deception system to distribute potentially large amounts of decoy information with the aim to detect nefarious acts as well as to increase the workload of an attacker to identify real information from bogus information, rather than providing unfettered access as broadly exists today. We developed a system to generate and place *decoy documents* within a file system. Our system generates decoy documents containing decoy credentials that are monitored (*e.g.*, Gmail credential monitoring) for misuse and stealthily embedded beacons that signal an alert when the document is opened.

To achieve the goal of wide spread deception we must consider methods to trap a wide variety of potential insiders with varying levels of sophistication. Toward this goal, we developed a proof-of-concept system we call D^3, the Decoy Document Distributor system. Samples of D^3 generated documents are presented in the Appendix. The contributions of this paper include:

- A novel set of generally applicable properties are proposed to guide the design and deployment of decoys and maximize the deception they induce for different classes of insiders who vary by their level of knowledge and sophistication.
- A large-scale automated creation and management system for deploying decoys that can detect the presence (and, in some cases, "identity") of malicious insiders, or at least indicate malicious insider activity. This provides a means for ordinary users to deploy honey documents without having to setup sophisticated honeypot systems and sensors.
- An offensive trap-based defense system is proposed to detect masqueraders and traitors, and to flood attackers with bogus exfiltrated information that they must analyze in order to find real information of value. Hence, our long term goal is to flood the miscreant marketplace with bogus information devaluing their quarry.
- A design of decoy information that combines a number of methods and monitors, both internal and external, to detect insider exploitation using a common and ubiquitous set of baited targets, ordinary looking documents.
  1. A watermark is embedded in the binary format of the document file to detect when the decoy is loaded in memory, or egressed in the open over a network.
  2. A "beacon" is embedded in the decoy document that signals a remote web site upon opening of the document indicating the malfeasance of an insider illicitly reading bait information.
  3. If these methods fail to detect an insider attack or an exfiltration of baited documents, the content of the documents contain bait and decoy information that is monitored as well. Bogus logins at multiple organizations as well as bogus and realistic bank information is monitored by external means.
- An easy to use system to broadly deploy decoys to ordinary users who are alerted by email when a decoy has been touched on their laptops and personal computers; no such system presently exists.

The reader is encouraged to visit the Decoy Document Distribution (D^3) web site to evaluate our technology developed to date at: http://www.cs.columbia. edu/ids/RUU/Dcubed[1].

2   Related Work

The use of deception, or decoys, plays a valuable role in the protection of systems, networks, and information. The first use of decoys (*i.e.*, in the cyber domain)

---
[1] Some features are restricted for internal use only.

has been credited to Cliff Stoll [29,23] and detailed in his novel "The Cuckoos Egg" [24], where he provides a thorough account of his crusade to catch German hackers breaking into Lawrence Berkeley Laboratory computer systems. Stoll's methods included the use of bogus networks, systems, and documents to gather intelligence on the German attackers who were apparently seeking state secrets. Among the many techniques waged, he crafted "bait" files, or in his case, bogus classified documents that really contained non-sensitive government information and attached "alarms" to them so that he would know if anyone accessed at them. To Stoll's credit, a German hacker was eventually caught and it was found that he had been selling secrets to the KGB.

Honeypots are effective tools for profiling attacker behavior. Honeypots are considered to have low false positive rates since they are designed to capture only malicious attackers, except for perhaps an occasional mistake by innocent users. Spitzner described how honeypots can be useful for detecting insider attack [22] and discusses the use of honeytokens [23] such as bogus medical records, credit card numbers, and credentials. In a similar spirit, Webb et al. [26] showed how honeypots can be useful for detecting spammers. In current systems, the decoy/honeytoken creation is a laborious and manual process requiring large amounts of administrator intervention. Our work extends these basic ideas to an automated system of managing the creation and deployment of these honeytokens.

Yuill et al. [29] extend the notion of honeytokens with a "honeyfile system" to support the creation of bait files, or as they define them, "honeyfiles." The honeyfile system is implemented as an enhancement to the Network File Server. The system allows for any file within user file space to become a honeyfile through the creation of a record associating a filename to userid. The honeyfile system monitors all file access on the server and alerts users when honeyfiles have been accessed. Their work does not focus on the content or automatic creation of files, but they do elicit some of the challenges of creating deceptive files (with respect to names) that we address in section 4.

In this paper, we introduce a set of properties of decoys to guide their design and maximize the deception they induce for different classes of insiders who vary by their level of knowledge and sophistication. To the best of our knowledge, the synthesis of these properties is indeed novel a contribution. Bell and Whaley [2] have described the structure of deception as a process of hiding the real and showing showing the false. They introduce several methods of hiding that include masking, repackaging, and dazzling, along with three methods of showing that include mimicking, inventing, and decoying. Yuill et al. [28] expand upon this work and characterize deceptive hiding in terms of how it defeats an adversary's discovery process. They describe an adversary's discovery process as taking three forms: direct observation, investigation based on evidence, and learning from other people or agents. Their work offers a process model for creating deceptive hiding techniques based on how they defeat an adversary's discovery process.

The decoy documents introduced in this paper utilize similar deception mechanisms as well as beacons to signal a remote detect and alert in real-time time when a decoy has been opened. Web bugs are a class of silent embedded tokens which have been used to track usage habits of web or email users [17]. Unfortunately, they have been most closely associated with unscrupulous operators, such as spammers, virus writers, and spyware authors who have used them to violate users privacy. Typically they will be embedded in the HTML portion of an email message as a non-visible white on white image, but they have also been demonstrated in other forms such as Microsoft Word, Excel, and PowerPoint documents [20]. When rendered as HTML, a web bug triggers a server update which allows the sender to note when and where the web bug was viewed. Animated images allow the senders to monitor how long the message was displayed. The web bugs operate without alerting the user of the tracking mechanisms. The advantage for legitimate advertisers is that this allows them to monitor advertisement effectiveness, while privacy advocates worry that this technology can be misused to spy on users' habits. Our work leverages the same ideas, but extends them to other document classes and is more sophisticated in the methods used to draw attention. In addition, our targets are insiders who should have no expectation of privacy on a system they violate.

3   Threat Model - Level of Sophistication of the Attacker

The insider seeks to identify and avoid the decoys and abscond with "real" information. We broadly define four monotonically increasing levels of insider sophistication and capability. Some will have tools available to assist in deciding what is a decoy and what is real. Others will only have their own observations and thoughts.

- **Low:** Direct observation is the only tool available. The adversary largely depends on what can be gleaned from a first glance. We strive to defeat this level of adversary with our beacon documents, even though decoys with embedded beacons may be distinguished with more advanced tools.
- **Medium:** A more thorough investigation can be performed by the insider; decisions based on other, possibly outside evidence, can be made. For example, if a decoy document contains a decoy account credential for a particular identity, an adversary may verify that the particular identity is real or not by querying an external system (such as www.whitepages.com). Such adversaries will require stronger decoy information possibly corroborated by other sources of evidence.
- **High:** Access to the most sophisticated tools are available to the attacker (*e.g.,* super computers, other informed people who have organizational information). The notion of the "Perfect Decoy" described in the next section may be the only indiscernible decoy by an adversary of such caliber.
- **Highly Privileged:** Probably the most dangerous of all is the privileged and highly sophisticated user. Such attackers might even be aware that the system is baited and will employ sophisticated tools to try to analyze, disable, and avoid decoys entirely. As an example of how defeating this level of threat

might be possible, consider the analogy with someone who knows encryption is used (and which encryption algorithm is used), but still cannot break the system because they do not have knowledge of an easy-to-change operational parameter (the key). Likewise, just because someone knows that decoys are used in the system does not mean they should be able to identify them. This is the principal– coming up with a scheme to satisfy it remains an open problem.

4   Generating and Distributing Bait

In order to create decoys to bait various levels of insiders, one must understand the core properties of a decoy that will successfully bait an insider.

4.1   Decoy Properties

We enumerate various properties and means of measuring these properties that are associated with decoy documents to ensure their use will be likely to snare an inside attacker. We introduce the following notation for these definitions.

Believable[2]: Capable of eliciting belief or trust; capable of being believed; appearing true; seeming to be true or authentic.

A good decoy should make it difficult for an adversary to discern whether they are looking at an authentic document from a legitimate source or if they are indeed looking at a decoy. We conjecture that believability of any particular decoy can be measured by adversary's failure to discern one from the other. We formalize this by defining a decoy believability experiment. The experiment is defined for the document space M with the set of decoys D such that $D \subseteq M$ and $M - D$ is the set of authentic documents.

The Decoy Believability Experiment: $\text{Exp}_{A,D,M}^{believe}$

- For any $d \in D$, choose two documents $m_0, m_1 \in M$ such that $m_0 = d$ or $m_1 = d$, and $m_0 \neq m_1$; that is, one is a decoy we wish to measure the believability of and the second is chosen at random from the set of authentic documents.
- Adversary A obtains m_0, m_1 and attempts to choose $\hat{m} \in \{m_0, m_1\}$ such that $\hat{m} \neq d$, using only information intrinsic to m_0, m_1.
- The output of the experiment is 1 if $\hat{m} \neq d$ and 0 otherwise.

For concreteness, we build upon the definition of "Perfect Secrecy" proposed in the cryptography community [12] and define a "perfect decoy" when:

$$\Pr[\text{Exp}_{A,D,M}^{believe} = 1] = 1/2$$

---

[2] For clarity, each property is provided with its definition gleaned from online dictionary sources.

The decoy is chosen in a believability experiment with a probability of 1/2 (the outcome that would be achieved if the volunteer decided completely at random). That is, a perfect decoy is one that is completely indistinguishable from one that is not. A benefit of this definition is that the challenge of showing a decoy to be believable, or not, reduces to the problem of creating a "distinguisher" that can decide with probability better than 1/2.

In practice, the construction of a "perfect decoy" might be unachievable, especially through automatic means, but the notion remains important as it provides a goal to strive for in our design and implementation of systems. For many threat models, it might suffice to have less than perfect believable decoys. For our proof-of-concept system described below, we generate receipts and tax documents, and other common form-based documents with decoy credentials, realistic names, addresses and logins, all information that is familiar to all users.

We note that the believable property of a decoy may be less important than other properties defined below since the attacker may have to open the decoy in order to decide whether the document is real or not. The act of opening the document may be all that we need to trap the insider, irrespective of the believability of its content. Hence, enticing an attacker to open a document, say one with a very interesting name, may be a more effective strategy to detect an inside attack than producing a decoy document with believable content.

Enticing: highly attractive and able to arouse hope or desire; "an alluring prospect"; lure.

Herein lies the issue of how does one measure the extent to which a decoy arouses desires, how well is it a lure? One obvious way is to create decoys containing information with monetary value, such as passwords or credit card numbers that have black market value [14,25]. However, enticement depends upon the attacker's intent or preference. We define enticing documents in terms of the likelihood of an adversary's preference; enticing decoys are those decoys that are chosen with the same likelihood. More formally, for the document space M, let P be the set of documents of an adversary's A preference, where $P \subseteq M$. For some value ϵ such that $\epsilon > 1/|M|$, an enticing document is defined by the probability

$$\Pr[m \to M | m \in P] > \epsilon$$

where $m \to M$ denotes m is chosen from M. An enticing decoy is then defined for the set of decoys D, where $D \subseteq M$, such that

$$\Pr[m \to M | m \in P] = \Pr[d \to M | d \in D]$$

We posit that by defining several general categories of "things" that are of "attacker interest", one may compose decoys using terms or words that correspond to desires of the attacker that are overwhelmingly enticing. For example, if the attacker desires money, any document that mentions or describes information that provides access to money should be highly enticing. We believe we can measure frequently occurring (search) terms associated with major categories of interest (*e.g.*, words or terms drawn from finance, medical information, intellectual property) and use these as the constituent words in decoy documents. To measure the effectiveness of this generative strategy, it should be possible to execute content searches and count the number of times decoys appear in the top 10 list of displayed documents. This is a reasonable approach also, to measuring how conspicuous, defined below, the decoys become based upon the attacker's searches associated with their interest and intent.

Conspicuous: easily visible; easily or clearly visible; obvious to the eye or mind; Attracting attention.

A *conspicuous* decoy should be easily found or observed. Conspicuous is defined similar to enticing, but conspicuous documents are found because they are easily observed, whereas enticing documents are chosen because they are of interest to an attacker. For the document space M, let V be the set of documents defined by the minimum number of user actions required to enable their view. We use a subscript to denote the number of user actions required to view some set of documents. For example, documents that are in view at logon or on the desktop (requiring zero user actions) are labeled V_0, those requiring one user action are V_1, etc. We define a "view", V_i of a set of documents as a function of a number of user actions applied to a prior view, V_{i-1}, hence

$$V_i = \text{Action}(V_{i-1}) \text{ where } V_j \neq V_i, j < i$$

An "Action" may be any command or function that displays files and documents, such as 'ls', 'dir', 'search.' For some value ϵ such that $\epsilon > 0$, a conspicuous document, d, is defined by the probability

$$\prod_{i=0}^{n} \Pr[V_i] > \epsilon$$

where n is the minimum value where $d \in V_n$. Note if d is on the desktop, V_0, $\Pr[V_0] = 1$ (*i.e.*, the documents in full view are highly conspicuous).

When a user first logs in, a conspicuous decoy should either be in full view on the desktop, or viewable after one (targeted) search action. One simple user action is optimal for a highly conspicuous decoy. Thus, a measure of conspicuousness may be a count of the number of search actions needed, on average, for a decoy to appear in full view. The decoy may be stored in the file system anywhere if a simple content-based search locates it in one step. But, this search act depends upon the query executed by the user. The query can either be a location (*e.g.*, search for a directory named "TAX" in which the decoy appears) or a content query (*e.g.*, using Google Desktop Search for documents containing the word "TAX.") In either case, if a decoy document appears after one such search, it is conspicuous. Hence, we may define the set P as all such files that can be found in some number of steps. But, this depends upon what search terms the attacker uses to query! If the decoy never appears because the attacker used the wrong search terms, the decoy is not conspicuous. We posit that the property of *enticing* is likely the most important property, and a formal measure to evaluate enticement will generate better decoys. In summary, an enticing decoy should be conspicuous to be an effective decoy trap.

Detectable; to discover or catch (a person) in the performance of some act: to detect someone cheating.

Decoys must ensure an alert is generated if they are exploited. Formally, this is defined for adversary A, document space M, and the set of decoys D such that $D \subseteq M$. We use $Alert_{A,d} = 1$ to denote an alert for $d \in D$. We say d is detectable with probability ϵ when

$$\Pr[d \to M : Alert_{A,d} = 1] \geq \epsilon$$

Ideally, ϵ should be 1.

We designed the decoy documents with several techniques to provide a good chance of detecting the malfeasance of an inside attack in real-time.

- At time of application start-up, the decoy document emits a beacon alert to a remote server.
- At the time of memory load, a host-sensor, such as an antivirus scanner, may detect embedded tokens placed in a clandestine location of the document file format.
- At the time of exfiltration, a NIDS such as Snort, or a stream event detection system such as Cayuga [5] may be used to detect these embedded tokens during the egress of the decoy document in network traffic where possible.
- At time of information exploitation and/or credential misuse, monitoring of decoy logins and other credentials embedded in the document content by external systems will generate an alert that is correlated with the decoy document in which the credential was placed.

This extensive set of monitors maximizes ϵ, forcing the attacker to expend considerable effort to avoid detection, and hopefully will serve as a deterrent to reduce internal malfeasance within organizations that deploy such a trap-based defense. In the proof-of-concept implementation reported in this paper, we focus our evaluation on the last item. We utilize monitors at our local IT systems, at Gmail and at an external bank.

Variability: The range of possible outcomes of a given situation; the quality of being subject to variation.

Attackers are humans with insider knowledge, even possibly with the knowledge that decoys are liberally spread throughout an enterprise. Their task is to identify the real documents from the potentially large cache of decoys. One important property of the set of decoys is that they are not easily identifiable due to some common invariant information they all share. A single search or test function would thus easily distinguish the real from the fake. The decoys thus must be highly varied. We define variable in terms of the likelihood of being able to decide the believability of a decoy given *any* known decoy. Formally, we define *perfectly variable* for document space M with the set of decoys D such that $D \subseteq M$ where

$$\Pr[d' \to D : \mathrm{Exp}_{A,D,M,d'}^{believe} = 1] = 1/2$$

Observe that under this definition an adversary may have access to *all* N previously generated decoys with the knowledge they are bogus, but still lack the ability to discern the N+1st. From a statistical perspective, each decoy is independent and identically distributed. For the case that an adversary can determine the N+1st decoy only after observing the N prior decoys, we define this as an *N-strong Variant*.

Clearly, a good decoy generator should produce an unbounded collection of enticing, conspicuous, but distinct and variable documents. They are distinct with respect to string content. If the same sentence appears in 100 decoys, one would not consider such decoys with repetitive information as highly variable; the common invariant sentence(s) can be used as a "signature" to find the decoys, rendering them distinguishable (and clearly, less enticing).

Non-interference: Something that does not hinder, obstructs, or impede.

Introducing decoys to an operational system has the potential to *interfere* with normal operations in multiple ways. Of primary concern is that decoys may pollute authentic data so that their legitimate usage becomes hindered by corruption or as a result of confusion by legitimate users (*i.e.,* they cannot differentiate real from fake). We define non-interference in terms of the likelihood of legitimate users successfully accessing normal documents after decoys are introduced. We use $\mathrm{Access}_{U,m} = 1$ to denote the success of a legitimate user U accessing a normal document m. More formally, for some value ϵ, the document space M, $\forall m \in M$ we define

$$\Pr[Access_{U,m} = 1] \geq \epsilon$$

on a system without decoys. Non-interference is then defined for the set of decoys D such that $D \subseteq M$ and $\forall m \in M$ we have

$$\Pr[Access_{U,m} = 1] = \Pr[Access_{U,m} = 1|D]$$

Although we seek to create decoys to ensnare an inside attacker, a legitimate user whose data is the subject of an attacker must still be able to identify their own real documents from the planted decoys. The more enticing or believable a decoy document may be, the more likely it would be to lead the user to confuse it with a legitimate document they were looking for. Our goal is to increase believability, conspicuous, and enticingness while keeping interference low; ideally a decoy should be completely non-interfering. The challenge is to devise a simple and easy to use scheme for the user to easily differentiate their own documents, and thus a measure of interference is then possible as a by-product.

Differentiable: to mark or show a difference in; constitute a difference that distinguishes; to develop differential characteristics in; to cause differentiation of in the course of development.

It is important that decoys be "obvious" to the *legitimate user* to avoid interference, but "unobvious" to the insider stealing information. We define this in terms of an inverted believability experiment, in which the adversary is replaced by a legitimate user. We say a decoy is differentiable if the legitimate user always succeeds. Formally, we state this for the document space M with the set of decoys D such that $D \subseteq M$ where

$$\Pr[\mathrm{Exp}_{U,D,M}^{believe} = 1] = 1$$

How might we easily differentiate a decoy for the legitimate user so that we maintain "non-interference" with the user's own actions and legitimate work? The remote thief who exfiltrates all of a user's files onto a remote hard drive may be perplexed by having hundreds of decoys amidst a few real documents; the thief should not be able to easily differentiate between the two cases. If we store a hundred decoys for each real document, the thief's task is daunting; they would need to test embedded information in the documents to decide what is real and what is not, which should complicate their end goals. For clarity, decoys should be easily *differentiable* to the legitimate user, but not to the attacker without significant effort. Thus, the use of "beacons" or other embedded content in the binary file format of a document, must be judiciously designed and deployed to avoid making decoys trivially differentiable for the attacker.

4.2   The Decoy Document Distributor (D^3) System

The D^3 web-based service generates and distributes decoy documents to registered users. The decoy properties guide the design of decoy templates in D^3 that are used to generate specific documents for download. The content of each decoy document includes several types of "bait" information such as online banking logins provided by a collaborating financial institution[3], login accounts for online servers, and web based email accounts. In our deployment we used Columbia University student accounts and Gmail email accounts as bait, but these can be customized to any set of monitored credentials. These decoy credentials are "bait" and are enticing targets for different types of adversaries [14,13].

4.3   Decoy Document Design

The primary goal of the trap based defense is to detect malfeasance. Since no system is foolproof, we propose that multiple overlapping signals be embedded in the decoy documents to ensure *detectability*. Any alert generated by the multiple decoys is an indicator that some insider activity has occurred. Since the attacker may have varying levels of sophistication, a combination of traps are used in

---

[3] By agreement, the institution request that its name be withheld.

decoy documents to increase the likelihood one will succeed in generating an alert. A sophisticated attacker may, for example, disable the internal beacon, or cut off network connections avoiding communication, disable or kill local host monitoring processes, or they may exfiltrate documents via a web-browser without opening them locally. The documents are designed with several means of detecting their misuse:

- embedded honeytokens, computer login accounts created that provide no access to valuable resources, and that are monitored when (mis)used;
- embedded honeytoken banking login accounts specifically created and monitored for this trap-based technology demonstration specifically to entice financially motivated attackers;
- a network-level egress monitor that alerts whenever a marker, specially planted in the decoy document, is detected (we are collaborating with Cornell to use Cayuga [5] for this purpose. Presently Snort may be used as simple signature detector as a proof-of-concept);
- a host-based monitor that alerts whenever a decoy document is "touched" in the file system such as a copy operation;
- an embedded "beacon" alerts a remote server at a site at Columbia that we call SONAR. The web site emits an email to the registered user who created and downloaded the decoy document.

The implementation of features are described below.

Honeytokens. This layer of defense is made up of "bait" information such as online banking logins provided by a collaborating financial institution, credit card numbers, login accounts for online servers, and web based email accounts. The primary requirement for bait is that it be detectable when (mis)used. For example, one form of bait that we use are usernames and passwords for Gmail accounts. D^3 is integrated with a variety of services to enable monitoring of these credentials once they are deployed as decoys. In the case of the Gmail accounts, custom scripts access *mail.google.com* to parse the bait account pages, gathering account activity information. The information includes the IP addresses for the previous 5 account accesses and the time. If there is any activity from IP addresses other than D^3's monitor, an alert is triggered with the time and IP of the offending host. Alerts are also triggered when the monitor cannot login to the bait account. In this case, we conclude that the account password was stolen (unless monitoring resumes) and the password changed unless other corroborating information (like a network outage) can be used to convince otherwise. In addition, some of our accounts have password monitors, allowing us to produce a seemingly unbounded collection of decoy variants for individual usernames.

In the case of financially motivated bait, we are beginning to use real credit card numbers in addition to banking login credentials. Many credit card providers offer "one-time-credit-card numbers" and other forms of Controlled Payment Numbers [18], which enable the generation of multiple credit card numbers for a single account. In the case of PayPal, single use credit card numbers can be generated with a predetermined balance. The D^3 monitor is being integrated with the PayPal APIs to automatically monitor the activity of the credit card numbers deployed through D^3. As is the case for all of the decoys, the benefit of deployment through D^3 is the automation, enabling their creation, monitoring, and distribution en masse.

Beacon Implementation. The highly sophisticated attacker will likely attempt to differentiate between a real document and a decoy by analyzing the binary file format prior to opening a file. This necessitates a design where beacon code and watermarks in decoy documents are hidden to avoid their easy identification. The attacker would surely avoid the decoys if they could easily identify them by a simple static test for an embedded beacon. The beacon code can be embedded in documents in a number of ways and made to appear statistically equivalent to its surrounding data using a blending technique called "spectrum shaping" (see [21,6]). Such obfuscation techniques are very hard to defeat [15].

Using common techniques developed for malware, beacons attempt to silently contact a centralized server with a unique token embedded within the document at creation time. The token is used to identify the decoy and document, the IP address of the host accessing the decoy document. Depending on the particular document type and the rendering environment used during viewing of the beacon document, some additional data may be collected.

The first proof-of-concept beacons have been implemented in MS Word and PDF and deployed through the D^3 web site. In the case of the MS Word document beacons, the examples rely on a stealthily embedded remote image that is rendered when the document is opened. The request for the remote image is a positive indication the document has been opened. In the case of PDF document beacons, the signaling mechanism relies on the execution of Javascript within the document. The D^3 site includes a tutorial guiding the user on how to generate, download, and enable the decoys' silent communication on hosts. It is important to point out that there are methods for disabling the beacon mechanism. In Section 5.2, we provide an evaluation of beacon robustness.

Embedded Marker Implementation. Beacon documents contain embedded markers that a host or network sensor may detect either when documents are loaded in memory or transmitted in the clear. The markers are constructed as a unique pattern of word tokens uniquely tied to the document creator. The sequence of word tokens is embedded within the beacon document's meta-data area or reformatted as comments within the document format structure. Both locations are ideal for embedding markers since most rendering programs ignore these parts of the document. The embedded markers can be used in Snort signatures for detecting exfiltration.

5   Evaluation

5.1   Masquerade Detection Using Decoy Documents as Bait

We have defined the general properties that decoys should have and discussed how we may measure these properties, but here we focus on the most important property: *detectability*. Under ideal testing conditions, decoy efficacy could be shown through deployment on true operational systems either within an enterprise environment, or on personal computers, by the number of attacks they are able to detect or thwart (they have a deterrence effect). However, given reasonable time limits, the infrequency of attacks within the insider threat model makes this approach impractical within a university environment. As we mentioned we are now seeking a larger user population to study and measure decoy generation over time.

Another approach to evaluation is a user study in which users are organized and asked to evaluate decoys based on each of the key decoy properties mentioned earlier. We take human evaluation to be the gold standard of evaluation since the human mind is the ultimate target of our decoys. That is, we wish to show how well our decoys can induce deception on human test subjects. One of the challenges of conducting a traditional user study lies in the logistics of obtaining volunteers. In our methodology, we attempt to reduce this challenge by leveraging external attackers to serve as participants in our study on masquerade detection. To do so, we "invite" attackers (or more accurately, bamboozle them) into our study by attracting them with a set of vulnerable systems on the university network, which also serve as our testing platform.

Our test platform is embedded within a honeynet [9]. It consists of several virtual machines running Linux and configured with Sebek [10] to capture attacker activities including commands and file references. In order to limit potential damage from system compromise and still allow for testing, we configured the honeynet to allow all incoming connections while restricting the number of outgoing connections.

The virtual machine hosts within the honeynet were configured with accounts and home directories for three decoy usernames. To make the environment as real as possible, genuine data from personal accounts on other systems were loaded into each of the home directories. We changed name references within the data to reflect those of the appropriate decoy users. In total, our phony user accounts contained 15 or more directories and 50-100 files. The hosts were then seeded with several of D^3's decoy files using the decoy distributor utility. The decoy files were generated to have conspicuous names such as "stolen passwords", "credit card", "private data", and "Gmail AccountInfo", but were distributed within the polluted home directories of the decoy accounts, making the environment as real as possible.

To lure test subjects into the study, our initial approach was to use attackers that attempt to gain internal access via password scanning. Password scanning attacks are common on the university network, where attempts on a typical machine are in the range of thousands per day. To enable attacker access, we conducted a short study to first determine the most common usernames and passwords (excluding those for root and actual users) used in these attempts. We created accounts with several of these usernames and passwords, to quickly learn that this breed of attacker was not going to suffice for our user study; their sole purpose seemed confined to creating zombies for botnets. While this may be a valid threat to study while evaluating decoys [7], allowing bots to operate on the university network poses too much risk.

In our second and more aggressive approach, we narrowed our recruitment effort to web forums and IRC channels with the expectation and hope that we would get fewer attacks involving botnets. In this approach, we selected several high volume forums to solicit volunteers and posted variations of invitations with messages that included hostnames, usernames, and passwords. The idea was to provide just enough innocent-looking information from a novice to lure people into our machines without providing direct evidence that we were conducting a deception-based experiment. Note that we deliberately omit the names of the forums used and the exact details of the messages, as this is an ongoing study.

While our methodology could, in theory, provide anyone with access to our test platform, by selectively choosing the location of postings and contents postings, we expected to recruit two primary classes of individuals:

- Legitimate and generally curious computer-savvy individuals. These users have no interest in extending privileges in an unauthorized way, but participate in the study out of curiosity, as there is no other incentive.
- Unscrupulous opportunistic hackers who attempt to extend their network access by whatever means afforded to them. These individuals are enticed by our posting as they see our machines as low "hanging fruit" in their targeting campaign.

In either case, we believe these individuals to be suitable candidates for our study (with one caveat mentioned later). Both classes of individuals can be used in measuring the enticement property of decoys. We measure this by examining the behavior exhibited in file access, both with respect to the particular files a user attempts to read and in the order in which the files are read. For example, if all users consistently read the same file first, we know the file must indeed be enticing.

In regards to indistinguishability of the decoys, we note that the content of these decoys contains bait information in the form of monitored credentials on real systems. Certainly, if our attackers take the time to use the decoy credentials, there is an implication that they must also be believable. More importantly though, if they use the credentials and we detect their use, we have also answered the most important question of – can we *detect* the attacker? Note that the first class of the individuals is by definition, not useful for this part of the study. That is, attempting to use credentials found on our machines is clearly an illicit activity, which they would not partake in. Unfortunately, we do not have a good way of distinguishing the two classes, so we get slightly skewed results.

Over the span of the first week[4], our hosts netted 20 unique users, determined anecdotally by source IP address and reinforced by the unique behaviors exhibited by each. The length of user sessions ranged from minutes to hours. The users that spent hours were focused on activities that included writing code and attempting to install other software, presumably with malicious intent. The

---

[4] Most attacks occurred within the first 8-hour period after posting.

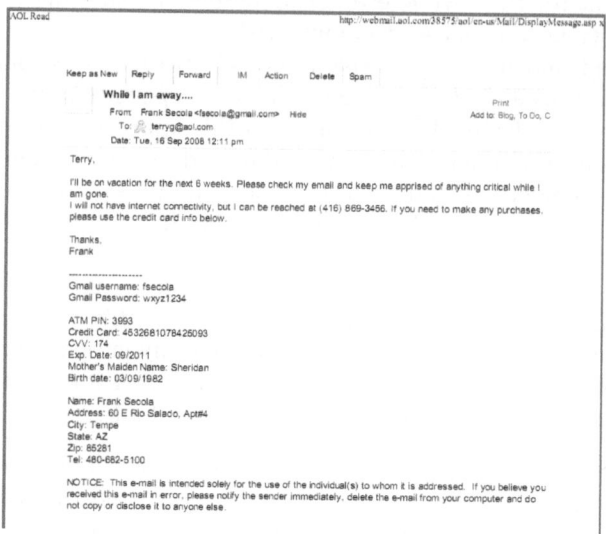

Fig. 1. Decoy email message with embedded Gmail account information

most common activities exhibited for users upon login included activities such as: determining who else was on the machine, checking and deleting the command history, and exploring the system (*i.e.,* reading the most *conspicuous* files). We note that in almost all cases, our decoys were among the first few files viewed by users upon initial login (as noted, viewing the history file was also popular). The most significant observations were made for 6 of the 20 unique users:

1. There were three attempts to use Gmail credentials that were contained in a decoy document, which triggered an alert on SONAR.
2. One attacker changed the password on a bogus Gmail account, which also triggered a SONAR alert.
3. There were at least two attempts to exfiltrate decoy files (with *scp* and *sftp*; one file, named "stolenpasswords", contained credentials to the university systems.
4. There was one attempt to use the university credentials contained in the "stolenpassword" file, which we were alerted to by the monitor that signaled an alert to SONAR.

We take these results as evidence that D^3/SONAR indeed has value as a defense against masqueraders. While only 5 of the 20 users sounded an alarm on SONAR, we emphasize that our methodology did include an unknown proportion of benign users. Furthermore, the focus of study was on masquerade detection; admittedly, we do not yet have a good way of evaluating our system on traitors, but this will be the focus of future work.

One flaw in our evaluation methodology that was revealed during testing was that we allowed users to make changes to the file system. We did this deliberately

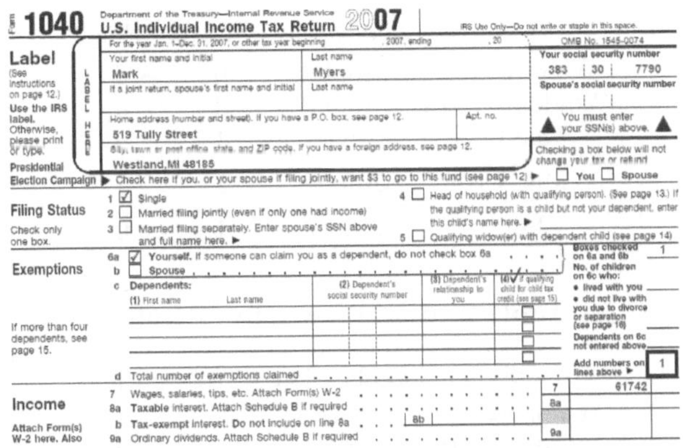

Form **1040** Department of the Treasury—Internal Revenue Service
U.S. Individual Income Tax Return 2007
IRS Use Only—Do not write or staple in this space.
For the year Jan. 1–Dec. 31, 2007, or other tax year beginning , 2007, ending , 20
OMB No. 1545-0074

Label (See instructions on page 12.) **Use the IRS label.** Otherwise, please print or type.
LABEL HERE
Your first name and initial: Mark — Last name: Myers
If a joint return, spouse's first name and initial — Last name
Home address (number and street). If you have a P.O. box, see page 12.: 519 Tully Street — Apt. no.
City, town or post office, state, and ZIP code. If you have a foreign address, see page 12.: Westland,MI 48185
Your social security number: 383 | 30 | 7790
Spouse's social security number
▲ You must enter your SSN(s) above. ▲

Presidential Election Campaign ▶ Check here if you, or your spouse if filing jointly, want $3 to go to this fund (see page 12) ▶
Checking a box below will not change your tax or refund. ☐ You ☐ Spouse

Filing Status
Check only one box.
1 ☑ Single
2 ☐ Married filing jointly (even if only one had income)
3 ☐ Married filing separately. Enter spouse's SSN above and full name here. ▶
4 ☐ Head of household (with qualifying person). (See page 13.) If the qualifying person is a child but not your dependent, enter this child's name here. ▶
5 ☐ Qualifying widow(er) with dependent child (see page 14)

Exemptions
6a ☑ **Yourself.** If someone can claim you as a dependent, do not check box 6a
b ☐ Spouse
c **Dependents:** (1) First name Last name — (2) Dependent's social security number — (3) Dependent's relationship to you — (4) ✓ if qualifying child for child tax credit (see page 15)
If more than four dependents, see page 15.
d Total number of exemptions claimed
Boxes checked on 6a and 6b: 1
No. of children on 6c who: • lived with you • did not live with you due to divorce or separation (see page 16)
Dependents on 6c not entered above
Add numbers on lines above ▶ 1

Income
Attach Form(s) W-2 here. Also
7 Wages, salaries, tips, etc. Attach Form(s) W-2 — 7: 61742
8a **Taxable** interest. Attach Schedule B if required — 8a
b **Tax-exempt** interest. Do not include on line 8a — 8b
9a Ordinary dividends. Attach Schedule B if required — 9a

Fig. 2. Decoy tax document with bogus user information

to increase the realism of the environment in the experiments. The problem this created was that it made decoy defense vulnerable to deletion (*e.g.*, several of our visitors executed wholesale deletion of files with "rm -rf *") . This poses a problem in our testing methodology, but not necessarily in practice. That is, the act of deleting files is in itself a detectable behavior that would alert monitors of suspicious behavior.

In this study, we omitted testing decoy documents with embedded beacons. The honeypots set up to attract remote attackers were stripped down Linux machines that had no installed applications necessary to open and render the decoy documents. We believe the value of beacon documents to be self-evident. We encourage the reader to visit and test the D^3 site, and participate in our planned longitudinal study. In the next section we describe tests of the beacon implementation on multiple hosts.

5.2 Beacon Implementation Tests

To test the robustness of the beacon implementations we tested them with the most common configurations of operating systems and document viewers. To this end, we contacted a random group of users across the Internet and sent them each two types of beacon documents along with a request that they open them as part of a benign experiment. The results of tests conducted on PDF and Word beacons are presented in Table 1 and 2 below. These results are a representative sample of real users across multiple hosts accessing the beacon documents. For the most part the beacon technology works well on the windows platform while not as well on Mac and Linux operating systems. The reason is that the default PDF reader is not Adobe's and does not execute Javascript embedded within the documents. Similarly, Word document beacons do not work when applications other than Microsoft Word (*e.g.*, OpenOffice or Google Docs) are used to open

Table 1. PDF Beacon Test Results

OS	Application	#Tests	#Pings
Windows XP	Adobe	6	6
Windows Vista	Adobe	4	4
Mac OS	Preview	1	0
Mac OS	Adobe	1	1
Ubuntu	Evince	1	0

Table 2. Word Beacon Test Results

OS	Application	#Tests	#Pings
Windows XP	Word	5	4
Windows XP	GoogleDocs	1	0
Windows Vista	Adobe	4	4
Mac OS	Word	2	2
Linux	OpenOffice	1	0

them. We are currently researching ways to address these limitations and will focus on them in future work.

6 Conclusions

Our work focuses on the study and creation of bait information with the aim of exposing or thwarting the exploitation of exfiltrated information by malicious insiders. As future work, we intend to explore how this approach might also be applicable in detecting accidental violations of policy, as a means of warning users and organizations about such violations. The benefit of using the proposed decoy document system for this purpose is that it can potentially operate without the privacy repercussions if a mistake is made; such a benefit differentiates the approach from traditional monitoring approaches. Another direction to explore is how to improve the believability of decoys documents. We are planning a series of user studies to help us determine how users treat different attributes of a document in a specific context, such as whether an attacker would find more believable a document purporting to contain tax information that is encrypted/protected with a weak (predictable) passphrase, compared to an unprotected version of the same document.

In conclusion, although the use of bait information and similar trap-based defenses is well known, most of those efforts have focused either on artifacts that are logically separate from the operational systems (*e.g.*, honeypots [22]) or on low-level snippets of information created manually (*e.g.*, fake database records [23]). The D^3 system is a scalable and automated trap-based defensive system that forces attackers to expend considerable effort to identify realistic useful information from purposely planted bogus information intended to deceive. Naturally, the probability of exposing a malicious insider with trap-based defense tactics increases with the amount of decoy information that is generated and disseminated. D^3 offers the novel service of automatically creating and managing decoy documents, enabling the throttling of bait based on the desired protection level or cost (*e.g., interference*) one is willing to pay.

Acknowledgments

This material is based upon work supported in part by the US Department of Homeland Security under Grant Award Number 60NANB1D0127 with the Institute for Information Infrastructure Protection (I3P), the Army Research Office (ARO) Under Grant Award W911NF-06-1-0151 - 49626-CI, and the National Science Foundation (NSF) under Grant CNS-07-14647. The I3P is managed by Dartmouth College. The views and conclusions contained in this document are those of the authors and should not be interpreted as necessarily representing the official policies, either expressed or implied, of the U.S. Department of Homeland Security, the I3P, ARO, NSF, or Dartmouth College.

We give special thanks to the Sandia National Laboratories Doctorate Study Program for supporting Brian Bowen and to Henner Mohr for his diligent effort and contributions to the development of the D^3 website and decoy document content.

References

1. Bell, D.E., LaPadula, L.J.: Secure Computer Systems: Mathematical Foundations, MITRE Corporation (1973)
2. Bell, J., Whaley, B.: Cheating and Deception. Transaction Publishers, New Brunswick (1982)
3. Butler, J., Sherri, S.: Security: Spyware and Rootkits. In: Login, December 2004, vol. 29(6) (2004)
4. Clark, D.D., Wilson, D.R.: A Comparison of Commercial and Military Computer Security Policies. In: IEEE Symposium on Security and Privacy, pp. 184–194 (1987)
5. Demers, A., Gehrke, J., Hong, M., Panda, B., Riedewald, M., Sharma, V., White, W.: Cayuga: A General Purpose Event Monitoring System. In: CIDR, pp. 412–422 (2007)
6. Detristan, T., Ulenspiegel, T., Malcom, Y., Von Underduk, M.S.: Polymorphic Shellcode Engine Using Spectrum Analysis. Phrack 11, 61–69 (2003)
7. Friess, N., Aycock, J.: Black Market Botnets. Department of Computer Science, University of Calgary, TR 2007-873-25 (July 2007)
8. Hoang, M.: Handling Today's Tough Security Threats. Symantec Security Response (2006)
9. The Honeynet Project, http://www.honeynet.org
10. The Honeynet Project, Know Your Enemy: Sebek, A Kernel based data capture tool (November 2003)
11. Honeypot Mailing List, Security Focus, http://www.securityfocus.com/archive/119

12. Katz, J., Yehuda, L.: Introduction to Modern Cryptography. Chapman and Hall CRC Press, Boca Raton (2007)
13. Kravets, D.: From Riches to Prison: Hackers Rig Stock Prices. Wired Blog Network (September 2008)
14. Krebs, B.: Web Fraud 2.0: Validating Your Stolen Goods. The Washington Post (August 20, 2008)
15. Li, W., Stolfo, S.J., Stavrou, A., Androulaki, E., Keromytis, A.: A Study of Malcode-Bearing Documents. In: Hämmerli, B.M., Sommer, R. (eds.) DIMVA 2007. LNCS, vol. 4579, pp. 231–250. Springer, Heidelberg (2007)
16. Maloof, M., Stephens, G.D.: ELICIT: A System for Detecting Insiders Who Violate Need-to-know. In: Kruegel, C., Lippmann, R., Clark, A. (eds.) RAID 2007. LNCS, vol. 4637, pp. 146–166. Springer, Heidelberg (2007)
17. McRae, C.M., Vaughn, R.B.: Phighting the Phisher: Using Web Bugs and Honeytokens to Investigate the Source of Phishing Attacks. In: Proceedings of the 40th Hawaii International Conference on System Sciences (2007)
18. Orbiscom, http://www.orbiscom.com/
19. Richardson, R.: CSI/FBI Computer Crime and Security Survey (2007)
20. Smith, R.M.: Microsoft Word Documents that Phone Home. Privacy Foundation (August 2000)
21. Song, Y., Locasto, M.E., Stavrou, A., Keromytis, A.D., Stolfo, S.J.: On the infeasibility of modeling polymorphic shellcode. In: Proceedings of the 14th ACM conference on Computer and communications security (CCS 2007), pp. 541–551 (2007)
22. Spitzner, L.: Honeypots: Catching the Insider Threat. In: Proceedings of ACSAC, Las Vegas (December 2003)
23. Spitzner, L.: Honeytokens: The Other Honeypot. Security Focus (2003)
24. Stoll, C.: The Cuckoo's Egg. Doubleday (1989)
25. Symantec. Global Internet Security Threat Report, Trends for July –December 2007 (April 2008)
26. Webb, S., Caverlee, J., Pu, C.: Social Honeypots: Making Friends with a Spammer Near You. In: Proceedings of the Fifth Conference on Email and Anti-Spam (CEAS 2008), Mountain View, CA (August 2008)
27. Ye, N.: Markov Chain Model of Temporal Behavior for Anomaly Detection. In: Proceedings of the 2000 IEEE Workshop on Information Assurance and Security, United States Military Academy, West Point, NY, June 2000, pp. 171–174 (2000)
28. Yuill, J., Denning, D., Feer, F.: Using Deception to Hide Things from Hackers: Processes, Principles, and Techniques. Journal of Information Warfare 5(3), 26–40 (2006)
29. Yuill, J., Zappe, M., Denning, D., Feer, F.: Honeyfiles: Deceptive Files for Intrusion Detection. In: Proceedings of the 2004 IEEE Workshop on Information Assurance, United States Military Academy, West Point, NY, June 2004, pp. 116–122 (2004)

MULAN: Multi-Level Adaptive Network Filter

Shimrit Tzur-David, Danny Dolev, and Tal Anker

The Hebrew University, Jerusalem, Israel
{shimritd,dolev,anker}@cs.huji.ac.il

Abstract. A security engine should detect network traffic attacks at line-speed. When an attack is detected, a good security engine should screen away the offending packets and continue to forward all other traffic. Anomaly detection engines must protect the network from new and unknown threats before the vulnerability is discovered and an attack is launched. Thus, the engine should integrate intelligent "learning" capabilities. The principal way for achieving this goal is to model anticipated network traffic behavior, and to use this model for identifying anomalies.

The scope of this research focuses primarily on denial of service (DoS) attacks and distributed DoS (DDoS). Our goal is detection and prevention of attacks. The main challenges include minimizing the false-positive rate and the memory consumption. In this paper, we present the MULAN-filter. The MULAN (MUlti-Level Adaptive Network) filter is an accurate engine that uses multi-level adaptive structure for specifically detecting suspicious traffic using a relatively small memory size.

1 Introduction

A bandwidth attack is an attempt to disrupt an online service by flooding it with large volumes of bogus packets in order to overwhelm the servers. The aim is to consume network bandwidth in the targeted network to such an extent that it starts dropping packets. As the packets that get dropped include also legitimate traffic, the result is denial of service (DoS) to valid users.

Normally, a large number of machines is required to generate volume of traffic large enough for flooding a network. This is called a distributed denial of service (DDoS), as the coordinated attack is carried out by multiple machines. Furthermore, to diffuse the source of the attack, such machines are typically located in different networks, so that a single network address cannot be identified as the source of the attack and be blocked away.

Detection of such attacks is usually done by monitoring IP addresses, ports, TCP state information and other attributes to identify the anomalous network sessions. The weakness of directly applying such a methodology is the large volume of memory required for a successful monitoring. Protocols that accumulate state information that grows linearly with the number of flows are not scalable.

In designing a fully accurate and scalable engine, one need to address the following challenges.

1. Prevention of Threats: The engine should prevent threats from entering the network. Threat prevention (and not just detection) adds difficulties to the engine, most of

Y. Chen et al. (Eds.): SecureComm 2009, LNICST 19, pp. 71–90, 2009.

which stem from the need to work at line-speed. This potentially makes the engine a bottleneck – increasing latency and reducing throughput.

2. Accuracy: The engine must be accurate. Accuracy is measured by false-negative and false-positive rates. A false-negative occurs when the engine does not detect a threat and a false-positive when the engine drops normal traffic.

3. Modeling the anticipated traffic behavior: A typical engine uses thresholds to determine whether a packet/flow is part of an attack or not. These thresholds are a function of the anticipated traffic behavior, which should reflect, as best as possible, actual "clean" traffic. Creating such a profile requires a continuous tracking of network flows.

4. Scalability: One of the major problems in supplying an accurate engine is the memory explosion. There is a clear trade-off between accuracy and memory consumption. It is a challenge to design a scalable engine using a relatively small memory that does not compromise the engine accuracy.

This paper presents the MULAN-filter. The MULAN-filter detects and prevents DoS/DDoS attacks from entering the network. The MULAN-filter maintains a hierarchical data structure to measure traffic statistics. It uses a dynamic tree to maintain the information used in identifying offending traffic. Each level of the tree represents a different aggregation level. The main goal of the tree is to save statistics only for potentially threatening traffic. Leaf nodes are used to maintain the most detailed statistics. Each inner-node of the tree represents an aggregation of the statistics of all its descendants.

Periodically, the algorithm clusters the nodes at the first level of the tree, it identifies the clusters that might hold information of suspicious traffic, for each node in such clusters, the algorithm retrieves its children and apply the clustering algorithm on the node's children. The algorithm repeats this process until it gets to the lower level of the tree. This way, the algorithm identifies the specific traffic of the attack and thus, this traffic can be blocked.

The MULAN-filter removes from the tree nodes that are not being updated frequently. This way, it maintains detailed information for active incoming flows that may potentially become suspicious, without exhausting the memory of the device on which it is installed.

The MULAN-filter uses samples. At the end of each sample it analyzes the tree and identifies suspicious traffic. When the MULAN-filter identifies a suspicious path in the tree, it examines this path to determine whether or not the path represents an attack, this may take a few more samples. As a result, there might be very short attacks, that start and end within few samples that the MULAN-filter will not detect. In [1], the authors conclude that the bulk of the attacks last from three to twenty minutes. By determining the duration of a sample to few seconds, our MULAN-filter detect almost all such attacks.

The MULAN-filter was implemented in software and was demonstrated both on traces from the MIT DARPA project [2] and on 10 days of incoming traffic of the Computer Science school in our university. Our results show that the MULAN-filter works at wire speed with great accuracy. The MULAN-filter preferably be installed a on a router, so the attacks are detected before they harm the network, but its design allows it to be installed anywhere.

2   Related Work

Detection of network anomalies is currently performed by monitoring IP addresses, ports, TCP state information and other attributes to identify network sessions, or by identifying TCP connections that differ from a profile trained on *attacks-free* traffic.

PHAD [3] is a packet header anomaly detector that models protocols rather than user behavior using a time-based model, which assumes that the network statistics can change rapidly, in a short period of time. According to PHAD, the probability, P, of an event occurring is inversely proportional to the length of time since it last occurred. $P(NovelEvent) = r/n$, where r is the number of observed values and n is the number of observations. PHAD assigns an anomaly score for novel values of $1/P(NovelEvent) = tn/r$, where t is the time since the last detected anomaly. PHAD detects $\sim 35\%$ of the attacks at a rate of ten false alarms per day after training on seven days on attack-free network traffic.

MULTOPS [4] is a denial of service bandwidth detection system. In this system, each network device maintains a data structure that monitors certain traffic characteristics. The data structure is a tree of nodes that contains packet rate statistics for subnet prefixes at different aggregation levels. The detection is performed by comparing the inbound and outbound packet rates. MULTOPS fails to detect attacks that deploy a large number of proportional flows to cripple the victim, thus, it will not detect many of the DDoS attacks.

ALPI [5] is a DDoS defense system for high speed networks. It uses a leaky-bucket scheme to calculate an attribute-value-variation score for analyzing the deviations of the values of the current traffic attributes. It applies the classical proportion integration scheme in control theory to determine the discarding threshold dynamically. ALPI does not provide attribute value analysis semantics; i.e., it does not take into consideration that some TCP control packets, like SYN or ACK, are being more disruptive.

Many DoS defense systems, like Brownlee et al. [6], instrument routers to add flow meters at either all, or at selected, input links. The main problem with the flow measurement approach is its lack of scalability. For example, in [6], if memory usage rises above a high-water mark they increase the granularity of flow collection and decrease the rate at which new flows are created. Updating per-packet counters in DRAM is impossible with today's line speed. Cisco NetFlow [7] solves this problem by sampling, which affects measurement accuracy. Estan and Varghese presented in [8] algorithms that use an amount of memory that is a constant factor larger than the number of large flows. For each packet arrival, a flow id lookup is generated. The main problem with this approach is in identifying large flows. The first solution they presented is to sample each packet with a certain probability, if there is no entry for the flow id, a new entry is created. From this point, each packet of that flow is sampled. The problem with that is its accuracy. The second solution uses hash stages that operate in parallel. When a packet arrives, a hash of its flow id is computed and the corresponding counter is updated. A large flow is a flow whose counter exceeds some threshold. Since the number of counters is lower than the number of flows, packets of different flows can result in updating the same counter, yielding a wrong result. In order to reduce the false-positives, several hash tables are used in parallel.

Schuehler et al. present in [9] an FPGA implementation of a modular circuit design of a content processing system. The implementation contains a large per-flow state store that supports 8 million bidirectional TCP flows concurrently. The memory consumption grows linearly with the number of flows. The processing rate of the device is limited to 2.9 million 64-byte packets per second.

Another solution, presented in [10], uses aggregation to scalably detect attacks. Due to behavioral aliasing the solution doesn't produce good accuracy. Behavioral aliasing can cause false-positives when a set of well behaved connections aggregate, thus mimicking bad behavior. Aliasing can also result in false negatives when the aggregate behavior of several badly behaved connections mimics good behavior. Another drawback of this solution is its vulnerability against spoofing. The authors identify flows with a high imbalance between two types of control packets that are usually balanced. For example, the comparison of SYNs and FINs can be exploited by the attacker to send spurious FINs to confuse the detection mechanism.

3   DoS Attacks

Denial of service (DoS) attacks cause service disruptions when too many resources are consumed by the attack instead of serving legitimate users. A distributed denial of service (DDoS) attack launches a coordinated DoS attack toward the victim from geographically diverse Internet nodes. The attacking machines are usually compromised zombie machines controlled by remote masters. Typical attacked resources include link bandwidth, server memory and CPU time. DDoS attacks are more potent because of the aggregated effect of the traffic converging from many sources to a single one. With knowledge of the network topology the attackers may overwhelm specific links in the attacked network.

The best known TCP DoS attack is the SYN flooding [11]. Cisco Systems Inc. implemented a TCP Intercept feature on its routers [12]. The router acts as a transparent TCP proxy between the real server and the client. When a connection request is made from the client, the router completes the handshake for the server, and opens the real connection only after the handshake is completed. If the amount of half-open connections exceeds a threshold, the timeout period interval is lowered, thus dropping the half-open connections faster. The real servers are shielded while the routers aggressively handle the attacks. Another solution is SYN cookies [13], which eliminates the need to store information per half open connection. This solution requires design modification, in order to change the system responses.

Another known DoS attack is the SMURF [14]. SMURF uses spoofed broadcast ping messages to flood a target system. In such an attack, a perpetrator sends a large amount of ICMP echo (ping) traffic to IP broadcast addresses, with a spoofed source address of the intended victim. The hosts on that IP network take the ICMP echo request and reply with an echo reply, multiplying the traffic by the number of hosts responding. An optional solution is to never reply to ICMP packets that are sent on a broadcast address [15].

Back [16] is an attack against the Apache web server in which an attacker submits requests with URL containing many front-slashes. Processing these requests slows down the server performance, until it is incapable of processing other requests. Sometimes this attack is not categorized as high rate DoS attacks, but we mention it since the MULAN-filter discovers it. In order to avoid detection, the attacker sends the front-slashes in separate HTTP packets, resulting in many ACK packets from the victim server to the attacker. An existing solution suggests counting the front-slashes in the URL. A request with 100 front-slashes in the URL would be highly irregular on most systems. This threshold could be varied to find the desired balance between detection rate and false alarm rate.

In all the above examples, the solutions presented are specific to the target attack and can be implemented just after the vulnerabilities are exploited. The MULAN-filter identifies new and unknown threats, including all the above attacks, before the vulnerability is discovered and the exploit is created and launched, as detailed later.

4   Notations and Definitions

- A *metric* is defined as the chosen rate at which the measurements by the algorithm are executed, for example, bit per second, packets per second etc.
- An L_n is the number of levels in the tree data structured used by the algorithm.
- *Sample value* is defined as the aggregated value that is collected from the captured packets in one sample interval.
- *Window interval* is defined as $m \times$ *sample interval*, where $m > 0$ and the *sample interval* is the length of each sampling.
- *Clustering Algorithm* is defined as the process of organizing *sample values* into groups whose members have "similar" values. A *cluster* is therefore a collection of *sample values* that are "similar" and are "dissimilar" to the *sample values* belonging to other clusters (as detailed later).
- *Cluster Info* is the number of samples in the cluster, the cluster mean and the cluster standard deviation, denoted by *C.Size*, *C.Mean* and *C.Std* respectively.
- *Anticipated Behavior Profile (ABP)* is a set of k Clusters Info, where k is the number of clusters.
- *Clusters Weighted Mean (WMean)* is the weighted mean of all clusters, alternatively, the mean of all samples.
- *Clusters Weighted Standard Deviation (WStd)* is the weighted standard deviation of all clusters, alternatively, the standard deviation of all samples.
- *High General Threshold (HGThreshold)* is $WMean + t_1 \times WStd$ and *Low General Threshold (LGThreshold)* is $WMean + t_2 \times WStd$, where $t_1 > t_2$.
- *Marked Cluster* is a cluster with mean greater than *LGThreshold*.

5   The MULAN-Filter Design

The MULAN-filter uses inbound rate anomalies in order to detect malicious traffic. The statistics are maintained in a tree-shaped data-structure. Each level in the tree represents an aggregation level of the traffic. For instance, the highest level may describe inbound packets rate per-destination, the second level may represent per-protocol rate for a specific destination and the third level hold per-destination port rate for a specific destination and protocol. Each node maintains the aggregated statistics of all its descendants. A new node is created only for packets with a potentially suspicious parent node. This way, for example, there is a need to maintain a detailed statistics only for potentially suspicious destinations, protocols or ports. Another advantage of using the tree is the ability to find specific anomalies for specific nodes. For example, one rate can be considered normal for one destination, but is anomalous for the normal traffic of another destination.

The MULAN-filter can be used in two modes, training mode and verification mode. The output of the training mode is the *ABP* and the thresholds. For each cluster C in the *ABP*, if $C.Mean > LGThreshold$, the cluster is denoted as a *marked cluster*. This information is used to compare the online rates in the verification mode process.

In order to calculate this information, the anticipated traffic behavior profile must be measured. There are two ways to measure such a profile: Either training a profile on identification of attack-free traffic, or by trying to filter the traffic from prominent bursts, which might indicate attacks and then creating the profile from the filtered traffic.

5.1   Anticipated Traffic Behavior Profile

In order to create the *ABP*, it is better to use an attack-free traffic. Alternative solutions strongly assume attack-free traffic, an assumption that may be impractical for several reasons. First, unknown attacks may be included in that traffic, so the baseline traffic is not attack-free. Furthermore, traffic profiles vary from one device to another, and unique attack-free training profiles need to be obtained for each specific device on which the system is deployed. Moreover, traffic profiles on any given device may vary over time. Thus, a profile may be relevant only at a given time, and may change a few hours later.

We propose a methodology in which anomalies are first identified, and then *refined*. The cleaner the initial traffic the more precise the initial state is, but our methodology works well with non-clean traffic. To achieve both goals, the algorithm aggregates per-*sample interval* statistics, creating a *sample value* for each such interval. At the end of each *window interval*, the algorithm employs a clustering algorithm in order to obtain a set of clustered *sample values*. If there are one or more clusters with significantly high mean values (3 standard deviations from *WMean*), the algorithm discovers the samples that are key contributors to the resulting mean values. The algorithm refines those samples by setting their value to the cluster mean value and then recalculates the clusters' means values. The "refinement" rule states that lower levels always override refinement of higher levels. This means that if the algorithm detects a high burst at one of the destinations and then detects a high burst at a specific protocol for that destination, it refines the node value associated with the protocol, which also impacts the value associated with the destination. The *refinement* process is performed at every *window interval* for maintaining a dynamic profile.

5.2   Data Structure

The MULAN-filter uses a tree-shaped data structure. The tree enables maintaining distinct statistics of *all* the relevant traffic. Traffic is considered relevant if there is a high probability that it contains an attack.

In our implementation example, there are three levels in the tree. The nodes at the first level hold statistics per-destination IP address, the nodes at the second level hold

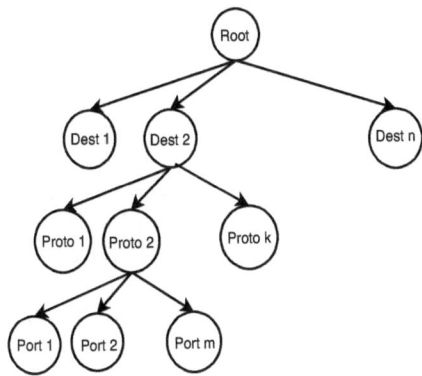

Fig. 1. The Tree

statistics per-protocol and the nodes at the third level hold statistics per-destination port (see Fig. 1). During the verification mode, when a *sample value* is calculated, the algorithm saves the aggregation for the first level. In our implementation, assume that the *sample value* is equal to SV and there are N_s packets that arrived during the *sample interval* with n different IP addresses. We define $Metric(Packet_j)$ to be the contribution of $Packet_j$ to SV, and SV_i to be the part of SV that is calculated from packets with IP_i in their header. Formally, $SV_i = \sum_{j, IP_i \in Packet_j} Metric(Packet_j)$, thus, $SV = \sum_i SV_i$, where $1 \leq j \leq N_s$ and $1 \leq i \leq n$. The tree structure is flexible to hold special levels for specific protocols, see Section 5.3.

In the verification mode, the tree is updated following two possible events. One is a completion of each *sample interval*. In this case, the algorithm compares SV to the clusters' means from the *ABP*. If the closest mean belongs to a *marked cluster*, a node for each IP_i is added to the first level in the tree. The second event at which the tree is updated may occur at packet's arrival. If the destination IP address in the packet header has a node in the first level, a node for the packet protocol is created at the second level, and so on. In any case, the metric's values along the path from the leaf to the root are updated. This way, each node in the tree holds the aggregated sum of the metric's values of its descendants.

A node that is not updated for long enough is removed from the tree. A node can not be removed unless it is a leaf, and it can become a leaf if all of its descendants have been removed. Thus, we focus only on nodes (or on traffic) that are suspected of comprising an attack; thus, saving on memory consumption.

5.3 Special Levels for Specific Protocols

Some protocols have special information in their header that can help the algorithm in blocking more specific data. Since our tree is very flexible in the data it can hold, we can add special levels for specific protocols. In our experiments we added a level for the TCP protocol that distinguishes between the TCP flags. This addition results in dropping only the SYN packets when the algorithm detects a SYN attack. The same can be done for the ICMP types in order to recognize the ECHO packets in the SMURF attack.

6   The Algorithm

Prior to implementing the algorithm, the following should be defined: depth of the tree, characteristics of the data aggregated at each level, sample interval, window interval, metrics to be measured, and $t1$ and $t2$ used for calculating *LGThreshold* and *HGThreshold*.

The MULAN-filter has been implemented in software. The input to our engine is taken both from the MIT DARPA project [2] and from the Computer Science school in our university. The input from MIT contains two stretches of five-days traffic (fourth and fifth week) with documented attacks and the input from the university contains 10 days of incoming traffic, this containing both real and simulated attacks that we injected.

The algorithm operates in two modes, the training mode and the verification mode. The training mode is generated at the end of a *window interval*. The input for this mode is a set of *N* samples and the output is *ABP* with indication of the *marked clusters*.

6.1   Training Mode

In order to create the *ABP*, the algorithm generates the K-means [17] clustering algorithm every *window interval* to cluster the *N sample values* into *k clusters*. For each cluster, the algorithm holds its size, mean and standard deviation, after which the algorithm can calculate the weighted mean, *WMean*, and the weighted standard deviation, *WStd*, and determine the value of *LGThreshold*. Since the samples may contain attacks, *LGThreshold* might be higher than it should. Therefore, for each cluster *C*, if *C.Mean > LGThreshold*, the algorithm retrieves *C*'s *sample values*. In our implementation, each *sample value* in the cluster holds *metric values* of IP addresses that were seen in that sample interval. For each sample, the algorithm gets the aggregation per IP address and generates new set of samples. The algorithm then generates K-means again, where the input is the newly created set. Running K-means on this set produces a set of clusters, a cluster with a high mean value holds the destinations with the highest metric value. The algorithm continues recursively for each level of the tree.

At each iteration in the recursion, the algorithm detects the high bursts and refines the samples in the cluster to decrease the bursts influence on the thresholds, see Section 5.1. As mentioned, the "refinement" rule states that lower levels always override refinement of higher levels. This means that if the algorithm detects a high burst at one of the destinations and then a high burst at a specific protocol for that destination, it refines the node value associated with the protocol, impacting on the value associated with the destination. When the refinement process is completed, the refined samples are clustered again to produce the updated *ABP* information and the *LGThreshold*. A cluster *C* is indicated a *marked cluster* if *C.Mean > LGThreshold*.

There can be cases in which the bursts that the algorithm refines represent normal behavior. In such cases *LGThreshold* may be lower than expected. Since the algorithm uses the training mode in order to decide whether to save information when running in verification mode, the only adverse impact of this refinement is in memory utilization, as more information is saved than actually needed. This is preferable to overlooking an attack because of a mistakenly calculated high threshold. The training mode algorithm is presented in Algorithm 1.

At each sample completion, the algorithm gets the *sample value* and finds the most suitable cluster from the *ABP*. In order for the profile to stay updated and for the clustering algorithm to be generated only at the end of each *window interval*, the cluster mean and standard deviation are updated by the new *sample value*.

6.2   Verification Mode

The verification mode starts after one iteration of the training mode (after one *window interval*). The verification mode is executed either on a packet arrival or following a sample completion.

To simplify the discussion, as a working example in this section we assume that the tree has three levels and the aggregation levels are IP address, protocol and port number in the first, second and third level, respectively.

On each packet arrival, the algorithm checks whether there is a node in the first level of the tree for the destination IP address of the packet. If there is no such node, nothing is done. If the packet has a representative node in the first level, the algorithm updates the node's metric value. From the second level down, if a child node exists for the packet, the algorithm updates the node's metric value, otherwise, it creates a new child.

At each sample completion, the algorithm gets the *sample value* and finds the most suitable cluster from the *ABP*. If the suitable cluster is a *marked cluster*, the algorithm adds nodes for that sample in the first level of the tree. In our example, the algorithm adds per-destination aggregated information from the sample to the tree. I.e. for each destination IP address that appears in the sample, if there is no child node for that IP address, the algorithm creates a child node with the aggregated metric value for that address (see Section 5.2).

The algorithm runs K-means on the nodes at the first level of the tree. Each sample value is per-destination aggregated information (SV_i with the notations from Section 5.2). As in the training mode, the clustering algorithm produces the set of *clusters info*, but in this case the algorithm calculates the threshold *HGThreshold*. If a cluster's mean is above the *HGThreshold*, a deeper analysis is performed. For each sample in the cluster (or alternatively, for each node at the first level), the algorithm retrieves the node's children and generates K-means again. The algorithm continues recursively for each level in the tree. At each iteration, the algorithm also checks the *sample values* in the cluster nodes. If a *sample value* is greater than *HGThreshold*, it marks the node as suspicious.

The last step is to walk through the tree and to identify the attacks. The analysis is done in a DFS manner. A leaf that has not been updated long enough is removed from the tree. Each leaf that is suspected too long is added to the black list, thus preventing suspicious traffic until its rate is lowered to a normal rate. For each node on the black-list, if its high rate is caused as a results of only a few sources, the algorithm raises an alert but does not block the traffic; If there are many sources, the traffic that is represented

Algorithm 1. Training Mode Algorithm

1: *packet* ⇐ *ReceivePacket*();
2: *UpdateMetricValue*(*sample*, *packet*);
3: **if** *End of Sample Interval* **then**
4:     *samples.AddSample*(*sample*);
5: **end if**;
6: **if** *End of Window Interval* **then**
7:     *UpdateTrainProfile*(*samples*);
8: **end if**.

UpdateTrainProfile(samples)

1: *clusters* ⇐ *KMeans*(*samples*);
2: *samples* ⇐ *Refine*(*clusters*, *samples*);
3: *ABP* ⇐ *BuildProfile*(*clusters*);
4: *LGThreshold* ⇐ *calcThreshold*(*ABP*);
5: **for all** *clusterInfo* ∈ *ABP* **do**
6:     **if** *clusterInfo.Mean* > *LGThreshold* **then**
7:         *setMarked*(*cluster*);
8:     **end if**;
9: **end for**.

by the specific path is blocked until the rate becomes normal. The verification mode algorithm is presented in Algorithm 2.

In addition of the above, to prevent attacks that do not use a single IP destination, like attacks that scan the network looking for a specific port on one of the IP addresses, the algorithm identifies sudden increase in the size of the tree. When such increase is detected, the algorithm builds a hash-table indexed by the source IP address. The value of each entry in the hash-table is the number of packets that were sent by the specific source. This way, the algorithm can detect the attacker and block its traffic (see Section 8). The algorithm maintains a constant number of entries and replaces entries with low values. The hash-table size is negligible and does not affect the memory consumption of the algorithm.

Since the algorithm detects anomalies at each level of the tree, it can easily recognize exceptions in the anomalies it generates. For example, if one IP address appears in many samples as an anomaly, the algorithm learns this IP address and its anticipated rate and adds it to an *exceptions list*. From this moment on, the algorithm compares this IP address to a specific threshold.

6.3   The Algorithm Parameters

In our simulation, the algorithm builds three levels in the tree. The first level holds aggregated data for the destination IP addresses, the second level holds aggregated data for the protocol for a specific IP address, and the third level holds aggregated data for a destination port for specific IP and port. Since we look for DoS/DDoS attacks, these levels are sufficient to isolate the attack's traffic.

At the end of each window interval the algorithm updates the *ABP* and, since the network can be very dynamic, we chose the window interval to be five minutes. The bulk

Algorithm 2. Verification Mode Algorithm

1: *packet* ⇐ *ReceivePacket*();
2: *UpdateMetricValue*(*sample*, *packet*);
3: *PlacePctInTree*(*packet*, *root*, 0);
4: **if** *End of Sample Interval* **then**
5:     *SetFirstLevel*(*sample*, *root*);
6:     *Verify*(*root.children*);
7:     *AnalyzeTree*(*root*);
8: **end if**.

PlacePctInTree(packet, node, level)

1: **if** *level* == *lastLevel* **then**
2:     *return*;
3: **end if**;
4: **if** *node.HasChild*(*packet*) **then**
5:     *child* ⇐ *node.GetChild*(*packet*);
6:     *child.AddToSampleValue*(*packet*);
7:     *PlacePctInTree*(*packet*, *child*, ++*level*);
8: **else**
9:     **if** *level* > 0 **then**
10:         *child* ⇐ *CreateNode*(*packet*);
11:         *node.AddChild*(*child*);
12:     **end if**;
13: **end if**.

SetFirstLevel(sample, root)

1: *cluster* ⇐ *GetClosestClusterFromABP*(*sample*);
2: *cluster.UpdateMeanAndStd*(*sample*);
3: **if** *MarkedCluster*(*cluster*) **then**
4:     *AddFirstLevelInfo*(*sample*);
5: **end if**.

Verify(nodes)

1: *clustersInfo* ⇐ *KMeans*(*nodes*);
2: *CalcThresholds*(*clustersInfo*);
3: **for all** *cluster* ∈ *clustersInfo* **do**
4:     **if** *cluster.Mean* > *LGThreshold* **then**
5:         **for all** *node* ∈ *cluster* **do**
6:             **if** *node.sampleValue* > *HGThreshold* **then**
7:                 *MarkSuspect*(*node*);
8:             **end if**;
9:             *Verify*(*node.children*);
10:         **end for**;
11:     **end if**;
12: **end for**.

AnalyzeTree(node)

1: **for all** *child* ∈ *node.children* **do**
2:     **if** *child.NoChildren*() **then**
3:         **if** *child.UnSuspectTime* > *cleanDuration* **then**
4:             *RemoveFromTree*(*child*);
5:         **end if**;
6:         **if** *child.SuspectTime* > *suspectDuration* **then**
7:             *AddToBlackList*(*child*);
8:         **end if**;
9:     **else**
10:         *AnalyzeTree*(*child*);
11:     **end if**;
12: **end for**.

of DoS/DDoS attacks lasts from three to twenty minutes, we have therefore chosen the sample interval to be five seconds. This way the algorithm might miss few very short attacks. An alternative solution for short attacks is presented in Section 6.5. A node is considered as indicating an attack if it stays suspicious for *suspect duration*; In our implementation the *suspect duration* is 18 samples. A node is removed from the tree if it is not updated for *clean duration*; In our implementation the *clean duration* is 1 sample. DoS/DDoS attacks can be generated by many short packets, like in the SYN attack example, thus, a bit-per-second metric may miss those attacks. In our implementation we use a packet-per-second metric.

The last parameters to be determined are $t1$ and $t2$ that are used for calculating *LGThreshold* and *HGThreshold*. These parameters are chosen using Chebyshev inequality. The Chebyshev inequality states that in any data sample or probability distribution, nearly all the values are close to the mean value, in particular, no more than $1/t^2$ of the values are more than t standard deviations away from the mean. Formally, if $\alpha = t\sigma$, the probability of an attribute length, can be calculated using the inequality:

$$p(|x - \mu| > \alpha) < \frac{\sigma^2}{\alpha^2}.$$

The Chebyshev bound is weak, meaning the bound is tolerant to deviations in the samples. This weakness is usually a drawback. In our case, since DoS/DDoS attacks are characterized by a very high rate, the engine has to detect just significant anomalies and this weakness of the Chebyshev boundary becomes an advantage. In our experiment we set $t1 = 1$ and $t2 = 5$.

Non-self-similar traffic may be found at the lower levels of the tree (per destination rate, per protocol rate etc.). Another problem at the lower levels is the reduced number of samples, complicating the ability to anticipate traffic behavior at these levels. In order to identify the anomalies at those levels, we introduce an alternative measurement model, see Section 6.4.

6.4   Modeling Non-self-similar Traffic

The MULAN-filter has to model anticipated traffic. There are two main challenges in modeling anticipated traffic: the complexity of network traffic, and its variance over time.

Bellovin [18] and Paxson [19] found that wide network traffic contains a wide range of anomalies and bizarre data that is not easily explained. Instead of specifying the extremely complex behavior of network traffic, they use a machine learning approach to model actual traffic behavior. Research by Adamic [20] and Huberman and Adamic [21] implies that this approach would fail since the list of observed values grows at a constant rate and is never completed, regardless of the length of the training period. However, Leland et al. [22] and Paxson & Floyd [23] show that this is not valid for many types of events, like measuring packets per second rate.

Non-self-similar traffic may be found at the lower levels of the tree (per destination rate, per protocol rate etc.). Another problem at the lower levels is the reduced number of samples, complicating the ability to anticipate traffic behavior at these levels. In order to identify the anomalies at those levels, an alternative measurement model should be introduced. Let N_c be the number of children of a node, and s be the sum of all sample values of the node children. If a "small" subset of N_c represents a "high percentage" of s, an anomaly is alerted. For example, consider a destination for which there are seven protocol nodes, of which six have sample values of approximately ten packets per second, and a seventh node has a sample value of 400 packets per second. This would result in a mean value of 65.7, with rather high standard deviation of 147.4. Using traditional models, it will be difficult to identify the seventh child as an anomaly. Using the proposed model, one child represents $\sim 87\%$ of all samples, so this node is identified as an anomaly.

6.5 Handling Short Attacks

MIT traces contain short DDoS attacks (some of them are 1 second long). An example from MIT traces is the SMURF attack. In the SMURF attack, the attacker sends ICMP 'echo request' packets to the broadcast address of many subnets with the source address spoofed to be that of the intended victim. Any machine that is listening on these subnets responds by sending ICMP 'echo reply' packets to the victim. Short attacks can exhaust a victim but usually cannot defeat it. Since our algorithm blocks the anomalies from entering the network, it declares an anomaly only after a node has being suspected for some time. By reducing the sample interval, our algorithm can easily detect the short attacks so an alert mechanism is added for them. As opposed to the common DoS or DDoS attacks, in order to exhaust a service, the rate of the short attacks must be significantly high so the anomaly will be much more conspicuous. Thus, in order to reduce the false-positives we use more stringent detection rules for the short attacks.

7 Optimal Implementation

The main bottleneck that might occur in our engine is the tree lookup, which is performed on arrival of each packet. Since the engine has to work at wire speed, software solutions might be unacceptable. We suggest an alternative implementation.

The optimal implementation is to use a TCAM (Ternary Content Addressable Memory) [24]. The TCAM is an advanced memory chip that can store three values for every bit: zero, one and "don't care". The memory is content addressable; thus, the time required to find an item is considerably reduced. The RTCAM NIPS presented in [25] detects signatures-based attacks that were drawn from Snort [26]. In the RTCAM solution, the patterns are populated in the TCAM so the engine detects a pattern match in one TCAM lookup. We can similarly deploy the MULAN filter in the TCAM. A TCAM of size M can be configured to hold $\lfloor M/w \rfloor$ rows, where w is the TCAM width. Let $|L_i|$ be the length of the information at level i in the tree. Assuming that there are m levels, w is taken to be $\sum_i |L_i|$, where $1 \leq i \leq m$. In our example, the IP address at the first level contains 4 bytes (for IPv4). An additional byte is needed to code the protocol at the second level. For the port number at the third level we need another two bytes. Thus, in our example $w = 7$. Since the TCAM returns the first match, it is populated as follows: the first rows hold the paths for all the leaves in the tree. A row for a leaf at level i, where $i < L_n$ is appended with "don't care" signs. After the rows for the leaves, we add

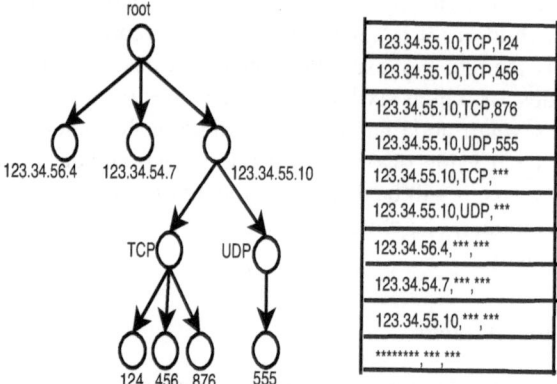

Fig. 2. TCAM Population

rows for the rest of the nodes, from the bottom of the tree up to the root. Each row for a non-leaf node at level l is appended with "don't care" signs for the information at each level $j \leq n$ such that $l < j$. The last row contains w "don't care" bytes, thus indicating that there is no path for the packet and providing the default match row.

Fig. 2 presents an example of the tree structure and the populated TCAM for that tree. As shown, each node (except the root) has a row in the TCAM. When a packet arrives, the algorithm extracts the relevant data, creates a TCAM key and looks for a TCAM match. Each row in the TCAM holds a counter and a pointer to the row of the parent node. When there is a TCAM match (except in the last row), the algorithm updates the counter at the matched row and follows the pointer to the parent node's row. The algorithm updates the counters for each row along the path from the node corresponding to the matched row to the row corresponding to the ancestor at the first level.

In our algorithm, there are only two places where the algorithm might add nodes to the tree, when nodes are set for the first level, and on packet arrival. In both cases, the algorithm adds leaves represented by the corresponding rows at the beginning of the TCAM. Similarly, when the algorithm"cleans" the tree, it removes leaves, again, handling the beginning of the TCAM. In order to easily update the TCAM while keeping the right order of the populated rows, the TCAM is divided into L parts, where L is the number of the levels in the tree.

The last obstacle our algorithm has to deal with is the TCAM updates. TCAM updates are done when adding nodes to the tree and when removing nodes from the tree. The TCAM can be updated either with a software engine or with a hardware engine. Using software engine is much simpler but is practical only when there is a low number of updates. Fig. 3 presents the average number of TCAM updates for each 100 packets of the incoming traffic of the Computer Science school. The figure clearly illustrates the creation of the tree. During the creation of the tree there are many insertions, thus the number of updates is relatively high.

Each value is an average of values of all the days of MIT traces. The total average update rate is ~ 1.5 updates for 100 packets, more than 99% of the values are below 50 updates, with a small number of scenarios when the engine has to make up to ~ 1700

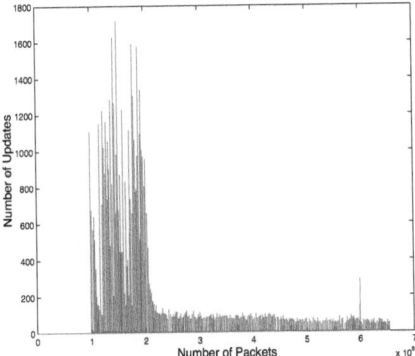

Fig. 3. TCAM Updates

TCAM updates. Today's enterprise network equipment supports hundreds of Giga bits per second of traffic and small and medium business devices handle $60 - 100$ Giga bits per second and above. One Giga interface supports 1.5 million packets per second, thus enterprise network devices need to deal with about 500 millions packets per second, and small and medium business need to deal with about 150 millions packets per second. A software engine will not be able to fulfil these requirements and thus is not acceptable. A hardware engine can achieve line speed rates. The available TCAM update speed with hardware engine is in the range of 3 to 4 nano seconds, which is $250,000,000$ to $330,000,000$ updates per second. In light of the rate of TCAM updates, it can be deduced that on average, one TCAM update is performed for every 67 packets. With a traffic rate of 500 million packets per second, the engine has to make $500M/67 \approx 7.5$ millions updates per second, which is significantly less than the available TCAM update rates limit. Even with 50 TCAM updates, the engine executes $500M/2 = 250$ millions updates per second which is still in range.

8   Experimental Results

The quality of performance of the algorithm is determined by two factors: scalability and accuracy. In order to analyze the performance of the algorithm, a simulation was implemented and tested with MIT DARPA traces and real traffic from our School of Computer Science.

8.1   Scalability

Demonstration of scalability of the algorithm requires analysis of the memory requirement at every stage of execution. We measured the number of nodes on each sample and we found an average size of the tree is 1028 nodes. This result is very encouraging since it is a very reasonable memory consumption.

Another major advantage of our algorithm is the fact that the increase in tree size is very moderate compared to the increase in the number of flows. This is clearly demonstrated in Fig. 4 (Note that the x axis is a logarithm scale). In general, for any number

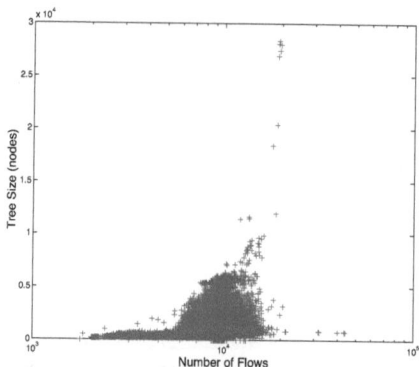

Fig. 4. Tree Size vs Number of Flows

of flows the tree size is below 10000 nodes. There are few cases where the size of the tree exceeds 30000 nodes, these cases occur when the traffic contains attacks. An optimization to the algorithm, thats prevent such cases is presented in Section 8.3.

Memory consumption is one of the major difficulties when trying to extract per-flow information in a security device. The main problem with the flow measurement approach is its lack of scalability. Memory consumption of algorithms presented in previous works is directly influenced by the number of flows, and in many cases the algorithm performance is affected. Cisco NetFlow [7] solves this problem by sampling, which affects measurement accuracy. Another work [8] develops algorithms that use an amount of memory that is a constant factor larger than the number of the large flows. The main problem in this approach is how to identify large flows. Two possible solutions were presented, both of which lack accuracy. In [4], the authors try to aggregate data by IP prefixes. For more than 1024 IPs, the data structure size does not fit in cache, so that the algorithm rates drop proportionally to the total memory consumption. In our engine, memory consumption does not grow linearly with the number of flows and the algorithm accuracy is therefore not affected.

8.2   Accuracy

Accuracy is measured by the false-negative and the false-positive rates. False-negative is a case in which the system does not detect a threat and false-positive is the case in which the system drops normal traffic. This section presents the accuracy results both on MIT DARPA traces and on the real traffic from our School of Computer Science.

MIT DARPA Traces. There are only two documented bandwidth attacks in the MIT DARPA traces, both are SYN-attacks. Our algorithm finds these attacks. In addition, our algorithm detects several other anomalous behaviors. The analysis indicates that in one of the days, there are many retransmissions packets and a large number of se-quential TCP-keep-alive packets, which is consistent with anomalous behavior. Another example is the back attack targeted at the Apache web servers by submitting requests with URL's containing many front-slashes. As the server tries to process these requests, it slows down and is unable to process other requests. In order to avoid detection, the attacker sends each front-slash in a different HTTP packet. The victim sends many TCP ACK packets back to the attacker. Since the engine compares traffic per destination (at the first level) it detects this traffic as anomaly. There is another case where our algorithm detects many TCP SYN, RST and FIN packets. In one of the SYN attacks, the source of the attack is an IP address within the network. As a result of the attack, the victim sends many TCP RST packets back to the attacker. Consequently, the engine detects two anomalies: the SYN packets to the victim and the RST packets to the attacker.

The School of Computer Science Traces. We analyzed the traffic in two modes. In the first mode, we ran the algorithm on the original data and we looked for real attacks. In the second mode we randomly added attacks to random destinations and verified that the algorithm detects the injected attacks.

In the first mode we found some very interesting anomalous behaviors. In one alert, the algorithm detects inbound scan on TCP, port 1433. In this attack, the attacker scans the network, looking for a Microsoft SQL Server installations with weak password protection and, if successful, looks to steal or corrupt data or use some features with SQL Server to compromise the host system. Another alert indicates a single source that scans the network for a listening HTTP server (scanning many IP addresses on port 80). One more interesting alert indicates an inbound scan on TCP, port 139. Such inbound scans are typically systems that are trying to connect to file shares that might be available on the system and therefore should be blocked. While most of this traffic is the result of worms or viruses, which can use open file shares to propagate, they can be also the result of malicious users attempt to connect to the victim. Once connected, they can download, upload or even delete or edit files on the connected file share.

The algorithm detected 4 exceptional IP addresses, all of them servers in the network. The algorithm generated 87 alerts, almost all of them are IP addresses that communicate with an IP address from the *exceptions list*. Since the *exceptions list* is a safe list of servers, these alerts were omitted from the final results. We were left with 24 alerts. There can be cases where a single host downloads a heavy file or backup heavy material etc. In such cases, there will be a high rate between a single host to a single destination. Our algorithm detects these channels as an attack. Since we don't want to prevent this legal traffic, our algorithm alerts these connections but it does not block the connection's traffic. Analysis of the results indicates that there are only 5 alerts containing more than one source. These 5 alerts are false positives and they were generated from the highest level in the tree, e.g. the alerts refer to IP addresses without indications for a specific protocol or port. In case of an attack on a specific service, the tree detects the attack also in lower levels, thus, an attack on this level may imply only some kind of network scan, i.e. a port scan. When an attacker tries to scan the network, the size of the tree significantly increases. Thus, by combining both anomalies, high rate on the highest level and the size of the tree, we can eliminate these false positives.

In the second mode we randomly injected DoS/DDoS attacks of different kinds, our algorithm found all of them. The injected attacks included the following attacks: ICMP flood, where a host is bombarded by many ICMP echo requests in order to consume its resources by the need to reply. Syn Attack, where random Syn packets are sent to the attacked host with intent to fill the number of open connections it can hold and therefore leave no free resources for new valid connections. DNS flood, roughly similar to ICMP flood, only more efficient against DNS servers as usually these requests require more time spent on the server side. Smurf, where the attacker spoofs many echo requests coming from the attacked host, and consequently the host is swamped by echo replies.

To reinforce our results, we compare the MULAN filter against LAD [27]. LAD is a triggered, multi-stage infrastructure for the detection of large-scale network attacks. In the first stage, LAD detects volume anomalies using SNMP data feeds. These anomalies are then used to trigger flow collectors and then, on the second stage, LAD performs analysis of the flow records by applying a clustering algorithm that discovers heavy-hitters along IP prefixes. By applying this multi-stage approach, LAD targets the scalability goal. Since SNMP data has coarse granularity, the first stage of LAD produces false-negatives. The collection of flow records on the second phase requires a buffer to hold the data and adds bandwidth overhead to the network, thus LAD uses a relatively high threshold that leads to the generation of false-negatives. One major difference between the MULAN filter and LAD is that LAD only supplies detection of attacks, which a network operator needs to process. This eases the implementation by two aspects; first, the attacks are not detected online and the second is the tolerance to false positives. The MULAN filter prevents the attacks with a negligible rate of false positives.

8.3   Controlling the Tree Size

In order to control the size of the tree in a way that it does not explode as it may do during scanning attacks, we added the following rule: When the algorithm detects an attack on any of the nodes in the tree, it stops adding children to that node until the node's rate falls below the threshold. As mentioned in Section 8.1, the reason the tree had $\sim 15,000$ nodes on that day is that two IP addresses received TCP traffic for many different ports. For each unique port, the algorithm created a node in the tree. With the above rule, when the anomalies are detected, the algorithm does not add more nodes for new ports, although it does update the counter at the parent node (in this example, the node that represents TCP). The algorithm resume adding children when the counter at the parent is reduced and the parent is no longer categorized as an anomaly. The tree size results after applying this optimization is presented in Table 1.

Table 1. Tree Size Results after Optimization

Day (W.D)	Packets Number	Average (Nodes)	Maximum (Nodes)
4.1	1,320,049	11.3	108
4.2	1,251,319	9.3	101
4.3	1,258,076	10.2	84
4.4	1,580,440	11.2	121
4.5	1,261,848	10.4	116
5.1	1,320,049	10.8	108
5.2	2,502,808	10.9	134
5.3	1,329,336	10.5	90
5.4	2,259,146	19.7	1,035
5.5	2,602,565	11.5	104

9   Discussion and Future Work

The engine presented in this paper detects DoS/DDoS attacks. We fully simulated and tested it with MIT DARPA traces and on real and recent traffic. There are two major advantages of our algorithm. One is the ability to save detailed information of the attacks while using a limited amount of memory. The second advantage is the fact that our engine finds all the attacks we expect it to find with a negligible number of false-positives. These two advantages were achieved by the use of a hierarchical data structure.

A future work can identify a way to generalize this algorithm so it can detect other types of attacks. One can create a state machine for each protocol, and identify patterns that repeat in the different state machines. Thus, the nodes in the tree will hold the state machine operations and suspicious behavior will be an anomaly from these operations.

Another algorithm could be developed for finding anomalies in different parts of a packet or a flow. For example, a normal pattern can be the number of HTTP headers, in which case, HTTP request with many headers (Apache2 attack) would be reported as an anomaly. Another example is addressing a Mailbomb attack in which the attacker sends many messages to a server, overflowing that server's mail queue and causing system crash. Each site has a different threshold of e-mail messages that can be sent by (or to) one user before the messages are considered a Mailbomb. Thus, a high rate detection engine might not discover this kind of attack. If the nodes in the tree will contain per protocol information, the algorithm will detect the unexpected number of emails.

References

1. Moore, D., Voelker, G.M., Savage, S.: Inferring internet denial-of-service activity. In: 10th Usenix Security Symposium, pp. 9–22 (2001)
2. Mit darpa project data set, http://www.ll.mit.edu/IST/ideval/index.html
3. Mahoney, M., Chan, P.: Phad: Packet header anomaly detection for identifying hostile network traffic. Technical report, Florida Tech., CS-2001-4 (2001)
4. Gil, T.M., Poletto, M.: MULTOPS: A Data-Structure for bandwidth attack detection. In: Proceedings of USENIX Security Symposium, pp. 23–38 (2001)
5. Ayres, P.E., Sun, H., Chao, H.J., Lau, W.C.: Alpi: A ddos defense system for high-speed networks. IEEE Journal on Selected Areas in Communications 24(10), 1864–1876 (2006)
6. Brownlee, N., Mills, C., Ruth, G.: Traffic flow measurement: Architecture, http://www.ietf.org/rfc/rfc2063.txt
7. Cisco netflow, http://www.cisco.com/en/US/products/sw/netmgtsw/ps1964/index.html
8. Estan, C., Varghese, G.: New directions in traffic measurement and accounting. In: Proceedings of the 2001 ACM SIGCOMM Internet Measurement Workshop, pp. 75–80 (2002)
9. Schuehler, D.V., Lockwood, J.W.: A modular system for FPGA-based TCP flow processing in high-speed networks. In: Becker, J., Platzner, M., Vernalde, S. (eds.) FPL 2004. LNCS, vol. 3203, pp. 301–310. Springer, Heidelberg (2004)
10. Kompella, R.R., Singh, S., Varghese, G.: On scalable attack detection in the network. IEEE/ACM Trans. Netw. 15(1), 14–25 (2007)
11. Cert coordination center: tcp syn flooding and ip spoofing attacks, http://www.cert.org/advisories/CA-1996-21.html

12. Eddy, W.M.: Cisco: Defenses against tcp syn flooding attacks,
    http://www.cisco.com/web/about/ac123/ac147/archived_issues/ipj_9-4/
    syn_flooding_attacks.html
13. Bernstein, D.J.: Syn cookies, http://cr.yp.to/syncookies.html
14. Cert coordination center: smurf ip denial-of-service attacks,
    http://www.cert.org/advisories/CA-1998-01.html
15. Ferguson, P., Senie, D.: Rfc 2827. network ingress filtering: Defeating denial of service attacks which employ ip source address spoofing,
    http://www.faqs.org/rfcs/rfc2827.html
16. Kendall, K.: A database of computer attacks for the evaluation of intrusion detection systems. Master Thesis, MIT Department of Electrical Engineering and Computer Science (1999)
17. MacQueen, J.B.: Some methods for classification and analysis of multivariate observations. In: Cam, L.M.L., Neyman, J. (eds.) Proc. of the fifth 5th Berkeley Symposium on Mathematical Statistics and Probability, vol. 1, pp. 281–297. University of California (1967)
18. Bellovin, S.M.: Packets found on an Internet. Technical report, Computer Communications Review (1993)
19. Paxson, V.: Bro: a system for detecting network intruders in real-time. Computer Networks 31(23-24), 2435–2463 (1999)
20. Adamic, L.A.: Zipf, power-laws, and pareto - a ranking tutorial,
    http://www.hpl.hp.com/research/idl/papers/ranking/ranking.html
21. Adamic, L.A., Huberman, B.A.: The nature of markets in the world wide web,
    http://www.hpl.hp.com/research/idl/papers/webmarkets/webmarkets.pdf
22. Leland, W.E., Taqq, M.S., Willinger, W., Wilson, D.V.: On the self-similar nature of Ethernet traffic. In: Sidhu, D.P. (ed.) ACM SIGCOMM, San Francisco, California, pp. 183–193 (1993)
23. Paxson, V., Floyd, S.: Wide area traffic: the failure of Poisson modeling. IEEE ACM Transactions on Networking 3(3), 226–244 (1995)
24. Arsovski, I., Chandler, T., Sheikholeslami, A.: A ternary content-addressable memory (tcam) based on 4t static storage and including a current-race sensing scheme. IEEE Journal of Solid-State Circuits 38(1) (2003)
25. Weinsberg, Y., Tzur-David, S., Anker, T., Dolev, D.: High performance string matching algorithm for a network intrusion prevention system (nips). In: High Performance Switching and Routing, HPSR 2006 (2006)
26. Snort, http://www.snort.org/
27. Sekar, V., Duffield, N., Spatscheck, O., Merwe, J.V.D., Zhang, H.: Lads: Large-scale automated ddos detection system. In: USENIX ATC, pp. 171–184 (2006)

Automated Classification of Network Traffic Anomalies

Guilherme Fernandes and Philippe Owezarski

LAAS - CNRS
Université de Toulouse
7 Avenue du Colonel Roche
31077 Toulouse, France
owe@laas.fr

Abstract. Network traffic anomalies detection and characterization has been a hot topic of research for many years. Although the field is very advanced in the detection of network traffic anomalies, accurate automated classification is still a very challenging and unmet problem. This paper presents a new algorithm for automated classification of network traffic anomalies. The algorithm relies on three steps: (i) after an anomaly has been detected, identify all (or most) related packets or flow records; (ii) use these packets or flow records to derive several distinct metrics directly related to the anomaly; and (iii) classify the anomaly using these metrics in a signature-based approach. We show how this approach can act as a filter to reduce the false positive rate of detection algorithms, while providing network operators with (additional) valuable information about detected anomalies. We validate our algorithm on two different datasets: the METROSEC project database and the MAWI traffic repository.

1   Introduction

The Internet has greatly grown in complexity, changing from a single best effort service to a multi-services network that is ever more demanding of guaranteed quality of service (QoS). Network traffic anomalies can seriously impact or disrupt the normal operation of networks. It is then vital that their identification and mitigation be quickly done by network administrators. A specific type, volume anomalies, is responsible for unusual modifications on network traffic volume characteristics (normally identified on the #packets, #bytes and/or #new flows). These anomalies can be caused by a myriad of events: from physical or technical network problems (e.g. outages, routers misconfiguration), to intentionally malicious behavior (e.g. denial-of-service attacks, worms related traffic), to abrupt changes caused by legitimate traffic (e.g. flash crowds, alpha flows). This diversity coupled with the great (natural) variability of normal Internet traffic volume [16], makes the identification and mitigation of these anomalies a very challenging task.

Despite these difficulties, constant progress has been made in network traffic anomaly detection. Methods have been created to detect anomalies in single-links and network-wide data, and techniques have been used to cope with the

Y. Chen et al. (Eds.): SecureComm 2009, LNICST 19, pp. 91–100, 2009.

high dimensionality of network traffic data (e.g. sketches [13][4] and principal components [11][12]). Algorithms for network traffic anomaly detection have evolved from only being able to signal an anomaly in time (e.g. [1][17]) to providing information about the actual flows that cause the anomaly [13][4]. This information is very valuable for network administrators that need to manually verify and mitigate potential anomalies, but is still not enough. Because of the characteristics of network traffic and the frequency of anomalies, it is not feasible to manually analyze (in real-time) all anomalies detected by state-of-the-art detection algorithms. Network operators need more information than just the anomalous flows to efficiently prioritize between detected anomalies.

Although there has been some effort to characterize network traffic anomalies, automated classification has not received much attention (a notable exception is [12]). Automated classification intends to add meaningful information to the alert of a detected anomaly. Ideally, the computed information can then be used to define the type of the anomaly or to at least help characterize the underlying cause. In this paper, we present a new algorithm for automated classification of network traffic anomalies. We show how the information obtained by further analyzing the identified anomalous flows can be used in a signature-based classification module to reliably characterize different types of anomalies (e.g. DDoS, network scans, attack responses). We also show how this approach provides the flexibility needed by network operators to understand and manipulate the classification process. We do a statistical validation for the automated classification of DDoS anomalies and discuss results obtained for other type of anomalies using two different datasets: the METROSEC project (see http://www.laas.fr/METROSEC) database and the MAWI traffic repository [2].

2 Related Work

The evolution of detection algorithms (see Section 1) has been followed by several studies on the characterization of network traffic anomalies. Barford et al. [1] used a wavelet-based signal analysis on single-link volume data to characterize four classes of network anomalies: outages, flash crowds, attacks and measurement failures. Lakhina et al. used the subspace method to characterize several types of network-wide anomalies based on traffic volume metrics [11] and on traffic features [12]. Prior work has also been directed to individual types of anomalies. For example, DoS and distributed DoS (DDoS) attacks received an in-depth analysis in [15][8][14]. Jung et al. [9] studied the differences on DDoS and flash crowds behavior from a web server perspective. We thoroughly use the knowledge of such previous work to convey different attributes of the traffic anomalies that are used by our classification module to reliably label the anomaly.

Previous work has proposed ways to (automatically) convey more information about network traffic (e.g. by creating and labeling clusters [12][5]) and to provide prioritization (e.g. by using heuristics such as unexpectedness [5]). Specific to network traffic anomalies, the unsupervised approach of [12] creates clusters based on how anomalies are represented in the entropy space of their traffic features (i.e. IP addresses and ports). Although all anomalies that belong to a specific cluster share a given characteristic, this approach is clearly not enough to uniquely classify an anomaly (as shown by their results). Closer to our work, Kim et al. [10] study how different types of DoS attacks and port scans behave, creating rules to detect and classify them based either on flow header information or on statistical analysis of the flow traffic. Our algorithm aims at general automated classification of network traffic anomalies which are *just being detected*.

3   Anomaly Classification

Our algorithm defines three steps for anomaly classification: (i) after an anomaly has been detected, identify all (or most) related packets or flow; (ii) use these packets or flow records to derive several distinct metrics directly related to the anomaly; and (iii) classify the anomaly using these metrics with a signature-based approach. These steps are based on the fact that much information is needed to reliably classify different types of anomalies and even to distinguish between subtypes, like the many types of DoS attacks. Since current detection algorithms are based on few parameters (i.e. traffic volume metrics or traffic features like IP address and ports), steps are necessary to obtain more information about the anomaly. Naturally, the best source of information are records on the packets or flows that actually cause the anomaly. From now on, we will refer only to packets traces, but similar results can be obtained using flow records.

To test our classification algorithm, we use a variation of the simple traffic volume anomaly detection algorithm presented on [6]. The detection algorithm can be explained as follows. Given a trace of duration T and a time-scale granularity of Δ (i.e. 30s throughout this paper), divide the trace in N slots where $N \in [1, T/\Delta]$. For each slot i obtain the data time series X of each traffic volume metric $\in \{\#packets, \#bytes, \#syn\}$. Obtain the *absolute deltoids* [3] P of X and calculate their standard deviation σ_p. For any p_i over the threshold $K * \sigma_p$, mark its slot as anomalous. Using the deltoids of the data time series is important to consider the variation over the amplitude of the curve instead of the variation of network traffic, as the latter is insignificant due to its natural high variability. Our choice of metrics is based on [11] (with #syn instead of #new flows), but the algorithm permits the use of any other data time series.

Detection of low intensity anomalies is important especially for DDoS anomalies [16] and for anomalies in highly aggregated traffic. To detect low intensity anomalies, we apply the detection algorithm to different aggregation levels at the same time. Aggregation is done based on destination IP address and a bit mask modifier for each packet. In this paper we use the following prefix sizes as aggregation levels /0 (i.e. whole traffic), /8, /16 and /24. As with any other detection algorithm, this increase in sensitivity generates a higher rate of false positives (i.e. normal traffic variations are considered anomalous). With the multi-level feature, the algorithm presented above is particularly sensitive to infrequent communications where only a few packets are seen for a given network/mask aggregation. Although this would generally make the algorithm unusable, we show how the classification process can be used as a filter to greatly reduce the number of false positives. The simplicity of the detection algorithm makes the next step (i.e. identification of corresponding packets and derivation of metrics) a straightforward task and permitted us to concentrate on the characterization of the anomalies.

3.1   Gathering Information

With the characterization of network traffic anomalies done in previous work [1][11][12], we see that different types of anomalies can affect volume metrics and traffic features, such as IP addresses and ports, in the same manner. This clearly shows that we cannot do reliable classification based only on these metrics, and further information needs to be identified. We then introduce the notion of anomaly *attributes*. An attribute is a feature that helps to characterize a specific anomaly (see Table 1). The classification module uses signatures based on attributes derived directly from the packets that compose the anomaly.

The detection algorithm that we use in this work makes it straightforward to get these packets. A detected anomaly is identified by its slot, network address and mask. We also know exactly why it was considered anomalous (i.e. the deltoid for one or more of the volume metrics was above the threshold). Using this information we then proceed to read all the packets in the corresponding slot that are destined to that network, so that we can find the responsible destination hosts (i.e. IP address/32). Our idea of responsible destinations is similar to the notion of dominant IP address range and/or port of [11]. In our algorithm, the set of responsible destinations is composed of all the destination hosts that appear in any of the possible combinations of minimum sets that would bring the anomaly's corresponding deltoid below a fraction of the original threshold. After identifying these hosts, we follow an equivalent approach to determine the responsible sources, ports and protocols. This notion could also be applied to any other traffic feature. Potentially, finding the packets (or flows) that compose an anomaly can be done with any detection algorithm that identifies the starting time and anomalous flows of the anomalies (e.g. [13][4]).

During the anomaly detection and responsible flows identification phases we compute the attributes shown in Table 1. Attributes *found* and *impactlevel* are specific to the detection algorithm we use in this work, but similar attributes should be available for other detection algorithms. The rest of the attributes are derived while identifying the responsible flows. This list is by no means absolute and can be extended. These attributes were the ones we identified as useful during this work and are justified in Section 3.2.

3.2   Classification

General Idea. The main objective of our algorithm is to automatically label network traffic anomalies while they are being detected. The vast number of

Table 1. Attributes derived from a given anomaly. p, b and s are for packets, bytes and syn respectively.

Attribute	Description
found{p,b,s}	If metric was anomalous, value of P, zero otherwise.
impactlevel{p,b,s}	# of anomalous parent aggregation levels due to this anomaly.
#respdest	Number of responsible destinations.
#rsrc/#rdst	Ratio of responsible sources to responsible destinations.
avg#rdstports	Average number of responsible destination/source ports.
avg#rsrcports	Average number of responsible source ports.
#rpkt/#rdstport	Ratio of number of packets to responsible destination ports.
#rpkt/#rsrc	Average number of packets of responsible sources.
bpprop	Average packet size (only packets of the anomaly).
spprop	Ratio of number of syn to number of packets of the anomaly.
samesrcpred	If a specific responsible source appears for the majority of dests.
samesrcportpred	If the majority of responsible sources use the same source ports.
oneportpred	If only one destination port dominated.
invprotopred	If packets using invalid protocol numbers or types dominated.
invalidpred	If the anomaly was mainly consisted of (other) invalid packets.
landpred	If most packets had the same source and destination IPs.
echopred	If most packets were of type ICMP Echo Request/Reply.
icmppred	If most packets were ICMP of any other type.
rstpred	If most packets were TCP with RST flag set.

different types of anomalies [11] and the variations of individual types make it necessary to create very specialized signatures to achieve low misclassification rates. To this extent, we define three types of signatures: (i) universal, (ii) strong and (iii) local. Universal signatures are rules that should never misclassify an anomaly independently of network characteristics. Strong signatures are expected to have low misclassification rates but usually rely on some kind of threshold (and thresholds are difficult to set). Local signatures are defined by network administrators specifically to their domain. Note that they can choose how to best label these anomalies and change thresholds to suit their needs.

We will now discuss the anomalies that we have studied and show some examples of how the attributes we have identified can be used to create strong or even universal signatures for them. The idea is to give the reader a better understanding of how automated classification can be done using these attributes and to show the expressiveness of our algorithm. New attributes and rules can certainly be identified by expert network administrators.

DoS Characterization. Denial-of-service (DoS) attacks are malicious attempts to negate access to network resources [15]. Distributed denial-of-service (DDoS) attacks are (flooding) DoS attacks which use multiple sources to cause much more damage while being hardly detectable. These attacks are extremely common [15][8] and can greatly reduce the QoS of a network even when it has enough resources to cope with the attack [16]. DDoS anomalies may greatly affect the time series of #packets, #flows or both [11][1], and the distributions of destination and source

Table 2. Examples of strong signatures used in this work. *gr* stands for the time series granularity and *sspp* is an abbreviation for the attribute *samesrcportspred*.

Id	Anomaly Type	Signature
1	ICMP Echo DDoS	#respdest == 1 and echopred and (#rpkt/#rdstport > 30*gr or #rsrc/#rdest > 15)
2	TCP SYN DDoS	#respdest == 1 and founds and spprop > 0.9 and oneportpred and #rpkt/#rdstport > 10*gr
3	Network Scan	#respdest > 200 and samesrcpred
4	SYN Port Scan	#respdest == 1 and #rsrc/#rdest == 1 and spprop > 0.8 and avg#rdstports > 5
5	Attack Response	#respdest == 1 and (rstpred or icmppred) and foundp > 20*gr and (not (impactlevelp == 3)) and (#rsrc/#rdest == 1 or sspp)

addresses and ports [12]. However, these characteristics are shared with other types of anomalies, and more detailed information is needed to create robust signatures for their automated classification.

Universal signatures for DDoS anomalies can be defined by analyzing the types of DDoS attacks that use packets which do not comply with the used protocol specification. For example, many attacks have been seen in the wild to use either minimum size IP packets (i.e. 40 bytes) [8], an invalid protocol (e.g. IP protocol 0 or 255 [15][8]), or using land packets for flooding (i.e. packets with the same source and destination IP) [4]. A simple and direct rule would be *if invalidpred or invprotopred or landpred then label as DoS* (see Table 1 for a description of the attributes used). Note that all the identification information (e.g. source(s) and destination IP and port, protocol, etc.) is given as part of the alert.

Creating universal signatures for DDoS anomalies generated by attacks that use compliant packets is very difficult. For this type of attacks we try to develop strong signatures using a rich variety of attributes. Table 2 shows some of the signatures used in this work. For example, the second signature of Table 2 classifies TCP SYN attacks destined to a specific service (*oneportpred*) with an average of 10 or more packets per second (*#rpkt/#rdstport*). It uses *founds* and *spprop* to verify that most of the packets that generate the anomaly have (only) the TCP SYN flag set.

Other Anomalies. We will now quickly go over the other type of anomalies and the most interesting attributes we have identified for each one. *Network scans* [14] are probing attempts to identify the availability of a specific service on many different machines. Network scans can be reliably characterized by a single source communicating with many destinations (i.e. attributes *#respdest* and *samesrcpred*). Stronger signatures can also use *bpprop, foundsyn, spprop, oneportpred* and *#rpkt/#rdstport* to improve accuracy and maybe lower the threshold for *#respdest*. *Port scans* are similar but concentrate on one destination to discover which services the host is running. They should create very little traffic but may have a noticeable impact on *#syn*. They are characterized by one source, one destination and multiple ports with few packets being used. Signature 4 of Table 2 shows an example for classifying TCP SYN port scans.

Flash crowds (FC) can be defined as a sudden surge of legitimate client requests for a resource. The distributed nature of FCs makes it difficult to distinguish them from DDoS attacks [9]. Attributes include *#rsrc/#rdst, oneportpred, foundsyn, foundpkts* and *#rpkt/#rsrc*, while also taking into consideration that they should only be detectable in higher granularities (i.e. > 5min). *Alpha flows* are unusual high-rate byte transfers from a single source to a single destination, having a strong impact in #bytes and #packets [11]. They also tend to use much bigger packets than DoS attacks. Normally, port information is used to identify known operations that create alpha flows (e.g. scheduled backups). Attributes include *impactlevelbytes, impactlevelpkts, #respdest, #rsrc/#rdst, bpprop* and *foundsyn*, and actual ports might be defined.

Finally, *attack response* anomalies are generated by victims of attacks (e.g. DDoS or scans). These response packets are normally either TCP packets with RST ACK, RST or SYN ACK flags set, or ICMP control packets [15]. The line between attack responses and low intensity DDoS anomalies is very thin, especially as these packets are known to be used in DDoS reflector attacks [8]. Signature 5 of Table 2 shows a unified signature for detecting responses to flooding attacks and to scanning attempts.

Local Signatures. The flexibility of being able to understand, add and modify the way that anomalies are classified is a key feature for the applicability of automated network traffic anomaly detection and classification on real networks. Network operators may modify (or disable) strong signatures (i.e. by changing thresholds and/or labels), and also develop local (i.e. domain specific) signatures. For example, instead of trying to separate attack responses from DDoS attacks that use TCP RST packets, a signature might be defined as *if #respdest == 1 and rstpred and impactlevelp > 2 then label as StrongRSTAnomaly*. The flexibility provided by this approach can also be used to reduce false positives of detection algorithms. The rationale is that a wide range of signatures can be defined to potentially cover most of the true anomalies and a default label — applied to any anomalies that did not match one of these signatures — could then be discarded by network operators. This reduces the detection rate of true anomalies but trades the false positive rate of the detection algorithm for the misclassifications of the signatures defined.

4   Validation

We use two datasets to validate our algorithm: the METROSEC project traces with artificially created anomalies and the MAWI traffic repository with anomalies seen on the wild. We concentrate on DDoS anomalies for their importance and multiformity. If we are able to successfully separate different DDoS anomalies from normal traffic and from other types of anomalies, it might follow that general automated classification of network traffic anomalies is possible. Note that because of space limitation, only the most significant results are presented. A full description of the validation process and results can be found in [7].

4.1   Data

The METROSEC traces consist of real traffic collected on the French operational network RENATER with simulated attacks performed using real DDoS attack tools. This dataset was created in the context of the METROSEC research project to, among other goals, study the nature and impact of anomalies on networks' QoS. This dataset has been used for validation by a number of different studies on anomaly detection (e.g. [17]). For the validation of our algorithm, we use 14 METROSEC traces containing DDoS attacks of intensities ranging from very low (i.e. 4-10% of the whole traffic) to very high (i.e. 87-92%). The attacks also vary in type (i.e. from TCP SYN flooding to Smurf attacks), number of attacking hosts (i.e. 1-4) and duration.

On the other hand, the MAWI dataset has real undocumented anomalies. It is composed of 15 minutes packets traces collected daily at 2PM from a Japanese network called WIDE since 1999 to present. These traces are provided publicly after being anonymized and stripped of their payload data (see http://mawi.wide.ad.jp/). Although these traces are undocumented, the authors of [4] started an effort to label anomalies found in this database. We randomly selected a total of 30 traces from 2001 to 2006 from which some had already been identified by [4] to contain DDoS anomalies. Using this second dataset is important to verify that our algorithm is not restricted to a single network nor to artificial attacks.

4.2   Methodology

The validation of our algorithm is divided in two parts. In the first part, a (proper) statistical validation is done using the METROSEC traces for the classification of DDoS anomalies. Different levels of sensitivity of the detection algorithm are used by varying its K parameter from 1.5 to 6. The classification signatures used are the same for all values of K, but only DDoS related signatures are considered. In the second part, the classification performance of our algorithm is tested for different types of anomalies (i.e. DDoS, port and network scan, and attack response) on both of the datasets presented in the previous section. A fixed K of 2 is used, and all the signatures are enabled (including the *same* DDoS signatures used in the first part). A granularity of 30 seconds and the levels of aggregation 0, 8, 16 and 24 are used in the detection algorithm for both parts.

4.3   Results and Discussion

The classification performance for the first part of our validation was very similiar for all values of K (i.e. the algorithm achieved a very high rate of correct classifications with a *very* small rate of misclassifications). The results obtained with K equal to 2 include 23 true positives (i.e. DDoS anomalies correctly classified), 2 false positives (i.e. non-DDoS anomalies misclassified as DDoS), 1 false negative (i.e. misclassified DDoS anomaly) and 455731 true negatives (i.e. non-DDoS anomalies classified as non-DDoS). Further analysis showed that one of the false positives was actually a real, unexpected DDoS ICMP reflector attack, and the attack responsible for the false negative was correctly classified in a subsequent anomaly.

The results for the second part of our validation were equally promising. On the METROSEC traces, the non-DDoS signatures found a total of 16 port scans, 13 attack responses and 2471 network scans. Manual analysis showed that all port scans and 10 attack responses were true positives. We were not able to identify the nature of the other 3 attack responses. Network scans were not manually analyzed, but the signature used (see Table 2) has a very low (if not inexistent) misclassification rate. Running the algorithm on the 30 fifteen minutes MAWI traces resulted in 22 DDoS, 4429 network scan, 5233 port scan and 72 attack response anomalies in a total of 2.5 million anomalies detected. Manual analysis and cross-referencing with the results of [4] revealed 19 true positives (of which 6 had not been detected by [4]), 3 false positives that might be ICMP reflector attacks, and 9 (known) false negatives. The false negatives were mainly due to the detection algorithm used, and are not a limitation of our classification approach or of the signatures used. Preliminary analysis of the other type of anomalies showed that many of them were due to worm scannings (and responses), with Sasser and Dabber variants being particularly common.

5   Conclusions

In this paper we presented a new approach for automated classification of network traffic anomalies. We defined an initial set of anomaly attributes and characterized different types of anomalies (e.g. DDoS, network scans, etc) using them. We showed how automated classification can be done (succesfully) using these attributes within a signature-based approach and leveraging on the capability of state-of-the-art detection algorithms to identify the anomalous flows. We evaluated our work using two very different sets of packets traces with real network traffic and several anomalies. The results obtained illustrate the expressiveness of our approach to differentiate between many types of DDoS anomalies and other anomalies (including normal traffic variations), and strongly hint that general automated classification is possible. On future work we intend to explore the subtleties of other types of anomalies and to see how state-of-the-art identification algorithms can be easily integrated to our classification approach.

Acknowledgment

This work has been done in the framework of the ECODE project funded by the European commission under grant FP7-ICT-2007-2/223936.

References

1. Barford, P., Kline, J., Plonka, D., Ron, A.: A signal analysis of network traffic anomalies. In: Internet Measurment Workshop, Marseille (November 2002)
2. Cho, K., Mitsuya, K., Kato, A.: Traffic data repository at the wide project. In: USENIX ATEC, San Diego, California (2000)

3. Cormode, G., Muthukrishnan, S.: What's new: finding significant differences in network data streams. IEEE/ACM Trans. Netw. 13(6), 1219–1232 (2005)
4. Dewaele, G., Fukuda, K., Borgnat, P., Abry, P., Cho, K.: Extracting hidden anomalies using sketch and non gaussian multiresolution statistical detection procedures. In: Workshop on Large-Scale Attack Defense (LSAD), Kyoto, Japan (2007)
5. Estan, C., Savage, S., Varghese, G.: Automatically inferring patterns of resource consumption in network traffic. In: ACM SIGCOMM, Karlsruhe (2003)
6. Farraposo, S., Owezarski, P., Monteiro, E.: A multi-scale tomographic algorithm for detecting and classifying traffic anomalies. In: IEEE ICC, Glasgow (June 2007)
7. Fernandes, G., Owezarski, P.: Automated classification of network traffic anomalies. LAAS Report No 08468 (2008)
8. Hussain, A., Heidemann, J., Papadopoulos, C.: A framework for classifying denial of service attacks. In: ACM SIGCOMM, Karlsruhe (2003)
9. Jung, J., Krishnamurthy, B., Rabinovich, M.: Flash crowds and denial of service attacks: Characterization and implications for cdns and web sites. In: WWW, Honolulu, Hawaii (May 2002)
10. Kim, M.-S., Kong, H.-J., Hong, S.-C., Chung, S.-H., Hong, J.: A flow-based method for abnormal network traffic detection. In: IEEE/IFIP Network Operations and Management Symposium, Seoul (April 2004)
11. Lakhina, A., Crovella, M., Diot, C.: Characterization of network-wide anomalies in traffic flows. In: Internet Measurement Conference, Taormina, Italy (2004)
12. Lakhina, A., Crovella, M., Diot, C.: Mining anomalies using traffic feature distributions. In: ACM SIGCOMM, Philadelphia (2005)
13. Li, X., Bian, F., Crovella, M., Diot, C., Govindan, R., Iannaccone, G., Lakhina, A.: Detection and identification of network anomalies using sketch subspaces. In: Internet Measurement Conference, Rio de Janeiro, Brazil (2006)
14. Mirkovic, J., Reiher, P.: A taxonomy of ddos attack and ddos defense mechanisms. SIGCOMM Comput. Commun. Rev. 34(2), 39–53 (2004)
15. Moore, D., Voelker, G.M., Savage, S.: Inferring internet denial-of-service activity. In: USENIX SSYM, Washington, DC (2001)
16. Owezarski, P.: On the impact of dos attacks on internet traffic characteristics and qos. In: ICCCN (October 2005)
17. Scherrer, A., Larrieu, N., Owezarski, P., Borgnat, P., Abry, P.: Non-gaussian and long memory statistical characterizations for internet traffic with anomalies. IEEE Trans. Dependable Secur. Comput. 4(1), 56–70 (2007)

Formal Analysis of FPH Contract Signing Protocol Using Colored Petri Nets

Magdalena Payeras-Capellà, Macià Mut-Puigserver, Andreu Pere Isern-Deyà,
Josep L. Ferrer-Gomila, and Llorenç Huguet-Rotger

Departament de Matemàtiques i Informàtica, Universitat de les Illes Balears
{mpayeras,macia.mut,andreupere.isern,jlferrer,l.huguet}@uib.es

Abstract. An electronic contract signing protocol is a fair exchange protocol where the parties exchange their signature on a contract. Some contract signing protocols have been presented, and usually they come with an informal analysis. In this paper we use Colored Petri Nets to formally verify the fairness and the resistance to five previously described attacks of FPH contract signing protocol. We have modeled the protocol and the roles of the signers, a trusted third party, malicious signers as well as the role of an intruder. We have proven that the protocol is resistant to typical attacks. However, we have detected three cases where the protocol generates contradictory evidences. Finally, we have explained which should be the behavior of an arbiter to allow the resolution of these conflicting situations.

Keywords: contract signing protocol, Coloured Petri Nets, formal verification.

1 Introduction

Contract signing procedures, certified electronic mail or electronic purchases are good examples of fair exchange protocols. A fair exchange of values always provides an equal treatment to all users, and, at the end of the execution of the exchange, all parties have the element that wished to obtain, or the exchange has not been solved successfully (in this case, nobody has its expected element). These protocols make use of non-repudiation services, so they have to produce evidences to guarantee non-repudiation services. In case of dispute an arbiter has to be able to evaluate the evidences and take a decision in favor of one party without any ambiguity. Contract signing protocols allow the signature of a previously accorded contract by two or more signers. The fair exchange protocol ensures that at the end of the exchange all the signers have the signed contract or none of them have it. Fair exchange protocols often use Trusted Third Parties (TTPs) helping users to successfully realize the exchange. Several electronic contract signing protocols have been presented, with TTPs involved in different degrees. Among them there are a few proposals where the exchange can be finished in only three steps. Micali's protocol [1] and FPH protocol [2] are both efficient protocols with 3 messages in the exchange protocol. These protocols differ in the resolution protocol as well as in the elements exchanged in the three steps. However, they have another common aspects like the use of an off-line TTP, called optimistic approach. This concept of *optimistic protocol* was introduced in [3] by Asokan et al. In an optimistic fair exchange protocol the TTP only intervenes in case of problems to guarantee the fairness of the exchange.

Y. Chen et al. (Eds.): SecureComm 2009, LNICST 19, pp. 101–120, 2009.
ⓒ Institute for Computer Science, Social-Informatics and Telecommunications Engineering 2009

Bao et Al. described [4] three attacks to Micali's protocol and proposed an improved protocol. Recently, Sornkhom and Permpoontanalarp [5] have applied a formal method to analyze the security of Micali's protocol by using Colored Petri Nets (hereinafter CPNs). This method allows the demonstration of the vulnerability of Micali's protocol to the three attacks described by Bao. Additionally, the method has been used to find two new attacks to Micali's protocol.

In this paper we have created a new model for the formal analysis of FPH protocol, similar to that used by Sornkhom and Permpoontanalarp but adapted to the features of the present analysis. Once the protocol is modeled, we can formally prove the behavior of the protocol in case of malicious users. Our first goal is to prove the fairness of this protocol; first we will do that in case of malicious signers, and then we have modeled a malicious intruder.

We have organized the paper as follows, in Section 2 we summarize FPH protocol with its security characteristics. Section 3 includes the description of the simulation model using CPNs. Section 4 presents the analysis of the protocol and the results obtained in different execution scenarios. Finally, section 5 includes the conclusions and describes future applications of the simulation model.

2   FPH Contract Signing Protocol

2.1   Ideal Features of a Contract Signing Protocol

Practical solutions for contract signing require of the existence and possible involvement of a TTP. To obtain efficiency, three objectives are usually pursued:

- To reduce the involvement of the TTP.
- To reduce the number of messages to be exchanged.
- Possible implication of the TTP should not require expensive operations, neither the storage of high volume of information.

The first objective has been achieved in some proposals. They are the optimistic solutions [3,6,7,8,9] and the TTP are not involved in every protocol run. Regarding the number of messages to be exchanged, [6] states that three is the minimum number of messages for a contract signing protocol. Protocols for contract signing have to provide evidence to parties to prove, at the end of the exchange, if the contract is signed and the terms of the contract. Some additional properties have to be achieved in optimistic protocols [7,9]:

- *Effectiveness*: if the parties behave correctly the TTP will not be involved;
- *Fairness*: no party will be in advantageous situation at any stage of a protocol run;
- *Timeliness*: parties can decide when to finish a protocol run;
- *Non-repudiation*: parties can not deny their actions;
- *Verifiability of the third party*: if the TTP misbehaves, all damaged parties will be able to prove it.

In this section we describe the FPH protocol that will be formally evaluated in next sections. This protocol achieves the previous requirements.

2.2 Description of FPH Contract Signing Protocol

It is assumed that both (A)lice and (B)ob have already agreed on a plaintext contract C before the exchange. Then they sign the contract using the protocol. The channel used among the signers is an unreliable channel, so it cannot be assumed that the messages sent through this channel arrive to their recipient. The channel between a signer and the TTP is a resilient channel, that is, the messages will eventually arrive to their recipient but the time of the arrival cannot be predicted. The originator, A, and the recipient, B, will exchange non-repudiation evidence directly. Only in case they cannot get the expected items from the other party, the TTP will be invoked, by initiating *cancel* or *finish sub-protocols*. The notation and elements used in the protocol description are in Table 1 while the *exchange sub-protocol* is described in Table 2.

Table 1. Elements

X, Y	Concatenation of two messages X and Y
$H(X)$	Collision-resistant one-way hash function of message X
$S_i(X)$	Digital signature on message X with the private key, or signing key, of i (using some hash function, H(), to create a digest of X)
$i \rightarrow j: X$	i sends message X to j
$M=\{A,B,C\}$	Message containing the contract to be signed, C, the originator, A(lice), and the recipient, B(ob)
$h_A = S_A(M)$	Signature of A on the contract M
$h_B = S_B(M)$	Signature of B on the contract M
$ACK_A = S_A(h_B)$	Signature of A on h_B; acknowledgement that A knows that the contract is signed, and is part of the necessary evidence for B
$ACK_T = S_T(h_B)$	Signature of the TTP on h_B; this is an equivalent acknowledgement to which A should have sent
$h_{AT} = S_A[H(M), h_A]$	Evidence that A has requested TTP's intervention
$h_{BT} = S_B[H(M), h_A, h_B]$	Evidence that B has requested TTP's intervention
$h'_B = S_T(h_B)$	Signature of the TTP on h_B to prove its intervention

Table 2. Exchange sub-protocol

1. $A \rightarrow B$:	M, h_A
2. $B \rightarrow A$:	h_B
3. $A \rightarrow B$:	ACK_A

If the protocol run is completed, the originator A will hold non-repudiation (NR) evidence, h_B, and the recipient B will hold non-repudiation evidence, h_A and ACK_A. So the protocol meets the effectiveness requirement. If it is not the case, A or B, or both, need to rectify the unfair situation by initiating the *cancel* or *finish sub-protocol*, respectively, so that the situation returns to a fair position.

If A "says" (A could be trying to cheat or being in a wrong conception of the exchange state) that she has not received message 2 from B, A may initiate the *cancel sub-protocol* (Table 3).

Table 3. Cancel sub-protocol

IF (*finished=true*)	1'. $A \to T$:	$H(M), h_A, h_{AT}$
	2'. T:	retrieves h_B
	3'. $T \to A$:	h_B, h'_B
ELSE	2''. $T \to A$:	$S_T(\text{"cancelled"}, h_A)$
	3''. T:	Stores *cancelled=true*

In the *cancel sub-protocol*, the TTP will verify the correctness of the information given by A. If it is not the case, the TTP will send an error message to A. Otherwise, it will proceed in one of two possible ways. If the variable *finished* is *true*, it means that B had previously contacted with the TTP (see paragraph below), and the TTP had given the NR token to B, ACK_T. Now it has to give the NR token to A. So, it retrieves this stored NR token, h_B, and sends it to A, and a token to prove its intervention, h'_B. But if B had not previously contacted with the TTP, the TTP will send a message to A to cancel the transaction, and it will store this information (*cancelled = true*) in order to satisfy future petitions from B. Whatever case, now, we are again in a fair situation.

Table 4. Finish sub-protocol

IF	(*can-*	2'. $B \to T$:	$H(M), h_A, h_B, h_{BT}$
celled=true)		3'. $T \to B$:	$S_T(\text{"cancelled"}, h_B)$
ELSE		3''. $T \to B$:	ACK_T
		4''. T:	stores *finished=true* and h_B

If B "says" that he has not received message 3, B may initiate the *finish sub-protocol* (Table 4). In the *finish sub-protocol*, the TTP will verify the correctness of the information given by B. If it is not the case the TTP will send an error message to B. Otherwise, it will proceed in one of two possible ways. If the variable *cancelled* is *true*, it means that A had previously contacted with the TTP (see paragraph above). The TTP had given a message to A to cancel the transaction, and now it has to send a similar message to B. Otherways, the TTP will send the NR token, ACK_T, to B. In this case the TTP will store the NR token, h_B, and will assign the value *true* to the *finished* variable, in order to satisfy future petitions from A. Again, whatever case, now, we are in a fair situation.

As a conclusion, the protocol is fair and we have not made timing assumptions (the protocol is asynchronous).

2.3   Informal Analysis of Fairness and Non-repudiation of FPH Protocol

After a protocol run is completed (with or without the participation of the TTP), disputes can arise between participants. We can face with two possible types of disputes: repudiation of A (B claims that the contract is *signed*) and repudiation of B (A claims that the contract is *signed*).

An external arbiter (not part of the protocol) has to evaluate the evidence held and brought by the parties to resolve these two types of disputes. As a result, the arbiter will determine who says the truth. The arbiter has to know who is the originator and who is the recipient; remember that the contract, M, contains this information.

In case of repudiation of A, B is claiming that he received the signature on the contract M from A. He has to provide the following information to an arbiter: M, h_A and ACK_A or ACK_T. The arbiter will check if h_A is A's signature on M, and if it is positive the arbiter will assume that A had sent her signature to B. Then, the arbiter will check if ACK_A is A's signature on h_B, or it will check if ACK_T is TTP's signature on h_B. If this verification is positive, the arbiter will assume that either A or the TTP had sent an acknowledgement to B. Therefore, the arbiter will side with B. Otherwise, if one or both of the previous checks fails, the arbiter will reject B's demand. If the evidence held by B proves he is right, and A holds a message like $PR_T[H("cancelled", h_A)]$, it means that the TTP or A had acted improperly.

In case of repudiation of B, A is claiming that B had signed the contract M. She has to provide the following information to an arbiter: M and h_B. The arbiter will check if h_B is B's signature on M, and if it is positive the arbiter will assume that B had received M and h_A, and that he is committed to obtain the acknowledgement, ACK_A or ACK_T. If the previous verification fails, the arbiter will reject A's demand. If the verification is positive, the arbiter should interrogate B. If B contributes a *cancel* message, it means that B contacted with the TTP, and the TTP observed that A had already executed the *cancel sub-protocol*. For this reason the TTP sent the *cancel* message to B. Now it is demonstrated that A has tried to cheat. Therefore, the arbiter will reject A's demand, and the arbiter will side with B. If B cannot contribute the *cancel* message, the arbiter will side with A.

As a conclusion, the protocol meets the non-repudiation requirement. Moreover, the protocol also fulfils the property of verifiability of the TTP [2]. This informal analysis doesn't cover all the possible situations derived of the execution of the protocol. It will be completed with a formal verification of the protocol (included in Section 4) resulting from the use of the model based on Petri Nets described in Section 3.

3   Description of the Model Used for the Formal Analysis of Fair Exchange Protocols

3.1   Colored Petri Nets

CPN (Colored Petri Nets) is a discrete-event modeling language combining Petri Nets with a programming language called standard ML [10]. Petri Nets are capable to provide the interaction between processes and the programming language is used for the definition and manipulation of the data types. So, CPN can be used as a formal method to analyze distributed systems and communication protocols. A CPN model is an executable model representing the states of the system and the transitions that can cause a change of the state of the system. CPN contains four kinds of components:

- *Places*. They represent the system state at a given time. The places change from the activation of the transitions.
- *Transitions*. They are the actions which implies a state change.
- *Arcs*. They are the links between places and transitions.

– *Color sets.* The tokens that move through the states and transitions have a value, called color.

The global system state, after firing an event, is called *marking.* So, a *marking* is like a photo of the state of the system after each event. One of the tools that implement CPN is CPNTools [10]. This is the tool we have used in this work. When the model is designed, we can submit a simulation process in order to generate the state space. The state space is the set of markings between initial and final event. Therefore, we extract a complete definition of the system behavior along its execution.

3.2 General Assumptions and Methodology

In order to use Petri Nets to model the protocol, a number of general assumptions are made:

– Each party in the model has a unique identifier.
– Each party already knows the public keys of the others.
– Cryptographic algorithms used in the model are secure.
– The messages sent between the TTP and any party will always be delivered to the intended destination without modification (resilient channel).

The methodology followed to analyze the fairness of the protocol is:

– Build the model
    – Declare color sets (*colsets*) to represent messages and elements in the protocol.
    – Create top-level net to model the parties.
    – Create entity-level net to model the behavior of each party.
      Create process-level net for each entity-level.
    – Declare functions and variables that will be used in the model.
– Generate the state space
    – Set up initial marking for each party.
    – Generate the state space of the model using CPNTools.
– Create query functions to search for attack states.
– Extract attack scenarios using paths between states if attacks are found.

3.3 Description of the Model

In our model, based on Sornkhom and Permpoontanalarp's model [5], we have four key parties: Alice (A), Bob (B), Intruder (I) and TTP. While the TTP is strictly honest, the other parties can take the role of a malicious party. A and B, in their malicious role (A_m and B_m respectively), can stop the exchange or they can contact to the TTP in many different steps and this way, they could try to cheat the other party. I is a malicious party who can acts as an observer, like a man in the middle, and moreover he can deploy many other tasks: *drop, store, forward* or *modify* messages in transit sent by any party involved in the exchange.

In order to model the *drop* and *stop* events made by malicious parties (e.g. A_m, B_m or I), the model has a mechanism to inform about these events to the other involved parties. When an event occurs, a message is immediately sent by the party who drops the message or stops the exchange to the other parties involved. This assumption helps us to avoid the use of a timeout on each party. When an event message is received, the party could act contacting the TTP or maybe stopping the exchange depending of which is the current protocol step.

Another important consideration is: messages between the TTP and any other party of the model will always be delivered to the intended destination without any modification.

With the provided data, we are able to build an scenario that can be used to model the protocol using different attack sessions, where each session can involve an initiator (e.g. A or A_m) and a responder (e.g. B or B_m). Note that I and TTP are implicitly present in every session trace. So, we can deploy four sessions: (A, B), (A_m, B), (A, B_m) and (A_m, B_m), where (X, Y) denotes which party is the initiator (X) and the responder (Y). In this paper, we won't consider parallel session attacks where malicious parties can be involved in multiple and concurrent sessions, and this task will be deployed in further works.

The architecture of the model can be divided into three big blocks, using a top-down technique: top, entity and process levels. All messages sent by any party are a combination of source, destination and a protocol message as a payload.

The top level scheme (Fig. 1) shows basic interaction with all parties involved in the protocol and the message flow between these parties. In the top level we can see the contents of each party's database, which contains the protocol messages sent and received by each party. Finally, we can see and control the content of the session. The

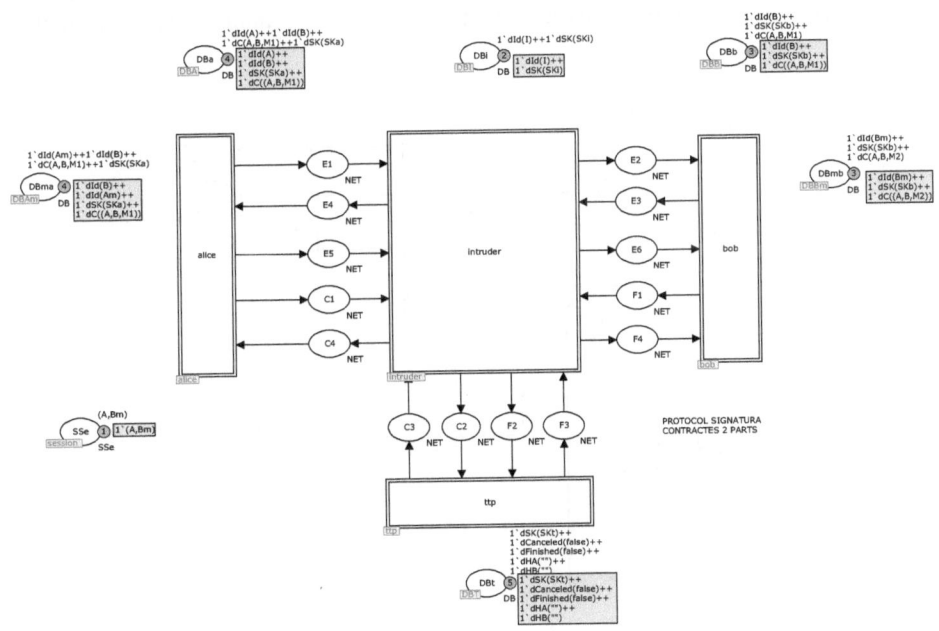

Fig. 1. Top level scheme

variable controls that will be used to distinguish the roles of the parties involved in the protocol execution (e.g. honest or malicious role). Moreover, in Fig. 1 we can see as the messages always are intercepted by I in their transit between parties.

The entity level shows us a more detailed model of the protocol and denotes all the steps each party can execute. In Fig. 2 we can see the entity level of A and her two roles. Transitions TA_1 to TA_4 are the transitions corresponding to her honest role, and TAm_1 to TAm_4 are the transitions of the malicious role. The first transitions of A, TA_1 and TAm_1, are to generate the first protocol message and send it to B. The transitions TA_2 and TAm_2 are to receive and verify the second message sent by B and send to B the third message. TA_3 and TAm_3 have the responsibility to contact the TTP using the *cancellation sub-protocol*, and the last transitions TA_4 and TAm_4 are to receive the response from the TTP. Note that the selection of the transitions that will be executed is done by the session configuration which tells if the party is honest or malicious.

B's entity level, as it is shown in Fig. 3, like A's entity level, implements the honest (TB_1 to TB_3) and malicious (TBm_1 to TBm_3) roles of B. TB_1 and TBm_1 are to receive and verify the first message of the protocol and they also send the second message, while TB_2 and TBm_2 are to receive and verify the third protocol message and, if it is needed, these transitions are able to contact the TTP. At last, TB_3 and TBm_3 are to receive the response from the TTP.

The process level implements all the actions deployed by the users and specifies how the relations between the entities are. The actions deployed by each process are atomic, e.g. only one process can be executed at the same time. This can be done by a unique token, which is shared between all parties of the model. It is captured by each party when a process starts, and it will be released when the process ends. Moreover, each process level is controlled by a session flow control mechanism. This mechanism

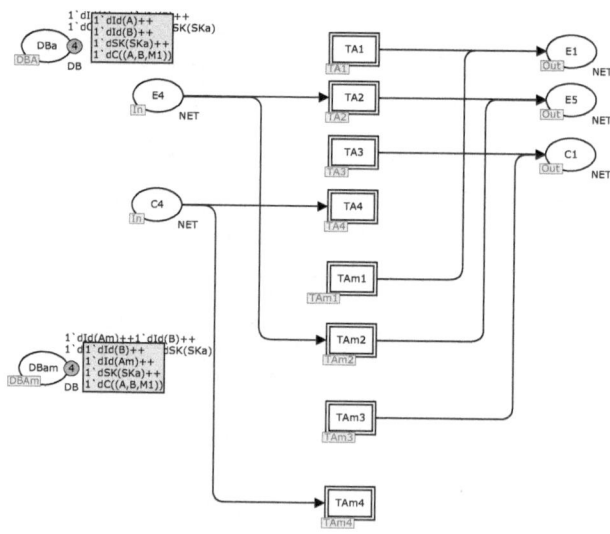

Fig. 2. A's entity level

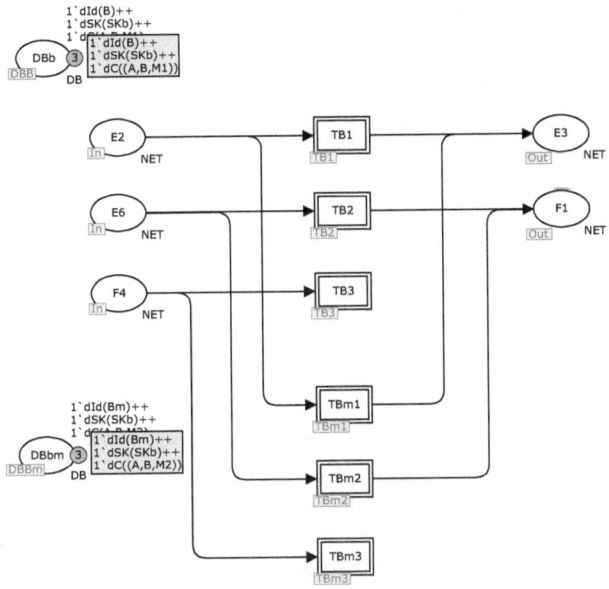

Fig. 3. *B*'s entity level

is defined like a token which passes through parties and at every step, they change it contents. This token controls the order in that actions will be done. For example, it controls that a message generation should be executed after the verification step.

3.4 Query Functions

In order to extract attack scenarios from state spaces we have developed a set of query functions, such that of Fig. 4 to find special contents in each party database. The main function is *SearchCommitsTerminalNodes(ack,id)*, where *ack* is the element or commit we would like to search in the database of the *id* party. This function returns a list of markings which fulfill some conditions. The function is build around the use of standard query function *PredNodes(p1,p2,p3)*. The first parameter is another custom query function named *SearchCommits(ack,id)*, where *ack* and *id* have the same use as in the previous query. This function is capable to take up the contents of the desired database *id* and tell us if the *ack* is in the database. The second parameter is to choose only markings which are leaf markings, e.g. terminal markings, which are markings that contain a complete execution of the protocol. The last parameter, *NoLimit*, tells the query should walk all markings and return all results.

The main query can be used to analyze the fairness property. In order to do this, we apply the query function against the parties involved in the exchange, depending of the session, to search the desired commit. The function will return a list of terminal markings. The analysis of this list will tell us if the exchange is fair or not.

```
fun SearchCommits( ack:DB, id:Id ) : Node list
 = PredAllNodes(
      fn n =>
      let
         val dba = Mark.Top'DBa 1 n
         val dbb = Mark.Top'DBb 1 n
         val dbi = Mark.Top'DBi 1 n
         val dbam = Mark.Top'DBma 1 n
         val dbbm = Mark.Top'DBmb 1 n
      in
         if (id=A) then
            cf( ack , dba ) > 0
         else if (id=B) then
            cf( ack , dbb ) > 0
         else if (id=Am) then
            cf( ack, dbam ) > 0
         else if (id=Bm) then
            cf( ack, dbbm ) > 0
         else (* id=I *)
            cf( ack , dbi ) > 0
      end
   )

fun SearchCommitsTerminalNodes( ack:DB, id:Id ) : Node list
 = PredNodes ( (SearchCommits(ack,id)) ,
               fn n => (Terminal n) andalso (FullyProcessed n),
               NoLimit)
```

Fig. 4. Search query functions developed in order to search commits into the party's databases

4   Formal Analysis of FPH Contract Signing Protocol

4.1   Evaluation of the Vulnerability to Previously Defined Attacks

Until today, several attacks to contract signing protocols have been described. Bao et Al. [4] found three attacks to Micali's ECS1 protocol (Table 5). Later Sornkhom and Permpoontanalarp [5] found two new attacks to the same protocol. The consequence of these attacks is the loss of fairness. For this reason, we have used the model based on CPN described in last section to evaluate the resistance of FPH protocol to all these attacks.

Micali's ECS1 protocol (Table 5) and FPH protocol are similar, so we will use the same notation to describe them. Moreover, we will use $E_X(Y)$ to denote the encryption using the public key of X of the message Y. A is committed to the contract, C, as an initiator if B has both $S_A(C, Z)$ and M where $Z = E_{TTP}(A, B, M)$ and M is a random. On the other hand, B is committed to C as a responder if A has both $S_B(C, Z)$ and $S_B(Z)$.

Now we are going to describe the five attacks to Micali's protocol and apply them to FPH protocol, then we will use the model to prove both the fairness and its resistance to these attacks.

Table 5. Micali's ECS1 protocol definition

	$A \rightarrow B$: $\quad S_A(C, Z)$
	$B \rightarrow A$: $\quad S_B(C, Z), S_B(Z)$
IF (Both signatures are valid)	$A \rightarrow B$: $\quad M$
IF (B receives valid M such that $Z = E_{TTP}(A, B, M)$)	The exchange is completed
ELSE	$B \rightarrow$ TTP: $\quad A, B, Z, S_B(C, Z), S_B(Z)$
	TTP $\rightarrow A$: $\quad S_B(C, Z), S_B(Z)$
	TTP $\rightarrow B$: $\quad M$

Bao's First Attack. A is a malicious initiator and sends a false element in step 1. In Micali's protocol this attack (Table 6) can be done if A sends a false Z where $Z = E_{TTP}(A, B, M)$. In this case, A can always obtain B's commitment but B will not have A's commitment. This attack is possible because B cannot verify the elements received in step 1.

Table 6. Bao's First Attack Trace

$A \rightarrow B$:	$S_A(C, Z)$ where $Z = E_{TTP}(A, B, M)$
$B \rightarrow A$:	$S_B(C, Z), S_B(Z)$
$A \rightarrow B$:	Nothing
$B \rightarrow$ TTP:	$A, B, Z, S_B(C, Z), S_B(Z)$
TTP $\rightarrow A$:	Nothing
TTP $\rightarrow B$:	Nothing

In order to detect the attack on the model, we have generated a session with A_m (A acting maliciously) and B, as we can see on Fig. 5. In this attack, A_m builds a false contract M_2 and she sets an arbitrary initiator (X) and arbitrary responder (Y). The first query searches h_A element in A_m's database, finding four cases, corresponding to markings 20, 21, 22 and 37. The second query searches the same element, h_A, in B's database and as we can see, B never has this element. This is because the verification stage fails and B never stores the received message. The two last queries search the response of the TTP into A_m's database, and we can see that A_m only receives a *cancel* message (marking 37) and she never obtains the NR evidence from the TTP.

Then, FPH protocol is not vulnerable against Bao's *first attack*, because B verifies the elements received in step 1 and in case of *attack 1* he doesn't send the message of step 2. Then A will not send message 3. If A tries to contact the TTP, the TTP will send a cancellation proof and stores *cancelled=true*. B will not contact the TTP because he doesn't have any valid element from A.

Fig. 5. First attack query results, A_m database contents and session configuration

Bao's Second Attack. *A* **conspires with another initiator** *A'* **and changes her identity in step 1.** In Micali's protocol this attack (Table 7) can be done if *A* conspires with *A'* and sends a false *Z* where $Z = E_{TTP}(A', B, M)$. In this case, malicious *A* can always obtain *B*'s commitment on a contract between *B* and *A'*, but *B* will not have anything. This attack is possible because *B* cannot verify the identity of *A* in the element received in step 1.

Table 7. Bao's Second Attack Trace

$A \rightarrow B$:	$S_A(C, Z)$ where $Z = E_{TTP}(A', B, M)$
$B \rightarrow A$:	$S_B(C, Z), S_B(Z)$
$A \rightarrow B$:	Nothing
$B \rightarrow TTP$:	$A, B, Z, S_B(C, Z), S_B(Z)$
$TTP \rightarrow A$:	Nothing
$TTP \rightarrow B$:	Nothing

The *second attack* can be detected in the model using the same session configuration (A_m, B) as the *first attack*, but using a different contract. In this case, we have built a false contract with a confabulated initiator (X), the initial receiver (B) and the previously accorded plain contract (M_1). As we can see in Fig. 6, the query results are the same as in the *first attack*, the second function never returns any result because *B* never builds message 2. Then, if we search the TTP's response on A_m's database, we can see A_m never obtains the NR and she only could have a cancellation proof.

So, FPH protocol is not vulnerable against this attack. *B* verifies the elements received in step 1 and in case of *attack 2* he doesn't send the message of step 2, as in *attack 1*. Then *A* will not send message 3 and the exchange will be stopped and *A* will not obtain *B*'s commitment. If *A* tries to conclude the exchange contacting the TTP, she will receive a cancellation proof. On the other side, *B* will not contact the TTP because he doesn't want to finish the exchange because he knows that the element sent in step 1 is false and, moreover, he hasn't sent any element.

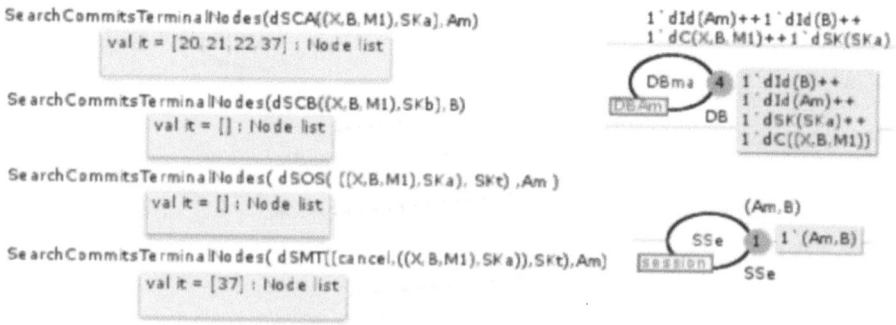

Fig. 6. Second attack query results, A_m database contents and session configuration

Table 8. Bao's Third Attack Trace

A → B:	$S_A(C, Z)$ where $Z = E_{TTP}(A', B, M)$
B → TTP:	$Z, S_B(C', Z), S_B(Z)$ for a false contract C'
TTP → A:	$S_B(C', Z), S_B(Z)$
TTP → B:	M

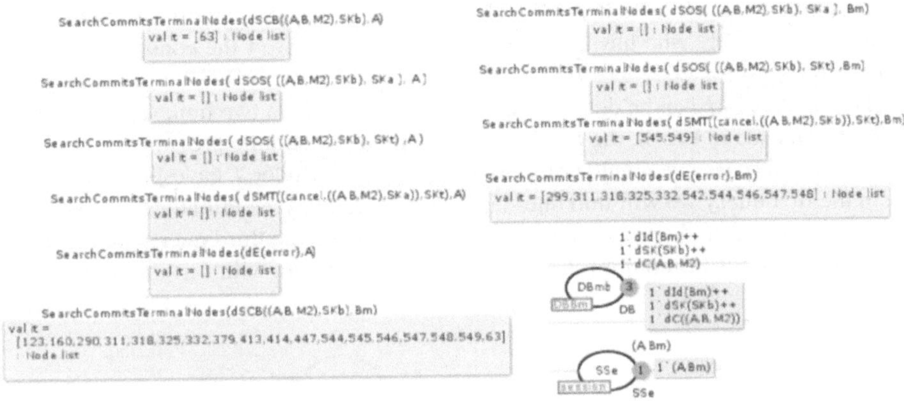

Fig. 7. Third attack query results, B_m database contents and session configuration

Bao's Third Attack. Malicious B contacts the TTP and requests the resolution with a false contract. In Micali's protocol this attack (Table 8) can be done if B, after the reception of a valid message in step 1, contacts the TTP to start the resolution of the exchange. In this request B includes a fake contract. In this case, malicious B always gets A's commitment on the original contract, but A obtains B's commitment on the false contract (selected by B). This attack is possible because A cannot request the resolution of the exchange and obtains from the TTP the elements resulting of the resolution started by B.

The *third attack* can be verified with the model using a session configuration where A is the honest initiator and B_m is the malicious responder, e.g. (A, B_m). B_m builds a contract containing a false plain text (M_2) but using the real initiator and responder. As we can see in Fig. 7, when B_m receives the first message, he changes its contents by setting a different plain contract (M_2). Then, we have searched if a false h'_B sent by B_m is into A's database and, effectively, it is in marking 63. Although A stores the message, she verifies it and she decides it is wrong and she doesn't generate the third message. Then A can contact the TTP, but she would ask for the original real contract using the *cancellation sub-protocol* and the TTP will send a cancellation proof to A. Finally, we can search the TTP's responses in B_m's database and we can see that he never obtains the alternative proof. Moreover, he can only obtain the cancellation proof and an error message because the TTP's verification fails.

In FPH protocol, however, when A receives a false $h'_B = S_B(M', A, B)$ in step 2, she detects the attack, stops the exchange and contacts the TTP. If B has contacted the TTP in first place and the request contained a false h_B, the TPP has been able to detect that h_A and h_B are not related with the same contract. Then, when A sends a resolution request, the TTP will send her a cancellation proof, so the contract will not be signed. If A contacts the TTP in first place, she will obtain a cancellation proof.

Fourth Attack. An Attacker eavesdrops B's commitment. The *fourth attack* was described in [5] and it is possible because Micali's protocol has an incomplete definition on B's commitment. The message $(S_B(C, Z), S_B(Z))$ is the evidence to prove that B has committed himself to contract C with any initiator. The evidence is not linked to the initiator, so anybody who has it can claim to be an initiator of the contract committed by B.

The *fourth attack* can be detected using a session between two honest parties, A and B. As we can see in Fig. 8, the databases of A and B contain the previously committed contract. In this case, we would search states where an intruder, I, eavesdrops messages. So, in the first query we will find one state where I changes the initiator of the contract. This message is found on B's database and finally, using the third query, we can prove how B never builds his commit, h_B, over the wrong contract with I as initiator.

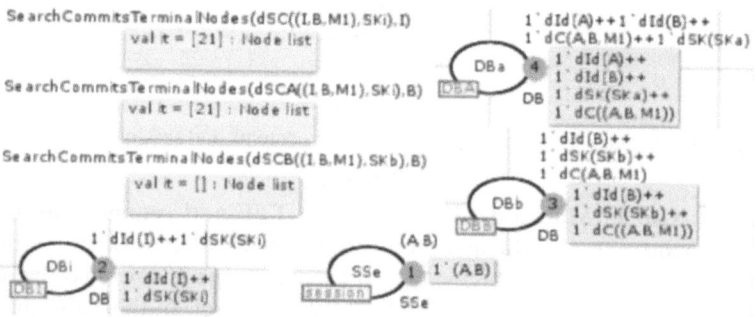

Fig. 8. Fourth attack query results, parties' database contents and session configuration

In contrast to Micali's protocol, FPH protocol has linked B's commitment to the contract. The evidence is the message $h_B = S_B(M)$, however holding this evidence is not enough for anyone to prove that B has committed himself to contract C. Because FPH protocol specifies that M contains the contract to be signed, C, and it indicates who is the originator, A, and who is the recipient, B. Thus, FPH protocol is resistant to this attack.

Fifth Attack. Swapping the initiator and the responder role. In the *fifth attack* (Table 9) described in [5] a malicious A can get B's commitment on a contract between B as an initiator and any conspired party A_r as a responder. But B will not get anything. In order to perform the attack, A involves B in the protocol so as to exchange the commitments on a contract. But A build a fake item Z with the identity of B as the initiator and a conspiring party A_r as the responder: $Z = E_{TTP}(B, A_r, M)$. Finally A will give $S_B(C, Z)$ and M to A_r. The TTP can't send anything to A and B because item Z doesn't fulfill the protocol specifications. Now, A_r can successfully claim B's commitment on the contract as an initiator and B doesn't have any kind of evidence.

Table 9. Fifth Attack Trace

A → B:	$S_A(C, Z)$ where $Z = E_{TTP}(B, A_r, M)$
B → A:	$S_B(C, Z), S_B(Z)$
A → B:	Nothing
B → TTP:	$A, B, Z, S_B(C, Z), S_B(Z)$
TTP → A:	Nothing
TTP → B:	Nothing

The *last attack* reconstruction (Fig. 9) uses a session composed by A_m and B. In this case, A_m changes the contents of the contract, swapping party's roles but she uses the previously committed contract (M_1). The application of the query functions against the model (Fig. 9) is the same as in the *first* and *second attacks*. The step 1 searches in A_m's database the first element h'_A and then the second query searches the second message into B's database. As it is shown, B does not build it. Finally, the third and fourth queries try to search responses sent by the TTP into A_m database. As we can see, she will only obtain a cancellation proof (marking 37).

However, in FPH protocol, as we have already explained, B verifies the item received at the step 1 of the protocol. Thus, if A has made improper changes in the message, B will detect it. Then, he will not continue and he will not send the message of step 2. Therefore, the attack described here will not be successful.

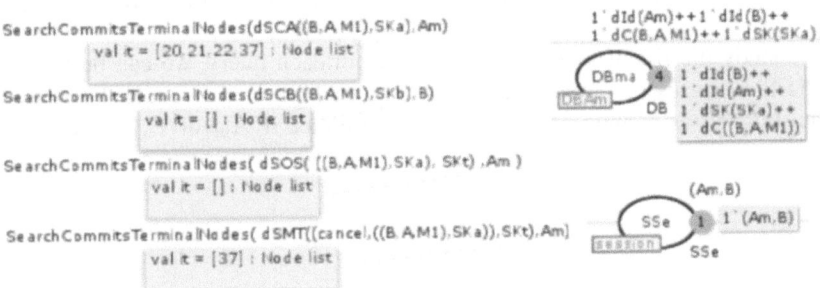

Fig. 9. Fifth attack query results, A_m database contents and session configuration

4.2 Fairness Analysis

In this section we will describe some conflicting situations in FPH protocol where the signers have contradictory evidences (see Section 2.3). The evidences generated by this protocol are not transferable, and an arbiter must contact both signers to solve a dispute, know the final state of the exchange and guarantee non repudiation. This property has been described in [11] and is called weak fairness. In this formal analysis of the fairness of the protocol we will prove that the arbiter can solve all kinds of conflicting situations derived from the execution of the protocol.

In [2], we described a conflicting situation where A can obtain NR evidence from B (h_B) and a *cancel* message from T, while B obtain NR evidence from A (h_A, ACK_A). A can do it, for instance, invoking the *cancel sub-protocol* after the end of the *exchange sub-protocol*. It seems that *A* can affirm that the contract is signed or is not signed (*cancelled*), depending on her usefulness, while *B* possesses NR evidence that will prove that the contract is signed. We have detected this situation in the formal analysis and we have called it *case 1*.

Moreover, thanks to our model we have discovered two more conflicting situations. The first one (we will call it *case 2*) is produced when a malicious *A* invokes the *cancel sub-protocol* after the end of the *exchange sub-protocol* (as in *case 1*) and then a malicious *B* executes it, too. It seems that *A* and *B* can state that the contract is signed or is not signed (*cancelled*), depending on her usefulness.

The last conflicting case we have detected (*case 3*) is achieved when the exchange is stopped after the step 2. In this case *A* has the NR evidence from *B* while *B* does not have the NR evidence from *A*. Both parties can contact the TTP. If *B* contacts in first place the TTP will send him the NR evidence and the contract will be signed. Instead, if *A* contacts in first place, the TTP will cancel the exchange and then *A* would have NR evidence from *B* (h_B) and a *cancel* message from *T*, while *B* obtains the *cancel* message.

The three conflicting situations are found in the model deploying a session composed by a malicious initiator or both a malicious initiator and a malicious responder, e. g. (A_m, B_m). This way, all possible behaviors of both parties are contemplated. Using the already known query functions, as shown in Fig. 10, we have searched into each party's database the desired commits. In this case, we have searched the second and third messages of the *exchange sub-protocol* and all the responses received from the TTP.

If we study the list of markings obtained from each query, we can build Table 10 with the three cases previously described. For each case, we denote the state of the contract (signed as S and cancelled as C) and either if A_m or B_m have contacted,

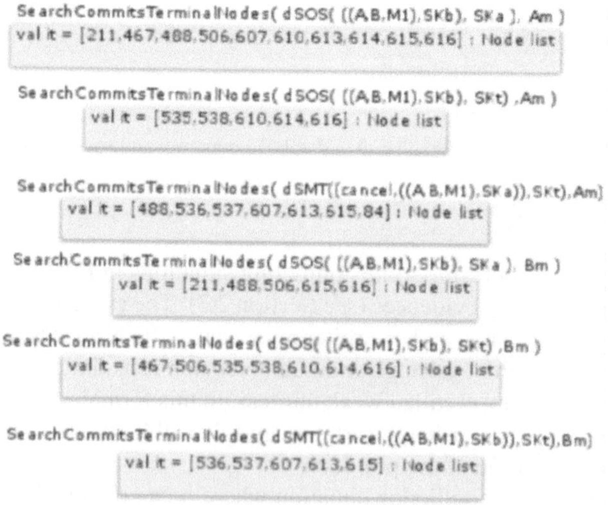

SearchCommitsTerminalNodes(dSOS(((A,B,M1),SKb), SKa), Am)
val it = [211,467,488,506,607,610,613,614,615,616] : Node list

SearchCommitsTerminalNodes(dSOS(((A,B,M1),SKb), SKt) ,Am)
val it = [535,538,610,614,616] : Node list

SearchCommitsTerminalNodes(dSMT[(cancel,((A,B,M1),SKa)),SKt),Am]
val it = [488,536,537,607,613,615,84] : Node list

SearchCommitsTerminalNodes(dSOS(((A,B,M1),SKb), SKa), Bm)
val it = [211,488,506,615,616] : Node list

SearchCommitsTerminalNodes(dSOS(((A,B,M1),SKb), SKt) ,Bm)
val it = [467,506,535,538,610,614,616] : Node list

SearchCommitsTerminalNodes(dSMT[(cancel,((A,B,M1),SKb)),SKt),Bm]
val it = [536,537,607,613,615] : Node list

Fig. 10. Query functions results over the model with (A_m, B_m) session

Table 10. List of the markings corresponding to the three cases with contradictory evidences

Case	Marking	A_m has NR	B_m has NR	A_m contacts TTP	B_m contacts TTP
1	488	S & C	S	Yes (M)	No
2	615	S & C	S & C	Yes (M)	Yes (M)
3	607	S & C	C	Yes (M)	Yes (H)
3	613	S & C	C	Yes (M)	Yes (H)

Table 11. Scenarios without contradictory evidences

Marking	A_m has NR	B_m has NR	A_m contacts TTP	B_m contacts TTP
84	C	Nothing	Yes (H)	No
211	S	S	No	No
467	S	S	No	Yes
506	S	S & S (by TTP)	No	Yes
535	S (by TTP)	S (by TTP)	Yes (H)	Yes (H)
536	C	C	Yes (H)	Yes (H)
537	C	C	Yes (H)	Yes (H)
538	S (by TTP)	S (by TTP)	Yes (H)	Yes (H)
610	S & S (by TTP)	S (by TTP)	Yes (M)	Yes (H)
614	S & S (by TTP)	S (by TTP)	Yes (M)	Yes (H)
616	S & S (by TTP)	S & S (by TTP)	Yes (M)	Yes (M)

maliciously (M) or honestly (H), the TTP. As shown in Table 10, using the model we
have located the three cases where A_m and B_m have contradictory evidences although
we have detected four possible scenarios, because *case 3* could appear twice.

As we can see, *case 1* happens on marking 488, when A_m obtains the NR evidence
from B_m (so A has evidence that the contract is *signed*), but she contacts the TTP
in order to cancel the exchange. This is a malicious behavior, because A_m shouldn't
contact the TTP to cancel an exchange that is already finished. The TTP sends A_m the
cancel message and then, A_m could affirm that the contract is *signed* or *cancelled*. B_m
receives the NR from A_m and he doesn't need to contact the TTP.

In *case 2*, corresponding to marking 615, the *exchange sub-protocol* ends successfully for each party, but A_m contacts the TTP, after the transfer of NR evidence to B_m,
in order to obtain a *cancel* message. Once B_m receives the NR evidence from A_m,
she also contacts the TTP and he obtains a *cancel* message. So, A_m and B_m have a
malicious behavior because they contact the TTP when they shouldn't.

Case 3 is detected twice on the model. In both scenarios, A_m obtains the NR from
B_m but B_m never receives the third message. In each scenario, A_m executes maliciously the *cancellation sub-protocol* and she receives a *cancel* message from the TTP.
In the first scenario, corresponding to marking 607, A_m decides to maliciously stop the
exchange and she doesn't send the third message to B_m. In the other hand, marking
613 is the result of a *drop* event of the third message by an intruder, *I*. From the point
of view of B_m, both scenarios are the same, and he will contact to the TTP in order to
resolve the situation obtaining a *cancel* message for each scenario.

In addition to the conflicting cases, there are other cases detected by the model where there aren't contradictory evidences but, in some cases, each party could have repeated proofs because they may contact the TTP when the protocol is successfully ended. Table 11 displays the scenarios without contradictory evidences.

The most interesting cases displayed on Table 11 are markings 84 and 616. In marking 84, B_m has nothing from A_m because an intruder I has executed a drop event on the first message. B_m cannot execute the *finalization sub-protocol* because he doesn't have any valid element from A_m. In the other hand, A_m resolves the contract executing the *cancellation sub-protocol* obtaining a *cancel* message from the TTP. The second marking, 616, is the case where A_m and B_m act maliciously contacting the TTP when the *exchange sub-protocol* ends successfully. It is similar to case 615, but this time, B_m contacts in first place the TTP, obtaining the corresponding NR evidence. Then, if A_m tries to cancel, the TTP sends a NR evidence that states that the contract is *signed*.

In order to solve these conflicting situations an arbiter must always contact both parties, and in case of contradictory evidences, we have established that he must act as follows:

- *Case 1*: A can state that the contract is *signed* or *cancelled*, but B possesses NR evidence that will prove that the contract is *signed*. If A tries to use the *cancel* message she will be proving she is a cheating party, so the arbiter will side with B.
- *Case 2*: As in *case 1*, an arbiter will contact both parties in case of contradictory evidences. If B shows NR evidence that will prove that the contract is *signed*, the arbiter can state that the contract is *signed*, and if B shows that the contract is *cancelled* the arbiter will state that the contract is *cancelled*. This way, due to the fact that A is always the first cheating party, if the arbiter sides always with B, the protocol will discourage A to act fraudulently.
- *Case 3*: Once again, A has acted fraudulently, and if the arbiter sides with B he will state that the contract is *cancelled*.

As a conclusion, we have detected the previously defined conflicting situation and we have discovered two additional cases. All the cases are due to the fraudulent behavior of A. To solve these situations, an arbiter must contact both parties and in case of conflict he must always side with B. This way, the protocol will be fair in all cases and moreover the fraudulent behavior of the parties is discouraged.

5   Conclusions and Future Work

In this paper we have formally analyzed, using a formal method (Petri Nets), an efficient contract signing protocol, FPH protocol [2], known as one of the solutions involving only three messages, as Micali's protocol. But, while Micali's protocol has been flawed (three attacks were found by Bao et.al. and two more attacks were found by Sornkhom and Perpoomtanalarp), FPH protocol is not vulnerable to any of these attacks due to its features. We have evaluated FPH protocol using a model that assumes that all the signers can be dishonest and an intruder can also attack the exchange, and we have proven the resistance to all these attacks using the model.

We have evaluated all the possible situations involving malicious users and intruders, and in all cases the exchange ends in a fair situation. Moreover, we have also detected that there are three cases in where, although the exchange is fair, one of the signers (or both) can have contradictory evidences. For these reason, although the exchange is fair, we cannot say that the proofs generated by the protocol are transferable, because both parties have to be interrogated by an arbiter to know the final state of the exchange. Finally, we have created a set of rules to determine the role of the arbiter in order to achieve fairness even when contradictory evidences are presented.

With the model created to evaluate the vulnerability of the protocol to previously described attacks and prove the fairness of the protocol we will be able, in a near future, to formally analyze other properties of the protocol, such as the verifiability of the TTP and also try to model more complex protocols such as a multiparty contract signing protocol. Moreover, we will adapt the model to work with new attack scenarios, like confabulated attacks using data from two different signature sessions. In parallel, we will work in the improvement of the model in order to include more control over the intruder's behavior and some other enhancements.

Acknowledgement

This work is partially supported by MEC and FEDER under projects: "Seguridad en la Contratación Electrónica basada en Servicios Web" (CICYT TSI2007-62986) and ARES "Grupo de Investigación Avanzada en Seguridad y Privacidad de la Información" (Consolider - Ingenio CSD2007-004). We would like to thank Yongyuth Permpoontanalarp for his useful comments and support during the development of this work.

References

1. Micali, S.: Simple and Fast Optimistic Protocols for Fair Electronic Exchange. In: Proceedings of 21st Symposium on Principles of Distributed Computing, pp. 12–19 (2003)
2. Ferrer-Gomila, J., Payeras-Capellà, M., Huguet-Rotger, L.: Efficient Optimistic N-Party Contract Signing Protocol. In: Davida, G.I., Frankel, Y. (eds.) ISC 2001. LNCS, vol. 2200, pp. 394–407. Springer, Heidelberg (2001)
3. Asokan, N., Shunter, M., Waidner, M.: Optimistic Protocols for Fair Exchange. In: 4th ACM Conference on Computer and Communications Security, pp. 7–17 (1997)
4. Bao, F., Wang, G., Zhou, J., Zhu, Z.: Analysis and Improvement of Micali's Fair Contract Signing Protocol. In: Wang, H., Pieprzyk, J., Varadharajan, V. (eds.) ACISP 2004. LNCS, vol. 3108, pp. 176–187. Springer, Heidelberg (2004)
5. Sornkhom, P., Permpoontanalarp, Y.: Security analysis of micali's fair contract signing protocol by using coloured petri nets. In: 9th ACIS Int. Conference on Software Engineering, Artificial Intelligence, Networking and Parallel/Distributed Computing, pp. 329–334 (2008)
6. Ferrer-Gomila, J.L., Payeras-Capellà, M.M., Huguet-Rotger, L.: Optimality in asynchronous contract signing protocols. In: Katsikas, S.K., López, J., Pernul, G. (eds.) TrustBus 2004. LNCS, vol. 3184, pp. 200–208. Springer, Heidelberg (2004)
7. Asokan, N., Shoup, V., Waidner, M.: Asynchronous Protocols for Optimistic Fair Exchange. In: IEEE Symposium on Research in Security and Privacy, pp. 86–99 (1998)
8. Garay, J.A., Jakobsson, M., MacKenzie, P.: Abuse-free optimistic contract signing. In: Wiener, M. (ed.) CRYPTO 1999. LNCS, vol. 1666, p. 449. Springer, Heidelberg (1999)

9. Zhou, J., Deng, R., Bao, F.: Some remarks on a fair exchange protocol. In: Imai, H., Zheng, Y. (eds.) PKC 2000. LNCS, vol. 1751, pp. 46–57. Springer, Heidelberg (2000)
10. Jensen, K., Kristensen, L.M., Wells, L.: Coloured Petri Nets and CPN Tools for Modelling and Validation of Concurrent Systems. Intenationals Journal on Software Tools for Technology Transfer, 213–254 (2007)
11. Kremer, S., Markowitch, O., Zhou, J.: An Intensive Survey of Fair Non-Repudiation Protocols. Computer Communications 25, 1606–1621 (2002)

On the Security of Bottleneck Bandwidth
Estimation Techniques

Ghassan Karame, David Gubler, and Srdjan Čapkun

Department of Computer Science
ETH Zürich, Switzerland
karameg@inf.ethz.ch, dgubler@student.ethz.ch, capkuns@inf.ethz.ch

Abstract. Several wide-area services are increasingly relying on bottleneck bandwidth estimation tools to enhance their network performance.
Selfish hosts have, therefore, considerable incentives to fake their bandwidths in order to increase their benefit in the network. In this paper,
we address this problem and we investigate the vulnerabilities of current
bottleneck bandwidth estimation techniques in adversarial settings. We
show that finding "full-fledged" solutions for the multitude of attacks on
the end-to-end bandwidth estimation process might not be feasible in the
absence of trusted network components; we discuss solutions that make
use of such trusted components. Nevertheless, we discuss other possible
solutions that alleviate these threats without requiring trusted infrastructure support and we evaluate the effectiveness of our proposals on
PlanetLab nodes.

Keywords: Security, Bandwidth Estimation, Bandwidth Shapers.

1   Introduction

Bottleneck bandwidth measurements are gaining increasing importance in many
wide-area Internet systems and services including multicast trees [1], content
distribution and peer-to-peer (P2P) systems [3]. Bottleneck bandwidth refers
to the maximum throughput that a path can provide to a flow, when there is
no other competing traffic load. Recently, bottleneck bandwidth estimation has
attracted significant interest in the literature. This is mainly due to the fact that
the performance and Quality-of-Service of most Internet services are based on
their bandwidth capacities.

Several tools for bottleneck bandwidth estimation (e.g., Nettimer [4], Pathchar
[5], pchar [6], bprobe [8], pathrate [9], Sprobe [10], etc.) have been proposed and
evaluated both by simulations and empirically over a number of Internet paths.
These techniques can be mainly classified in two categories [11]: the one-packet
and the packet-pair technique. Both techniques are well understood and can
provide accurate estimates under certain conditions. In both techniques, probe
packets are exchanged between the *verifier* (or the sender) and the *prover* (or
the receiver) to extract estimates of the network bandwidth characteristics.

Y. Chen et al. (Eds.): SecureComm 2009, LNICST 19, pp. 121–141, 2009.
© Institute for Computer Science, Social-Informatics and Telecommunications Engineering 2009

To measure bandwidth in a scalable way, current bandwidth measurement tools push the estimation functionality to the end-hosts. This renders them vulnerable to a wide range of security threats as *trust* is pushed to end-hosts that are more likely to be compromised than core/edge routers. Due to the increasing reliance on bandwidth estimation in current Internet services, untrusted hosts have considerable incentives to abuse this trust and fake their bottleneck bandwidth claims in order to increase their advantage from these services (e.g., free-riding in P2P networks [3]). Indeed, current measurement techniques are often at odds with "security" when deployed in adversarial settings. A malicious host can abuse the operation of these techniques in numerous ways to claim an inflated and/or deflated bandwidth: an untrusted host can make use of bandwidth shapers or can delay its probe packets to claim any bandwidth of its choice. By inflating its bandwidth claims, an untrusted host is likely to be delegated high priority in the network. For example, the untrusted host can be chosen as a super-peer in a P2P network [12] or a recommended server in content distribution networks based on the highest-capacity path. Similarly, untrusted provers might claim lower bandwidths to reduce their contribution in the network.

While some proposals (e.g., [10], [42], etc.) recommend the deployment of bottleneck bandwidth estimation tools across Internet hosts, we argue that the easy and accurate realization of attacks against current bottleneck bandwidth estimation techniques raises serious concerns about the suitability of their deployment. A thorough evaluation of these techniques in adversarial settings should therefore precede any prospective large-scale deployment.

Previous work [10], [11], [13], [14], [16], [17], [18] focused on evaluating the performance of bandwidth estimation techniques and did not address their security vulnerabilities. In this paper, we address this problem and we analyze the major security threats against current bottleneck bandwidth estimation techniques. We also investigate the impact of available software – such as traffic shapers – on the bandwidth estimation process. We demonstrate the effect, feasibility and the accuracy of these attacks on PlanetLab nodes [43]. Another important aim of this work is to extract relevant lessons about the security prospects of existing bottleneck estimation techniques and to hint application designers on the choice of a bandwidth estimation technique that better satisfies their desired level of assurance in the measurements. To the best of our knowledge, this is the first work that investigates the security vulnerabilities of bandwidth measurements in adversarial settings.

Our findings suggest that "full-fledged" solutions against the multitude of attacks on the bandwidth estimation process might not be feasible without requiring functionality from trusted network components; namely, since measurements are conducted end-to-end, fully mitigating delay-attacks against bandwidth estimation emerges as a challenging research problem. Remote attestation by trusted network components represents one of the few viable options to prevent such attacks. In this work, we discuss the viability and the effectiveness of this proposal in securing bandwidth measurements. We further propose and analyze several other solutions and heuristics that do not require any infrastructural support and we demonstrate that these schemes counter a large subset of attacks on current bandwidth estimation techniques.

The rest of the paper is structured as follows: Section 2 briefly overviews current bandwidth estimation techniques. Section 3 compiles the list of security threats against bottleneck bandwidth estimation techniques. In Section 4, we briefly discuss a solution to thwart these attacks based on remote attestation by trusted network components. In Section 5, we propose a set of techniques that do not require infrastructure support and we evaluate their effectiveness on PlanetLab nodes. In Section 6, we discuss possible insights in the design space of secure bottleneck bandwidth measurements. We conclude the paper in Section 7.

2   Bottleneck Bandwidth Estimation

The bottleneck bandwidth B_{min} of a path is the maximum rate that the path can provide to a flow from the source to the sink. B_{min} is determined by the *minimum* link capacity in the path. In what follows, we outline the operation of the two major bottleneck bandwidth estimation techniques: the one-packet technique and the packet-pair technique.

The One-Packet Technique. The one-packet technique relies on the assumption that a packet's traversal time across a path can be computed as the sum of its transmission and propagation delays, as follows:

$$t_l^j = t_0^j + \sum_{i=0}^{l-1}(\frac{S_j}{B_i} + d_i), \qquad (1)$$

where t_l^j is the traversal time of packet j through l links, t_0^j is the sending time of packet j, S_j is the packet size, B_i is the bandwidth of link i and d_i is the latency of link i.

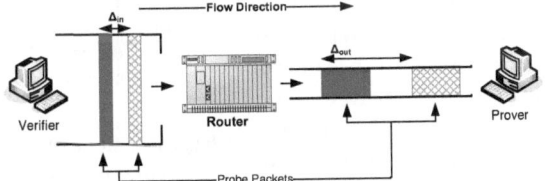

Assuming that the transmission delay is linear with respect to the packet size, it is highly likely that if the verifier transmits a large number of packets of *variable* size, at least one will have negligible queuing delay, and therefore the minimum round-trip time (RTT) values of these pack-

Fig. 1. Packet-Pair technique: The temporal spacing between the packets after the bottleneck link is inversely proportional to the bandwidth. The narrow part of the pipe represents the bottleneck link.

ets will form a line whose slope is the inverse of the link bandwidth to the prover [11]. This technique produces an estimate of the bandwidth at each hop in the path; the bottleneck bandwidth is then computed as the minimum value of the estimated link bandwidths. Note that the one-packet technique can only measure the download bandwidth (i.e., from the verifier to the prover). Examples of tools using the one-packet technique are Pathchar [5] and Clink [7].

The Packet-Pair Technique: Here, the verifier sends *two* back-to-back *large* packets of equal size to the prover. Once the prover receives these packets, it issues back its reply packet-pairs; the verifier then estimates the prover's *download* bandwidth by measuring the time dispersion between the reply packet-pairs [10]. Similarly, to estimate the prover's *upload* bandwidth, the prover sends two *large* packets adjacently in time to the verifier. The intuition behind the packet-pair technique is that when two large packets of the same size are sent back-to-back, it is highly likely that their queuing occurs at the bottleneck link of capacity B_i. Once the bottleneck link is traversed, the temporal spacing Δ_{out} between the two packets remains constant (Figure 1) and is inversely proportional to the bottleneck bandwidth [11]. Assuming FIFO queuing, the dispersion Δ_{max} after the packet-pair traverse H hops is as follows:

$$\Delta_{max} = \max_{i=0...H}\left(\frac{S}{B_i}\right) = \frac{S}{\min_{i=0...H}(B_i)} = \frac{S}{B}, \tag{2}$$

where B_i is the bandwidth of link i, S is the packet size and B is the bottleneck bandwidth of the path.

Several implementations of the packet-pair technique exist such as Nettimer [4], Pathrate [9] and Sprobe [10].

2.1   The Need for Secure Measurements

Bottleneck bandwidth measurements have the potential to solve considerable problems in applications and areas such as network management, end-to-end admission control, routing and traffic engineering [1], P2P networks, content distribution architectures [2], etc.. Selfish hosts might, therefore, have considerable incentives to fake their bandwidth claims

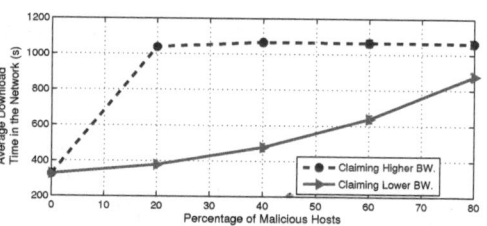

Fig. 2. Effect of malicious hosts on the average download time in a multicast binary tree application. Here, the num. of hosts is 1000 and the resource size is 3 MB. Each data point is averaged over 100 runs.

and increase their profit from these applications; by claiming higher bandwidths, selfish hosts are likely to be assigned higher priority in the network. Alternatively, hosts might claim lower bandwidths to limit their contribution in the network. This renders "secure" bandwidth measurement a crucial task nowadays.

For instance, in multicast distribution architectures, the download performance of hosts is highly affected by the organization of the nodes in the tree;

one slow peer located near the root of the tree can significantly impact the resource distribution time in the network [2]. In a prototype simulation that we have conducted[1], we investigate the effect of selfish hosts faking their bandwidths in an exemplary multicast binary-tree architecture. We assume a realistic bandwidth distribution amongst the nodes derived from the findings in [3]. As shown in Figure 2, selfish hosts can considerably affect the average resource download times in the entire network by claiming incorrect bandwidths. This effect is even more detrimental when hosts claim higher bandwidths than they actually have; the average download time over all peers in the network almost *quadruples* when only 20% of peers over-report their bandwidths.

3   Bandwidth Manipulation Attacks

In this section, we investigate delay-based attacks along with the major security threats against current bottleneck bandwidth estimation techniques.

3.1   System and Attacker Model

Our system consists of a *verifier* and a *prover*, connected by a *network*. The verifier *measures* and *verifies* the bottleneck bandwidth of the path to an untrusted prover. Here, we assume that the verifier actively probes the prover by issuing probe packets. The prover echoes its reply probe packets to the verifier. The latter estimates the bandwidth of the prover by extracting packet arrival times according to either the one-packet or the packet-pair technique. We focus on *bottleneck* bandwidth measurement and we assume that the application making use of the bandwidth measurement requires that the prover *cooperates* with the verifier during this process (otherwise it would be difficult to securely estimate its bandwidth). We limit our analysis to those applications that require an accurate estimate of the bottleneck bandwidth to the prover for their correct operation. For instance, while bandwidth manipulation attacks can be tolerated in BitTorrent [15], such attacks might affect the performance of the entire network in routing services, content distribution networks, multicast architectures, etc..

We further assume that the verifier uses a high-speed connection; therefore, its bandwidth will not affect the bottleneck bandwidth of the path to the prover. We do acknowledge that current bandwidth estimation tools can result in rather large estimation errors, however we assume that *enough* probe packets are exchanged to abstract away the effects of noisy measurements.

Untrusted provers constitute the core of our *internal* attacker model; by an untrusted prover, we refer to a host that is involved in bandwidth measurements, however it is not trusted by the verifier to correctly execute the measurement protocols. We assume that untrusted provers need to inflate/deflate their bandwidth claims by a considerable amount ($> 200\%$) to increase their profit in the network.

---

[1] Simulation details are omitted due to lack of space.

An external attacker Eve can equally compromise routers on the path between the verifier and the prover. By compromising routers, Eve can delay the exchanged probes to alter the bandwidth estimated by the verifier. Eve can also re-route probe packets through another bottleneck link to influence the conducted measurements.

3.2 Attacks on Current Techniques

Bandwidth measurement tools were developed without prior security considerations as they rely on ICMP/TCP implementations at end-hosts and do not guarantee any form of source nor destination authentication. An external attacker can *spoof* the IP [26] of the prover and issue back ICMP replies on its behalf; the measured bandwidth would be that of the attacker. The adversary could also re-route the probes to hosts at its disposal ([19], [20]) to claim a bandwidth of her choice (*sybil attack* [27]). In what follows, we analyze the detrimental impact of **delay attacks** on bottleneck bandwidth measurements.

Delay & Rushing Attacks on the One-Packet Technique: An untrusted prover can intentionally *delay* its reply packets to convince the verifier of a bandwidth claim of its choice (Figure 3). Given a set $\mathbb{S}$ of the variable-sized packets used in the one-packet technique, the prover can claim *lower* bandwidths $B_{claimed}$ than its genuine bandwidth B_{auth} by introducing a delay Δ_j to all packets $j \in \mathbb{S}$ of size $S_j > S_i$, where i is the *smallest* packet in $\mathbb{S}$, as follows:

$$B_{claimed} = \frac{S_j - S_i}{RTT_j - RTT_i + \Delta_j} \qquad (3)$$

$$\Delta_j = (S_j - S_i) \cdot (\frac{1}{B_{claimed}} - \frac{1}{B_{auth}}), \qquad (4)$$

Here, RTT_j denotes the *smallest* round trip time of probe j from the verifier to the prover. Note that the prover can equally claim a *higher* bandwidth by delaying probes $j \in \mathbb{S}$ of size $S_j < S_k$, where k denotes the *largest* packet in $\mathbb{S}$, by $\Delta_j = (S_k - S_j) \cdot (\frac{1}{B_{auth}} - \frac{1}{B_{claimed}})$.

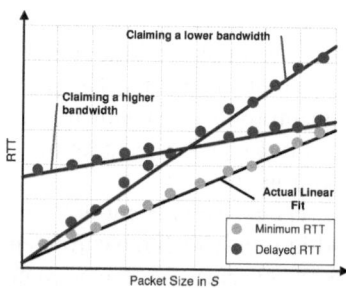

Fig. 3. Delay Attacks on the One-Packet technique

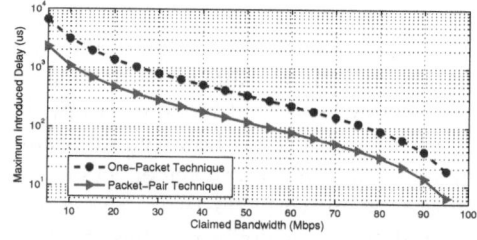

Fig. 4. Maximum delay required to fake bandwidth claims in the one-packet and the packet-pair technique. $B_{auth} = 100$ Mbps, the probe size ranges from 58 bytes to 1500 bytes. The path contains 5 link-layer hops.

In Equation 4, we assume that there are no intermediate hops on the path between the verifier and the prover. In practice, the untrusted prover has to further take into account the delays caused by the intermediate hops. This could be achieved by repeatedly applying Equation 4 for all link-layer hops in the desired link as follows:

$$\Delta_j = \sum_{i=1}^{H} \left((S_j - S_i) \cdot \left(\frac{1}{B_{claimed}} - \frac{1}{B_{auth}} \right) \right), \tag{5}$$

where H is the total number of link-layer hops in the measured path. Delay attacks can be very hard to detect given the unnoticeable delay that they introduce (Figure 4). Note that this attack is not only restricted to untrusted provers; a *rogue router* (compromised by Eve) can equally trick the verifier into accepting a fake bandwidth claim by introducing appropriate delays to the packet traversal time.

An untrusted prover can also predict the *Identifier* and *Sequence Number*[2] fields in the ICMP echo request packets and "rush" its reply by sending *specially crafted* ICMP echo replies ahead of time. In this way, an attacker can claim a smaller RTT which translates to a different bottleneck bandwidth measurement. A combination of these rushing and delay attacks could even reduce the maximum delay Δ_j that needs to be introduced to fake bandwidth claims.

Packet-Attraction and Repulsion attacks on the Packet-Pair Technique. In current implementations of the packet-pair technique [10], the verifier sends large back-to-back TCP SYN packets and awaits the corresponding TCP RST packets from the prover. Assuming that the prover immediately replies to the probe requests, this time dispersion will also be reflected in the difference of TCP SYN packet arrival times. By intentionally delaying the second reply probe, an untrusted prover increases the time dispersion between the packet-pairs and consequently the verifier would assume the existence of a smaller bottleneck link on the path to the prover. The required delay Δ is computed as follows:

$$B_{claimed} = \frac{S}{\Delta_{dispersion} + \Delta} \tag{6}$$

$$\Delta = S \cdot \left(\frac{1}{B_{claimed}} - \frac{1}{B_{auth}} \right) \tag{7}$$

where $\Delta_{dispersion}$ is the genuine dispersion between the packet-pairs, Δ denotes the additional delay between the packet-pairs, S is the size of the probes, $B_{claimed}$ is the fake claimed bandwidth of the prover and B_{auth} is the genuine bandwidth of the prover. As shown in Figure 4, Δ is considerably small – even for the largest probe size of 1500 bytes – compared to the delay required in the one-packet technique. This suggests that delay attacks are indeed more challenging to detect in the packet-pair technique when compared to the one-packet technique (Section 5.2).

Similarly, an untrusted prover or a rogue router can claim a smaller time dispersion between packet-pairs and consequently a higher download bottleneck

---

[2] Generally, the *Sequence Number* field in the ICMP echo request is incremental and therefore can be easily predicted.

bandwidth. The prover can delay its reply till both TCP SYN packets are received before sending its packet-pair replies with a time dispersion of its choice. Since RST packets are typically small in size, they will not queue at the bottleneck link. In this way, the prover can successfully claim a higher bandwidth than its genuine physical one.

At first glance, one might consider that these attacks can only be mounted by sophisticated attackers. However, this intuition is *not* correct. While a sophisticated user is able to manipulate his interface to temporarily delay all reply probes, less powerful provers can cause the same effect by using bandwidth shapers as shown in the following section.

3.3 Demonstration of Delay Attacks

In what follows, we demonstrate the feasibility of delay attacks on the one-packet and packet-pair techniques. Our findings are depicted in Figures 5 and 6. In our plots, *target bandwidth* refers to the bottleneck bandwidth claimed by an untrusted prover and *measured bandwidth* denotes the bottleneck bandwidth estimate extracted by the verifier. We rely on 10 and 100 Mbps symmetric physical connections deployed on three paths: **Path1** where both the verifier and the prover hosts (running Ubuntu *v.* 7.04 with 1 GB of RAM) are both located in Switzerland, **Path2**[3] where the verifier and the prover (host running Debian with 2 GB of RAM) are located in Switzerland and Germany, respectively, and **Path3** where the verifier is located in Switzerland and the prover is located in Illinois, USA. The prover runs RedHat Linux with 320 MB of RAM. Each data point in our plots is averaged over 1000 measurements.

One-Packet Technique. We created a prototype tool based on *Pathchar* [5] that delays the prover's reply packets (Equation 5). We used probe sizes ranging from 58 bytes to 1514 bytes (Ethernet headers included). Our application replaces the kernel's TCP/IP stack by a raw socket and uses an iptable rule to drop all replies issued by the kernel; it then sends back the reply probes with the desired delay.

As shown in Figures 5(a) and 5(c), an untrusted prover can claim any bandwidth of its choice in the one-packet technique by appropriately introducing small – almost unnoticeable – delays before issuing its replies (Figure 5(b)). Given the impact of small delays, the accuracy of the bandwidth claims can be further increased by accounting for the prover's PCI bus delays (Figure 5(c)).

Packet-Pair Technique: We demonstrate delay attacks on Sprobe [10] using an application that modifies the prover's networking interface and an open-source traffic shaper.

The cumulative distribution functions (CDF) of the conducted measurements (Figure 6(a)) suggests that these attacks – whether originating from a modified

---

[3] We were not able to conduct one-packet experiments on Path2 due to the fact that intermediate routers were blocking the ICMP probes.

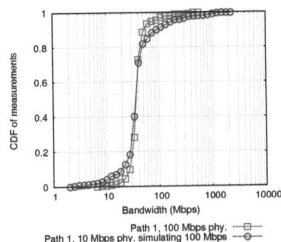 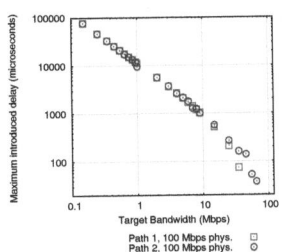 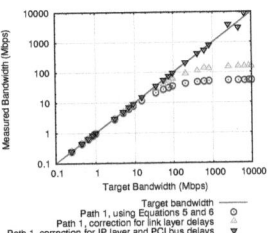

(a) Claiming 100 Mbps bandwidth on a 10 Mbps connection.

(b) Maximum introduced delay in claiming a lower bandwidth over a 100 Mbps downlink connection.

(c) Impact of link-layer and PCI bus delays on measurements over a 100 Mbps downlink connection.

Fig. 5. Delay attacks on the One-Packet technique

application or from bandwidth shapers – are almost *statistically* indistinguishable at the verifier's side from *authentic* bandwidth measurements, which renders them very hard to detect.

Our analysis in Section 3.2 is further validated in Figure 6(b). Indeed, the prover can claim a bandwidth of its choice irrespective of its actual physical download bottleneck[4]. These attacks can be equally achieved by bandwidth shapers (Figure 6(b)). We further investigate the effect of bandwidth shapers on bandwidth estimation in Section 5.2.

4   Trusted Infrastructure Support for Bandwidth Measurement

To the best of our knowledge, it is hard, if not impossible, for the verifier to *fully* ensure that the remote provers did not intentionally introduce delays before issuing their replies. Although some schemes were proposed in the scope of securing link quality measurement [40] and RTT measurements [41], they assume that the prover does not have incentives to mount delay-based attacks; this is not the case in bandwidth estimation scenarios. An intuitive solution to thwart this problem is to use tamper-resistant hardware [39] to prevent hosts from tampering with their network interface. However, this hardware comes at a high cost.

Remote attestation by trusted network components emerges as one of the few workable alternatives to fully securing bottleneck bandwidth measurements. In what follows, we briefly outline a scheme that makes use of trusted edge-routers and we show that our solution effectively mitigates delay attacks against bandwidth estimation. In Section 5, we discuss several other alternatives to partially alleviate these attacks without requiring infrastructural support.

---

[4] Note that Path2 featured considerable cross-traffic during the measurements, which explains the estimate errors in the plots.

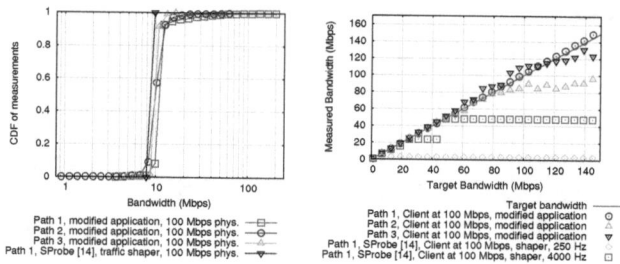

(a) Claiming 10 Mbps on a (b) Packet-delay attacks on a
100 Mbps uplink connection. 100 Mbps downlink connection.

Fig. 6. Delay attacks on the Packet-Pair technique

As shown in Figure 7, we assume in our analysis that the bottleneck links reside between the outer-most edge-routers and the end-hosts. Sample experiments on PlanetLab [43] nodes confirm that this is a reasonable assumption. We further assume that edge-routers are trusted by all entities and can timestamp, generate and authenticate packets.

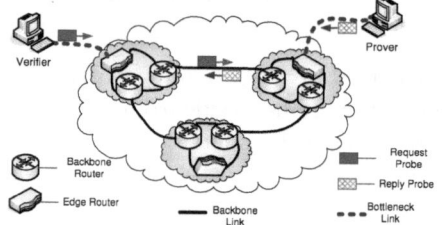

Fig. 7. Bottleneck bandwidth measurements using trusted edge routers

Our scheme for securing bottleneck bandwidth measurements unfolds as follows: when the verifier wishes to measure the bottleneck bandwidth of the path to a prover, it sends along that path a request packet containing the IP address of the prover and the type of bandwidth measurement of interest (upload and/or download). Upon reception of the latter packet, the edge-router connected to the prover measures the capacity of the bottleneck link it shares with the prover and sends its measurement results to the verifier. The verifier can validate the authenticity of the measurement results since they come enclosed with the signature of the edge-router. The edge-router estimates the bottleneck bandwidth of the link it shares with the prover as follows:

- **Upload Bandwidth Measurement.** Similar to the packet-pair technique, the prover sends two large back-to-back packets to the edge-router. Since the latter is located on the other side of the bottleneck link, it can verify that no additional delay Δ (Equation 7) was introduced between the packet-pairs (the edge-router measures the time delay between the last bit of the first packet and the first bit of the second packet is negligible). By doing so, the edge-router is certain that both packets queued at the bottleneck link. It then measures the time dispersion between the packets to estimate the bottleneck link of the path to the prover according to the packet-pair technique.

- **Download Bandwidth Measurement.** To measure the downlink bottleneck of the prover, the edge-router can estimate the time it needs to upload a
packet-pair on the path to the prover. Since the bottleneck link is shared by
both the prover and the edge-router and assuming a high transmission rate,
the latter's upload throughput corresponds to the download capacity[5] of the
bottleneck link.

5 "Best-Effort" Solutions for Current Bandwidth Estimation Techniques

In Section 4, we showed that by relying on trusted network components, security threats against bottleneck bandwidth measurements can be fully mitigated.
Given the current architecture of the Internet, we do acknowledge, nevertheless,
that relying on trusted infrastructure might constitute a rather "bulky" proposal
nowadays. In this section, we discuss and evaluate several other "best-effort"
countermeasures that do not require trusted infrastructure support.

5.1 Mitigating Spoofing and Rushing Attacks

Bottleneck bandwidth measurement tools can make use of lightweight authentication protocols to counter impersonation attacks. Furthermore, the verifier
can use pseudo-random functions to generate its request probes such that they
cannot be predicted by the provers and require that the reply probes are correlated in content to its request probes. Alternatively, the verifier can make use of
distance bounding protocols [30] or can require that the prover authenticates the
received pseudo-random probes using a shared key. Thus, the probability that
the prover correctly rushes its replies before receiving the request probes can be
made satisfactorily negligible ($O(2^{-k})$ for k-bit probes).

Note that the time required to authenticate each request probe is negligible
compared to the probes' propagation times. For example, the time required to
encrypt a 1500 bytes message with a 256 bit key using the AES implementation
in the Crypto++ library on an Intel Core 2 1.83 GHz processor running Windows XP is 19 μs [21]. We implemented a variant of the Sprobe tool [10] in which
the prover is required to encrypt (using AES) the request probes and we have
conducted sample bandwidth measurements on Paths 1 and 2 using this application. Our findings in Figure 8 show that the accuracy of the measurements is
preserved, which makes AES-based authentication suitable for integration within
current bandwidth estimation techniques.

5.2 Alleviating Delay Attacks

In what follows, we discuss some techniques to alleviate delay attacks on bandwidth estimation.

---

[5] Note that the edge-router can estimate the full capacity of the bottleneck link since
it can ensure that no downlink traffic is present at the time of the measurements.

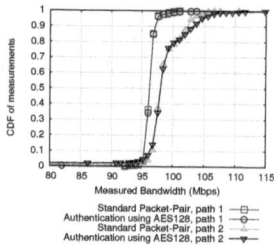

Fig. 8. Effect of Authentication on measurements

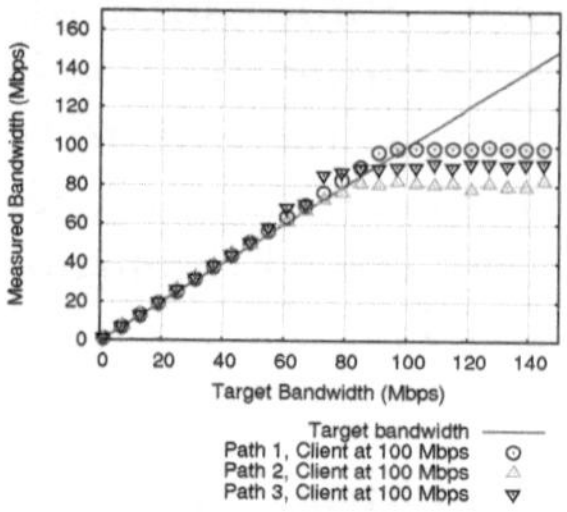

Fig. 9. Attacks on a 100 Mbps uplink connection

1) Mitigating Bandwidth Inflation Attacks in the Packet-Pair Technique: Given large probe sizes, the packet-pair technique ensures that the upper bound on the *upload* bandwidth that an *untrusted prover* can claim is bounded by its physical bottleneck bandwidth. This is depicted in Figure 9. In fact, the lower bound on the time dispersion between large packet-pairs is determined by the queuing on the bottleneck link. Even if the untrusted prover manipulates the transmission times of its reply probes, they will queue at the bottleneck link with high probability. Given this, the only viable strategy to claim a higher bandwidth would be to send each of the packet-pairs using different paths. However, this requires accurate knowledge of the network status; in practice, the attacker will only succeed with negligible probability. Note that an untrusted prover can also distribute its authentication credentials to other hosts under its control. In this case, the upper limit on the claimed bandwidth is bounded by the highest physical bandwidth of all the compromised hosts.

2) "Reference" Round-Trip Times: Theoretically, delay attacks can be alleviated if the verifier knows an estimate of the RTT to the prover. The verifier can acquire RTT estimates via offline measurements or from online servers that perform RTT measurements around the globe (e.g., [31]).

For instance, in the one-packet technique, if the verifier knows a reference RTT for a median-size probe packet in the set $\mathbb{S}$ of the variable probe sizes, then the bandwidth range that a prover can claim is bounded by the accuracy ϵ of the estimated reference RTT as follows:

$$\frac{B_{auth} \times (S_j - S_i)}{\epsilon \cdot B_{auth} + (S_j - S_i)} \leq B_{claimed} \leq \frac{B_{auth} \times (S_k - S_j)}{(S_k - S_j) - \epsilon \cdot B_{auth}},$$

where $B_{claimed}$ is the bandwidth claimed by the untrusted prover, B_{auth} is the genuine bottleneck bandwidth on the prover's side, S_i is the size of the smallest probe packet used in the variable-size probing set $\mathbb{S}$ and S_k is the size of the largest probe packet in $\mathbb{S}$.

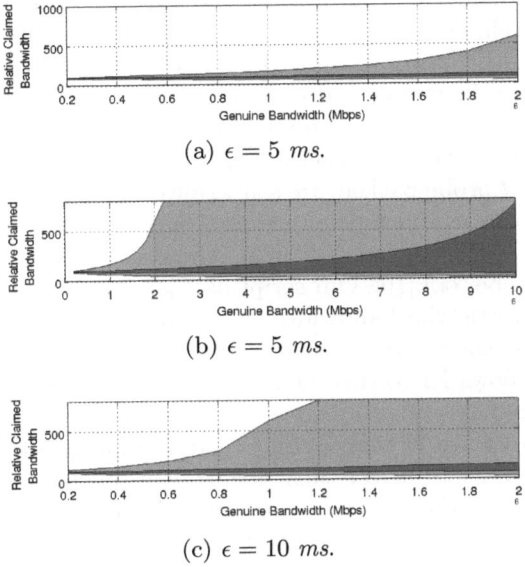

(a) $\epsilon = 5\ ms$.

(b) $\epsilon = 5\ ms$.

(c) $\epsilon = 10\ ms$.

Fig. 10. Range of achievable bandwidth claims for $\epsilon = 5\ ms$ and $\epsilon = 10\ ms$. The dark and light areas represent the achievable claims in the one-packet technique and the packet-pair technique, respectively.

Similarly, in the packet-pair technique, the *download* bandwidth that an untrusted prover can claim is equally bounded by the accuracy ϵ of the estimated RTT:

$$\frac{S}{\Delta_{dispersion} + \epsilon} \leq B_{claimed} \leq \frac{S}{\Delta_{dispersion} - \epsilon},$$

where S is the request probe packet size, $\Delta_{dispersion}$ is the time dispersion between the request probe packet-pairs originating at the bottleneck link and ϵ is the acceptable deviation in time from the reference RTT.

We investigate the benefits of this approach in Figure 10 for estimation errors $\epsilon = 5$ ms and $\epsilon = 10$ ms from the reference RTT. Given the variability of RTTs in current networks and the error ϵ in estimating the reference RTT, this technique can only *limit* the range of false claims (within 20 % of the genuine bandwidth[6]) in the case where the genuine bottleneck bandwidths are modest (typically < 10 Mbps). Our findings also show that this technique is not well-suited to upper-bound fake bandwidth claims in the packet-pair technique. This is due to the fact that the time dispersion between packets $\Delta_{dispersion}$ is comparable to typical values of ϵ, even when dealing with small bandwidths. It can, however, significantly lower-bound the claims of modest-bandwidth hosts.

---

[6] This is rather acceptable compared to the estimation errors resulting from current bandwidth estimation tools.

3) Detecting Bandwidth Shapers in the Packet-Pair Technique: Bandwidth and traffic shapers (e.g., NetLimiter [28], NetEqualizer [29], HTB [32]) provide a simple mechanism to limit the amount of data a host transmits and accepts by delaying incoming and outgoing packets to match a specified rate limit. Due to their mode of operation, bandwidth shapers *cannot* alter the measurements conducted by the one-packet technique since they cannot limit the rate at which *individual* probe packets are sent. However, they present themselves as effortless routines to conduct delay attacks on the packet-pair technique. We implemented a prototype shaper script on the prover's side and we studied its impact on the Sprobe tool [10]. Our script uses iptable rules and the HTB traffic shaper [32] to throttle the bandwidth of the prover on the fly. We have also conducted upload bandwidth measurements on Sprobe using NetLimiter [28] running on a Windows XP kernel. Our measurements were conducted on **Path 1** (refer to Section 3.3).

Our findings in Figure 11 suggest that current implementations of bandwidth shapers allow a verifier to *detect* their deployment on the prover's side. In fact, bandwidth shapers can only receive, store, and release packets whenever a system timer interrupt occurs [32]. This suggests that the maximum rate at which a *pair of packets* can be sent is bounded by the timer frequency of the underlying operating system: $B_{max} = S \cdot F_{sys}$, where B_{max} is the maximum achievable bandwidth claim, S is the packet size and F_{sys} is the system timer frequency. Furthermore, the achievable time dispersions between a packet-pair $T_{nominal}$ are inversely proportional to the system timer frequency F_{sys}. The achievable bandwidth claims are therefore computed as follows:

$$B_{claimed} = \frac{S \cdot F_{sys}}{i}, \forall i \in \mathbb{N}^*, \tag{8}$$

which explains the step-wise curves obtained in Figure 11. In most Linux systems, it is however possible to increase the system frequency through kernel re-compilation. As shown in Figure 11(a), a prover can achieve a higher upper bound on the claimed bandwidth by re-compiling its kernel to operate at a higher timer frequency.

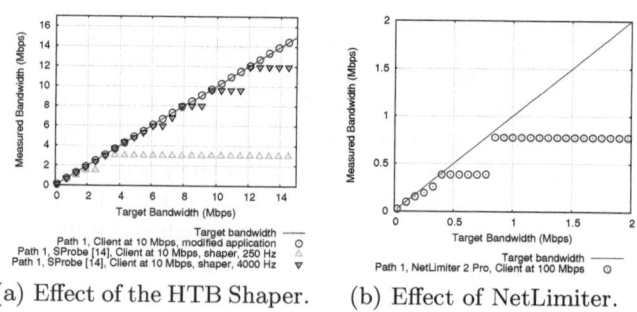

(a) Effect of the HTB Shaper.   (b) Effect of NetLimiter.

Fig. 11. Effect of Bandwidth/Traffic Shapers

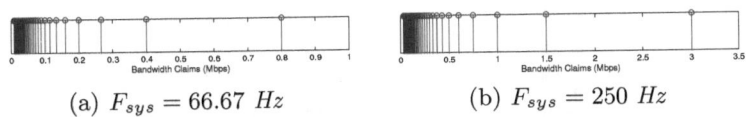

(a) $F_{sys} = 66.67\ Hz$ (b) $F_{sys} = 250\ Hz$

Fig. 12. Achievable bandwidth claims by traffic shapers in the packet-pair technique. A spike at value X indicates that bandwidth X can be achieved.

By comparing the time dispersions of the reply packet-pairs with $T_{nominal}$ for typical system frequencies, a verifier can suspect the presence of bandwidth shapers on the provers' side and can rule out the resulting estimate. This is especially true for small bandwidths (Figure 12).

We validate this claim via extensive measurements on 200 PlanetLab provers. To truthfully represent Internet nodes, we chose the PlanetLab nodes whose bandwidth distribution follows the distribution in the current Internet [38]. In accordance with the findings in [3], we assume that 40 % of the provers are selfish and make use of bandwidth shapers to vary their bandwidth claims over time. Untrusted provers can claim both higher (inflate) or lower (deflate) bandwidths by factors ranging from 1 to 10; as suggested in [3], we assume that high-bandwidth provers claim higher bandwidths with probability 0.1 and lower bandwidths with probability 0.9. Low-bandwidth provers claim higher bandwidth capabilities with probability 0.9.

In our experiments, we compute the median B_{med} of the measured bandwidths (typically 10 packet-pairs to remove noisy measurements) and we compare it to the *closest* bandwidth that a shaper can achieve B_{min} as follows: we assume a normal distribution[7] around B_{min} and we compute the probability P that B_{med} is within a threshold number of standard deviations ($n \cdot \sigma$) of B_{min}:

$$P = \frac{1}{\sigma\sqrt{\pi}} \int_{B_{min}-n\sigma}^{B_{min}+n\sigma} e^{-\frac{(x-B_{min})^2}{2\sigma^2}}, \tag{9}$$

Our results are illustrated in Figure 13; a significant fraction (76 %) of provers that use bandwidth shapers were correctly identified, which confirms the feasibility of bandwidth shaper detection in the packet-pair technique. Since the granularity of bandwidths that a bandwidth shaper can produce is small for low bandwidths, it can emulate a large number of low bandwidths (Figure 12). Given this, and in the presence of cross-traffic, the claims of low-bandwidth provers (e.g., modem users) can be easily mis-judged to be originating from bandwidth shapers. This explains the false negatives obtained in identifying honest provers.

Note that some Linux built-in traffic shapers (Linux kernel v. 2.6.23) do not rely on system timer interrupts. Thus, other techniques will be needed to detect them in the future. Windows-based shapers (e.g., NetLimiter [28]) can, however, still be detected using the aforementioned method.

---

[7] Experimental results conducted on 200 different Internet paths confirm this assumption.

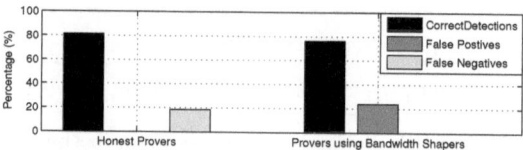

Fig. 13. Bandwidth Shaper Detection conducted on 200 different PlanetLab hosts

4) Verifying Bandwidth Capability. The block transfer method is a conventional mean to measure the *available* bandwidth (the unused capacity) of a path. Here, the verifier asks the prover to download/upload a full block of data and computes the data transfer rate to estimate the residual bandwidth of the path to the prover. Several applications make use of this technique to estimate the *available* bandwidth of a path (e.g., BitTorrent [15]). In the block transfer method, the upper bound on the claimed bandwidth is guaranteed to match the genuine bottleneck bandwidth. This allows the verifier to *identify* whether the bottleneck

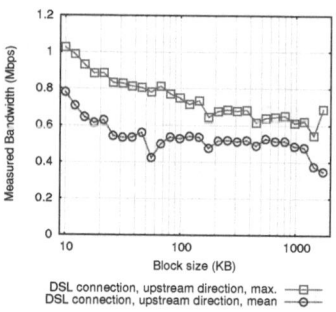

Fig. 14. Tradeoffs in the Block-Transfer method

bandwidth claimed by the prover can be achieved in practice. However, this solution incurs significant overhead in the network and depends on the traffic load in the path. Nevertheless, our findings suggest the existence of a potential trade-off with respect to the required data transfer size. We conducted block transfer measurements from a high-speed verifier to a prover connected to the Internet by a 0.8 Mbps upload connection. Figure 14 depicts the tradeoff between the size of the transfer block and the accuracy of the bandwidth estimated by the verifier. Indeed, even blocks of moderate size (50-100 KB) result in an indicative bandwidth estimate, which might justify the use of this method to filter out suspicious bandwidth claims.

5) Reverse-Resolve DNS Names. By resolving a prover's IP address into its Domain Name Server (DNS), the verifier might deduce the prover's type of Internet connection and detect false bandwidth claims. For example, if the prover's DNS name contains the string "dsl", it is highly likely that it has a DSL Internet connection [10]. We evaluated the viability of this proposal through extensive experiments on *1,000,000* randomly chosen IPs. We classified the obtained DNS names depending on whether they contain the strings: "dsl", "cable", "dial", "isdn", "WLAN" and "T1" or "T3". Our findings indicate that 34 % of the IPs leak their host's bandwidth information[8] (Figure 15). This information can be used by the verifier to detect discrepancies in the measured bandwidth. For in-

---

[8] Our results could be further improved given better knowledge of the local providers specific to each country (e.g., AT&T for DSL in the USA).

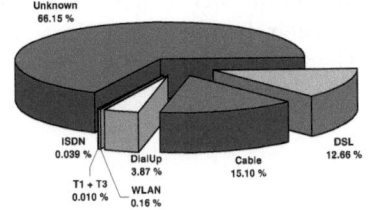

	Detection Rate	False Positives
Honest Provers	81.4 %	18.6 %
Bandwidth Shapers	75.6 %	21.2 %
Provers with a Modified Interface	68.1 %	31.9 %
Overall	**74.5 %**	**25.5 %**

Fig. 15. Parsing results of the DNS names of 1,000,000 randomly chosen IPs around the globe

Fig. 16. Detection Results on 200 Planet-Lab Nodes

stance, if the verifier measures a 5 Mbps download bandwidth while the prover's DNS name is "smartuser.dialup.com", then it is highly likely that there was an attack on bandwidth estimation.

6) Additional Heuristics: Statistical outlier detection [22], [23], [24] can also be used to prevent untrusted hosts from faking their bandwidth claims. Using outlier detection methods, correlations between different measurements can be identified and discrepancies can be detected. Furthermore, it is often the case that various performance metrics implicitly exhibit well-defined correlations [22], [44], which might allow the verifier to detect inconsistencies. For example, it is unlikely that a host having a 5 Mbps download bandwidth will have a 10 Mbps upload bandwidth.

To evaluate the viability of this proposal, we refined the PlanetLab experiment described in Section 5.2-2. We assume that a dedicated server periodically monitors the bandwidth claims of hosts and keeps *history* of the recorded measurements. We consider the following setting: 20 % of the hosts modify their networking interface to fake their bandwidth claims, 20 % of the hosts make use of bandwidth shapers and the remaining 60 % are "honest" hosts. We use a combination of bandwidth shaping detection (described earlier) and outlier detection based on the Z-score test to identify malicious hosts that fake their bandwidth claims.

$$Z = \frac{\sqrt{n}(X - \mu)}{\sigma}, \tag{10}$$

Here, n is the number of measurements per host, μ is their mean and σ is the standard deviation from the mean. If the P-value of the Z-score is above a threshold value (0.05), the host is considered to be malicious.

Our findings (Table 16) suggest almost 75 % of the fake claims were successfully detected; most of those detections correspond to provers that use bandwidth shapers and/or that vary their bandwidth claims over time. Larger detection rates could be achieved by incorporating additional techniques, such as reverse-resolve DNS names and reputation-based approaches [33], [34], [36], [37] in the detection process. In the latter approach, each host can be associated with a *reputation value* that indicates how trustworthy it is. Interacting hosts measure their respective bandwidth and form an opinion about each other. Malicious hosts, claiming incorrect bandwidths or varying their bandwidth claims, will be associated with low reputation values and, therefore, will not be chosen in subsequent interactions.

6 Discussion and Outlook

A great deal of lessons can be extracted from the operation of current bottleneck bandwidth estimation tools. We therefore hope that our findings hint application designers on the design of secure bandwidth measurement tools:

"Security" Features of Current Techniques. Till recently, the accuracy and the overhead of bandwidth estimation techniques have highly influenced the decision of application designers to choose a certain estimation technique (e.g., one-packet, packet-pair) given the requirements of their applications [10], [11]. However, "security" is another important factor that needs to be taken into account to ensure consistent bandwidth measurements. Although the design of current techniques cannot give "clear-cut" security guarantees, our findings suggest that some techniques are likely to perform better than others in different adversarial settings. On one hand, the packet-pair technique cannot prevent untrusted provers from inflating nor deflating their download bandwidth claims. Although it can successfully deflate its upload bandwidth, an untrusted prover cannot inflate its upload bandwidth claims given large probes in the packet-pair technique. An important observation here is that bandwidth deflation/inflation attacks on the packet-pair technique can be achieved by bandwidth shapers, and thus can be easily realized by untrusted provers. On the other hand, delay attacks on the one-packet technique require more sophisticated users, capable of altering their networking interface, since bandwidth shapers cannot affect the measurements in the one-packet technique. Fortunately, bandwidth manipulation attacks mounted by modest-bandwidth provers might be successfully mitigated in both techniques (with the exception of download bandwidth inflation attacks) if the verifier knows an estimate of the RTT to the prover.

Active and Cooperative Measurements. Some previous work [10] argues that bandwidth estimation tools should be designed to work in uncooperative environments in order to scale to a large number of hosts. Although this is indeed a desirable property, we find support for uncooperative environments rather unrealistic. In fact, with the proliferation of "de-facto" security applications, such as home firewalls, probing techniques based on uncooperative TCP/UDP and ICMP functionality find less applicability in the near future as they are likely to be considered hostile by the end-hosts. Some routers already *filter* ICMP packets due to their potential malevolent use [11]. Therefore, support for cooperative measurements is inevitable in the near-future [10]. Furthermore, end-to-end security would impair the use of bandwidth monitoring tools in passive and uncooperative environments as it involves active end-host cooperation for source authentication; cooperative environments present themselves as vital "playgrounds" for secure end-to-end bandwidth monitoring in the current Internet.

Network Measurements as "First Class Citizens". Current measurement tools do not take into account the impact of untrusted hosts on bandwidth measurements. Given the current trends in designing a "clean-slate" future Internet, our findings indirectly motivate the need for a *secure* next-generation Internet. Since network measurements are gaining paramount importance in monitoring the performance of the Internet, *secure* infrastructural support for network measurements becomes rather a necessity. As shown in Section 4, by pushing functionality from end-hosts back to *dedicated* and *trusted* network components, several security threats can be eliminated. Performance "awareness" is another desirable design property for next-generation Internet. Dedicated network components could in the future construct and store bandwidth and latency "maps" of Internet hosts. This would indeed eliminate the need for active probing-based end-to-end *insecure* measurement tools.

7 Conclusions

In this paper, we analyzed and demonstrated the major security vulnerabilities of current bottleneck bandwidth estimation techniques. Given the increasing reliance on bandwidth estimation tools in current Internet services, these vulnerabilities might affect the performance of all the applications that make use of these tools. Another important aim of this work is to extract relevant lessons about the security prospects of existing bottleneck estimation techniques and to hint application designers on the choice of a bandwidth estimation technique that better suits their applications. Our findings suggest that it is very hard, if not impossible, to *fully* counter all security challenges against existing tools without requiring functionality from trusted network components. More specifically, delay attacks pose serious challenges to the consistency of bandwidth measurements. Nevertheless, we proposed other possible solutions and heuristics – that do not require infrastructural support – to mitigate attacks on existing tools and we showed via extensive measurements on PlanetLab nodes that they can alleviate a significant fraction of attacks on current bottleneck bandwidth measurement techniques.

Acknowledgments

The authors would like to thank the anonymous reviewers for their helpful suggestions and feedback.

References

1. Ratnasamy, S., McCanne, S.: Inference of Multicast Routing Tree Topologies and Bottleneck Bandwidths using End-to-end Measurements. In: Proceedings of IEEE INFOCOM (1999)
2. Schiely, M., Renfer, L., Felber, P.: Self-Organization in Cooperative Content Distribution Networks. In: Proceedings of NCA (2005)

3. Saroiu, S., Gummadi, P., Gribble, S.: A Measurement Study of Peer-to-Peer File Sharing Systems. In: MMCN (2002)
4. Lai, K., Baker, M.: Nettimer: A Tool for Measuring Bottleneck Link Bandwidth. In: USITS (2001)
5. Jocobson, V.: Pathchar (1997),
   http://www.caida.org/tools/taxonomy/perftaxonomy.xml#pathchar
6. Math, B.: pchar (1999),
   http://www.caida.org/tools/taxonomy/perftaxonomy.xml#pchar
7. Clink: a tool for estimating Internet link characteristics,
   http://allendowney.com/research/clink/
8. Carter, R.: Cprobe and bprobe Tools (1996),
   http://cs-people.bu.edu/carter/tools/Tools.html
9. Dovrolis, C.: pathrate (2001),
   http://www.cis.udel.edu/~dovrolis/bwmeter.html
10. Sariou, S., Gummadi, P., Gribble, S.: SProbe: A Fast Technique for Measuring Bottleneck Bandwidth in Uncooperative Environments. In: Proceedings of INFOCOM (2002)
11. Lai, K., Baker, M.: Measuring Link Bandwidths Using a Deterministic Model of Packet Delays. In: ACM SIGCOMM (2000)
12. KaZaA, http://www.kazaa.com/
13. Strauss, J., Katabi, D., Kaashoek, F.: A Measurement Study of Available Bandwidth Estimation Tools. In: IMC (2003)
14. Hu, N., Li, L., Mao, Z., Steenkiste, P., Wang, J.: A Measurement Study of Internet Bottlenecks. In: Proceedings of INFOCOM (2005)
15. BitTorrent, http://www.bittorrent.org/protocol.html
16. Carter, R., Crovella, M.: Measuring Bottleneck Link Speed in Packet-Switched Networks. In: Performance Evaluation (1996)
17. Dovrolis, C., Ramanathan, P., Moore, D.: What do packet dispersion techniques measure? In: Proceedings of INFOCOM (2001)
18. Prasad, R., Dovrolis, C., Murray, M., Claffy, K.: Bandwidth estimation: metrics, measurement techniques, and tools. IEEE Network (2003)
19. Revealed, the Internet's Biggest Security Hole,
    http://blog.wired.com/27bstroke6/2008/08/revealed-the-in.html
20. More on BGP Attacks,
    http://blog.wired.com/27bstroke6/2008/08/how-to-intercep.html
21. Speed Comparison of Popular Crypto Algorithms,
    http://www.cryptopp.com/benchmarks.html
22. Walters, A., Zage, D., Nita-Rotaru, C.: A Framework for Mitigating Attacks Against Measurement-Based Adaptation Mechanisms in Unstructured Multicast Overlay Networks. ACM/IEEE Transactions on Networking (2007)
23. Soule, A., Salamatian, K., Taft, N.: Combining Filtering and Statistical Methods for Anomaly Detection. In: Proceedings of IMC (2005)
24. Snader, R., Borisov, N.: EigenSpeed: Secure Peer-to-peer Bandwidth Evaluation. In: Proceedings of IPTPS (2009)
25. Savage, S., Cardwell, N., Wetherall, D., Anderson, T.: TCP Congestion Control with a Misbehaving Receiver. Computer Communication Review (1999)
26. Harris, B., Hunt, R.: TCP/IP security threats and attack methods. Computer Communications (1999)
27. Douceur, J.: The sybil attack. In: Druschel, P., Kaashoek, M.F., Rowstron, A. (eds.) IPTPS 2002. LNCS, vol. 2429, p. 251. Springer, Heidelberg (2002)

28. NetLimiter, http://www.netlimiter.com/
29. NetEqualizer, http://www.netequalizer.com/
30. Brands, S., Chaum, D.: Distance-bounding protocols. In: Helleseth, T. (ed.) EUROCRYPT 1993. LNCS, vol. 765, pp. 344–359. Springer, Heidelberg (1994)
31. The CAIDA DNS root/gTLD RTT Dataset,
    https://data.caida.org/datasets/dns/root-gtld-rtt/
32. HTB Traffic Shaper, http://luxik.cdi.cz/~devik/qos/htb/
33. Kamvar, S., Schlosser, M., Garcia-Molina, H.: The EigenTrust Algorithm for Reputation Management in P2P Networks. In: WWW (2003)
34. Sears, W., Yu, Z., Guan, Y.: An Adaptive Reputation-based Trust Framework for Peer-to-Peer Applications. In: NCA (2005)
35. Damiani, E., Vimercati, S., Paraboschi, S., Samarati, P.: Managing and Sharing Servents' Reputations in P2P Systems. IEEE Transactions on Knowledge and Data Engineering (2003)
36. Dimitriou, T., Karame, G., Christou, I.: SuperTrust: A Secure and Efficient Framework for Handling Trust in Super Peer Networks. In: Proceedings of ACM PODC (2007)
37. Karame, G., Christou, I., Dimitriou, T.: A Secure Hybrid Reputation Management System for Super-Peer Networks. In: Proceedings of IEEE CCNC (2008)
38. OECD, Broadband Growth and Policies in OECD Countries,
    http://aui.es/IMG/pdf_Informe_OCDE_Banda_Ancha_en_el_Mundo.pdf
39. Jin, H., Lotspiech, J.: Forensic Analysis for Tamper Resistant Software. In: Proceedings of ISSRE (2003)
40. Zeng, K., Yu, S., Ren, K., Lou, W.: Towards Secure Link Quality Measurement in Multihop Wireless Networks. In: Globecom (2008)
41. Courtay, O., Karroum, M., Duran, A.: Method and Devices for Secure Measurements of Time-Based Distance Between Two Devices. Patent no. WO/2006/136278 (2006)
42. Barford, P.: Measurement as a First Class Network Citizen. White Paper,
    http://pages.cs.wisc.edu/~pb/sngi_whitepaper.pdf
43. PlanetLab, http://www.planet-lab.org/
44. Jiang, G., Cybenko, G.: Temporal and spatial distributed event correlation for network security. In: American Control Conference (2004)

An Eavesdropping Game with SINR as an Objective Function

Andrey Garnaev[1] and Wade Trappe[2]

[1] St. Petersburg State University, Russia
agarnaev@rambler.ru
[2] WINLAB, Rutgers University, USA
trappe@winlab.rutgers.edu

Abstract. We examine eavesdropping over wireless channels, where secret communication in the presence of an eavesdropper is formulated as a zero-sum game. In our problem, the legitimate receiver does not have complete knowledge about the environment, i.e. does not know the exact values of the channels gains, but instead knows just their distribution. To communicate secretly, the user must decide how to transmit its information across subchannels under a worst-case condition and thus, the legal user faces a max-min optimization problem. To formulate the optimization problem, we pose the environment as a secondary player in a zero-sum game whose objective is to hamper communication by the user. Thus, nature faces a min-max optimization problem. In our formulation, we consider signal-to-interference ratio (SINR) as a payoff function. We then study two specific scenarios: (i) the user does not know the channels gains; and (ii) the user does not know how the noise is distributed among the main channels. We show that in model (i) in his optimal behavior the user transmits signal energy uniformly across a subset of selected channels. In model (ii), if the user does not know the eavesdropper's channel gains he/she also employs a strategy involving uniformly distributing energy across a subset of channels. However, if the user acquires extra knowledge about environment, e.g. the eavesdropper's channel gains, the user may better tune his/her power allocation among the channels. We provide criteria for selecting which channels the user should transmit on by deriving closed-form expressions for optimal strategies for both players.

1 Introduction

Security is one of the most prominent problems surrounding wireless communications, largely due to the broadcast nature of the wireless medium, which facilitates eavesdropping. Although much of the work in confidentiality for wireless systems has focused on cryptographic solutions, which necessitate key management, there has been a recent movement towards exploring new security mechanisms for wireless systems. There has been an effort by the wireless research community to develop new forms of confidential communication that

Y. Chen et al. (Eds.): SecureComm 2009, LNICST 19, pp. 142–162, 2009.
© Institute for Computer Science, Social-Informatics and Telecommunications Engineering 2009

exploit the fading characteristics of the wireless channel to achieve secret communications through appropriate coding constructions[1,2,3,4,6,7]. Such work has largely built upon prior information-theoretic work of [8,9,10,11], where the notion of secrecy capacity was introduced to describe the rate at which a sender could communicate in an information-theoretically confidential manner in the presence of an eavesdropper. Recent results have sought to incorporate modern communication system design, and take advantage of the many degrees of freedom available in a dynamic wireless fading environment. For example, it is possible to use multiple subcarriers in order to provide a large number of parallel subchannels, as is utilized in OFDM transceivers (which is becoming a de facto physical layer strategy for many existing and emerging wireless systems, including 802.11g and WiMax), and the underlying frequency selectivity induced by multipaths can provide a diversity advantage. Recent results related to secret communication over independent, parallel channels has been reported in [4,5].

In the basic formulation of confidential communication, we have three entities: Alice, Bob and Eve. Alice seeks to communicate secretly with Bob while in the presence of an eavesdropper Eve. In this formulation, there are two sets of channels of interest, first is the channel from Alice to Bob, and second is the channel from Alice to Eve. Using G as a generic representation for the Alice to Bob channel, and H as a generic representation for the Alice to Eve channel, a natural question that arises is how secret communication rates can be characterized under different assumptions regarding which entities know the states of various channel states. The results of [4], for example, were formulated for the case of complete channel state information where Alice, Bob and Eve all have perfect knowledge of the CSI for channels G and H. For complete CSI it has been shown that the secrecy capacity for a collection of independent parallel channels can be solved through appropriate water-filling of the channel differences between G and H.

Unfortunately, the case of complete CSI is not representative of what one would expect to face in an adversarial setting, where the eavesdropper is not likely to reveal its presence. Instead, incomplete CSI cases are more appropriate but, for the most part, have not been considered in the literature. Generally, it is reasonable to assume that the receiver has knowledge of the state of the channel from the transmitter. Hence, we are interested in cases where Alice does not have complete knowledge of G or H. In this paper, we examine the problem of secret communication over fading channels for several specific cases of incomplete CSI.

To address the problem of how the sender can best communicate secretly to a legitimate receiver while having varying levels of knowledge about the corresponding channel states, we formulate the problem of secret communication as a zero-sum game. Here, the user must decide how to transmit information across which subchannels under a worst-case condition, while we pose the environment as a secondary player in a zero-sum game whose objective is to hamper successful communication by the user. We consider signal-to-interference ratio (SINR) as a payoff function since, in the regime of low SINR, this objective is an approximation to the secrecy rate.

We begin the paper in Section 2 by presenting our three entities (Alice, Bob and Eve), and providing a description of the basic communication model that we will use throughout this paper. In the sections that follow, we examine several distinct cases where different assumptions are placed on how well Alice or Eve know the channel gains. Throughout the paper we present conclusions that can be drawn from theoretically formulating the eavesdropping problem in a game-theoretic scenario. We provide proofs in the Appendix.

2   Problem Overview

Alice seeks to communicate secretly with Bob, while in the presence of a potential (passive) eavesdropper, Eve. We consider a communication system involving n independent subchannels, as might arise in an OFDM system. Letting Alice's transmitted signal on channel i be X_i, then Bob's received signal is

$$Y_i = \sqrt{g_i} X_i + W_i^{AB}, \tag{1}$$

while Eve receives the signal

$$Z_i = \sqrt{h_i} X_i + W_i^{AE}. \tag{2}$$

We may collect Alice's channel input as $X^n = [X_1, \cdots, X_n]$, and similarly define Bob's received signals as Y^n, and Eve's as Z^n. In the communication literature, the channel gains g_i and h_i may follow many different distributions and one of the most common is the Rayleigh fading model, where g_i and h_i follow an exponential distribution with an average channel gain $E[g_i]$ or $E[h_i]$ capturing distance-dependent attenuation and shadowing. In general the Alice-to-Bob channel and Alice-to-Eve channel will have different average characteristics, i.e. in general $E[g_i] \neq E[h_i]$. Further, we note that the W_i^{AB} and W_i^{AE} are additive noise terms that (unless noted otherwise) have been normalized appropriately (relative to the main Alice-to-Bob channel gains g_i) to have unit variance.

In [4], it was shown under the complete CSI assumption, that the secrecy capacity of the system of n independent channels for Alice-Bob-Eve can be expressed as $C_n(\mathbf{g}, \mathbf{h}, \mathbf{P}^*) = \sum_{i=1}^n C_{AWGN}(g_i, h_i, P_i)$, where $C_{AWGN}(g_i, h_i, P_i)$ is the secrecy capacity for an additive white Gaussian noise channel model, and was given by Leung-Yan-Cheong and Hellman in [13]. Further, $\mathbf{P}^*$ is the optimal power allocation across the n subchannels and corresponds to waterfilling appropriately by considering the relative differences between $\mathbf{g}$ and $\mathbf{h}$.

3   Optimization with SINR as the Objective Function

In this section, we formulate the secret communication problem as an optimization problem. As noted earlier, Alice would like to transmit information through n channels, and to do this she must allocate power $P = (P_1, \ldots, P_n)$ across these channels, where

$$P_i \geq 0 \text{ for } i \in [1, n] \tag{3}$$

and

$$\sum_{i=1}^{n} P_i = \bar{P}. \tag{4}$$

Here $\bar{P} > 0$ denotes the signal total power budget she may transmit. Up to a normalization factor, Alice's payoff is given as follows

$$v(P) = \sum_{i=1}^{n} \left[\ln\left(1 + g_i P_i\right) - \ln\left(1 + h_i P_i\right) \right]_+ \tag{5}$$

where g_i and h_i are the corresponding fading channel gains of the main (Alice to Bob) and eavesdropper (Alice to Eve) channels. The individual secrecy rate terms $\ln\left(1 + g_i P_i\right) - \ln\left(1 + h_i P_i\right)$ are generally unwieldy, and as a useful approximation, we may instead define a more convenient payoff function, which we shall refer to as the SINR payoff. The SINR payoff for Alice is given as follows

$$v(P) = \sum_{i=1}^{n} g_i P_i - \sum_{i-1}^{n} h_i P_i. \tag{6}$$

SINR has been considered in non-eavesdropping communication scenarios. Specifically, it has been used as an objective function in the power control game in [16], [17] and [15]. In [16], the Braess paradox in the context of the power control game has been studied and in [17] all users have a single common channel and choose between several base stations, while in [15] jamming and cooperative scenarios are considered. Lastly, we note that in the regime of low SINR the present objective serves as an approximation to the secrecy rate.

Since the payoff is linear in P the optimal power strategy assigns transmission power across the channels by placing a preference to channels with greater difference between the channel gains of the main and eavesdropper channels, $g_i - h_i$. Namely, the following result holds.

Theorem 1. *The optimal power allocation strategy, P, for Alice for the secret communication optimization problem with SINR as the payoff, under condition (3) and (4), is given as follows*

$$P_i \begin{cases} = 0 & \text{for } i \in [1, n] \backslash I_*, \\ \geq 0 \text{ such that } \sum_{i \in I_*} P_i = \bar{P} & \text{for } i \in I_*, \end{cases}$$

where $I_* = \{ i \in [1, n] : g_i - h_i = \max\{g_j - h_j : j \in [1, n]\}\}$ is the maximal difference between fading channel gains of the main (Alice to Bob) and eavesdropper (Alice to Eve) channels. The payoff corresponding to this strategy is $v = \bar{P} \max\{g_j - h_j : j \in [1, n]\}$.

Now look at the problem assuming that Alice has fixed the power allocation, i.e. the vector $P = (P_1, \ldots, P_n)$ satisfying (3) and (4), yet the environmental parameters are not completely known, i.e. Alice does not know the exact values of the channels gains. To capture this assumption, we shall further assume that Alice knows the best case scenario for the main and eavesdropper channel gains, but does not know the precise values of any instantaneous realization. Hence, we assume that the gains g_i for the main subchannel i is given by

$$g_i = g_i^0 - G_i, \tag{7}$$

where g_i^0 is the best possible channel gain, and G_i reflects additional degradation of the channel that might arise from fading or other factors. For analysis, we assume that Alice knows that the degradation G_i is such that

$$G_i \geq 0 \text{ for } i \in [1, n] \tag{8}$$

and that she knows an (ensemble) characterization of this degradation across all n subchannels

$$\sum_{i=1}^{n} G_i = \bar{G}, \tag{9}$$

where $\bar{G} > 0$ thus corresponds to the total main channel perturbation.

Similarly, we assume that Alice has imprecise knowledge of the gains of the eavesdropper subchannel i, given by

$$h_i = h_i^0 + H_i, \tag{10}$$

where h_i^0 is (best, and hence smallest) possible channel gain and is known to the user. However, as before, about the perturbation of this channel gain, H_i, she knows only that it is such that

$$H_i > 0 \text{ for } i \in [1, n] \tag{11}$$

and

$$\sum_{i=1}^{n} H_i = \bar{H}, \tag{12}$$

where $\bar{H} > 0$ is the total eavesdropper's channels perturbation known to Alice.

The payoff is then given as follows

$$v((G, H)) = \sum_{i=1}^{n} (g_i^0 - G_i)P_i - \sum_{i=1}^{n} (h_i^0 + H_i)P_i$$
$$= \sum_{i=1}^{n} \xi_i^0 P_i - \sum_{i=1}^{n} (G_i + H_i)P_i, \tag{13}$$

where ξ_i is the difference between fading channel gains of the main (Alice to Bob) and eavesdropper (Alice to Eve) channels i, namely,

$$\xi_i^0 = g_i^0 - h_i^0, \quad i \in [1, n]. \tag{14}$$

We will assume that $g_i^0 > h_i^0$ for $i \in [1, n]$, so $\xi_i^0 > 0$ for $i \in [1, n]$. The following result allows Alice to quantify the worst payoff she could have, as she would like to minimize (13) for any admissible (G, H).

Theorem 2. *Let $I_{max} = \{i \in [1, n] : P_i = P_{max}\}$ where $P_{max} = \max_{j \in [1,n]} P_j$. Then the optimal strategy (G, H) is given as follows*

$$G_i \begin{cases} = 0, & i \in [1, n] \backslash I_{max}, \\ \geq 0 \text{ such that } \sum_{j \in I_{max}} G_j = \bar{G}, & i \in I_{max}, \end{cases} \qquad (15)$$

$$H_i \begin{cases} = 0, & i \in [1, n] \backslash I_{max}, \\ \geq 0 \text{ such that } \sum_{j \in I_{max}} H_j = \bar{H}, & i \in I_{max}. \end{cases} \qquad (16)$$

The payoff corresponding to this strategy is $v = \sum_{i=1}^{n} \xi_i^0 P_i - P_{max}(\bar{G} + \bar{H})$.

4 An Eavesdropping Game with Unknown Gains

We continue our analysis of the situation where Alice does not know the exact values of the channels gains, as described previously. Alice faces the problem of allocating power so that information can be transmitted under the worst-case conditions or, in other words, Alice faces a maxmin problem. To address this question we draw upon game theory since we may consider Alice as a player in a game, while we may model the environment (nature) as a second player with a goal opposite to Alice's, namely, to hamper information transmission by Alice (by selecting channel states so as to benefit the eavesdropper Eve)[1]. Thus, nature faces a minmax problem and the optimal strategies of the players for the maxmin and minmax problems will coincide with each other.

We assume that the gains of the main channel i is given by (7) and the gains of the eavesdropper channel i is given by (10). The strategy for the environment is governed by appropriately selecting (G, H), which consists of two components: $G = (G_1, \ldots, G_n)$ – the main channel's degradations about g; and $H = (H_1, \ldots, H_n)$ – the eavesdropper's channel degradations about h, as per the conditions (8), (9) and (11), (12). Alice's P is given by satisfying (3) and (4). The SINR payoff for Alice is given as follows

$$v(P, (G, H)) = \sum_{i=1}^{n} (g_i^0 - G_i)P_i - \sum_{i=1}^{n} (h_i^0 + H_i)P_i. \qquad (17)$$

Both players know the values of g_i^0, h_i^0, $i \in [1, n]$ as well as $\bar{P}$, $\bar{G}$ and $\bar{H}$. We consider the situation as a zero-sum game Alice versus nature with Alice's payoff as (36), while the payoff to nature is $-v(P, (G, H))$.

We will look for the value of the game v and the optimal strategies P^* of Alice and (G^*, H^*) for nature. Recall that optimal strategies and the value of the game satisfy the conditions:

$$v(P, (G^*, H^*)) \leq v := v(P^*, (G^*, H^*)) \leq v(P^*, (G, H))$$

for any strategies P and (G, H) for the players (Alice and nature).

---

[1] We note, contrary to intuition, Eve is not the second player in our formulation, but is a passive beneficiary of the strategy employed by the environment.

Note that the payoff (17) of the game by (14) can be rewritten in the following equivalent form

$$v(P, (G, H)) = \sum_{i=1}^{n} \xi_i^0 P_i - \sum_{i=1}^{n} (G_i + H_i) P_i. \tag{18}$$

Without loss of generality we can assume that the channels are arranged in such a way that

$$\xi_1^0 \geq \xi_2^0 \geq \ldots \geq \xi_n^0 > 0. \tag{19}$$

We introduce the following auxiliary notation,

$$\varphi_k := \sum_{i=1}^{k} (\xi_i^0 - \xi_k^0) \text{ for } k \in [1, n]. \tag{20}$$

It is clear that the sequence φ_k, $k \in [1, n]$ is increasing since the following relations hold:

$$\varphi_{k+1} - \varphi_k = \sum_{i=1}^{k+1} (\xi_i^0 - \xi_{k+1}^0) - \sum_{i=1}^{k} (\xi_i^0 - \xi_k^0) = (\xi_k^0 - \xi_{k+1}^0) k \geq 0,$$

and $\varphi_1 = 0$. For this game we can prove the following result describing the optimal strategies as well as the value of the game.

Theorem 3. *(a) Let*

$$\bar{G} + \bar{H} \geq \varphi_n, \tag{21}$$

then the value of the game is given by

$$v = \frac{\bar{P}}{n} \left(\sum_{i=1}^{n} \xi_i^0 - \bar{G} - \bar{H} \right). \tag{22}$$

Alice's optimal strategy P^* assigns power uniformly across all the n channels, i.e.

$$P_i^* = \bar{P}/n \text{ for } i \in [1, n]. \tag{23}$$

Nature's optimal strategy (G^*, H^*), meanwhile, involves assigning the eavesdropper and main channel components H^* and G^* to equalize the difference in quality between the fading channel gains of the main (Alice to Bob) and eavesdropper (Alice to Eve) channels, namely, H^* satisfies (11) and (12), G^* satisfies (8) and (9) and

$$G_i^* + H_i^* = \frac{1}{n} \left(\bar{H} + \bar{G} - \sum_{j=1}^{n} (\xi_j^0 - \xi_i^0) \right), \tag{24}$$

say,

$$G_i^* = \frac{\bar{G}}{n(\bar{G} + \bar{H})} \left(\bar{H} + \bar{G} - \sum_{j=1}^{n} (\xi_j^0 - \xi_i^0) \right), \tag{25}$$

$$H_i^* = \frac{\bar{H}}{n(\bar{G} + \bar{H})} \left(\bar{H} + \bar{G} - \sum_{j=1}^{n} (\xi_j^0 - \xi_i^0) \right) \tag{26}$$

for $i \in [1, n]$.

(b) Let

$$\bar{G} + \bar{H} < \varphi_n.$$

Then, there is a $k_* \in [1, n-1]$ such that

$$\varphi_{k_*} \leq \bar{G} + \bar{H} < \varphi_{k_*+1}. \tag{27}$$

The value of the game is given as follows

$$v = \frac{\bar{P}}{k_*} \left(\sum_{i=1}^{k_*} \xi_i^0 - \bar{G} - \bar{H} \right).$$

Alice's optimal strategy P^* assigns power equally among the first k_* channels, i.e.

$$P_i^* = \begin{cases} \bar{P}/k_* & \text{for } i \in [1, k_*], \\ 0 & \text{for } i \in [k_* + 1, n]. \end{cases} \tag{28}$$

Nature's optimal strategy (G^*, H^*) assigns G^* only to the main channel components unused by Alice, while H^* and G^* are assigned across the eavesdropper's subchannels so as to equalize the k_* best differences in quality between fading channel gains of the main (Alice to Bob) and eavesdropper (Alice to Eve) channels. Namely, H^* satisfies (11) and (12), G^* satisfies (8) and (9),

$$G_i^* = H_i^* = 0 \text{ for } i \in [k_* + 1, n] \tag{29}$$

and

$$G_i^* + H_i^* = \frac{1}{k_*} \left(\bar{H} + \bar{G} - \sum_{j=1}^{k_*} (\xi_j^0 - \xi_i^0) \right), \tag{30}$$

say,

$$G_i^* = \frac{\bar{G}}{k_*(\bar{G} + \bar{H})} \left(\bar{H} + \bar{G} - \sum_{j=1}^{k_*} (\xi_j^0 - \xi_i^0) \right), \tag{31}$$

$$H_i^* = \frac{\bar{H}}{k_*(\bar{G} + \bar{H})} \left(\bar{H} + \bar{G} - \sum_{j=1}^{k_*} (\xi_j^0 - \xi_i^0) \right) \tag{32}$$

for $i \in [1, k_*]$.

5   Either the Eavesdropper's Channels Gains or the Main Channels Gains Are Unknown

In this section, we first consider the case where Alice does not know the exact values of gains of the eavesdropper's channels, but she does have full knowledge about the main (Alice to Bob) channel gains. The payoff for Alice is

$$v(P, H) = \sum_{i=1}^{n} g_i^0 P_i - \sum_{i=1}^{n} (h_i^0 + H_i) P_i. \tag{33}$$

Nature's strategy thus consists only of appropriately selecting the eavesdropper's channels component H while satisfying (11) and (12). For this case we can prove the following result, which basically states that in order to harm Alice (and thus help Eve), nature has to spoil equalizing k channels with the largest gains differences, while Alice has to assign power uniformly across these k channels.

Theorem 4. *The value of the game is given as follows*

$$v = \frac{\bar{P}}{k} \left(\sum_{i=1}^{k} \xi_i^0 - \bar{H} \right).$$

where

$$k = \begin{cases} n & \text{for } \varphi_n \leq H, \\ k_* : \varphi_{k_*} \leq \bar{H} < \varphi_{k_*+1} & \text{for } \varphi_n > H. \end{cases}$$

Alice's optimal strategy P^* has her using an equalizing strategy among the k best channels. Namely,

$$P^* = \begin{cases} \bar{P}/k, & i \in [1, k], \\ 0, & \text{otherwise.} \end{cases} \tag{34}$$

Nature's optimal strategy H^* involves equalizing the k best channels. Namely,

$$H_i^* = \begin{cases} \frac{1}{k} \left(\bar{H} - \sum_{j=1}^{k} (\xi_j^0 - \xi_i^0) \right), & i \in [1, k] \\ 0, & \text{otherwise.} \end{cases}$$

If Alice does not know the exact values of the gains of the main subchannels, while she has full knowledge about eavesdropper's channel gains, then the payoff to Alice is given as follows

$$v(P, G) = \sum_{i=1}^{n} (g_i^0 - G_i) P_i - \sum_{i=1}^{n} h_i^0 P_i. \tag{35}$$

Theorem 5. *The value of the game is given as follows*

$$v = \frac{\bar{P}}{k} \left(\sum_{i=1}^{k} \xi_i^0 - \bar{G} \right).$$

where

$$k = \begin{cases} n & \text{for } \varphi_n \leq G, \\ k_* : \varphi_{k_*} \leq \bar{G} < \varphi_{k_*+1} & \text{for } \varphi_n > G. \end{cases}$$

Alice's optimal strategy P^* has her using an equalizing strategy among the k best channels. Namely,

$$P^* = \begin{cases} \bar{P}/k, & i \in [1, k], \\ 0, & \text{otherwise.} \end{cases}$$

Nature's optimal strategy G^* involves equalizing the k best channels. Namely,

$$G_i^* = \begin{cases} \frac{1}{k}\left(\bar{G} - \sum_{j=1}^{k}(\xi_j^0 - \xi_i^0)\right), & i \in [1, k] \\ 0, & \text{otherwise.} \end{cases}$$

Let us demonstrate some numerical results showing how information about the channels impacts the value of the eavesdropping game we have formulated. Suppose there are five subchannels, $n = 5$, and ξ_i is given by an exponential law, namely, let $\xi_i = 4\kappa^{i-1}$ for $i \in [1, n]$ and $\kappa = 0.7$. We examine the value of the game and the number of channels employed to communicate for the two cases: (1) with unknown gains as in Section 4, (2) with unknown eavesdropper channels gains. For both plots we will assume that $\bar{P} = 3$ and $\bar{G} \in [1, 7]$ and $\bar{H} = 1$. However, for the second case we assume that $\bar{H}$ is uniformly distributed across the subchannels H_i. In Table 1 we present the value of the game for different values of k. Of course, when the players use all the five channels then the value of the two cases of the eavesdropping game coincide, which occurs for large $\bar{G}$ (in this example, $\bar{G} = 7$). If $\bar{G}$ is small (equals 1) then having extra information about the channels (the second case) allows her to improve her SINR (and hence secrecy) payoff by a factor of roughly 1.5.

Table 1. The value of the game and k for two plots

G	Case 1	k	Case 2	k
1	1.587	3	2.400	1
2	1.283	4	1.800	2
3	1.033	4	1.320	3
4	0.818	5	0.987	3
5	0.618	5	0.683	4
6	0.418	5	0.433	4
7	0.218	5	0.218	5

6   The Worst Case for the Main Gains Are Known

To show that the optimal strategies essentially depend on the information the players have, in this section we slightly change the formulation of the game to assume that the worst possible values for the main channels gains are known (instead of the best possible values), and then demonstrate the impact that such a change has on the optimal strategies. We assume that the SINR payoff for Alice is given as follows

$$v(P,(G,H)) = \sum_{i=1}^{n}(g_i^0 + G_i)P_i - \sum_{i=1}^{n}(h_i^0 + H_i)P_i, \qquad (36)$$

where now g_i^0 is the worst possible value for the main subchannel i's gain.

For this game we can prove the following result describing the optimal strategies as well as the value of the resulting eavesdropping game:

Theorem 6. *(a) Let (21) hold. Then the value of the game is given by*

$$v = \frac{\bar{P}}{n}\left(\sum_{i=1}^{n}\xi_i^0 - \bar{H} + \bar{G}\right). \qquad (37)$$

Alice's optimal strategy P^* assigns power uniformly across all n subchannels, i.e. by (23). Nature's optimal strategy (G^*, H^*), meanwhile, involves assigning the eavesdropper channel component H^* to equalize the eavesdropper channels, while assigning the main channel component G^* uniformly across subchannels,

$$G_i^* = \bar{G}/n, \qquad (38)$$

$$H_i^* = \frac{1}{n}\left(\bar{H} - \sum_{j=1}^{n}(\xi_j^0 - \xi_i^0)\right) \qquad (39)$$

for $i \in [1, n]$.

(b) Let $\bar{H} < \varphi_n$. Then, there is a $k_* \in [1, n-1]$ such that (27) holds. Also, let

$$A < 0, \qquad (40)$$

where

$$A := \bar{G} - \frac{1}{k_*}\sum_{i=k_*+1}^{n}\left(\sum_{j=1}^{k_*}(\xi_j^0 - \xi_i^0) - \bar{H}\right)$$

$$= \bar{G} - \frac{1}{k_*}\left((n - k_*)\sum_{j=1}^{k_*}\xi_j^0 \qquad (41)\right.$$

$$\left. - k_*\sum_{j=k_*+1}^{n}\xi_j^0 - \bar{H}(n - k_*)\right).$$

Then the value of the game is $v = \bar{P}\left(\sum_{i=1}^{k_*}\xi_i^0 - \bar{H}\right)/k_*$. Alice's optimal strategy P^* assigns power equally among the first k_* channels, i.e. it is given by (28).

Nature's optimal strategy (G^*, H^*) assigns G^* only to the main channel components not used by Alice, while H^* is assigned across the eavesdropper's subchannels so as to equalize the quality of the k_* best channels for Alice. Namely,

$$
G_i^* \begin{cases} = 0, & i \in [1, k_*], \\ \leq \frac{1}{k_*}\left(\sum_{j=1}^{k_*}(\xi_j^0 - \xi_i^0) - \bar{H}\right) & \\ \text{such that } \sum_{j=k_*+1}^{n} G_i^* = \bar{G}, & i \in [k_* + 1, n], \end{cases} \tag{42}
$$

$$
H_i^* = \begin{cases} \frac{1}{k_*}\left(\bar{H} - \sum_{j=1}^{k_*}(\xi_j^0 - \xi_i^0)\right), & i \in [1, k_*], \\ 0, & i \in [k_* + 1, n]. \end{cases} \tag{43}
$$

(c) Let $\bar{H} < \varphi_n$ and $A \geq 0$. The value of the game is given by (37). Alice's optimal strategy P^* is given by (23). Nature's optimal strategy (G^*, H^*) assigns H^* according to (43), and equalizes the quality of the k_* best channels, while component for the main channel G^* is assigned to supplement all the channels until they have an equal level, as follows

$$
G_i^* = \begin{cases} \frac{A}{n}, & i \leq k_*, \\ \frac{A}{n} + \frac{1}{k_*}\left(\sum_{j=1}^{k_*}(\xi_j^0 - \xi_i^0) - \bar{H}\right), & i > k_*. \end{cases} \tag{44}
$$

Since the inequality

$$
\frac{\bar{P}}{n}\left(\sum_{i=1}^{n} \xi_i^0 - \bar{H} + \bar{G}\right) < \frac{\bar{P}}{k_*}\left(\sum_{i=1}^{k_*} \xi_i^0 - \bar{H}\right)
$$

is equivalent to

$$
k_* \bar{G} < (n - k_*)\sum_{j=1}^{k_*} \xi_j^0 - k_* \sum_{j=k_*+1}^{n} \xi_j^0 - \bar{H}(n - k_*)
$$

or, by (41), to $A < 0$, we can summarize the result of Theorem 6 about the value of the game in the following statement.

Theorem 7. *The value of the game is given as follows: if $\varphi_n > \bar{H}$, then*

$$
v = \max\left\{\frac{\bar{P}}{n}\left(\sum_{i=1}^{n} \xi_i^0 - \bar{H} + \bar{G}\right), \frac{\bar{P}}{k_*}\left(\sum_{i=1}^{k_*} \xi_i^0 - \bar{H}\right)\right\}.
$$

We now present some numerical results to illustrate the implications of Theorem 6 and 7. As before, suppose there are five subchannels, $n = 5$ and ξ_i is given by the exponential law, namely, let $\xi_i = 4\kappa^{i-1}$ for $i \in [1, n]$ and $\kappa = 0.7$. We compare how the optimal strategies change around the switching point A. In Table 2 we put together the optimal strategies for nature when $\bar{G} = 1$, $\bar{P} = \{3, 4\}$ corresponding to the values of the game 5.76 and 4.855. In spite of the fact that $k_* = 3$ for both cases, there is a switching point between $\bar{P} = 3$ and $\bar{P} = 4$ since for the first case

Table 2. The optimal strategies for the *nature* player in the example eavesdropping game

$H^* \& G^*(\bar{P})$	1	2	3	4	5
$H^*(3)$	2.08	0.88	0.04	0	0
$G^*(3)$	0	0	0	≤ 0.446	≤ 0.858
$H^*(4)$	2.413	1.213	0.373	0	0
$G^*(4)$	0.032	0.032	0.032	0.246	0.658

$A = -0.507$ and for the second case $A = 0.159$. In the case $\bar{P} = 3$ a variety of G components is possible that do not use the first three channels. For example, it could be any $G^* = (0, 0, 0, G_4^*, G_5^*)$ such that $G_4^* \leq 0.446$, $G_5^* = 0.858$, $G_4^* + G_5^* = 1$. Meanwhile, in the case $\bar{P} = 4$ the G^* component uses all the channels.

7   The Optimization Problem with Unknown Noise and Eavesdropper's Channel Gains

In this section, we relax the assumptions about the noise term (W_i^{AB} from Section 2), and consider the situation where Alice does not know how the noise is distributed among the main (Alice to Bob) subchannels. For example, the noise power may not be uniform across subchannels. To reflect this case, we assume that the main channels gains are given by

$$g_i = 1/(N_i^0 + N_i) \text{ for } i \in [1, n],$$

where N_i^0 is a constant part of the noise level in the main channel i and N_i is a variable component for which Alice knows only the total perturbation $\bar{N}$, which satisfies

$$\sum_{i=1}^{n} N_i = \bar{N} \tag{45}$$

and

$$N_i \geq 0 \text{ for } i \in [1, n]. \tag{46}$$

We note that this is representation allows us to reflect the variable noise terms directly in the channel gains g_i. For example, low levels of noise (i.e. small N_i^0 and N_i) leads to a correspondingly large subchannel gain g_i, which implies that the ith subchannel is good.

Assume that Alice has fixed the power allocation strategy for signal transmission, i.e. the vector $P = (P_1, \ldots, P_n)$ satisfying (3) and (4), but the parameters for the environment are not completely known, i.e. Alice does not know how the noise is distribution for Eve, or the values the eavesdropper's channels gains. The payoff is given as follows

$$v((N, H)) = \sum_{i=1}^{n} \frac{P_i}{N_i^0 + N_i} - \sum_{i=1}^{n} (h_i^0 + H_i) P_i. \tag{47}$$

Alice would like to know what the worst payoff she could have, so, she would like to minimize (47) by (N, H).

Since the payoff is linear in H and concave in N, the strategy (N^*, H^*) is the optimal one if and only if there is ν, such that

$$\frac{P_i}{(N_i^0 + N_i^*)^2} \begin{cases} = \nu & \text{for } N_i^* > 0, \\ \leq \nu & \text{for } N_i^* = 0, \end{cases} \tag{48}$$

$$H_i^* \begin{cases} \geq 0 & \text{for } P_i = P_{max}, \\ = 0 & \text{otherwise.} \end{cases} \tag{49}$$

Then the optimal H^* is given by (16) and the optimal N^* is of the form

$$N_i^* = N_i(\nu) = \left[\sqrt{P_i/\nu} - N_i^0\right]_+ \quad \text{for } i \in [1, n],$$

where $\nu = \nu_*$ is the unique positive root of the equation

$$\sum_{i=1}^n \left[\sqrt{P_i/\nu} - N_i^0\right]_+ = \bar{N}.$$

The payoff corresponding to (N^*, H^*) is given as follows

$$v = \sqrt{\nu_*} \sum_{N_i(\nu_*)>0} \sqrt{P_i} - \sum_{i=1}^n h_i^0 P_i - \bar{H} P_{max}.$$

8 The Game with Unknown Noise in the Main Subchannels

In this section we consider the situation where there is unknown noise in the main subchannels, and examine this case from game-theoretical position. There are two players: Alice and nature. Alice has to transmit the total power $\bar{P}$ using strategy P satisfying (3) and (4). Recall that nature's objective is to harm Alice-to-Bob communication, and thus in this case nature's strategy consists only of a jamming component N satisfying (45) and (46), i.e. nature introduces noise to the main subchannels. The payoff to Alice is given as follows

$$v(P, N) = \sum_{i=1}^n \frac{P_i}{N_i^0 + N_i} - \sum_{i=1}^n h_i^0 P_i \tag{50}$$

The payoff to nature is $-v(P, N)$.

In the following theorem we find the value of the game and the optimal strategies for the players. In particular, we show that nature should hamper precisely the same channels that Alice employs. The optimal strategy for nature is a water filling strategy, but from an adversarial point of view.

Theorem 8. *The value of the game is $\omega_* \bar{P}$ where ω_* is the unique root in $[-\min_i h_i^0, \infty)$ of the water filling equation*

$$H_N(\omega) := \sum_{i=1}^n \left[\frac{1}{h_i^0 + \omega} - N_i^0 \right]_+ = \bar{N}. \qquad (51)$$

The optimal nature's strategy is given by

$$N_i^* = N_i(\omega) = \left[\frac{1}{h_i^0 + \omega} - N_i^0 \right]_+, \quad i \in [1, n]. \qquad (52)$$

The optimal Alice's strategy is given as follows

$$P_i^* = \begin{cases} \bar{P} \dfrac{1/(h_i^0 + \omega_*)^2}{\displaystyle\sum_{j: N_j(\omega_*) > 0} (1/(h_j^0 + \omega_*)^2)} & \text{if } N_i(\omega_*) > 0. \\ 0 & \text{otherwise.} \end{cases}$$

Remark 1. *It is interesting that the optimal strategy for nature does not take into account the power of signal Alice has to transmit but only the parameters of the environment, which is quite reasonable because nature is Alice's rival.*

As a numerical example we consider five channels $n = 5$ case. Let N_i^0 and h_i^0 are given by the same exponential law, namely, $N_i^0 = h_i^0 = \kappa^{i-1}$ for $i \in [1, n]$ where $\kappa = 0.5$. Also, let $\bar{P} = 1$ and $\bar{N} = 0.5$ In Table 3 the value of the game and the players' optimal strategies are given as a function of κ. For $\kappa = 0.1$ these strategies use four out of the five subchannels, for $\kappa = 0.8$ they use two subchannels, and for intermediate values these strategies use three subchannels.

Table 3. The value of the game and the optimal strategies of the players

κ	v		1	2	3	4	5
0.1	6.315	N	0.000	0.047	0.142	0.154	0.157
		P	0.000	0.230	0.248	0.258	0.263
0.2	5.314	N	0.000	0.000	0.140	0.176	0.184
		P	0.000	0.000	*0.321*	0.336	0.344
0.3	4.655	N	0.000	0.000	0.114	0.182	0.204
		P	0.000	0.000	0.319	0.336	0.345
0.4	3.858	N	0.000	0.000	0.083	0.187	0.229
		P	0.000	0.000	0.316	0.336	0.347
0.5	3.056	N	0.000	0.000	0.052	0.189	0.258
		P	0.000	0.000	0.312	0.337	0.351
0.6	2.345	N	0.000	0.000	0.025	0.189	0.286
		P	0.000	0.000	0.306	0.338	0.356
0.7	1.764	N	0.000	0.000	0.006	0.186	0.307
		P	0.000	0.000	**0.298**	0.339	0.363
0.8	1.314	N	0.000	0.000	0.000	0.183	0.317
		P	0.000	0.000	0.000	0.478	0.522

It is interesting to note that for $\kappa \in [0.2, 0.7]$ the maximal difference is 11% (it is accentuated in bold font) from the uniform strategy, and arises right before switching to using smaller number of channels and smallest in 1% (accentuated in italic font right after the switching point).

9 Conclusion

Recently, there has been increasing interest in using the properties of the physical layer in a wireless system to support security (specifically, confidentiality) objectives. The basic principle behind this new form of confidentiality is to take advantage of conditions where the main Alice-to-Bob channel is better than the adversarial channel Alice-to-Eve. One fundamental challenge facing the formulation of such physical layer secrecy is understanding the implications of varying assumptions for what knowledge the participants (Alice, Bob and Eve) have in the secret communication. In this paper we have examined the problem of eavesdropping over fading channels, where the problem of secret communication in the presence of an eavesdropper is formulated as a zero-sum game. In our problem, the legitimate receiver does not have complete knowledge about the environment, i.e. does not know the exact values of the channels gains. Rather, we consider that the receiver has some partial knowledge characterizing the channel, such as its distribution. The transmitter's task then involves deciding how to transmit its information across which subchannels. We have posed this problem as an optimization problem, where the environment acts as a secondary player in a zero-sum game whose objective is to hamper successful communication by the user. In our formulation, we have chosen to use signal-to-interference ratio (SINR) as the payoff function, due to the tractability it provides, but note that at low SINR our objective function approximates the secrecy capacity. We have studied a variety of scenarios where different assumptions are placed on the amount of knowledge that the transmitter, Alice, has in the eavesdropping game. In the case where Alice does not know the gains for the various subchannels, then the best strategy is to distribute energy equally across a subset of selected channels. On the other hand, if Alice does not know the eavesdropper's channel gains, then Alice should also employ a strategy involving uniformly distributing energy across a subset of channels. However, if the user acquires extra knowledge about environment, e.g. the eavesdropper's channel gains, then we show how Alice may better tune her power allocation among the channels.

References

1. Li, X., Chen, M., Ratazzi, E.P.: Space-time transmissions for wireless secret-key agreement with information-theoretic secrecy. In: Proc. IEEE SPAWC 2005, June 2005, pp. 811–815 (2005)
2. Koorapaty, H., Hassan, A.A., Chennakeshu, S.: Secure Information Transmission for Mobile Radio. IEEE Trans. Wireless Commun., 52–55 (July 2003)
3. Hero, A.E.: Secure Space-Time Communication. IEEE Trans. Info. Theory, 3235–3249 (December 2003)

4. Li, Z., Yates, R., Trappe, W.: Secrecy Capacity of Independent Parallel Channels. In: Allerton Conference on Communication, Control, and Computing (2006)
5. Li, Z., Xu, W., Miller, R., Trappe, W.: Securing wireless systems via lower layer enforcements. In: WiSe 2006: Proceedings of the 5th ACM workshop on Wireless security, pp. 33–42 (2006)
6. Liang, Y., Poor, H.V., Shamai, S.: Secure Communication over Fading Channels. IEEE Transactions on Information Theory, Special issue on Information Theoretic Security 54(6), 2470–2492 (2008)
7. Gopala, P., Lai, L., El Gamal, H.: On the secrecy capacity of fading channels. IEEE Trans. Inform. Theory (accepted for publication)
8. Wyner, A.: The wire-tap channel. Bell. Syst. Tech. J. 54(8), 1355–1387 (1975)
9. Csiszár, I., Körner, J.: Broadcast channels with confidental messages. IEEE Trans. on Inf. Theory 24(3), 339–348 (1978)
10. Maurer, U.M., Wolf, S.: Information-theoretic key agreement: From weak to strong secrecy for free. In: Preneel, B. (ed.) EUROCRYPT 2000. LNCS, vol. 1807, pp. 351–368. Springer, Heidelberg (2000)
11. Bennett, C., Brassard, G., Crepeau, C., Maurer, U.M.: Generalized privacy amplification. IEEE Trans. on Information Theory 41, 1915–1923 (1995)
12. Van Dijk, M.: On a special class of broadcast channels with confidential messages. IEEE Trans. on Information Theory 43(2), 712–714 (1997)
13. Leung-Yan-Cheong, S.K., Hellman, M.: The gaussian wire-tap channel. IEEE Transactions on Information Theory 24(4), 451–456 (1978)
14. Altman, E., Avrachenkov, K., Garnaev, A.: A jamming game in wireless networks with transmission cost. In: Chahed, T., Tuffin, B. (eds.) NET-COOP 2007. LNCS, vol. 4465, pp. 1–12. Springer, Heidelberg (2007)
15. Altman, E., Avrachenkov, K., Garnaev, A.: Transmission power control game with SINR as objective function. In: Altman, E., Chaintreau, A. (eds.) NET-COOP 2008. LNCS, vol. 5425, pp. 112–120. Springer, Heidelberg (2009)
16. Altman, E., Kamble, V., Kameda, H.: A Braess Type Paradox in Power Control over Interference Channels. In: Physicomnet workshop, Berlin, April 4 (2008)
17. Ji, H., Huang, C.-Y.: Non-cooperative uplink power control in cellular radio systems. Wireless Networks 4, 233–240 (1998)

A   Appendix

Proof of Theorem 3. Since the payoff is linear in P, G and H, the strategies P^*, (G^*, H^*) for Alice and nature are in equilibrium (so, these strategies are the best response to one another) if and only if there are ω, ν_G and ν_H such that

$$P_i^* \begin{cases} \geq 0 & \text{for } \xi_i^0 - G_i^* - H_i^* = \omega, \\ = 0 & \text{for } \xi_i^0 - G_i^* - H_i^* < \omega, \end{cases} \tag{53}$$

$$G_i^* \begin{cases} \geq 0 & \text{for } P_i^* = \nu_G, \\ = 0 & \text{for } P_i^* < \nu_G, \end{cases} \tag{54}$$

$$H_i^* \begin{cases} \geq 0 & \text{for } P_i^* = \nu_H, \\ = 0 & \text{for } P_i^* < \nu_H. \end{cases} \tag{55}$$

(a) Let P^* be given by (23). Then, by (54) and (55), $\nu_G = \nu_H = \bar{P}/n$ and any strategy (G, H) is the best response one for (23), in particular, the strategy given by (29). Let H^* and G^* be given by (29). Then, by (21) they present a strategy and $\xi_i^0 - G_i^* - H_i^* = \omega$ for $i \in [1, n]$ where $\omega = \left(\sum_{j=1}^n \xi_j^0 - \bar{H} - \bar{G}\right)/n$. Then, by (53), any strategy for Alice is the best response strategy to nature's strategy given by (29). This proves (a).

(b) Let P^* be given by (28). Then, by (54) and (55), $\nu_G = \nu_H = \bar{P}/k_*$ and any strategy for nature (G, H) satisfying the following conditions is the best response for (28).

$$H_i = 0 \text{ and } G_i = 0 \text{ for } i \in [k_* + 1, n]. \tag{56}$$

By (27), (G^*, H^*) given by (29) and (30) is a strategy which satisfies to (56). So, (G^*, H^*) is the best response one for (28). Let (G^*, H^*) be given by (29) and (30). Then

$$\xi_i^0 - G_i^* - H_i^* \begin{cases} = \omega, & i \in [1, k_*], \\ \leq \omega, & i \in [k_* + 1, n], \end{cases}$$

where $\omega = \left(\sum_{j=1}^{k_*} \xi_j^0 - \bar{H} - \bar{G}\right)/k_*$. So, (28) is the best response to (29) and (30) by (53).

Proof of Theorem 6. Since the payoff is linear in P, G and H, the strategies P^*, (G^*, H^*) for Alice and nature is in equilibrium (so, these strategies are the best response each other) if and only if there are ω, ν_G and ν_H such that

$$P_i^* \begin{cases} \geq 0 & \text{for } \xi_i^0 + G_i^* - H_i^* = \omega, \\ = 0 & \text{for } \xi_i^0 + G_i^* - H_i^* < \omega, \end{cases} \tag{57}$$

$$G_i^* \begin{cases} \geq 0 & \text{for } P_i^* = \nu_G, \\ = 0 & \text{for } P_i^* > \nu_G, \end{cases} \tag{58}$$

$$H_i^* \begin{cases} \geq 0 & \text{for } P_i^* = \nu_H, \\ = 0 & \text{for } P_i^* < \nu_H. \end{cases} \tag{59}$$

(a) Let P^* be given by (23). Then, by (58) and (59), $\nu_G = \nu_H = \bar{P}/n$ and any strategy (G, H) is the best response one for (23), in particular, the strategy given by (38) and (39). Let H^* be given by(39). It is clear that for this H^* (11) holds and, by (39), (19) and (21),

$$H_j^* = \frac{1}{n}\left(\bar{H} - \sum_{j=1}^n (\xi_j^0 - \xi_i^0)\right) \geq \frac{1}{n}\left(\bar{H} - \varphi_n\right) \geq 0 \text{ for } j \in [1, n].$$

So, (11) also holds and H^* is the eavesdropper's channel component arising in nature's strategy to harm the secrecy of communication between Alice and

Bob. It is clear that G^* given by (38) satisfies (8) and (9) and for G^* and H^* holds the following relation: $\xi_i^0 + G_i^* - H_i^* = \omega$ for $i \in [1, n]$ where $\omega = \left(\sum_{j=1}^{n} \xi_j^0 - \bar{H} + \bar{G} \right) / n$. Then, by (57), any strategy for Alice is the best response strategy to nature's strategy given by (38) and (39). This proves (a).

(b) Let P^* be given by (28). Then, by (58) and (59), $\nu_G = 0$ and $\nu_H = \bar{P}/k_*$ and any nature's strategy (G, H) satisfying the following conditions is the best response for (28).

$$
\begin{aligned}
H_i &= 0 \text{ for } i \in [k_* + 1, n], \\
G_i &= 0 \text{ for } i \in [1, k_*].
\end{aligned}
\tag{60}
$$

Let H^* be given by (43). It is clear that for this H^* (11) holds. Also, by (22), (19) and (21)

$$
H_j^* = \frac{1}{k_*} \left(\bar{H} - \sum_{j=1}^{k_*} (\xi_j^0 - \xi_i^0) \right) \geq \frac{1}{k_*} (\bar{H} - \varphi_{k_*}) \geq 0 \text{ for } j \in [1, k_*].
\tag{61}
$$

So, for H^*, (12) also holds and it is the eavesdropper's channel components for a strategy employed by nature. By (19) and (27),

$$
\sum_{j=1}^{k_*} (\xi_j^0 - \xi_i^0) - \bar{H} \geq \sum_{j=1}^{k_*} (\xi_j^0 - \xi_{k_*+1}^0) - \bar{H}
\tag{62}
$$
$$
= \varphi_{k_*+1} - \bar{H} \geq 0 \text{ for } j \in [k_* + 1, n].
$$

Thus, for G^* given by (42), (9) holds. Then, by (40), it is the main channels component of a strategy by nature. It is clear that H^* and G^* satisfy (60). Therefore, they present the best response to (28).

Let G^* and H^* be given by (43) and (42). Then

$$
\xi_i^0 + G_i^* - H_i^* \begin{cases} = \omega, & i \in [1, k_*], \\ \leq \omega, & i \in [k_* + 1, n], \end{cases}
$$

where $\omega = \left(\sum_{j=1}^{k_*} \xi_j^0 - \bar{H} \right) / k_*$. So, (28) is the best response to (43) and (42).

(c) Let P^* be given by (23), then any strategy (G, H) is the best response for (23), in particular to the strategy given by (43) and (44).

Let G^* and H^* be given by (43) and (44). Then, by (22), (19), (21) and (61), H^* are the eavesdropper channel components of a strategy by nature. Also, as $A > 0$, then, by (62), G^* corresponds to the main channel components for a strategy employed by nature. Then $\xi_i^0 + G_i^* - H_i^* = \omega$ for $i \in [1, n]$ where

$$
\omega = \frac{A}{n} + \frac{1}{k_*} \left(\sum_{i=1}^{k_*} \xi_i^0 - \bar{H} \right).
$$

Thus, any Alice strategy is the best response for (H^*, G^*), such as the strategy given by (23), and $\omega \bar{P}$ is the value of the game. Then, since

$$
\frac{A}{n} + \frac{1}{k_*}\left(\sum_{i=1}^{k_*}\xi_i^0 - \bar{H}\right)
$$

$$
= \frac{1}{n}\left(\bar{G} - \frac{(n-k_*)\sum_{j=1}^{k_*}\xi_j^0 - k_* \sum_{j=k_*+1}^{n}\xi_j^0 - \bar{H}(n-k_*)}{k_*}\right)
$$

$$
+ \frac{1}{k_*}\left(\sum_{i=1}^{k_*}\xi_i^0 - \bar{H}\right) = \frac{1}{n}\left(\sum_{i=1}^{n}\xi_i^0 + \bar{G} - \bar{H}\right).
$$

the value of the game is given by (37). This completes the proof of Theorem 6.

Proof of Theorem 8. Since the payoff is linear in P and concave on N, the strategies P^*, N^* of Alice and nature is in equilibrium (so, these strategies are the best response to each other) if and only if there are ω, ν such that

$$
P_i^* \begin{cases} \geq 0 & \text{for } \dfrac{1}{N_i^0 + N_i^*} - h_i^0 = \omega, \\ = 0 & \text{for } \dfrac{1}{N_i^0 + N_i^*} - h_i^0 < \omega, \end{cases} \tag{63}
$$

$$
\frac{P_i^*}{(N_i^0 + N_i^*)^2} \begin{cases} = \nu & \text{for } N_i^* > 0, \\ \leq \nu & \text{for } N_i^* = 0. \end{cases} \tag{64}
$$

Thus, by (64), if $P_i^* = 0$ then $N_i^* = 0$. It is reasonable to look for the optimal nature strategy in a subclass of strategies which hamper only the channels employed by Alice to transmit the signal, so for the strategies that have $P_i^* > 0$ then $N_i^* > 0$. Then, by (63), the optimal strategy N^* is of the form

$$
N_i^* = N_i(\omega) = \left[\frac{1}{h_i^0 + \omega} - N_i^0\right]_+, \tag{65}
$$

where $\omega = \omega_*$ is the unique root in $[-\min_i h_i^0, \infty)$ of the following water filling equation

$$
H_N(\omega) := \sum_{i=1}^{n}\left[\frac{1}{h_i^0 + \omega} - N_i^0\right]_+ = \bar{N}. \tag{66}
$$

By (64) and (65) we have that the Alice's optimal strategy is of the form

$$
P_i^* = P_i(\nu) = \begin{cases} \dfrac{\nu}{(h_i^0 + \omega_*)^2} & \text{if } N_i(\omega_*) > 0, \\ 0 & \text{otherwise} \end{cases} \tag{67}
$$

and $\nu = \nu_*$ can be found as the unique root of $H_P(\nu) := \sum_{i=1}^n P_i(\nu) = \bar{P}$. Thus,

$$\nu_* = \frac{\bar{P}}{\sum_{j:N_j(\omega_*)>0}(1/(h_j^0 + \omega_*)^2)}.$$

It is clear that the strategies defined by (65) and (67) satisfies the conditions (63) and (64). That is why they are the optimal ones. This completes the proof of Theorem 8.

Ensemble: Community-Based Anomaly Detection for Popular Applications

Feng Qian, Zhiyun Qian, Z. Morley Mao, and Atul Prakash

University of Michigan, Ann Arbor MI 48109, USA
{fengqian,zhiyunq,zmao,aprakash}@umich.edu

Abstract. A major challenge in securing end-user systems is the risk of popular applications being hijacked at run-time. Traditional measures do not prevent such threats because the code itself is unmodified and local anomaly detectors are difficult to tune for correct thresholds due to insufficient training data.

Given that the target of attackers are often popular applications for communication and social networking, we propose *Ensemble*, a novel, automated approach based on a trusted community of users contributing system-call level local behavioral profiles of their applications to a global profile merging engine. The trust can be assumed in cases such as enterprise environments and can be further policed by reputation systems, *e.g.*, by exploiting trust relationships inherently associated with social networks. The generated global profile can be used by all community users for local anomaly detection or prevention. Evaluation results based on a malware pool of 57 exploits demonstrate that Ensemble is an effective defense technique for communities of about 300 or more users as in enterprise environments.

1 Introduction

End-user systems can be difficult to secure for a variety of reasons. They are typically unmanaged: users download software, browser bugs, *etc.* In this paper, we focus on defending against a class of attacks in which popular applications are hijacked at run-time. In the past, this has led to wide-spread attacks such as the Skype worm [14] spread using Skype and buffer overflows in Outlook email clients to execute arbitrary code [7]. Traditional measures, such as anti-virus scanners [5], do not prevent such threats because the application code itself is unmodified. Prior work indicates that system-call level profiling [23,33,37] may help detect such attacks early but a significant barrier is a lack of sufficient training data to ensure low false positive rates.

In this paper, we present *Ensemble*, a novel unsupervised anomaly detection approach based on the idea of a trusted community of users contributing system-call level *local profiles* of an application to a common merging engine. The merging engine generates a *global profile* that captures the possible space of normal run-time behaviors of an application. The global profile can be used to detect or prevent anomalies in application behavior at each end-host in real time. The promise of this approach is that it helps overcome the problem of a lack of sufficient training data at each host and can be largely automated. The challenges are making such a system efficient, overcoming the differences in profiles due to factors such as variations in installation directories or hardware, and identifying the appropriate information to collect in profiles.

Y. Chen et al. (Eds.): SecureComm 2009, LNICST 19, pp. 163–184, 2009.
© Institute for Computer Science, Social-Informatics and Telecommunications Engineering 2009

The underlying hypothesis of Ensemble is that, *as the number of local profiles increases, the aggregate global profile tends to converge, thus revealing the normal behavior of the target application.* Most applications in our experiments were found to satisfy this property, though we also identified types of applications that would be exceptions. This paper makes the following contributions.

Handling diversity in execution environments. Various factors impact community-based profiling, *e.g.,* the same application at different hosts may be installed in different directories, run with different amount of memory, and use different number of CPUs. All these can cause variations in the system call traces with their parameters. We determined the types of data to use for generating behavioral profiles to handle these variations, while keeping profiles compact and representative of the application.

Analysis of the relationship between the community size and false positive rates. We first applied community-based anomaly detection to a community of 12 users using a normal, clean instant messaging application. The detailed system-call level data were sampled for 50 minutes during 5 hours with each local profile generated based on one minute of sampled data. We found that high false positive rates to be of significant concern, just as with single-host profiling using system calls. A testbed of virtual machines was subsequently used to study the impact of scaling up the system to a larger user community. We found that the techniques, in general, tend to become much more effective with larger community size. Significant reduction in false positive rates was observed after reaching approximately 300 users.

Techniques to reduce data transfer by sharing summary data generated by profiling applications. We show that while each host collects detailed system-call level data [23,26,36] for local analysis, it only needs to send a modest amount of local profile data per application (approximately, 4-5 KB/sec) to a common server to create community profiles.

A general interface. Our system provides a useful abstraction of a general interface for any target application to be protected. Multiple applications can subscribe to the Ensemble service.

Ensemble is currently implemented in user space in Windows. We used Detour library [27] by Microsoft Research to intercept system calls for target applications. For improved efficiency, as discussed in §4.2, Ensemble can be implemented as a service in the OS kernel. The rest of the paper is organized as follows: §2 overviews the related work; §3 describes the overall model of Ensemble; §4 details our implementation; and §5 evaluates the system experimentally. Finally, §6 discusses limitations before concluding in §7.

2 Related Work

Our work improves on existing work in the area of anomaly detection by exploring the applicability of community-based profiling to generate detailed run-time behavior profiles of applications at the system call level. Below we highlight some of the related approaches in malware detection and containment.

Anomaly Detection. One of the first studies on anomaly detection for applications was done by Forrest *et al.* [23,26,36]. They executed an application multiple times with different inputs to collect system call sequences and then used those to form the baseline behavior of the program. Any significant deviation from the baseline was considered as an anomaly. Many of the follow-up studies [16,24,21,25,37,33,20] incorporate machine learning techniques such as hidden-Markov model and neural networks. Later studies examined the inclusion of system call arguments [13] and call stack information [22]. Generating a common model from different runs is a non-trivial problem. In [16], Ballardie and Crowcroft explore several representative models, including frequency-based models, a data-mining approach, and a finite state machine approach.

All these above approaches can suffer from high false positive rate. The data collection process is typically manual or may take a long time to cover most normal behavior. If the application's normal behaviors are not adequately captured, unobserved normal behavior is likely misclassified as abnormal. While better machine learning algorithms [25,33] can help, one fundamental problem in making these schemes practical is the difficulty in getting sufficient training data to capture comprehensive application behavior.

Our work builds on the approaches in the above systems. The primary contribution is to show that if a large user community sharing their training data with an IDS at a fine-grained level, behavioral profiles can be generated that are much more complete and accurate than local profiles. One of the challenges we examined in extending the techniques to a community environment is that not just the inputs, but the operating environment for the software can be different. In our experiments, we allowed applications to be installed in random directories on various systems with diverse hardware configuration and varying workload imposed by other applications. We extend existing algorithms for combining profiles to handle likely variations.

Community-based Systems. The concept of "application community" [2] has been proposed to collaboratively diagnose and respond to attacks by generating appropriate configuration patches and filters. The goal is to generate a community-specific situation awareness gauge to predict imminent attacks. But it does not focus on anomaly detection as in our work to help prevent attacks.

A similar concept of "collaborative learning for security" [19] is applied to automatically generate a patch to the problematic software without affecting application functionality. However, the detectors used are static detectors without training, and the ways in which the community is utilized are limited to gathering detailed execution constraints in the binary, distributing the generated patch, and letting the user community evaluate them.

Companies, such as Symantec [12], Microsoft, and Google also leverage the notion of a community to help identify malware programs or spam emails [4] from user based feedback. Vigilante [17] and Sweeper [34] try to contain Internet worms by automatically detecting exploits. Both enable a user community to share their antibodies to prevent and stop future attacks from Internet worms.

In other application contexts, the concept of community has also been explored. Peer-Pressure [35] utilizes it to automatically detect and troubleshoot misconfigurations by assuming that most users in the community have the correct configuration. The Gamma System [32] was proposed to split the monitoring task among community users, enabling minimally intrusive program analysis and software evolution. Similarly, Cooperative Bug Isolation [31,30] leverages the community to do "statistical debugging" based on the feedback data automatically generated by community users.

In contrast to the above body of work, our work examines the effectiveness of applying the notion of community at a much finer-grained level. Instead of just combining binary feedback or signatures of worms, we integrate run-time behavioral profiles, consisting of system calls and associated parameters, of applications across a community of heterogeneous users. This allows us to extend anomaly detection to additional classes of software applications.

Signature based anti-virus (AV) software. In this approach, a user typically uses a signature database of known attacks, resulting in the advantage of negligible false positives. Unfortunately, it is difficult to maintain signatures covering new attacks. A study by Oberheide *et al.* [28] found that commercial AV software has a detection rate ranging from only 54.9% to 86.6% for attacks that occurred in the previous year. More importantly, the AV software had significantly poorer detection rates for more recent malware samples. This implies that anomaly based detection is still indispensable.

Behavior-based intrusion detection systems (IDS). These systems rely on pre-defined rules to detect anomalies in the run-time system behavior. They can better detect zero-day attacks that attempt to evade code-based signatures. But, getting the rules right can be difficult and therefore the rules tend to be relatively coarse-grained. For example, by default, McAfee VirusScan Enterprise 8.5i [5] Access Protection rule blocks outbound port 25 to filter malicious email programs. However, to get normal email applications to work, 42 popular email clients, such as outlook.exe and thunderbird.exe [11], are exempt. Note these applications are often the ones exploited.

3   Methodology

In this section, first we present high-level methodologies used in *Ensemble*, then explain them in detail in §3.1 to §3.3.

The goal of Ensemble is to detect application misbehavior, particularly caused by zero-day attacks. As the start point of our approach, we generate a *local profile* for each application instance. A *profile* is a summary of target application's inter-process communications and its behavior that can result in persistent changes (changes that survive across reboots) to the file system, the registry, network, and other system settings. They are abstracted from system call traces. Statistically, it can be seen as representative data points in the sample space containing all possible state changing behavior of the target application.

We envision that a large number of community users feed local profiles of an application to a central server, which periodically aggregates them into a *global profile*, depicting the application's normal behavior as a baseline. The global profile serves as a *classifier* that identifies anomalies using collected local profiles as training data.

To detect and prevent intrusion, we monitor the application behavior and compared it with the global profile continuously. An alarm is triggered when the application is about to perform an operation that does not match the global profile. The user can be alerted or the system can be configured to directly block the operation. Next we investigate several important challenges of our methodology.

3.1 Profile Generation

Local profiles. A local profile is generated from raw system call traces [26]. In Windows, system calls are undocumented, thus we use Windows API calls in our prototype. For simplicity we ignore a set of APIs that do not modify host file system or network state such as graphics and user interface API that are unlikely abused or even if abused will likely be visible through other APIs we monitor. Also, we only focus on operations executed by the target application given the profile is for a particular application, with the exception of the process dependency, as discussed below.

Global profiles. A global profile is distilled from multiple local profiles. We develop a taxonomy for APIs in terms of functionality (process dependency, file access, network access, *etc.*). For each category, corresponding records in local profiles are aggregated by key attributes (Table 1). An example of aggregating File Access category is shown in Table 2.

Table 1. Key attributes for primary categories in global profiles

Category	Key Attributes
Process Dependency	Src Process Name/Image Hash, Dst Process Name/Image Hash, Type $\in$ {Fork, Hook, File...}
File Access	Filename, Type $\in$ {Read, Write}
Registry Access	Registry key, Type $\in$ {Read, Write}
Network Connection	Remote IP, Remote Port, Protocol $\in$ {TCP, UDP, other}

Table 2. Example: aggregate records in local profile (a) into global profile (b)

(a) Local profiles

Profile ID	Filename	Bytes accessed	Type
1	a.dat	10	read
1	a.dat	15	read
1	b.dat	10	read
2	b.dat	10	read

(b) Global profiles

Filename	Type	Count by profiles
a.dat	read	1
b.dat	read	2

Among all the categories, the process dependency [29] depicts the interaction among processes of the target application and other processes. A local profile contains two types of dependencies: indirect and direct dependency. Indirect dependency, such as a file dependency (Process A writes file F, which is then read by Process B), requires an object (*e.g.*, a file or an IP address) as an intermediary. It is synthesized by correlating multiple API calls. Direct dependency, such as a fork dependency, takes place without an intermediary. It can be inferred from a single API call.

3.2 The Environment Diversity Challenge

For categories other than process dependency, the simplified methodology illustrated in Table 2 has limitations. For example, for a text processor, different users edit different files, thus the file access category is not aggregatable if naively using the filename as the key attribute. Similarly, a P2P client may talk to random IP addresses, leading the aggregation in the global profile to be a set of IP addresses each with very few occurrences. We apply two methods to address this challenge.

First, we use predefined rules to normalize the path and file names. For example, `c:\Documents and Settings\Alice\a.dat` is normalized to *USER-DOC*`\`
`a.dat`. This also helps protect the privacy of community users.

Second, our main solution is *Stack Signature*, which describes the stack history of the calling thread for each API call. The key idea is that the "random" events of the same functionality of a program such as sending a message or making a VoIP call in Skype, should be associated with a fixed set of execution paths that can be represented by call stacks. Based on this assumption, we introduce Stack Signature, a compact version of call stack. A Stack Signature is calculated by iterating all stack frames of the current thread and XORing their return addresses. In the case of recursive calls, return addresses occurring multiple times are counted once.

In a global profile, the relationship between stack signatures and objects (*e.g.*, filenames and IP addresses) can be characterized by a weighted bipartite graph, whose vertices are divided into two disjoint sets X and Y, where X is the set of stack signatures and Y is the set of objects. There is an edge $e : x \rightarrow y \in E$ where $x \in X$ and $y \in Y$, if and only if an event accessing object y has stack signature x in at least one local profile. Each element in X, Y and E has a weight, indicating its occurrence frequency in terms of the number of local profiles. Except for the process dependency which is fairly stable, we introduce stack signatures and use bipartite graphs as the data abstraction for all other categories.

We observe many such cases in our experiments. For example, at stack signature `0x61AE46F8`, QQ [8] – an instant messaging application may receive data from at least 64 different servers such as 121.14.*.*, 219.133.*.*, 58.61.*.*, via port 8000. All servers are found at Guangdong, China, where the headquarter of QQ is located. The size of received data is always a multiple of 10240 bytes.

3.3 Anomaly Detection

As described at the beginning of this section, Ensemble clients periodically pull the global profile from the server. The anomaly detection and prevention are performed continuously. Before each operation monitored by Ensemble is executed, the API call is intercepted and compared with the global profile using the following comparison algorithm.

1. **Threshold-based process dependency anomaly detection.** If a process dependency D is detected (*e.g.*, a *fork* or *file* dependency), we locate its frequency $f(D) = \frac{\text{\# of local profiles containing } D}{\text{\# of local profiles}}$ in the global profile, if $f(D) < th_{PD}$, where th_{PD} is a threshold, then D is regarded as abnormal.

2. **Stack signature analysis.** If the operation to be executed by the target application falls into other categories in Table 1, then its stack signature x is calculated, its object y is identified, and $e : x \rightarrow y$ is matched against the bipartite graph $B_G = \{X_G, Y_G\}$ in the global profile. Let the frequency of e and x in B_G be $f(e)$ and $f(x)$, respectively. (*i.e.*, $f(e) = \frac{\text{\# of local profiles containing } e}{\text{\# of local profiles}}$). Let the degree of x in B_G be $d(x)$. We also introduce thresholds th_e, th_x and deg_x. We determine whether e is an abnormal action by several tests searching for the predictable relation of the objects accessed by stack signatures.

Test 1. Does a fixed stack signature always access a fixed object? (*e.g.*, The program reads a constant configuration file) Formally, if $f(e) > th_e$, then e passes the test and no further tests are needed.

Test 2. Does a fixed stack signature always access different objects? (*e.g.*, A file editor may open different files) Formally, if $f(x) > th_x$ and $d(x) > deg_x$, then e passes the test and no further tests are needed. This handles the "the Environment Diversity Challenge."

Some challenges arise, as we observe that in multiple executions of the same application, a single object may be accessed by different stack signatures forming one or more clusters. Figure 1 is an example of reading file `ServUCert.key` in 1,305 executions by Serv-U 5.0.0.0 (a commercial FTP server). The stack signatures form a cluster ranging from `0x1019A500` to `0x1019A5FF`. We conjecture two reasons: (1) The locality of object access. The same object is often accessed at close-by instruction addresses. For example, the code in Figure 2 is common in C programs. The consecutive calls of `fread` satisfy the locality principle. (2) The accumulation of varieties. A

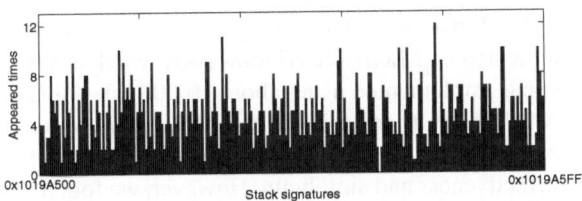

Fig. 1. Frequency of accessing `ServUCert.key` from different stack signatures in 1305 local profiles

```
FILE * ifs = fopen("data.dat", "r");
fread(&para1, sizeof(para1), 1, ifs);
if (para1 == 1) fread(&para2, sizeof(para2), 1, ifs);
/* read other parameters */
fclose(ofs);
```

Fig. 2. Sample code of reading a file

signature is calculated by XORing return addresses of n stack frames with each frame having a variety of k_i, the total variety can be as large as $\prod_{i=1}^{n} k_i$.

Motivated by the above observation, we add two additional tests to reduce false positives.

Test 3. Does a cluster of stack signatures access a fixed object? We define a cluster by a window centering at x: $X_{win} = \{z \in X_G \| |z - x| \leq winSize\}$. Formally, if $\sum_{z \in X_{win}} f(e' : z \to y) > th_e$, then $e : x \to y$ passes this test.

Test 4. Does a cluster of stack signatures access different objects? Formally, if $\sum_{z \in X_{win}} f(z) > th_x$ and $\sum_{z \in X_{win}} d(z) > deg_x$, then e passes this test. It is a further generalization of Test 3.

Test 3 and 4 may introduce false negatives; however, they are expedient alternatives in the situation where the number of samples is limited. Ideally, when the global profile contains a large enough sample space, Test 3 and 4 can be replaced by Test 1 and 2, respectively, since the range of stack signatures is finite. Figure 3 illustrates four patterns in the global profile, corresponding to the above four tests.

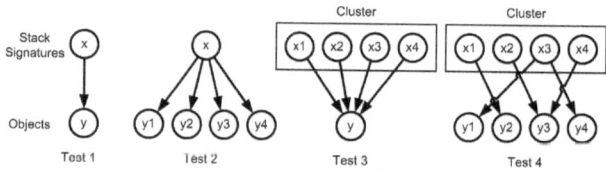

Fig. 3. Four API invocation patterns

4   Implementation

The architecture of our Ensemble prototype is illustrated in Figure 4. It is designed to perform online anomaly detection using continuously updated global profiles and generated local profiles. Existing work is mostly evaluated in Linux environments while our system is implemented on Microsoft Windows XP, which is a more common attack target. Our prototype is implemented using about 10,000 lines of C++ code.

In our design, we initially tried to implement Ensemble by using system call sequences (N-gram previously proposed [23,26,36]) as the representation of local profiles, due to its claimed effectiveness and simplicity. However, we found that N-gram has surprisingly low convergence speed for Windows API sequences in terms of obtaining the model of application's normal behaviors, likely due to a much larger sample space than in Linux (the number of Windows APIs is 6 times the number of Linux syscalls). We estimate two reasons for such big discrepancy: first, there are distinct difference between Unix/Linux system calls and Windows APIs; second, modern applications are becoming more and more complicated. System calls may be a too find-grained characterization of program behavior. Note that a lot of researchers apply N-gram algorithm on virus or malwares, whose binary sizes are much less than legitimate applications. Therefore, instead we resort to the simpler frequency-based model as described in §3.1 that has a faster convergence behavior.

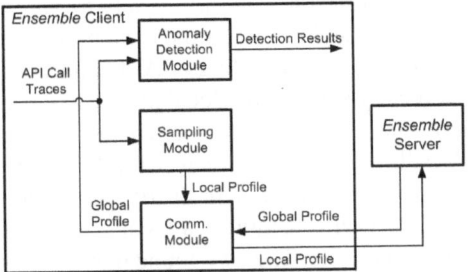

Fig. 4. The Ensemble Architecture

4.1   Generating Profiles and Anomaly Detection

We used the *Detour* Library [27] to monitor and log 106 APIs calls related to file system (26), registry (8), file mapping (6), messages (8), thread (4), process (8), network (13), pipe (6), hook (3), clipboard (3), system time (6), DNS (2), handle management (2) and user accounts management (11), most of which are Windows specific. To the best of our knowledge, they cover most APIs that can cause inter-process communications, or result in persistent changes to the file system, the registry, the network, and other system settings. Note that it is fairly easy to include new APIs to the framework. We generate stack signatures using the `StackWalk64` function in Windows Debugging Library.

Given the raw API traces and their stack signatures, the local profiles are generated as described in §3.1 (for process dependency) and §3.2 (for other categories). We implemented seven categories for profiles. (1) process dependency, (2) file access, (3) directory access, (4) registry access, (5) network connection, (6) DNS, and (7) IP prefix access. For (1), we handle 4 types of direct process dependencies: send message, set hook, create/terminate/suspend process (thread) and write/read/alloc/dealloc process memory, and 8 types of indirect dependencies: files, registry, file mapping, network, named pipes, anonymous pipes, system time and clipboard. The transformation from API traces to other categories (*e.g.*, file access, network access) is trivially done by translating API parameters.

The global profile is generated by grouping various local profiles. Except for the process dependency, which is represented by a table like Table 2(b), other categories are represented using bipartite graphs (stack signature → object names).

Our anomaly detection algorithm described in §3.3 is very efficient. For process dependency, the dependency inference and frequency look up is $O(1)$ in run time using hash tables. For other categories using bipartite graphs, the computational complexity for Tests 1 and 2 is $O(1)$; while Test 3 and 4 are also $O(1)$ given that the window size is a small constant.

4.2   Operational Model

Finally, we present an overview of Ensemble's operational model. At each client, Ensemble is running as a system service and is transparent to the target application. CAPTCHA is used when subscribing or unsubscribing Ensemble services to prevent tampering from bots.

When the application is running, the *Ensemble sampling module* periodically logs its API calls with stack signatures[1] and generates the local profile (*e.g.*, every 3 hours, one local profile is generated from 1-min sampling of API call traces). The *Ensemble communication module* periodically submits the local profile to the server, and also fetches the global profile from it. The *Ensemble Anomaly Detection Module* keeps monitoring target application's API calls and matching them with the global profile. If an alarm is triggered, the requested operation is denied, or the decision is left to the user.

Initially our anomaly detection is sampled: a local profile is generated periodically and compared with the global profile. Then we found that even if the anomaly detection is performed continuously, the extra overhead is acceptable (less than 2%), given that in most cases, the applications' API calls are not invoked in a "bursty" manner.

The Ensemble server can be maintained either on a large scale (*e.g.*, by the application vendor), or on a small scale (*e.g.*, within an enterprise network). Its tasks include collecting local profiles, generating the global profile and other management functionalities. Ideally, each version of the application should have its own global profile. Depending on the specific application, one global profile may also characterize several versions with minor differences.

4.3   Limitations of the Prototype

Our current prototype has the following limitations which are not fundamental to our design. At the client side, the sampling module is implemented at the user level, using a third-party library. For future work we plan to move the entire system into Windows kernel. At the server side, in order to prevent pollution of global profiles, we plan to investigate the use of reputation systems that establish trust among community users. Currently, we envision our system to be mainly deployed in enterprise environments where trust can be assumed.

The latest Windows Vista adopts Address Space Load Randomization (ASLR) technique [1], which hampers the functionality of Stack Signatures. We can address this problem by using the relative offset of the return address from the module's start address, together with the module signature. We plan to explore this as future work.

5   Evaluation and Experiments

In this section, we systematically evaluate Ensemble. First we describe a small-scale deployment for a community of 12 users (§5.1). Based on the negative results due to the limited size of the community, we introduce our testbed and target applications used for experiments (§5.2), then analyze the generated local profiles (§5.3) and the resulting global profiles (§5.4). Next, we measure false positives (§5.5) and estimate false negatives using a recent malware collection (§5.6). Finally we present the performance evaluation of our system (§5.7).

5.1   Small Scale Real Deployment

We deployed Ensemble among 12 real users, using *Windows Live Messenger* (MSN) as the target application. All users were using Win XP SP2 but with different software

---

[1] To capture process dependency, some APIs called by other processes also need to be logged.

and hardware configurations. Before the experiment, we manually upgraded their MSN to the same version (2008 Build 8.5.1302.1018) and ensured the systems are virus-free. Users were not familiar with technical details of Ensemble, and were told to use MSN as usual. For each user, we collected 50 API call traces, each lasting 1 minute, during a 5-hour period. We used this dataset to evaluate false positives.

We used 5-fold cross validation on 600 traces to evaluate false positives. For each trace in the test group, if any API call triggered a false alarm, then the local profile was counted as one false positive. For the parameters in §3.3, we empirically set $th_e = 1\%, th_x = 1\%, deg_x = 10, winSize = 4KB$ (We tried different parameters such that $th_e < 2\%, th_x < 2\%, deg_x < 20$, and obtained similar results). We found that the false positive rates were too high to be accepted (greater than 30% for file access and registry access). The reason is that 12 users are not sufficient to form a community to cover diverse application behavior.

5.2 Experimental Infrastructure

To test the impact of a larger community, we created an automated testbed to simulate a community environment. The idea is simple: to execute the target application multiple times on the testbed. In each execution, a local profile is created and fed to the global profile generator, as if it was submitted by a real community user. Then we use the global profile to test against normal and abnormal behaviors and evaluate false positives and negatives. We have two design goals for the testbed.

- **Diverse User Behaviors.** Random user actions are injected during each trial. The distribution of the randomness should roughly conform to that of a real community.
- **Diverse System Environment.** During each trial, the system environment should also vary to simulate hardware and software variations in a real community. For example, a VoIP client may adjust its voice encoding strategy according to available network bandwidth, leading to different local profiles.

We manually created a Finite State Machine (FSM) for each target application to describe most of its main functionalities from an end user's perspective. FSM can be generated in a more automated fashion by combining user traces and adding some perturbation to include additional usage behavior. Despite the manual effort, FSM based representation for understanding application usage, even approximate, can aid in generating more diverse usage scenarios for a given application. Figure 5 is a simplified FSM for MSN. In each automated execution, the testbed partially iterates the FSM based on a Markov chain model, which characterizes the popularity of application's different functionalities. Each state transition $S_x \rightarrow S_y$ in the FSM represents a user action. A weight is assigned to e indicating the probability that the next state is S_y given the current state is S_x. For example, in Figure 5, "Login" is the initial state where the user starts the application. The probability that the user successfully logs in ($\frac{10}{1+2+10} = 77\%$) is much higher than the probability that the user enters an invalid ID or password (8%).

The testbed not only randomly chooses the action, but also executes some actions with randomness. For instance, it is able to operate an instant messenger by selecting a random user and chatting with him/her via random text messages, emotion icons, handwritings or Flash winks. In another example, the "make phone call" action in Skype is carried out by dialing a number from 3000 toll-free numbers we collected.

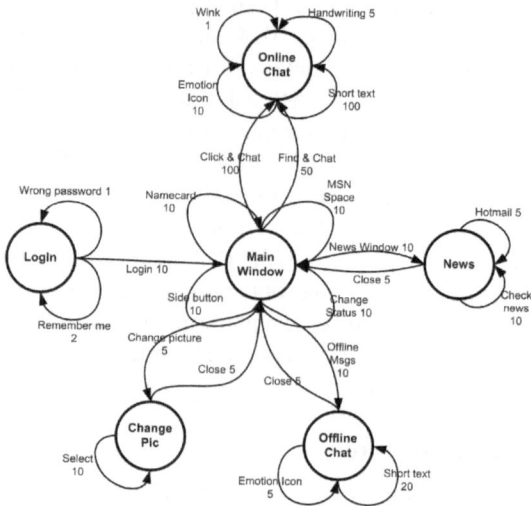

Fig. 5. A simplified finite state machine of MSN. Labels on edges indicate state transition probability.

We admit that our approach contains subjective elements and thus may not perfectly simulate a community environment. However, a community itself is a set of subjective users and has a tendency to change from time to time. Also, we will show in §5.3 the heavy-tailed distribution of simulated users' behaviors, which are usually the case in a real community.

To tackle the system environment randomness, the testbed automatically changes the hardware/software configurations for each trial. All experiments were conducted on virtual machines (VMware 6.0.2) for ease of management. The varied configuration includes memory, number of processors, installed software, existing running processes, system workload, firewall settings, system time, network bandwidth, DNS server, *etc.*

The testbed includes a FSM script parser, an action executor that maintains the state synchronization and sends mouse/keyboard input to the target application, a configuration manipulator that changes the system environment and a communicator that communicates with the Ensemble kernel. The testbed is built using about 3,000 lines of C++ code.

We chose four applications running on Microsoft Windows XP SP2 as our initial target applications: *Skype* 3.5.0.239; *Windows Live Messenger* (MSN) 2008 Build 8.5.1302.1018; *Tecnet QQ* [8] (2007 Beta 4, 7.0.374.204), an ICQ client with typically more than 30 million daily online users in China; *Serv-U* [9] (5.0.0.0), a commercial FTP server. These applications were selected due to their popularity and past history of attacks targeting them.

5.3 Local Profiles

Table 3 shows the number of local profiles, sampling times and API log sizes of local profiles of each target application. The sampling time was set to conform to a Gaussian

Table 3. Statistics of local profiles

Target App	# of local profiles	Sample Time (Mean)	Sample Time (Std Dev)	API Trace Size (Mean)	LP Size (Mean)
Skype	550	60 secs	5 secs	3.40MB	0.20MB
MSN	1298	75 secs	5 secs	1.17MB	0.09MB
QQ	1118	60 secs	5 secs	1.18MB	0.09MB
Serv-U	1305	45 secs	5 secs	0.23MB	0.03MB

Table 4. Statistics of global profiles

Target App	Process Dependency	File Read	File Write	Dir Read	Dir Write	Reg Read	Reg Write	Connections	IP Prefixes	DNS Query
Skype	8	209	237	178	208	4,587	328	135,844	115,864	0
MSN	10	2,884	244	795	90	54,506	2,749	6,417	554	0
QQ	4	6,549	8,029	6,541	8,021	59,491	229	11,867	9823	10,691
Serv-U	1	2,609	835	305	7	146	0	23,295	2	1

distribution. The sampling process started either at or after the application starts, and stopped either at or before the application terminates. The entire collection of local profiles lasted for one week.

As mentioned, we created randomness during each trial to simulate different user behavior in the community. Thus each "user" may explore a different subset of the application functionalities. Figure 6 illustrates the distribution of FSM patterns for Skype, MSN and QQ. A pattern defines the states iterated by the testbed in a single trial. If there are n possible states in FSM, then there exists $2^n - 1$ possible patterns $(0, 0, ..., 0, 1), ..., (1, 1, ..., 1, 1)$. For pattern $(a_1, a_2, ..., a_n)$, $a_i = 1$ iff the i-th state is visited at least once in a trial. The heavy-tailed distributions in Figure 6 demonstrate the diversity of user behaviors generated by our testbed, as well as the similarity of most users' behaviors. Although this may not exactly match the actual user behavior, we believe our method adds sufficient randomness to closely approximate general user activities.

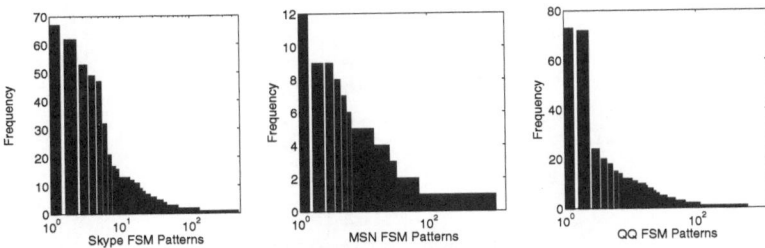

Fig. 6. FSM Pattern distribution for Skype (474 patterns), MSN (1137 patterns) and QQ (584 patterns). The X-axis is log-scaled.

5.4   Global Profiles

Table 4 presents statistics of global profiles. The numbers in the table are the numbers of process dependencies and, for other categories, the number of edges in the bipartite graphs.

The process dependency categories of QQ, MSN and Skype are shown in Figures 9(a), 10, and 11(a), respectively. Only parts with solid line represent the observed dependencies; while the dotted lines indicate detected misbehavior (§5.6). The percentage on the edge denotes its occurrence frequency. The size of bipartite graphs is usually much larger.

Figure 7 shows examples of the bipartite graphs. For each subfigure, the upper part X is the set of stack signatures; the lower part Y is the set of objects (registry keys, directory names, *etc.*), which are represented by a number (object ID). The numbers in square brackets are the frequencies.

- Subfigure (a) is a common case where a fixed stack signature accesses a fixed object. For example, stack signature 0x7BF74721 always reads 3 registry keys:
  \REGISTRY\MACHINE\SOFTWARE\Classes\QQCPHelper...
  \REGISTRY\MACHINE\SOFTWARE\Classes\CLSID\23752AA7...
  \REGISTRY\MACHINE\SOFTWARE\Classes\CLSID\23752AA7...
- Subfigure (b) illustrates a random event problem. For each trial, Stack signature 1814742014 (0x6C2AC3FE) writes different registry keys under
  \REGISTRY\MACHINE\SOFTWARE\Classes\CLSID\ and
  \REGISTRY\MACHINE\SOFTWARE\Classes\TypeLib\.
- Subfigure (c) illustrates the slight variation of stack signatures, as explained in §3.3. We can observe two clusters of stack signatures in subfigure (c): 4582218??, 1819194???. Both clusters access the user cookie directory *USER-DOC*\cookies.

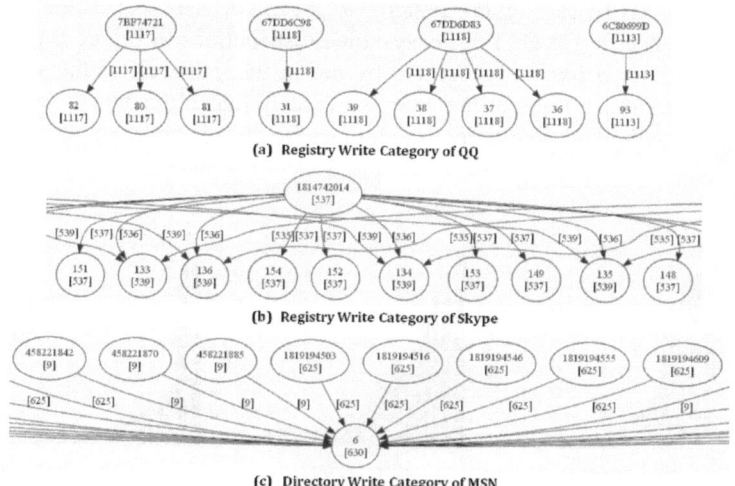

(a) Registry Write Category of QQ

(b) Registry Write Category of Skype

(c) Directory Write Category of MSN

Fig. 7. Examples of bipartite graphs. From top to bottom: (a) Registry write category of QQ (b) Registry write category of Skype (c) Directory write category of MSN.

5.5   False Positives

We used the same methodology (5-fold cross-validation) and the parameters as in the real deployment (§5.1) to evaluate the false positives for the testbed. In Table 5, the column "LPs" indicates the number of local profiles in the test group; the columns "Worst" and "Best" indicate the highest and lowest number of false positives (traces that contain at least one API call that triggers the false alarm), respectively, in 10 independent experiments (each experiment has 5 passes).

Table 5. Coarse-grained false positives (counting the number of local profiles)

Target App Category	Skype			MSN			QQ			ServU		
	LPs	Worst	Best	LPs	Worst	Best	LPs	Worst	Best	LPs	Worst	Best
Process Dependency	110	0	0	262	0	0	226	1	0	196	0	0
File Read	110	0	0	262	0	0	226	0	0	261	0	0
File Write	110	0	0	262	0	0	226	0	0	261	0	0
Directory Read	110	0	0	262	0	0	226	0	0	261	0	0
Directory Write	110	0	0	262	0	0	226	0	0	261	0	0
Registry Read	110	0	0	262	4	2	226	1	0	261	0	0
Registry Write	110	0	0	262	1	0	226	0	0	0	0	0
Connections		N/A		262	4	2	226	1	0	261	0	0
IP Prefixes		N/A		262	0	0	226	0	0	261	0	0
DNS Query	0	0	0	0	0	0	226	0	0	261	0	0

Table 6 presents a fine-grained false positive measurement. Similar as above, we employed 5-fold cross-validation and the experiment was repeated for 10 times using the same parameters. In Table 6, the column "Avg E" denotes the average number of API calls[2] in the test group, which were fed into Ensemble Anomaly Detection Module; the columns "Worst" and "Best" indicate the highest and lowest numbers of API calls that are mistakenly detected as abnormal, respectively.

For Skype and ServU, no false positives were observed. For MSN and QQ, although their fine-grained false positives of Registry Read and Connections categories were slightly higher even when the false positive rate converges (shown in Figure 8), the mistakenly detected API calls concentrated in a few local profiles (Upon manual inspection of the logs, it was highly possible that during the generation of these local profiles, the application terminated unexpectedly.). Ideally, if they were indeed application's natural behaviors, then as the pool of training data becomes larger, the initial "strange" behaviors will become normal, and the large size of training data is exactly the advantage of a community.

When we were testing Skype, it produced unacceptable false positive rates for network-related behavior (two categories whose false positives labeled as "N/A" in Table 5 and Table 6). Upon manual inspection, we found that the stack signatures from network related APIs were almost uniformly distributed in the entire address space, and

---

[2] To be precise, "Avg E" is the number of process dependencies or the number of edges in the bipartite graph.

Table 6. Fine-grained false positives. (counting the number of edges in PDGs or bipartite graphs)

Target App Category	Skype Avg E	Worst	Best	MSN Avg E	Worst	Best	QQ Avg E	Worst	Best	ServU Avg E	Worst	Best
Proc. Dep.	498	0	0	2203	0	0	844	1	0	196	0	0
File Read	13271	0	0	31650	0	0	40578	0	0	6290	0	0
File Write	1938	0	0	3623	0	0	40138	0	0	3473	0	0
Dir Read	10214	0	0	22292	0	0	39903	0	0	2758	0	0
Dir Write	1650	0	0	2711	0	0	40114	0	0	1810	0	0
Reg Read	43398	0	0	611294	55	37	415532	1	0	23943	0	0
Reg Write	33639	0	0	25441	1	0	23805	0	0	0	0	0
Connections	N/A			23398	12	4	18074	11	0	7194	0	0
IP Prefixes	N/A			17974	0	0	16385	0	0	516	0	0
DNS Query	0	0	0	0	0	0	17085	0	0	258	0	0

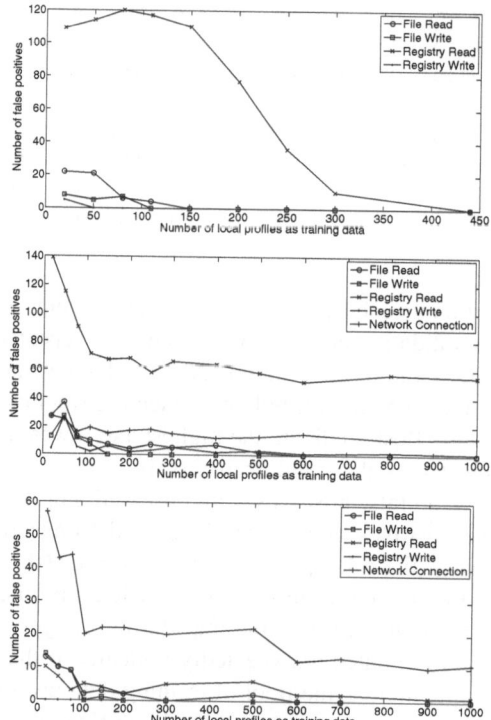

Fig. 8. Convergence of fine-grained FP as local profiles increase. (Top: Skype; Middle: MSN; Bottom: QQ)

the dumped stack frames were also abnormal. Based on our estimation, Skype may employ some obfuscation techniques to protect their code against reverse engineering [10]. In summary, we believe that the false positives of Ensemble are acceptable.

Furthermore, we used 600 API call traces obtained in real deployment to test against the global profile generated by 1,298 MSN local profiles from the testbed. We obtained false positive rates of 0% (process dependency), 6% (file read), 4% (file write), 2% (directory read), 1% (directory write), 11% (registry read), 6% (registry write), 9% (connections) and 3% (IP prefixes), using the metric in Table 5. Upon manual inspection, the main cause of false positives was the incompleteness of our FSM model, in which some use cases such as video chat were not covered.

We also measured the relationship between the community size and the false positive rate using a 5-fold cross-validation, and presents the results using the worst case (the highest number of false positives in 10 independent experiments). As shown in Figure 8 for three applications, it is clear that the fine-grained false positive rate significantly decreases with increasing number of local profiles, and converges to a stable value (We discussed the high false positives of QQ and MSN earlier in this section). A real active community is believed to have orders of magnitude of more local profiles submitted by users, thus ensuring a low false positive rate.

5.6 False Negatives

We evaluate false negatives on a total of 57 known malware programs and exploits for each target application by performing online comparison between the application behavior monitored in real time and the global profile, which was generated from local profiles described in Table 3. We used the same parameters as in the false positive evaluation.

Table 7 summarizes our selected malwares and exploits against target applications. They were selected from a malware collection obtained from honeypots, Web page crawling, and spam traps. It seems that these 57 malwares and exploits have somewhat common exploit techniques. However, we argue that the core merit of anomaly detection system is that, no matter how sophisticated an attack will be, as long as the application's behavior deviates from the baseline, the anomaly can be detected without prior knowledge.

For QQ, we tested 27 password stealer trojans, all of which were detected by Ensemble. Figure 9 shows a representative case. The trojan process (1180.EXE) sets a keyboard hook to QQ.EXE and tries to log users' keystrokes. The trojan also caused abnormal file accesses: KERNEL32.DLL and ISIGNUP.SYS. The latter was extracted by the trojan.

We attempted two buffer overflow exploits using the Metasploit framework [6] against Serv-U. Both exploits were detected by Ensemble. One exploit caused ServU to spawn a command line shell, which could be remotely controlled by the attacker. Another exploit

Table 7. Our malware/exploit collection used in false negative evaluation

Target App	# of Malwares/Exploits	Descriptions
Skype	3	Worm
MSN	25	Worm, password trojan
QQ	27	Password trojan
Serv-U	2	Buffer overflow exploits

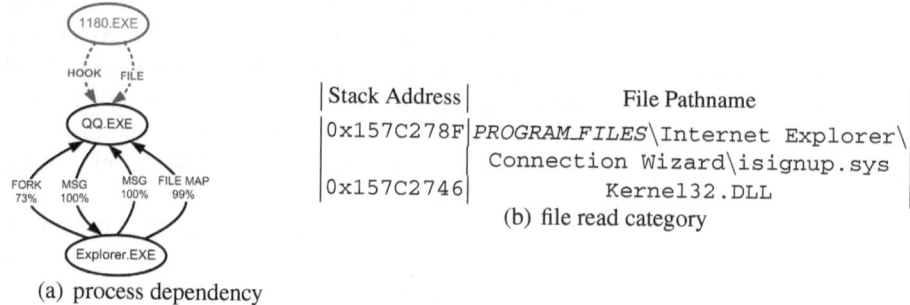

Stack Address	File Pathname
0x157C278F	*PROGRAM_FILES*\Internet Explorer\ Connection Wizard\isignup.sys
0x157C2746	Kernel32.DLL

(b) file read category

(a) process dependency

Fig. 9. Anomaly detection results of the QQ trojan

Table 8. Anomaly detection results of the Serv-U buffer overflow exploit (unusual file and network access)

Stack Signature(s)			Object Type	Object Name
6607A2DC	6606A17F		File Read	*IE_TEMP*\Content.IE5\H0SBCDN6\putty.exe
112CF1F2	660AC700	660AC7D1	File Write	*IE_TEMP*\Content.IE5\H0SBCDN6\putty.exe
11201534	11211697		File Write	*SYSTEM32*\a.exe
6606A17F	6607A2DC		Dir Read	*IE_TEMP*\Content.IE5\H0SBCDN6\
11211697	11201534		Dir Write	*SYSTEM32*\
660AC7D1	660AC700	112CF1F2	Dir Write	*IE_TEMP*\Content.IE5\H0SBCDN6\
60814BDC	17A77DFF		Connection	193.201.200.66:80 TCP
1B772B23	1B7729D0		IP Prefix	193.201.200.0/23

(Omitted: 106 registry read edges and 26 registry write edges)

made ScrvU to download a file and execute it. The exploit was constructed in Metasploit by providing a URL pointing to an executable file (in our experiment, the downloaded executable was putty.exe, which was then renamed to a.exe and executed). In Table 8, a series of events before the execution of a.exe were clearly revealed by failing to match abnormal edges with bipartite graphs in the global profile.

For MSN, we tested 25 worms that hijack MSN to send out malicious contents to the user's contacts. In one example shown in Figure 10, the malware process with a long file name tried to modify registry keys and files that MSN read later.

Skype consists of Skype.exe and SkypePM.exe. We tested three worms that abused the Skype API to send malicious links to deceive receivers to click them. Since the Skype API on Windows is implemented using the message mechanism, Ensemble detected the worm named StWinsDat.exe that sent messages to Skype.exe, as shown in Figure 11. Ensemble also detected that Skype read the file StWinsDat.exe from two stack addresses that never appeared in the global profile.

As part of the real-deployment in §5.1, we manually executed 25 MSN worms on 3 real machines with different configurations. All abnormal behaviors were detected by Ensemble. Furthermore, it seems that all above anomalies can be covered by the process dependency category. However, we argue that other categories are necessary. For one reason, it is possible that some attacks can happen without process dependency (*e.g.*, anomalies caused by network packets such as Apache-Knacker exploit [3]).

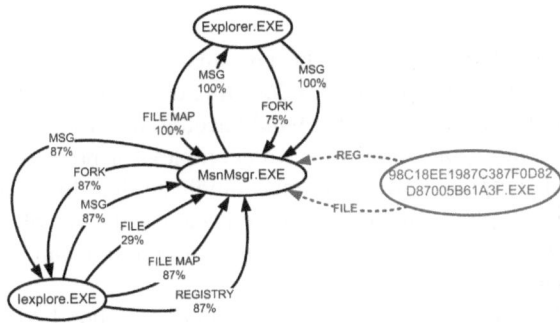

Fig. 10. Anomaly detection results of the MSN worm (process dependency)

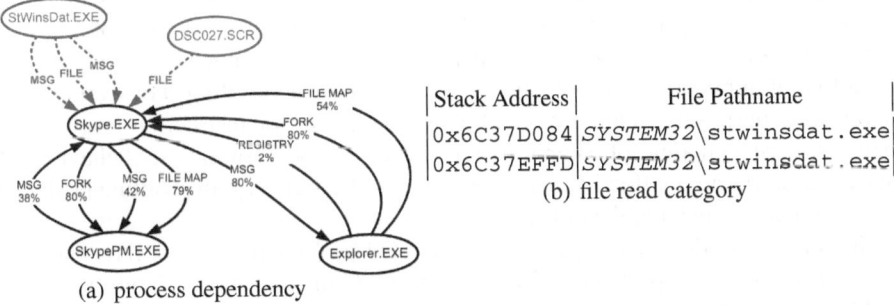

(a) process dependency

Fig. 11. Anomaly detection results of the Skype worm

Furthermore, as shown in Figure 9(b), Figure 11(b) and Table 8, other categories provide more detailed information about the anomaly.

5.7 Performance Evaluation

Using four target applications mentioned above, we measured the overhead of our prototype in terms of time and space. The evaluation was done on a commodity Dell Inspiron 530 PC (2.33G Core2 Duo CPU, 2GB memory, with WinXP SP2 installed). We believe that the overall overhead is acceptable. Extra delay incurred by local profile collection is less than 15%. Note that this happens infrequently (*e.g.*, 1 minute per 3 hours), and Ensemble does not collect local profiles for two applications simultaneously. Extra overhead caused by anomaly detection is less than 2%. The logging size of API traces is less than 0.25 MB/min per application. The global profile size is less than 10MB per application. Like software update, the Ensemble server can transfer a "patch" of the new version of the global profile, with a much smaller size.

6 Limitations of Ensemble

While we found Ensemble's approach to be a promising direction for addressing a difficult problem of using run-time profiles for detecting code injections and other run-time anomalies, we also noted limitations that would need to be addressed in the future.

We expect that some applications to be too complex for profiles to converge using limited system-call sampling. Our experiments indicate that this is the case for complex plug-in enabled applications such as IE and MS Word since plug-ins may behave differently from the original applications. Additional sampling and larger communities may help in such cases.

We plan to evaluate Ensemble in a real community with hundreds of users. Privacy concerns must be addressed, even though only summary data about system calls is exchanged with a server.

If a significant fraction of community of users mounted a coordinated attack to pollute the global profile, it is conceivable that the global profile can be corrupted. This is more likely in open communities, where sybil attacks [18] are possible. In closed communities as in enterprise environments, such attacks are much less likely.

Different applications may require different types of profiling. For example, if an application purposely randomizes addresses at function or instruction level (*e.g.*, the network access module of Skype mentioned in §5.5 to obfuscate its behavior), then stack signatures are ineffective. Alternative methods, such as path profiling [15], can be added to handle such applications.

In our design, the stack signature is generated by XORing unique return addresses of stack frames. The probability of collision is non-negligible in 32-bit OS, but very unlikely in 64-bit systems which are becoming increasingly popular.

6.1 Over-Generalization

Each application has a set of "normal behaviors" (true baseline). False negative may happen when the detector-defined normal behaviors go beyond the true baseline (*i.e.*, over-generalized) because the features or methods are not well-chosen or the model is not precise enough (*i.e.*, an imperfect detector). For almost all practical IDS, the detector-defined normal behaviors are broader than the true baseline, thus allowing mimicry attacks. This is a problem with any detectors not just ours. The aggregation process should not introduce much additional over-generalization. Consider the aggregation of local profiles whose diversities are caused by: *(i)* User randomness. Different users can generate different profiles but they mostly fall within true baseline assuming profiles are trusted (User randomness can be regarded as exercising different normal execution paths in the application). *(ii)* System environment randomness. We admit that different system environment may have different set of "normal behaviors". However, this should introduce limited over-generalization, if any at all. In the worst case, we can have separate aggregations/pools for different OSes and software versions as mentioned in §4.2.

6.2 Mimicry Attacks

A perfect detector should leave no opportunity for mimicry attacks which are due to over-generalization. Note that the aggregation process is independent of what features or approaches are used for anomaly detection. The existence of mimicry attack is mainly due to limitations in feature selection and detection techniques, not in profile aggregation. Our focus is to show that with a reasonable detector, how we can reduce false positives rather than making the features rich enough to eliminate the possibility for mimicry attacks.

7 Conclusions

We have described the design of Ensemble, an unsupervised anomaly detection and prevention system relying on a user community to detect or prevent anomalies in popular applications. Local behavioral profiles are combined into a global profile, which can be used to detect or prevent code-injection or behavior-modifying exploits. Hosts participating in Ensemble only need to contribute summary run-time profile data (about 0.5 MB) periodically. Ensemble addresses the problem of merging profiles from hosts that may have different operating environments. From evaluation based on 57 test exploits for four candidate applications, we found that the quality of global profiles, and the resulting false positive rate, significantly improves as the community size grows to approximately 300 users, demonstrating that the use of communities is a practical way to automatically generate behavioral profiles without much manual training, and the resulting behavioral profiles are effective for run-time anomaly detection and prevention.

References

1. Address space layout randomization, http://blogs.msdn.com/
2. Application Community, http://www.darpa.mil/
3. C. CAN-2003-0245. Apache apr-psprintf memory corruption vulnerability, http://web.nvd.nist.gov/
4. Gmail: We're working as a community, give your support!, http://news.softpedia.com/
5. McAfee Anti-virus software, http://mcafee.com/
6. Metasploit framework, http://www.metasploit.com
7. Microsoft Outlook Buffer Overflow in Processing TNEF Messages Lets Remote Users Execute Arbitrary Code, http://securitytracker.com/
8. QQ Instant Messenger, http://im.qq.com
9. Serv-U FTP Server, http://www.serv-u.com/
10. Should we be afraid of Skype, http://www.ossir.org/
11. VirusScan Enterprise 8.5i Access Protection rule blocks outbound SMTP mail on Port 25, https://knowledge.mcafee.com/
12. Malware flood driving new AV (December 2007), http://www.infoworld.com/
13. Kruegel, C., Mutz, D., Valeur, F., Vigna, G.: On the Detection of Anomalous System Call Arguments (2003)
14. Arak, V.: On the worm that affects Skype for Windows users (September 2007), http://share.skype.com/
15. Ball, T., Larus, J.: Efficient Path Profiling. In: 29th Annual IEEE/ACM International Symposium on Microarchitecture (1996)
16. Ballardie, T., Crowcroft, J.: Multicast-specific Security Threats and Counter-measures. In: Proc. of the IEEE Symposium on Security and Privacy (1999)

17. Costa, M., Crowcroft, J., Castro, M., Rowstron, A., Zhou, L., Zhang, L., Barham, P.: Vigilante: end-to-end containment of internet worms. In: SOSP (2005)
18. Douceur, J.R.: The Sybil Attack. In: Peer-To-Peer Systems: First International Workshop (2002)
19. Ernst, M.: Self-defending software: Collaborative learning for security, http://norfolk.cs.washington.edu/
20. Eskin, E.: Anomarly Detection over Noisy Data using Learned Probability Distributions. In: International Conference on Machine Learning (2000)
21. Eskin, E., Lee, W., Stolfo, S.J.: Modeling system calls for intrusion detection with dynamic window sizes. In: Proceedings of DARPA Information Survivability Conference and Exposition II (DISCEX II) (2001)
22. Feng, H.H., Kolesnikov, O.M., Fogla, P., Lee, W., Gong, W.: Anomaly Detection Using Call Stack Information (2003)
23. Forrest, S., Hofmeyr, S.A., Somayaji, A., Longstaff, T.A.: A Sense of Self for Unix Processes. In: IEEE Symposium on Security and Privacy (1996)
24. Ghosh, A., Wanken, J., Charron, F.: Detecting anomalous and unknown intrusions against programs. In: Proc. of the 1998 Annual Computer Security Applications Conference, ACSAC 1998 (1998)
25. Ghosh, A.K., Schwartzbard, A., Schatz, M.: Learning program behavior profiles for intrusion detection. In: Proceedings of the 1st conference on Workshop on Intrusion Detection and Network Monitoring, vol. 1 (1999)
26. Hofmeyr, S.A., Forrest, S., Somayaji, A.: Intrusion detection using sequences of system calls. Journal of Computer Security (1998)
27. Hunt, G., Brubacher, D.: Detours: Binary Interception of Win32 Functions. In: Proceedings of the 3rd USENIX Windows NT Symposium (1999)
28. Jon Oberheide, E.C., Jahanian, F.: CloudAV: N-Version Antivirus in the Network Cloud. In: Proceedings of 17th Usenix Security Symposium (2008)
29. King, S.T., Chen, P.M.: Backtracking intrusions. In: SOSP (2003)
30. Liblit, B., Naik, M., Zheng, A.X., Aiken, A., Jordan, M.I.: Public deployment of cooperative bug isolation. In: Proceedings of the Second International Workshop on Remote Analysis and Measurement of Software Systems, RAMSS (2004)
31. Liblit, B.R.: Cooperative bug isolation. PhD thesis, Berkeley, CA, USA, Chair-Alexander Aiken (2004)
32. Orso, A., Liang, D., Harrold, M.J., Lipton, R.: Gamma system: continuous evolution of software after deployment. SIGSOFT Softw. Eng. Notes 27(4) (2002)
33. Sekar, R., Dhurjati, M.D., Bollineni, P.: A Fast Automation-Based Method for Detecting Anomalous Program Behaviors. In: IEEE Symposium on Security and Privacy (2001)
34. Tucek, J., Newsome, J., Lu, S., Huang, C., Xanthos, S., Brumley, D., Zhou, Y., Song, D.: Sweeper: a lightweight end-to-end system for defending against fast worms. In: EuroSys. (March 2007)
35. Wang, H.J., Platt, J.C., Chen, Y., Zhang, R., Wang, Y.-M.: Automatic misconfiguration troubleshooting with peerpressure. In: OSDI (2004)
36. Warrender, C., Forrest, S., Pearlmutter, B.: Detecting Intrusions using System Calls: Alternative Data Models. In: IEEE Symposium on Security and Privacy (1999)
37. Yeung, D.-Y., Ding, Y.: Host-based intrusion detection using dynamic and static behavioral models. Pattern Recognition 36 (2003)

Using Failure Information Analysis to Detect Enterprise Zombies

Zhaosheng Zhu[1], Vinod Yegneswaran[2], and Yan Chen[1]

[1] Department of Electrical and Computer Engineering, Northwestern University
{z-zhu,ychen}@northwestern.edu
[2] Computer Science Laboratory, SRI International
vinod@csl.sri.com

Abstract. We propose failure information analysis as a novel strategy for uncovering malware activity and other anomalies in enterprise network traffic. A focus of our study is detecting self-propagating malware such as worms and botnets. We begin by conducting an empirical study of transport- and application-layer failure activity using a collection of long-lived malware traces. We dissect the failure activity observed in this traffic in several dimensions, finding that their failure patterns differ significantly from those of real-world applications. Based on these observations, we describe the design of a prototype system called Netfuse to automatically detect and isolate malware-like failure patterns. The system uses an SVM-based classification engine to identify suspicious systems and clustering to aggregate failure activity of related enterprise hosts. Our evaluation using several malware traces demonstrates that the Netfuse system provides an effective means to discover suspicious application failures and infected enterprise hosts. We believe it would be a useful complement to existing defenses.

1 Introduction

Due to the persistent and ubiquitous nature of the Internet's background radiation [35], modern enterprise networks have become relentless targets of attacks from a plethora of Internet malware including worms, self-propagating bots, spamming bots, client-side infects (drive-by downloads) and phishing attacks. Estimates on the number of malware instances released vary vastly (between ten of thousands to more than hundred thousand per month) depending on census methodologies [16, 31]. However, there is consensus that malware is becoming increasingly prevalent, sophisticated, and a formidable threat not just to network communications but also as a purveyor of data and identity theft. Network security analysts in today's enterprise networks rely primarily on a combination of network intrusion detection systems (NIDS) [36, 41] and antivirus (AV) systems to shield enterprise networks from this deluge of malware.

A NIDS passively monitors packets on the network wire and uses rules to discover suspicious activities, such as scans and exploit attempts, directed against systems in the network. Knowledge-based and behavior-based detection are two

Y. Chen et al. (Eds.): SecureComm 2009, LNICST 19, pp. 185–206, 2009.
© Institute for Computer Science, Social-Informatics and Telecommunications Engineering 2009

fundamental approaches to intrusion detection [14]. Knowledge-based intrusion detection systems [41] use signatures of well-known exploits and intrusions to identify attack traffic. However, reliable and accurate performance requires constant maintenance of the knowledgebase to reflect the latest vulnerabilities. In contrast, behavior-based intrusion detection techniques [27] compare current activity with a predefined model of normal behavior and flag deviants from known models as anomalies. A drawback with many behavioral approaches is the inherent difficulty of building robust models of normal behavior whose incompleteness results in high false alarm rates.

Contemporary AV software monitors end hosts by performing periodic system scans and real-time monitoring, checking existing files and process images with a dictionary of malware signatures that is constantly updated. Certain vendors also incorporate heuristic detection engines that identify infections based on static traits (*e.g.*, whether it is packed) or approximate behavioral profiles of known malware. Despite their ubiquity and sophistication, most AV systems have been shown to have unsatisfactory detection rates [10] especially in early days of an outbreak. Our experience at honeynets shows that the median day-zero detection rate for 30 AV vendors is around 82% [26]. The proliferation of the recent Conficker A and B worms offers further testament to the inefficacy of current AV systems. By leveraging a well-publicized Windows RPC vulnerability (MS08-67) [32], Conficker has successfully infected millions of hosts [17, 18], and in the early days of the outbreak only 3/39 AV engines were able to detect this binary as being malicious [43]. Like most malware, Conficker disables AV updates after infection, so subsequent signature updates by AV vendors were not particularly effective in curtailing this worm. In summary, the remarkable success of a scan-and-infect worm such as Conficker (seven years after Code Red I [13]), underscores why network security analysts need better tools to understand, react to, and cope with infections in their enterprises.

Our approach. In this paper, we introduce a new behavior-based approach to detect infected hosts within an enterprise network. Our objective is to develop a system that is independent of malware family and requiring no apriori knowledge of malware semantics or command and control (C&C) mechanisms. We devise an approach that is motivated by the simple observation that many malware communication patterns result in abnormally high failure rates. While prior efforts have tried to exploit this in the specific context of portscans [28] or studied types of failures [44, 39], we extend this to broadly consider a large class of failures in both transport and application levels. We have developed a prototype system called Netfuse that correlates network and application failures to detect infected hosts within enterprise networks. The event correlation engine of our system is inspired by prior systems such as BotHunter [23]. While BotHunter relies on exploit signatures from Snort, an important distinction of our approach is that it requires no specific knowledge of malware. Instead, Netfuse relies on application knowledge that it obtains from network protocol analyzers such as Wireshark [8] and L7 filters [5]. In some sense, Netfuse could be considered a behavior-based detection system whose model for malicious behavior is derived from underlying protocol analyzers. However, its novelty lies in its use of multipoint failure monitoring for support vector machine (SVM)-based classification of malware failure profiles. We believe that Netfuse could be a useful sensor input to BotHunter.

The Netfuse system has several integral components. First, it has a protocol failure analysis component that is built on the Wireshark protocol analyzer. It specifically analyzes transport failures (TCP RSTs, ICMP) and application failures on common ports TCP/25 (SMTP), TCP/80 (HTTP), UDP/53 (DNS) and TCP/6667 (IRC). Furthermore, it uses L7 filters to detect when common protocols are observed in nonstandard ports (*e.g.*, HTTP or DNS activity on a high-order port) and routes them to the appropriate Wireshark protocol handler. Second, it has a lightweight DNS monitor that monitors DNS activity between enterprise clients and the DNS server. Finally, it has a clustering and correlation component that aggregates alerts observed by the two sensors producing a condensed summary of failure activity that classify anomalous activity. For every IP with failure activity, it computes four different scores: (*i*) composite failure (*ii*) divergence (*iii*) persistence and (*iv*) failure entropy. This information is used by an SVM driven classification engine to detect suspicious hosts. Furthermore, a cluster summary is produced that aggregates suspicious hosts with similar failure profiles. The combination of these scores and clustering enables security analysts to easily comprehend failure patterns in the enterprise and quickly identify suspicious hosts in the network. We find that our approach is effective in isolating the presence of a vast majority of contemporary malware without specialized signatures.

Contributions. The contributions of our work are as follows:

1. We describe application-aware failure monitoring as a new approach for identifying infected hosts and uncovering anomalies in enterprise traffic.
2. We develop a prototype implementation of the Netfuse monitor using Wireshark and L7 filters. An important aspect of the implementation is multipoint failure monitoring.
3. We develop an SVM-based classifier to identify infected hosts.
4. We use multiple network traces of malware and benign traffic to evaluate detection rates and the false positive rate of Netfuse.

The remainder of the paper is organized as follows. In Section 2, we provide an analysis of network and application failures that motivate the Netfuse system. In Section 3, we introduce our Netfuse prototype implementation. In Section 4, we describe our classification and clustering algorithm. Then we describe our in-situ and online experiences with the Netfuse system and analyze results in Section 5. We survey related work in Section 6. We summarize our results and discuss future work in Section 7.

2   An Empirical Survey of Application Failure Anomalies

We explore reasons behind the occurrence of application failures in enterprise traffic. We begin with a case study analysis of the failure patterns of malware using over 30 long-lived malware (5-8 hour) traces. We then examine failure profiles of several normal applications that may cause failures similar to malware including webcrawlers, P2P software and popular video sites. Then we discuss the potential and implications of using protocol failure anomalies to detect misbehaving clients in the enterprise network.

In the following, we define the term *failure* to broadly refer to both network and application failures. Network failure corresponds to presence of packets, indicating transport-level failures such as TCP RSTs and ICMP unreachable messages in the trace. Application failures indicate higher-level protocol failures as shown in Table 1.

Table 1. Commonly observed protocol failure messages

Protocol	Layer	Failure Types
DNS	Application	NXDOMAIN (No such domain)
HTTP	Application	400 Bad Request, 404 Not Found, 403 Forbidden, 411 Length Required 500 Internal Server Server, 501 Not Implemented
FTP	Application	Transient Negative Completion reply Permanent Negative Completion reply
SMTP	Application	Domain service not available, mailbox unavailable Syntax error, command not implemented Machine does not accept mail, mailbox unavailable User not local, requested mail action aborted
IRC	Application	No such nick, No such server No such channel, Cannot send to channel

2.1 Malware Trace Analysis

The first part of our analysis is a study of application failure patterns observed in contemporary Internet malware. We started with a corpus of 32 different malware instances that we each executed in a controlled virtual machine (VM) environment for several hours. The sources of the malware include our honeynet [43], malicious email attachments, and the Offensive Computing website [6]. To obtain accurate and complete results of network interaction, it was necessary to collect long-lived traces and to allow the hosts to communicate with the outside world. We collected `tcpdump` traces of all network activity, and we analyze the failure patterns found in these traces below.

We find that contemporary malware instances generate a diverse set of failures, in both the transport and application levels. Interestingly, we find that these failures could be attributed to a small set of causes, *i.e.*, broken C&C channels, scanning and spam delivery attempts. Furthermore, the volume of failure activity seems to be strongly correlated with the volume of overall network activity. For example, scanners tend to generate a lot of flows, many of which generate transport failures. Likewise, many malware instances periodically retry failed communication attempts, which results in larger network traces with redundant activity.

Among the 32 malware instances, eight did not generate failures. These include two worms, three IRC botnets, and three spyware instances. As the three IRC bots contacted the server successfully and did not receive any MOTD commands from the server, there were no failures. Likewise, the well-behaved spyware binaries simply contacted a few active websites.

Table 2 illustrates the distribution of failures by protocol for each of the malware instances that generated transport or application failures. First, we note that 24/32 botnet and worm instances generate some sort of failure (either application or transport). We find that most of them (18/24 instances) trigger DNS failures. Furthermore, malware with spam capabilities (notably Storm) also tends to produce high volumes of SMTP failures. Finally, malware with P2P C&C channels and malware with scanning behavior are also associated with abnormally high ICMP failures. We examine the failure breakdowns within each protocol in greater detail below and provide explanations for their causes.

DNS failures. In our analysis, we found that 18 malware traces contained DNS failures. All of these were due to unresolved domain names or NXDOMAIN responses from the DNS server. In many cases, particularly for IRC bots, these arise because the C&C server gets taken down by ISPs or is otherwise blocked by law enforcement. While many well-behaved applications terminate connection attempts after a few failed tries, we find that malware tends to be remarkably persistent in its repeated attempts to contact its C&C server. We also observed that for certain malware, there is built-in redundancy in that they will query a set of domain names for the remote server. Although some domains do not resolve, C&C communication will still continue based on the successful DNS lookups.

Table 2. Failure profile summary (in hourly rates) of 24 malware instances

Malware	Class	DNS rate	HTTP rate	ICMP rate	SMTP rate	TCP rate
Look2me	Spyware	5				
Wsnpoem	Spyware	15				
Bobax	HTTP botnet	148				191
Kraken I	HTTP botnet	348				
AgoBot	IRC botnet	5312				
Gobot	IRC botnet			891		9539
Sdbot	IRC botnet	2188				
Sdbot II	IRC botnet	53				
Spybot I	IRC botnet	283				1506
Spybot II	IRC botnet	16				50
Spybot III	IRC botnet	16				
Wootbot	IRC botnet					275
Irc.Webloit	IRC botnet					477
Nugache	P2P botnet					291
Storm I	P2P botnet	26		5432	284	73
Storm II	P2P botnet			27151		
Allaple	Worm	9		33413		5738
Grum	Worm	60		160		31330
Kwbot	Worm	37				
Mytob	Worm	221		385		53
Netsky	Worm	51012				
Protoride	Worm	503				151
Virut	Worm	222	10	409		14
Weby	Worm		67			24

In fact, for some bots, such as Kraken, DNS failures could be considered part of normal behavior. This malware uses a dynamic C&C-based communication structure that constructs a new list of C&C rendezvous points each day. The fully qualified domain name (FQDN) of the C&C server is constructed from a dynamically generated hostname (based on the date) and one of the following four base domain names: .mooo.com, .dynserv.com, .dyndns.org, and .yi.org. As long as the botmaster and the malware use the same algorithm to generate domain names, it is very easy for the botmaster to change the C&C server names and IP addresses to evade detection. While resolutions for most of these DNS domains are expected to fail everyday, the botmaster simply has to register one of the daily domains when he wants to instruct the bots to perform a task. Hence, a lot of DNS lookup failures are observed in the trace. For example, our trace shows that the host received 1740 DNS failures in about 5 hours, which is highly anomalous for a normal host. A similar strategy is also adopted by the recent Conficker worms [18].

SMTP failures. In our analysis, we found that SMTP failures result from spamming behaviors. A typical example is the Storm botnet, which also uses SMTP to generate emails for spam as well as propagation. Hence, its trace includes a flurry of SMTP activity and a lot of failures. Certain SMTP servers immediately close the connection after the TCP handshake. Other failures occur early in the SMTP connection setup, most common reason being "550 Recipient address rejected: User unknown". In our traces, we found hundreds of SMTP failures from several email servers. But these failures were not persistent, *i.e.*, Storm does not retry a rejected username on the same SMTP server. In certain traces of the Storm botnet, this spam behavior stops after an hour, suggesting that certain malware instances do eventually learn from failures (albeit after a long time). We feel that any malware that generates spam is bound to produce such failures. Besides Storm, there were other malware instances that attempted to send spam email, *e.g.*, Bobax, but could not succeed in establishing communication with the remote SMTP server.

HTTP failures. We found that the HTTP failures in our traces could be attributed to two reasons: (1) sending mal-formed packets for DoS attacks and (2) querying for a configuration file that has since been removed from the control server. For example, malware Mimail.L sends the following request to the target HTTP server to launch a DoS attack: "GET / HTTP/1.0" to port 80, followed by 2048 bytes of data to port 80. As a result, it receives a flurry of "HTTP 400" errors from the server implying "Bad or Malformed HTTP request". Certain other failures are due to the missing files in controlling servers. For example, clients infected with the Weby malware will try to get a configuration file from several servers. Since this file is removed in the servers, it results in "HTTP 404/File not found" errors, which are quite persistent. In our 5-hour trace, there were 335 "HTTP 404/File not found" failures.

IRC failures. For botnets that use the IRC protocol for communication and control, the following failure modes are common. Sometimes, the channel is removed from a public IRC server, which results in IRC application failures like "no such channel". In certain other cases, the channel might be full due to too many bots, which would result in a "Cannot join channel" message.

TCP layer failures. We consider unproductive TCP flows *i.e.*, which do complete a TCP handshake and/or terminate the connection with RST prior to sending any payload. The prevalence of such unproductive flows (which also results from scanning behavior), is another characteristic of malware. For some malware instances, we observed that there were continuous TCP layer failures in certain ports. For example, some IRC botnet clients receive failures in the IRC port (TCP/6667) from the remote servers (either because the server has been taken down or because it is too busy). Certain Bobax clients receive failures in the SMTP (TCP/25) port from remote email servers because the client network has been blacklisted. While scanning is usually good evidence of malware, we find that persistent TCP failures from the same remote host could be another useful indicator of malware. For example, we observed that many IRC botnets generate TCP failures from being unable to contact a previously active C&C server that has since been taken down.

ICMP failures. In our analysis, we found that ICMP failures result from scanning behavior and communication patterns of P2P botnets such as Storm. As we discuss below, this is quite unlike normal P2P applications, such as BitTorrent and eMule, that generate few ICMP failures.

2.2   Failure Patterns of Normal Applications

The second part of our analysis studies failure patterns of normal applications. As studying failure patterns of all applications is outside the scope of this study, we focus on applications that one might typically expect to produce failure patterns similar to what was observed in the malware corpus that we analyzed. The goal of this study is to understand the degree to which malware failure patterns could be used to distinguish malware traffic from other benign enterprise traffic. Our Netfuse system uses these network failures as symptoms to detect suspicious hosts. Thus, these results could inform the feasibility and design of the Netfuse system and help us prioritize failure patterns that are used for detection. Specifically, we focus our investigation on three classes of applications, which at the first glance may cause similar failures: web crawler, P2P applications (BitTorrent, eMule), and online video service (youtube).

We collected several long-lived traces for each of these normal applications, in order to get a good understanding of the types of failures they generate. Table 3 provides a summary of these traces.

Webcrawler. webcrawlers, popularly known as webspiders or webrobots, are automated scripts that systematically scan all web-pages in a site looking for specific types of content. These are commonly used by search engines to build automated meta-data (indexes) of public web-pages, but are also used for mirroring websites, data mining, and by other web-based applications such as mashups

Table 3. Normal application trace summary

Type	Site	Size	Time	Pkts	# URLs
	news.sohu.com	3.1 GB	2 days	3577674	25334
Webcrawler	amazon.com	1.9 GB	2 days	2058630	23111
Mirror	bofa.com	144 MB	12 hours	186711	4141
	imdb.com	252 MB	16 hours	333583	8113
P2P	BitTorrent	6.1 GB	18 hours	7338627	n/a
P2P	eMule	1.3 GB	1 day	1982682	n/a
Video	youtube.com	16MB	2 hours	25498	n/a

Table 4. Failure profile summary (in hourly rates) for normal applications

Application Name	HTTP	ICMP	TCP
	rate	rate	# ports / rate
Web crawler(sohu)	1.4		1/0.4
Web crawler(amazon)			1/1.4
Web Crawler(imdb)	0.04		1/0.2
Web Crawler(bofa)	0.8		1/0.9
BitTorrent	0.6		382/333
eMule		68	839/370

and portals. Since webcrawlers have become very popular and they follow hyperlinks in an automated fashion, one might expect such systems to frequently stumble upon many failed links and generate HTTP failures. Hence, we pick them as the first class of application to study.

We used the default settings and -m (mirror) option in `wget` [3] that forces wget to act as a webcrawler, recursively following all links in a given site, until all the pages have been downloaded. We collected traces from crawling four popular websites in the US and China including `bofa.com`, `amazon.com`, `imdb.com` and `sohu.com`. Each crawl took 1-2 days and involved 144 MB to 3 GB of data transfer. We found that the webcrawler produced very few HTTP and transport failures. As an example, for the website `news.sohu.com`, there were only 18 transport layer (TCP) failures and 66 HTTP failures in 2 days. Other websites also show the similarly low failure patterns as shown in Table 4. As one might expect, we find that in webcrawlers, HTTP failures are restricted to "HTTP 404/File not found" messages.

P2P applications. We select two popular peer-to-peer (P2P) software programs for our analysis: BitTorrent and eMule. BitTorrent and eMule are P2P file sharing protocols used to transfer large amounts of data such as media files, software, and OS distributions. A single large file is broken up into pieces, which are replicated and distributed among a set of peers. In BitTorrent, the publisher of the file acts as the *first seed*, and every peer who downloads the data also uploads the content to other peers. A client wishing to download the file first obtains the meta-data file, called the *torrent*, which specifies where to download the pieces. Thus, a single HTTP request for a large file is translated into several small data requests to various peers in the network. eMule is similar in concept but implements a different protocol based on Kademlia [4].

Since the status of peers in both of these networks can dynamically change (from online to offline), we expect these P2P applications to have many failures. We used BitTorrent to download a popular Linux distribution (Fedora 10) and monitored the activity of this peer for one day. It turns out there were very few (11) ICMP failures and HTTP failures, but many TCP failures. Likewise, we used eMule to download another popular Linux distribution (Ubuntu) and monitored its activity for a day. It had many ICMP and TCP failures. An important difference between transport-level failure profiles of BitTorrent and the malware we analyzed is that for BitTorrent the TCP failures happen on a large set of ports. This did not occur in the malware traces, *i.e.*, failures were restricted to fewer ports and typically occurred in one or two ports. As an example, most TCP failures with the Storm worm were dominated by its activity on port 25/TCP (arising from its spam campaigns and unrelated to its P2P communication).

Online video service. YouTube.com is one of the largest and most popular websites that provide online video hosting service. Users can upload, view, and publicly share video clips. In this experiment, we collected traces by opening videos from youtube.com, and then keeping the browser open for several hours. In analyzing the trace, we found that there were no transport-layer failures. While we did find several "HTTP 304/Not Modified" errors, we did not find any other application-level failures. Since "HTTP 304/Not Modified" messages were not found in the malware traces, we infer that this might be an error code to be considered in a whitelist.

2.3   On the Potential of Failure Analysis to Uncover Suspicious Activities

We summarize the results of our exploratory empirical analysis on the utility of failure profile analysis. After our analysis of a collection of traces from both malware and benign applications, we find several notable differences in failure pattern between malware and normal applications that could be exploited in network-based detection systems.

1. Failures in malware occur frequently in both the transport and application levels. In general, failures are rare for normal applications, except for certain P2P protocols that can generate high volumes of transport failures. Thus *high volume* of failure traffic could be a useful indicator of malware.
2. DNS failures and in particular NXDOMAIN errors are common among malware applications and relatively infrequent in normal applications. Furthermore, these failures tend to *persist* (repeat with high frequency) in malware.
3. Failures in malware applications tend to be restricted to a few ports and often a few domains. Thus, malware failure patterns tend to have *low entropy*.

3   Architecture

In the prior section, we explored the possibility to using failure information to detect suspicious hosts in the enterprise network. Here, we describe the system framework and our prototype implementation of a system that realizes our ideas.

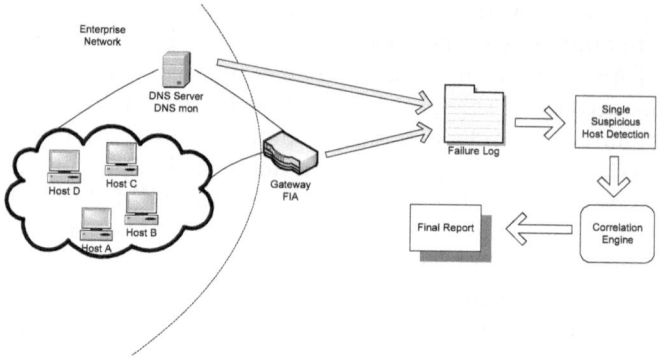

Fig. 1. Netfuse multi-point monitoring architecture

3.1   System Overview

As shown in Figure 1, Netfuse is composed of three parts: the failure information analysis (FIA) engine, DNSMon and the correlation engine. The FIA engine will typically be deployed on the perimeter of the enterprise networks. The major function of this component is to extract the failure information by looking at all packets that transit the enterprise gateway router. It will generate the failure information if any, by including both flow-level and application-level information (if available). The DNSMon system monitors interaction between enterprise clients and the local DNS server.

After the failure information is collected, the correlation engine implements a diagnostic algorithm to classify hosts according to their failure profiles and to group those suspicious hosts with similar failures. It then generates a classification report that identifies suspicious hosts based on four different criteria: failure volume, failure entropy, failure persistence, and failure uptick. We implemented our prototype FIA engine by modifying the wireshark network protocol analyzer. Our correlation engine is implemented in Python and uses a publicly available clustering package [1].

3.2   Building an FIA from Wireshark

Wireshark([8]) is an open-source network protocol analyzer that is based on libpcap library. Hence, Wireshark can analyze packets captured from a live network connection or read from a captured pcap trace file. It is distinguished by its flexible design that makes it easy to add dissectors for new protocols and built-in support for hundreds of popular protocols.

We modified Wireshark to automatically extract failure information. The failures we consider include transport-level and selected application-level protocols such as FTP, HTTP, SMTP, DNS, and IRC. For each ICMP failure, we record the error type and client IP address. For TCP failure, we record client and server IP addresses and corresponding port numbers. For DNS failures, we record the failure type, domain name, and client IP address. For FTP, IRC, HTTP and SMTP failures, we record the server IP address, error code, client IP address, and detailed failure information that may be helpful to an administrator. We also capture the packet associated with each failure message. We focus on these five protocols simply because they were the most popular in the enterprise traffic that we monitored. However, the design of Wireshark makes it straightforward to track failures in other protocols. Finally, as we are interested only in identifying potentially infected local hosts, we configure our system to only track inbound failure messages.

3.3 L7-Based Automatic Protocol Inference

One problem with Wireshark is that it does not have built-in protocol inference capability. It does not detect when a well-known protocol, *e.g.*, HTTP, is used in nonstandard ports. Wireshark expects each dissector to be tied to one or more ports and relies on the user to explicitly decode the packet by choosing a dissector when the packets are observed in unspecified ports. This is a fundamental limitation especially for malware analysis, as malware often transmits packets in nonstandard ports to evade monitoring systems.

To improve the fidelity of the FIA engine, we enhance Wireshark with L7 filter protocol signatures. L7-filter [5] is a classifier that can identify packets based on packet payload. It uses regular expressions to automatically classify packets as belonging to certain common protocols. We provide below examples of L7 protocol signatures for HTTP and IRC:

 – **HTTP Protocol:** http/(0.9|1.0|1.1)[1-5][0-9][0-9][\x09-\x0d-˜]*(connection:|content-type:|content-length:|date:)|post[\x09-\x0d-˜]*http/[01].[019]

 – **IRC Protocol:** ^(nick[\x09-\x0d]*user[\x09-\x0d]*:|user [\x09\x0d]*:[\x02-\x0d]*nick[\x09-\x0d]*\x0d\x0a)

We modified the connection struct in Wireshark to maintain a dissector tag for each connection. Every connection starts without any pre-specified dissectors. When a packet arrives, we first check to see if the connection has been allocated to a dissector. If not, we check to see if the packet matches one of the L7 filter signatures. If it finds a suitable dissector, then the connection struct is updated so future packets can be accelerated, bypassing the L7 regular expression check. Once the packet is parsed with the appropriate dissector, the output is examined for any failure messages that are stored in a log file. The FIA engine is installed as a monitor on the span port of the gateway router of the enterprise networks and logs inbound failure responses from remote servers. Figure 2 illustrates the modified Wireshark packet processing engine.

3.4 Multipoint Deployment

We begin with a simplified overview of a domain name lookup using the domain name service. As in our deployment, DNS servers are typically located inside the enterprise network. Local enterprise clients submit name resolution requests to the

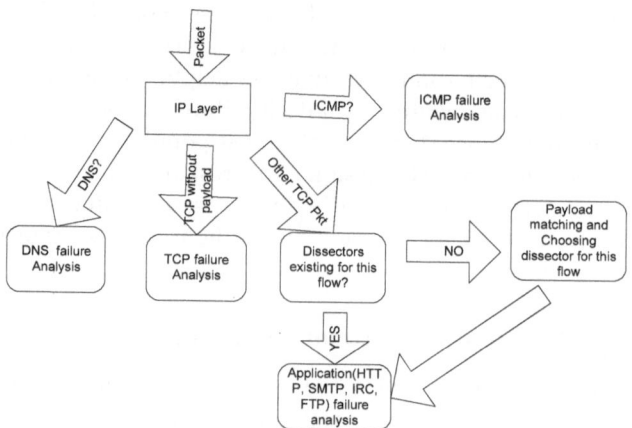

Fig. 2. Modified Wireshark packet processing engine

local DNS server (resolving name server). The resolving server checks its cache and if the name does not exist queries the authoritative name server on behalf of the local client. (The resolving server might have to query additional servers to obtain the name of the authoritative server for a specific domain.) Finally, the resolving name server responds back to the client with the appropriate IP address or NXDOMAIN if the name does not exist, or other type of DNS failure.

A side-effect of the hierarchical DNS system is that it poses additional challenges for any network-based monitoring system as monitoring the gateway only provides a view of the interaction between the resolving name server and external DNS servers. While suspicious domain lookups could be identified, they cannot typically be tracked back to the client that originated the name lookup. Netfuse addresses this problem by integrating an additional lightweight monitor (which we call DNSMon) that tracks activity between the local clients and the resolving name server. DNSMon produces regular alert logs that summarize DNS failure activity of all enterprise hosts. By combining DNSMon alerts with the data collected at the gateway monitor, we get a comprehensive log of network failure activity. Next, we describe how the Netfuse correlation engine processes this information to intelligently isolate suspicious enterprise hosts.

4   Correlation and Clustering Engine

Here, we first describe the algorithm that we implement for ranking suspicious hosts based on failure profiles. Next, we describe our algorithm for classifying groups of hosts with similar failure profiles. Finally, we discuss some techniques that we implemented for reducing false positives in our enterprise network.

Based on our empirical experience from analyzing malware traces, the current prototype system implementation is focused on failures that occur in the transport-layer and five application-layer protocols: HTTP, FTP, SMTP, DNS, and IRC. As Wireshark has dissectors for hundreds of protocols, it is not difficult to extend the system to support additional protocols. We now describe how our detection algorithm works based on failure input from these protocol analyzers.

4.1 Detecting Suspicious Hosts

The primary inputs to the diagnostic algorithm are failure logs obtained from the FIA engine and DNSMon described in Section 3. First, we classify and aggregate failure information based on host IP address, protocol, and failure type. Next, we compute the following four different scores for each host in the enterprise network with failure activity: (i) composite failure, (ii) failure divergence, (iii) failure persistence and (iv) failure entropy. The scores are each normalized to be in the range of 0 and 1. Finally, we use an SVM-based learning technique to classify suspicious hosts. We begin by describing the four scoring functions in greater detail.

Composite Failure Score. This score estimates the severity of the observed failures by each host based on volume. For every host, the failure profile can be represented as a vector$\{N_i\}$, where N_i represents the number of failures of the i_{th} protocol. We proceed as follows to compute the composite failure score for each host.

Step 1: In Section 2, we observed that malware tends to have a large number of failures. So the first step in our analysis is a filtering step that culls hosts with the fewest number of failures. Let α_i, β_i, and γ_i represent the number of application level failures, number of TCP RSTs and number of ICMP failures respectively of host i. Furthermore, let $\mu(\beta)$ and $\sigma(\beta)$ represent the average and standard deviation of TCP failures for a host. Likewise, let $\mu(\gamma)$ and $\sigma(\gamma)$ represent the average and standard deviation of ICMP failures for a host.

Specifically, we consider only hosts that satisfy either of the following three constraints: (1) $\alpha_i > \tau$ (where τ is a constant, set to be 15 for our experiments); (2) $\beta_i > \mu(\beta) + 2 * \sigma(\beta)$ (TCP RST count more than two standard deviations from the mean); or (3) $\gamma_i > \mu(\gamma) + 2 * \sigma(\gamma)$ (ICMP failure count more than two standard deviations from the mean). The final two constraints remove backscatter traffic [33], which artificially inflates the TCP RST and ICMP failure counts for IP addresses in the network.

Step 2: Next, we compute a composite score for each of the remaining hosts as follows: score($host_i$) $= \sum_{i=1}^{n} N_i/T_i$, where T_i is the total number of failures for i_{th} protocol across all hosts.

Step 3: Finally, we sort all the hosts according to the score calculated in the second step. Hosts with higher scores are more suspicious than hosts with lower scores.

Failure Divergence Score. The objective of the failure divergence score is to measure the degree of uptick in a host's failure profile. In particular, we would like to measure the delta between a host's current (daily) failure profile and past failure profiles. We expect that newly infected hosts would show a strong and positive divergence in their failure patterns while other hosts (clean hosts and those that have been infected for a while) would demonstrate a more stable failure profile.

To quantify this we adopt a well-known statistical forecasting technique, exponentially weighted moving averages (EWMA) [7], that uses a weighted moving average of past observations as the basis for predicting the failure profile for the next day. EWMA uses an exponential distribution to weigh recent observations more heavily than past observations and it is controlled by the parameter α, where α is the smoothing factor, and $0 \leq \alpha \leq 1$. In our measurements, we set α to be 0.5. We compute divergence as follows for each host in the network. Let E_{ijt} correspond to the expected number of failures for host i, on protocol j on day t. We compute E_{ijt} as shown in Figure 3. We then compare the actual value X_{ijt} with E_{ijt} by calculating the distance as follows: $1-(E_{ijt}-X_{ijt})/(E_{ijt}+X_{ijt})$. Finally, we normalize by dividing by the maximum divergence score across all hosts in that day to obtain a score in the range $[0,1]$.

$$E_{ij0} = X_{ij0} \tag{1}$$

$$E_{ijt} = \alpha X_{i,j,t-1} + (1 - \alpha)E_{i,j,t-1} \tag{2}$$

$$Dist_{it} = \sum_{j=1}^{n} 1 - \frac{E_{ijt} - X_{ijt}}{E_{ijt} + X_{ijt}} \tag{3}$$

$$Divergence_{it} = \frac{Dist_{it}}{\forall k \ max(Dist_{kt})} \tag{4}$$

Fig. 3. Simple exponential prediction model and divergence computation

Failure Entropy Score. The failure entropy score measures the degree of diversity in a host's failure profile. This is based on the insight derived from Section 2 that failures in many malware applications tend to have a high degree of redundancy, e.g., failures are often restricted to a few ports or domains such as in a bot that tries to repeatedly contact a C&C server that is currently inactive.

For TCP failures, we track entropy in the server distribution and host distribution of each client receiving TCP RST failures. For every server H_i, we record the number of N_i failures from it. We repeat the same for each server port P_i. For DNS failures, we track entropy in the domain names that are associated with failures. For each domain name D_i appearing in failure response, we record the number N_i. For HTTP, FTP IRC, and SMTP failures, we track entropy in the disribution of various failure types (e.g., HTTP/404) within each protocol and remote servers that issue the errors. For each host H_i and each error type E_i, we calculate the corresponding number N_i. We do not consider ICMP failures in the entropy computation.

For those protocols that have two distribution sets, we calculate the average entropy [2] for each set. We begin by computing weights for each host i and protocol j. Then, for each host i, we compute the significance (s) of protocol j as $s_{ij} = N_{ij}/\sum k = 1^n N_{kj}$ (i.e., number of failures of host i in protocol j divided by the total number of failures in protocol j across all hosts). The weight of protocol j for host i is simply its normalized significance $w_ij = s_ij \,/\, \sum_{k=1}^{n} s_i k$. The weighting function ensures that for each host, protocols that are responsible for a large portion of its failures will dominate its entropy value. Next, for each host i and protocol j, we calculate the entropy p_{ij}. The failure entropy score for the host is simply the weighted average entropy score, *i.e.*, $\sum_{i=1}^{n} w_i * p_i$.

Failure Persistence Score. The final score is failure persistence, which is motivated by the observation from our case study that malware failures tend to be long-lived. Prior approaches have used autocorrelation techniques to detect long-lived periodic behavior of malware additivity [24]. While we could leverage similar statistical approaches to measure persistent malware activity, we adopt a simpler approach to measure persistence. We simply split the time horizon into N parts (where N is set to 24 in our prototype implementation), and compute the percentage of parts where the failure happens. High failure persistence values provide yet another useful indicator of potential malware infections.

SVM-based Algorithm to Classify Suspicious Hosts. Support vector machines are a recent and well-studied family of supervised learning algorithms used for classification of multidimensional data. Given a training data set, SVMs work by building a hyperplane (or a predictor function) that efficiently seperates positive and negative examples. In our case, we are interested in the maximal margin classifier, *i.e.*, a hyperplane that separates positive and negative examples with maximal distance. In many environments, SVMs have been shown to outperform traditional linear classifiers. Indeed, we had a similar experience in testing different classifiers on our data set. For this system, we use a publicly available tool WEKA [9] to implement our SVM-based classification. The input to the system is a series of four-dimensional vectors where each vector corresponds to the four scores of a individual host. We train the system using a set of malware traces and clean traces for which we have ground truth. The classification problem is identifying the set of suspicious hosts in the network.

4.2   Detecting Failure Groups

After we get the result of suspicious hosts, we want to know whether they are infected by the same malware. For example, we want to know whether they belong to the same botnet. This information can help the network administrator rapdily assess what has happened inside the network. To enable this, we developed a clustering algorithm to detect failure groups which we discuss below. We begin by defining the scoring function that is used for comparing failure profiles.

Scoring Function. According to the description above, each type of failure can be represented as a set of (F_i, N_i), where F_i is the failure property and N_i is the number of failures with this property. Given this representation, we can define the similarity between two hosts as follows. The pseudocode for the algorithm is provided in Algorithm 1. For each protocol, the algorithm compares the number of failures for hosts i and j. The similarity score is incremented by protocol failure count of each host minus the difference between the larger and smaller failure count. It should be apparent that hosts with identical failure profiles would end

Let (F_i, N_i) be the set of one host, and (F_j, N_j) be the set of the other.
procedure Similarity$((F_i, N_i), (F_j, N_j))$
Let $sum = 0$ be the total number of failures of these two sets ;
Let $sim = 0$ be the number of failures that show similarity;
1 **foreach** (F_{ik}, N_{ik}) in set (F_i, N_i) **do**
2   **foreach** (F_{jl}, N_{jl}) in set (F_j, N_j) **do**
3   **if** $F_{ik} = F_{jl}$ **then**
4              $sim = sim + (N_{ik} + N_{jl} - abs(N_{ik} - N_{jl}))$
5              $sum = sum + N_{ik} + N_{jl}$
      end
  end
6 Return sim/sum;

Algorithm 1. Function to calculate similarity between two failure profiles

up with higher similarity scores. Finally, the similarity score is normalized by dividing by the total number of failures between the two hosts.

Clustering Method. The similarity metric enables us to cluster hosts into distinct groups based on their respective failure profiles. We apply hierarchical clustering based on Peter Kleiwig's publicly available clustering package [1]. The unique aspect of this tool is its flexibility, which lets us choose between seven different clustering algorithms. We chose Ward's minimum variance clustering method, which is widely used for hierarchical clustering. The clustering generates a dendrogram that illustrates similarity among hosts in the network based on their failure profiles. Then instead of fixing a threshold to cut them into clusters, we implement Silhouette Validation Method [37] to find the optimal cut index.

5 Evaluation

To evaluate the performance of Netfuse, we conducted comprehensive tests to measure its detection and false positive rates. The traffic that we use includes five traces shown in Table 5: three malware trace sets and two clean traces from a research institute network, which we refer to as the institute trace. First, we built a model from the training trace. Then to test the classification performance, we use traces from different malware sets and mix them with the institute traces.

Table 5. Training and testing data set

	5-day Institute Trace	12-day Institute Trace
Malware Trace I	Training	Testing
Malware Trace II		Testing
Malware Trace III		Testing

1. Malware Trace I: We reuse 24 traces from Table 2 which we combine with clean traces to build the classification model.

2. Malware Trace II: This data set contains five malware families that are not included in the training set (Peacomm, Kraken, Rbot, Mimail and Bifros) and three malware instances represented in the training set. We created a VMware-based virtual machine (VM) environment running eight isolated Windows XP virtual machines, infecting each with a different malware instance. We let these systems run for 10 hours and collected traces of all their network activity. We repeated the experiment three times collecting a total of 24 traces (three per malware). We use this trace to evaluate the classification system and the clustering component.

3. Malware Trace III: This data set contains more than 5,000 malware traces that were obtained from a sandnet. This corpus is particularly attractive because it represents a large and diverse collection of malware. However, a deficiency of sandnet traces is that the malware binaries are often run only for a short period and many of them do not generate any network activity. From this large corpus, we downselected 242 longer running traces based on duration and trace size.

4. Benign Institute Trace: We deployed our system online in the research institute network and continuously ran it for over three weeks. The network is rigorously monitored by NIDSs and has more than one hundred systems (mix of Linux, Macs and Windows PCs). Being a relatively small, well-administered network with a diverse mix of traffic makes it a good candidate for evaluating false positives. We use two traces from this network (a 5-day trace for training and a 12-day trace for testing). In our analysis of clients that generate many failures, we stumbled upon a group of misconfigured Tor nodes that are part of another project. These hosts are grouped together by the clustering engine and classified as benign by the SVM classifier.

5.1 Classification and Detection Results

We will first describe the training process. Then we use the built model to test the performance of our system, including detection rate and false positives.

Training Process. In the training process, we use the SVM algorithm to build a classification model. First, we combine malware trace I with the 5-day institute trace to construct the input data set. Intuitively, a larger training set implies a more accurate model. An example of a rule generated by the SVM algorithm is $-4.266 \times$ (normalized) divergence_score $-0.042 \times$ persistence_score $+0.664 \times$ entropy_score $+0.561 \times$ failure_score $+ 1.8486$. For our evaluation the detection rate for training is 97.2% and the false positive rate is 0.3%.

Performance Evaluation. To measure the detection performance and estimate false positive rates, we mixed different malware traces I, II, and III with 12 days of institute traces. We then processed them through the Netfuse classifier, which took under one hour to process the failure logs for 12 days. In each case, we counted the number of malware traces that were identified (true-positives)

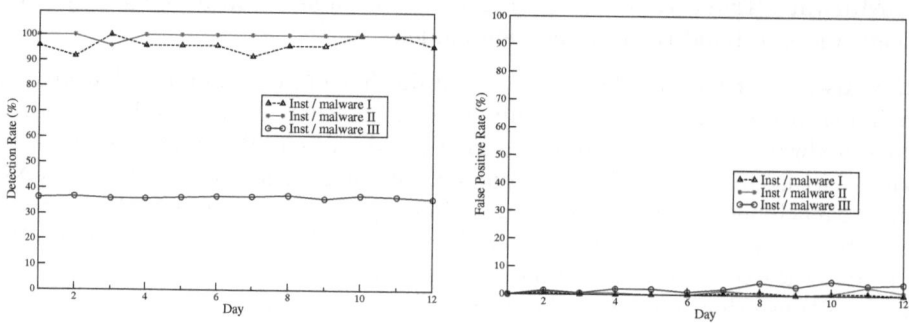

Fig. 4. Detection (left) and False Positive (right) rates on Institute/malware I,II,III mixture traffic

and the number of benign clients that were classified incorrectly. The results are shown in Figure 4. The detection rate is more than 92% for traces I and II. For trace 3, the detection rate varies between 35% and 40%, *i.e.*, around 90/242 malware instances detected. The lower detection rate for trace 3 could be attributed to two reasons. First, the trace set includes many types of malware, including adware that often have traffic profiles similar to benign applications. Second, the traces are quite short (around 15 minutes long). Despite this, Netfuse is able to detect over a third of the malware without any specialized signatures. The false positive rate is consistently lower than 5%.

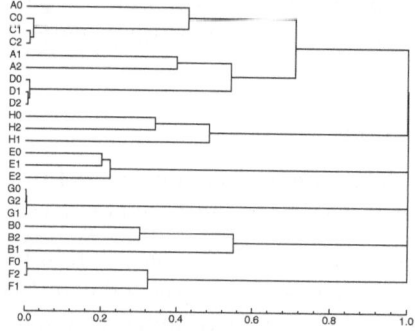

Table 6. Malware clustering summary

Bot Trace	Packets	Clustered	Accuracy
Peacomm	999905	3/3	100%
Bifrose	30635	3/3	100%
Mimail	279962	3/3	100%
Kraken	49505	3/3	100%
Sdbot	312796	3/3	100%
Spybot	79750	3/3	100%
Rbot	1175083	3/3	100%
Weby	9000	3/3	100%

Fig. 5. Malware clustering dendrogram

Clustering Results. After we identify suspicious hosts, we group them according to their failure profiles to simplify analysis of the network administrators. We use malware trace II to test the clustering engine. As shown in Table 6, we find that in all cases the clustering is quite robust. The corresponding dendrogram is provided in Figure 5, where 24 hosts are infected with eight malware instances listed A-H.

6   Related Work

Over the last three years, botnets have become one of the hot areas in networking and security research. In [40] Rajab *et al.* use a multifaceted approach to conduct a comprehensive study on the prevalence of IRC botnets. Dagon *et al.* ([12]) use DNS sink-hole redirection to measure botnet properties and develop a diurnal model for botnet propagation. In [21], Grizzard *et al.* study the structure of botnets and discuss how the single point weakness will force botherders to a P2P structure using the Storm botnet as an example. Vogt [42] *et al.* discuss a recent trend toward smaller botnets and raise the threat of superbots, *i.e.*, an army of distributed botnets that can be coordinated to act as a single network. More recently, Holz *et al.* discuss the emerging threat of fast-flux service networks [25]. Bayer *et al.* [11] propose a scalable algorithm to cluster the malware according to the host behavior profiles.

Inspired by these measurement and modeling studies, there has been a considerable research thrust in building better botnet detection systems. The Rishi [20] system detects IRC botnets by matching IRC bot nickname patterns. BotHunter was the first system to use dialog correlation to detect botnet activity. BotSniffer uses spatio-temporal correlation to detect botnet C&C activity [24]. The BotMiner system [22] combines clustering techniques with heuristics developed by BotHunter and BotSniffer to classify malware based on both malware activity patterns and C&C patterns. The motivation for Netfuse and its correlation approach bears certain similarities to these systems. However, these systems fundamentally differ from Netfuse in that they ignore application-layer failures and focus on successful communication patterns of bots.

Others have developed machine-learning approaches to detect botnets [30, 19]. Bayesian network classifiers are used in [30]. In this paper, authors use machine learning techniques to distinguish between non-IRC traffic, botnet IRC traffic and non-botnet IRC traffic. A different framework, which uses an entropy classifier and a machine-learning classifier, to detect chat bots is provided in [19]. It shows that message sizes and inter-message delays are sufficient to differentiate humans from chat bots. We consider these efforts complementary to our system. Statistical traffic anomaly detection techniques have also been demonstrated to have the potential of identifying botnet-like activity. The exPose system [29] uses statistical rule-mining techniques to extracting significant communication patterns and identify temporally correlated flows, such as worm flows. Threshold random walk is a well-known algorithm that uses hypothesis testing to identify portscanners and Internet worms [28].

Finally, we are also informed by traffic characterization studies such as Pang *et al.* [34] and efforts to automate characterization of enterprise use patterns [15]. A comprehensive analysis of DNS query traffic and its use in identifying network anomalies is provided in [38]. While our system is tuned toward the botnet detection problem, Netfuse could be easily extended to be used as a traffic characterization tool.

7 Conclusion

We propose failure information analysis as a new paradigm for detecting application-layer failures and suspicious activities in the enterprise. We are motivated by the goal of automatically discovering infected hosts in the enterprise. We use an empirical analysis case study to highlight certain differences in bot-like malware and production enterprise traffic that could be exploited to identify infection activity. Using this framework, we develop a prototype system called Netfuse that has three integral components: FIA, DNSMon and the correlation engine. The correlation engine uses four different scores (composite failure, divergence, failure entropy, and failure divergence) to classify suspicious hosts and a clustering component aggregates hosts with similar failure profiles to simplify analysis. We evaluate the system using several malware traces. Our evaluation and analysis shows that Netfuse is an efficient and effective system for discovering embedded malware. In future work, we plan to address the problem of adapting Netfuse to deal with knowledgeable adversaries.

Acknowledgements

This material is based on work supported by the Army Research Office under Cyber-TA Grant No. W911NF-06-1-0316 and by the National Science Foundation Grant No. CNS-0716612. This work was also partially supported by DOD (Air Force of Scientific Research) Young Investigator Award FA9550-07-1-0074 and a grant from NU-Motorola Center for Seamless Communication. Any opinions, findings, and conclusions or recommendations expressed in this material are those of the authors and do not necessarily reflect the views of the funding sources. We also wish to acknowledge help from Michael Hodgsett, Matt Jonkman and the useful feedback received from our Securecomm reviewers.

References

1. Data clustering, http://www.let.rug.nl/~kleiweg/clustering/
2. Entropy, http://en.wikipedia.org/wiki/Information_entropy
3. Gnu wget, http://www.gnu.org/software/wget/
4. Kademlia, http://en.wikipedia.org/wiki/Kademlia
5. L7-filter: Application Layer Packet Classifier for Linux,
   http://l7-filter.sourceforge.net/
6. Offensive Computing, Community Malicious code research and analysis,
   http://www.offensivecomputing.net/
7. Simple Exponential Smoothing,
   http://en.wikipedia.org/wiki/Exponential_smoothing
8. Wireshark: The World's Most Popular Network Protocol Analyzer,
   http://www.wireshark.org/
9. WEKA-Machine Learning Software in Java (2008),
   http://weka.wiki.sourceforge.net/Primer-?token=
   2b7a093d07966047b281eeec0da1b9fd

10. Bailey, M., Oberheide, J., Andersen, J., Mao, Z.M., Jahanian, F., Nazario, J.: Automated classification and analysis of internet malware. In: Kruegel, C., Lippmann, R., Clark, A. (eds.) RAID 2007. LNCS, vol. 4637, pp. 178–197. Springer, Heidelberg (2007)
11. Bayer, U., Comparetti, P.M., Hlauscheck, C., Kruegel, C., Kirda, E.: Scalable, behavior-based malware clustering. In: Network and Distributed System Security Symposium, NDSS (2009)
12. Dagon, D., Zou, C., Lee, W.: Modeling botnet propagation using time zones. In: Network and Distributed System Security Symposium, NDSS (2006)
13. Moore, D., Shannon, C., Brown, J.: Code-Red: A case study on the spread and victims of an Internet worm. In: Proceedings of the Internet Measurement Workshop (2002)
14. Debar, H.: An Introduction to Intrusion Detection Systems. In: Proceedings of Connect (2000)
15. Estan, C., Savage, S., Varghese, G.: Automatically Inferring Patterns of Resource Consumption in Network Traffic. In: Proceedings of ACM SIGCOMM (2003)
16. F-Secure. Kapersky Security Bulletin 2008: Malware Evolution January - June 2008 (2008), http://www.viruslist.com/analysis?pubid=204792034
17. F-Secure. Calculating the Size of the Downadup Outbreak (2009), http://www.f-secure.com/weblog/archives/00001584.html
18. Fitzgerald, P.: Downadup: Geolocation, Fingerprinting and Piracy (2009), https://forums.symantec.com/t5/Malicious-Code/Downadup-Geo-location-Fingerprinting-and-Piracy/ba-p/380993
19. Gianvecchio, S., Xie, M., Wu, Z., Wang, H.: Measurement and classification of humans and bots in internet. In: USENIX Security (2008)
20. Goebel, J., Holz, T.: Rishi: Identify bot contaminated hosts by irc nickname evaluation. In: Hot Topics in Understanding Botnets (HotBots) (2007)
21. Grizzard, J.B., Sharma, V., Nunnery, C., Kang, B.B.: Peer-to-peer botnets: Overview and case study. In: Hot Topics in Understanding Botnets (HotBots) (2007)
22. Gu, G., Perdisci, R., Zhang, J., Lee, W.: Botminer: Clustering analysis of network traffic for protocol- and structure-independent botnet detection. In: Proceedings of the 17th USENIX Security Symposium (2008)
23. Gu, G., Porras, P., Yegneswaran, V., Fong, M., Lee, W.: BotHunter: Detecting malware infection through IDS-driven dialog correlation. In: Proceedings of 16th USENIX Security Symposium (2007)
24. Gu, G., Zhang, J., Lee, W.: Botsniffer: Detecting botnet command and control channels in network traffic. In: Proceedings of the 15th Annual Network and Distributed System Security Symposium, NDSS 2008 (2008)
25. Holz, T., Gorecki, C., Rieck, K., Freiling, F.C.: Measuring and detecting fast-flux service networks. In: NDSS (2008)
26. SRI International. Malware Threat Center (2008), http://mtc.sri.org
27. Javitz, H., Valdes, A.: The SRI IDES statistical anomaly detector. In: Proceedings of IEEE Symposium on Research in Security and Privacy (1991)
28. Jung, J., Paxson, V., Berger, A.W., Balakrishnan, H.: Fast portscan detection using sequential hypothesis testing. In: Proceedings of the IEEE Symposium on Security and Privacy (2004)
29. Kandula, S., Chandra, R., Katabi, D.: What's going on? Learning communication rules in edge networks. In: Sigcomm (2008)

30. Livadas, C., Walsh, R., Lapsley, D., Strayer, W.T.: Using machine learning techniques to identify botnet traffic. In: Proc. IEEE LCN Workshop on Network Security, WoNS 2006 (2006)
31. Trend Micro. Trend Micro Threat Roundup and Forecast - 1H 2008 (2008), http://us.trendmicro.com/us/threats/enterprise/security-library/threat-reports/index.html
32. Microsoft. Microsoft Security Bulletin MS08-067 – Critical (2008), http://www.microsoft.com/technet/security/Bulletin/MS08-067.mspx
33. Moore, D., Voelker, G.M., Savage, S.: Inferring internet denial-of-service activity. In: Proceedings of the 10th Usenix Security Symposium (2001)
34. Pang, R., Allman, M., Bennett, M., Lee, J., Paxson, V., Tierney, B.: A first look at modern enterprise traffic. In: IMC (2005)
35. Pang, R., Yegneswaran, V., Barford, P., Paxson, V., Peterson, L.: Characteristics of Internet background radiation. In: Proceedings of the 4th ACM SIGCOMM Internet Measurement Conference (2004)
36. Paxson, V.: Bro: A system for detecting network intruders in real-time. In: Proceedings of the 7th USENIX Security Symposium, San Antonio, TX (January 1998)
37. Rousseeuw, P.: Silhouettes: a graphical aid to the interpretation and validation of cluster analysis. Journal of Computational and Applied Mathematics 20 (1987)
38. Plonka, D., Barford, P.: Context-aware clustering of dns query traffic. In: Proceedings of ACM Internet Measurement Conference (2008)
39. Plonka, D., Barford, P.: Context-aware Clustering of DNS Query Traffic. In: Proceedings of the 8th ACM SIGCOMM Internet Measurement Conference (2008)
40. Rajab, M.A., Zarfoss, J., Monrose, F., Terzis, A.: A multifaceted approach to understanding the botnet phenomenon. In: Proceedings of the 6th ACM SIGCOMM Internet Measurement Conference (2006)
41. Roesch, M.: The SNORT Network Intrusion Detection System (2002), http://www.snort.org
42. Vogt, R., Aycock, J., Jacobson Jr., M.J.: Army of botnets. In: Network and Distributed System Security Symposium, NDSS (2008)
43. Yegneswaran, V., Porras, P., Saidi, H., Sharif, M., Narayanan, A.: SRI's Multiperspective Malware Infection Analysis Page (2009), http://www.cyber-ta.org/releases/malware-analysis/public/
44. Zdrnja, B., Brownlee, N., Wessels, D.: Passive Monitoring of DNS Anomalies (2007)

Dealing with Liars: Misbehavior Identification via Rényi-Ulam Games*

William Kozma Jr. and Loukas Lazos

The University of Arizona, Electrical and Computer Engineering Dept. Tucson,
Arizona, 85712
{wkozma,llazos}@ece.arizona.edu

Abstract. We address the problem of identifying misbehaving nodes
that refuse to forward packets in wireless multi-hop networks. We map
the process of locating the misbehaving nodes to the classic Rényi-Ulam
game of 20 questions. Compared to previous methods, our mapping allows the evaluation of node behavior on a per-packet basis, without the
need for energy-expensive overhearing techniques or intensive acknowledgment schemes. Furthermore, it copes with colluding adversaries that
coordinate their behavioral patterns to avoid identification and frame
honest nodes. We show via simulations that our algorithms reduce the
communication overhead for identifying misbehaving nodes by at least
one order of magnitude compared to other methods, while increasing the
identification delay logarithmically with the path size.

1 Introduction

Multi-hop networks, such as wireless ad-hoc, sensor, and mesh networks rely
on collaboration among network nodes to provide reliable data services. If the
destination is not within the communication range of the source, data has to be
relayed by intermediate nodes. Implicit in this relay process is the assumption
that intermediate nodes are willing to forward traffic other than their own.

However, a fraction of nodes may not conform to the specifications of collaborative routing protocols. Sophisticated users can misconfigure their devices
to behave in a selfish manner and drop relay traffic, in order to save energy
resources [8, 9, 34]. Moreover, in hostile environments, an adversary may compromise several nodes and configure them to misbehave. It has been shown that
even a small fraction of misbehaving nodes refusing to relay packets, can lead to
a significant drop in the overall network performance [6, 7, 20, 21]. In this paper,
we address the problem of *developing resource-efficient methods for identifying
nodes that refuse to collaborate in relaying packets*. We define resource efficiency
in terms of the communication overhead associated with the identification of all
misbehaving nodes along a routing path.

Previously proposed solutions addressing routing misbehavior can be classified to reputation-based systems [6, 7, 21], credit-based systems [8, 9, 16, 34],

---

* This research was supported by BAE systems, and Connection One (an I/UCRC
NSF/industry/university consortium).

and acknowledgment-based systems [1, 2, 18, 20, 23]. A common element in all these solutions is the evaluation of node behavior on a *per-packet* basis. This approach provides a fine granularity in quantifying the behavior of nodes and low delay in identifying the misbehaving ones. However, it expends energy (in the form of receptions or transmissions) on a per-packet basis. For example, in acknowledgment-based systems, packets must be acknowledged two or more hops upstream [2, 1], thus consuming energy and bandwidth.

We develop a communication-efficient solution that allows the per-packet evaluation of behavior while not incurring the per-packet overhead. Nodes themselves are responsible for monitoring the packets they receive and forward to the next hop. When misbehavior is observed on a particular path, the source requests from nodes along the path to commit to a proof of the packets they receive and forward via an audit process (similar to [18, 1]). Although misbehaving nodes may lie about the packets they forward, the source combines multiple audit replies from honest nodes to identify the misbehaving ones.

Our Contributions: We map the problem of misbehavior identification to the classic Rényi-Ulam game of 20 questions [29, 32]. Rényi-Ulam games have been extensively used in various contexts including error correction codes [3], selecting, sorting, and searching in the presence of errors [25, 30, 31], to name a few. We develop communication-efficient algorithms for locating misbehaving nodes, based on different versions of Rényi-Ulam games. Our mapping allows the per-packet evaluation of node behavior without incurring the per-packet communication overhead. Furthermore, our formulation addresses colluding adversaries who coordinate their attacks to avoid identification and frame honest nodes.

The remainder of the paper is organized as follows. In Section 2, we present related work. In Section 3, we state the problem and our model assumptions. In Section 4, we map the misbehavior identification problem to Rényi-Ulam games and develop two auditing (searching) strategies. In Section 5, we present an efficient method for constructing audits. In Section 6, we compare the performance of our algorithms to previously proposed schemes. In Section 7, we conclude.

2   Related Work

Previously proposed methods for addressing the misbehavior problem can be classified into three categories: (a) credit-based systems, e.g., [8, 9, 16, 34], (b) reputation-based systems, e.g., [14, 7, 13, 21, 6, 22], and (c) acknowledgment-based systems, e.g., [1, 2, 20, 23].

Credit-Based Systems: Credit-based systems [8, 34, 9, 16] are designed to provide incentives for forwarding packets in the form of credit payments. Nodes accumulate credit that can be later used to pay for sending their own traffic. Buttyan et al. [8, 9] proposed a scheme in which a *nuglet counter* is used to tabulate the amount of credit accumulated at each node. To prevent tampering with the accumulated credit, the nuglet counter is implemented in tamper proof hardware. Zhong et al. [34] proposed Sprite, in which nodes collect *receipts* for the packets they forward which can be later exchanged for credit in a Credit Clearance Service (CCS). Jakobsson et al. [16] used cryptographic payment tokens that are attached to all packets and managed by a virtual bank. In credit-based systems a misbehaving node can drop relayed traffic if it is not interested in routing its own packets. Moreover, colluding nodes can agree to forward their own flows to accumulate credit while dropping all other flows. Finally, credit-based systems favor well connected nodes to boundary ones.

Reputation-Based Systems: Reputation-based systems [6,7,13,21,22,14] rely on building a reputation metric for each node according to its behavioral pattern. Buchegger et al. [6,7] proposed the *CONFIDANT* scheme, in which neighboring nodes monitor the behavior of their peers via overhearing. A similar monitoring method was proposed by Marti et al. [21]. In building the reputation metric, monitoring nodes usually overhear the transmission and reception of messages on a per-packet basis, thus operating their radio in promiscuous mode. Ganeriwal et al. [13] used a Bayesian model to map binary ratings into reputation metrics. He et al. [14] proposed SORI, which monitors neighboring nodes using a watchdog mechanism and propagates collected information to nearby nodes, thus relying on both first- and second-hand evaluations. Michiardi et al. [22] proposed CORE, where nodes combine reports from other nodes and task-specific monitoring to assign reputation metrics.

Node monitoring becomes complex in cases of multi-channel networks or nodes equipped with directional antennas. Neighboring nodes may be engaged in parallel transmissions in orthogonal channels thus being unable to monitor their peers. Moreover, operating in promiscuous mode requires up to 0.5 times the amount of energy for transmitting a message [12], thus making message overhearing an energy expensive operation.

Acknowledgment-Based Systems: Acknowledgment-based systems [1, 2, 20, 23] rely on the reception of acknowledgments to verify that a message was forwarded to the next hop. Liu et al. [20] proposed the 2ACK scheme, where nodes explicitly send acknowledgments two hops upstream to verify cooperation. Packets that have not yet been verified remain in a cache until they expire. A value is assigned to the quantity/frequency of unverified packets to determine misbehavior. The 2ACK scheme is susceptible to collusion of two or more consecutive nodes. Furthermore, colluding nodes can frame honest ones by claiming not to receive the acknowledgments. Padmanabhan et al. [23] proposed a method based on traceroute in which the source probes the path with pilot packets indistinguishable from data packets. Finally, Awerbuch et. al. [1] proposed an ACK-based scheme relying on a binary search process to identify a single misbehaving link. As with previous schemes, node collusion is not considered.

In our previous work [18], we proposed REAct, a reactive misbehavior identification scheme relying on audits. In REAct, the destination periodically sends acknowledgments to the source indicating the performance on the route. In the case of a performance drop, the source initiates a series of random audits to identify the misbehaving nodes. Nodes in the path in question provide a proof of the packets they forward to the next hop using Bloom filters. REAct reduces the communication overhead for identifying misbehaving nodes due to the compact representation of its audits. However, REAct does not address collusion.

3   Network and Adversarial Models

Network Model: We assume a multi-hop ad hoc network where nodes collaboratively relay traffic according to an underlying routing protocol such as DSR [17] or AODV [26]. The path P_{SD} used to route traffic from a source S to a destination D is assumed to be known to S. This is true for source routing protocols such as DSR. If DSR is not used, P_{SD} can be identified through a traceroute operation. For simplicity, we number nodes in P_{SD} in ascending order, i.e., n_i is *upstream* of n_j if $i < j$.

We assume that the source and destination collaboratively monitor the performance of P_{SD}. The destination periodically reports to the source critical metrics such as throughput or delay. If a misbehaving node drops the periodic updates as part of its misbehavior pattern, the source interprets the lack of updates as misbehavior. Likewise, the destination explicitly alerts the source in case the performance in P_{SD} is restored. These alerts are used to pause the misbehavior identification process and account for: (a) temporal variations of performance due to traffic or intermittent connectivity, and (b) random behavioral patterns of the misbehaving nodes. We initially consider a quasi-static network in which P_{SD} does not change during the misbehavior identification process. This is later relaxed, allowing changes in P_{SD} due to node mobility.

We assume that the integrity, authenticity, and freshness of critical control messages can be verified using resource-efficient cryptographic methods. For example, a public key cryptosystem realized via computationally-efficient elliptic curve cryptography may be used to verify the authenticity and integrity of messages while providing confidentiality [19]. Note that such cryptosystems require the existence of a trusted certificate authority (CA) for initialization (issuance of keys and certificates) as well as revocation of users via a certificate revocation list (CRL). Several methods have been proposed for the distributed implementation of a CA [11, 28, 33]. Alternatively, methods based on symmetric keys can be used to protect critical messages [15, 24, 27].

Adversarial Model: We assume that a set M of misbehaving nodes exist in a path of length $k \geq |M|$. Misbehaving nodes can be located anywhere in P_{SD}. The source and destination have a mutual interest in communicating, thus misbehavior of S and D is not considered. Misbehaving nodes are aware of the mechanism used for misbehavior identification. The goal of misbehaving is twofold; degrade throughput in P_{SD}, and remain undetected. We consider two models with respect to the behavioral pattern of nodes in M.

Independently misbehaving nodes: In this model, nodes in P_{SD} misbehave independently without coordinating their packet dropping patterns. Misbehavior is modeled after an ON/OFF process in which nodes alternate between dropping packets and behaving honestly. The duration of the misbehaving/behaving period is exponentially distributed with parameters μ_1, μ_2.

Colluding nodes: Colluding nodes share information with respect to the misbehavior identification process. For example, one misbehaving node can notify another of any actions of the source. Information sharing is achieved either in-band via the exchange of encrypted messages, or through an out-of-band coordination channel. Based on collective knowledge, the colluding nodes coordinate their behavioral patterns to avoid identification or frame honest nodes. In this model, we assume that colluding nodes are controlled by a single entity.

4 Misbehavior Identification

4.1 Motivation and Problem Mapping

The behavior monitoring mechanisms in previously proposed schemes operate on a per-packet basis, either with acknowledgments [1, 2, 20, 23], or message overhearing [6, 7, 21]. To reduce this overhead, we request nodes to self-evaluate the set of packets they forward to the next hop. In this self-evaluation process, honest nodes faithfully report the set of packets they received and forwarded, while misbehaving nodes may lie regarding packets they dropped.

We map the process of identifying lies to Rényi-Ulam searching games [29,32], that have been used for recovering an unknown value in the presence of errors. Using our mapping to Rényi-Ulam games, we develop novel misbehavior identification methods that are collusion resistant. We first provide a brief background on Rényi-Ulam games and then describe our mapping.

Background on Rényi-Ulam Games: Rényi-Ulam games are searching games independently proposed by Rényi [29] and Ulam [32]. These games involve two players; a questioner and a responder. The responder selects a secret value ω from a finite search space Ω. The questioner attempts to determine ω by asking at most q questions to which the responder is allowed up to ℓ lies. Before starting the game, the players agree on: (a) the search space Ω, (b) the number of questions q, (c) the number of lies ℓ, and (d) the mode of interaction between the players. The format of the questions can be classified into three categories: (a) bit questions, (b) cut questions, and (c) membership questions. Bit questions are defined as "Is the ith-bit of ω equal to 1?" Cut questions are defined as, for some $y \in \Omega$, "Is $\omega \leq y$?" Membership questions are defined as, for some subset $A \subseteq \Omega$, "Is $\omega \in A$?" The same questioning format is assumed for the entire game.

Two modes are possible for the interaction between the players; batch mode and adaptive mode. In batch mode, the questioner submits all questions to the responder at the same time. The responder is therefore able to review all questions before answering. In adaptive mode, the questioner asks questions one at a time. The questioner can adapt its strategy based on all previous answers. The questioner wins the game if it determines ω after at most q questions. Else, the responder wins. The questioner is said to have a "winning strategy" if it can find ω after at most q questions, independent of ω, or how the responder lies.

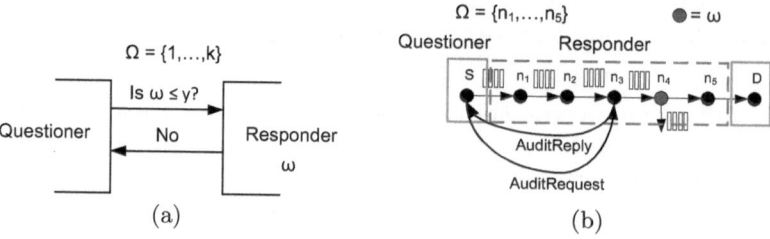

Fig. 1. (a) A generic Rényi-Ulam game. (b) Misbehavior identification mapped to a Rényi-Ulam game.

Mapping to Rényi-Ulam Games: In our mapping of misbehavior identification to Rényi-Ulam games, the role of the questioner is assumed by the source and destination, while the role of the responder is assumed by P_{SD}. The search space is defined as the set of nodes in P_{SD}, i.e., $\Omega = \{n_1, \ldots, n_k\}, k = |P_{SD}|$. The responder selects $\omega \in \{1, \ldots, k\}$, corresponding to the node n_ω in P_{SD} which is misbehaving. The source's goal is to determine n_ω, i.e., to locate the misbehaving node. Questions submitted by the questioner correspond to audits performed by the source to nodes in P_{SD}.

When responding to an audit, nodes state the set of packets forwarded to the next hop. The source combines one or more audits to construct bit, cut, or membership questions. The responder lies when a misbehaving node lies with respect to the packets forwarded to the next hop. For example, a node lies by either claiming to forward all packets received when in reality it drops them, or claiming to have forwarded no packets indicating they were dropped somewhere upstream. The location of the misbehaving nodes in P_{SD} is mapped to the placement of such lies by the responder. Note that since the responder is a single entity controlling the lies (i.e. location of misbehaving nodes and response to audits), our mapping implicitly assumes collusion. Figures 1(a) and 1(b) show the mapping of the misbehavior identification problem to a Rényi-Ulam game.

In our game, an honest node will always respond faithfully to an audit, thus a lie can only occur if a misbehaving node is audited. By adaptively selecting the nodes to be audited, the source can gather sufficient honest replies to identify nodes in M. If each node in P_{SD} is audited at most one time, the number of possible lies is limited to $\ell = |M|$. If nodes are audited multiple times, the number of lies depends on the exact auditing strategy. We now present two adaptive auditing strategies inspired by Rényi-Ulam games.

4.2 Rényi-Ulam Inspired Auditing Strategies

Let X_i denote the set of packets forwarded by a node n_i to the next hop. For example, the source sends packets X_S to the destination, and nodes n_i, n_j forward packets X_i, X_j respectively. In the absence of misbehavior in P_{SD} and assuming no packet loss $X_S = X_i = X_j$. In reality, some portion of the packets may be lost due to the wireless channel conditions or congestion, and hence $X_S \approx X_i \approx X_j$.

Definition 1. *A link (n_i, n_{i+1}) is defined as misbehaving if its two incident nodes n_i, n_{i+1} provide conflicting claims with respect to the packets forwarded to the next hop, i.e., $|X_i \cap X_{i+1}| \ll |X_i|$.*

Proposition 1. *At least one node incident to a misbehaving link is misbehaving.*

Proof. By contradiction. Assume that both nodes n_i, n_{i+1} of a misbehaving link are honest. Hence, the set of packets X_{i+1} forwarded by n_{i+1} to the next hop is approximately equal to the set of packets X_i, forwarded by n_i to n_{i+1}, i.e., $|X_i \cap X_{i+1}| \approx |X_i|$. This contradicts the definition of a misbehaving link.

Definition 2. *A simultaneous audit is defined as auditing two or more nodes with respect to the same set of packets X_S, sent from S to D via P_{SD}.*

Corollary 1. *The link between two behaving nodes n_i, n_{i+1} cannot be identified as misbehaving, when n_i, n_{i+1} are simultaneously audited.*

Proof. By Proposition 1, at least one misbehaving node is incident to any misbehaving link. Hence, two behaving adjacent nodes cannot be incident to a misbehaving link. The simultaneous audit requirement ensures that the dropping pattern of any misbehaving node upstream of behaving node n_i has the same effect on the packets observed by n_i, n_{i+1}. Thus packets forwarded by n_i are also forwarded by n_{i+1}, i.e., $|X_i \cap X_{i+1}| \approx |X_i|$.

Note that the converse of Corollary 1 is not true. For two nodes n_i, n_{i+1} for which $|X_i \cap X_{i+1}| \approx |X_i|$, we cannot conclude that both nodes are honest. Two colluding nodes may be incident to a link, and thus claim similar audit replies regardless of the packets forwarded.

Adaptive Audits with Cut Questions. We now show how the source can identify misbehaving nodes using an adaptive strategy and cut questions. Cut questions can be implemented by auditing one node at a time. These questions are of the form, "Is the misbehaving node upstream of n_i?", where n_i is the audited node. Assume there exists a single continuously misbehaving node n_M in P_{SD}. Define the set of nodes suspicious of misbehavior as $\mathcal{V} = \{n_1, \dots, n_k\}$. If $n_i \in \mathcal{V}$ is audited and replies with X_i such that $|X_S \cap X_i| \ll |X_S|$, the source concludes that all nodes downstream of n_i are behaving honestly, and therefore $n_M \le n_i$. This is true since either n_i is honest in which case it never received packets in X_S indicating an upstream misbehaving node, or n_i is the misbehaving node lying about its audit reply. If n_i replies that $|X_S \cap X_i| \approx |X_S|$, the source concludes that all nodes upstream of n_i are honest, and therefore $n_M \ge n_i$. This is true, since if any node upstream of n_i was the misbehaving one, n_i would not have received packets in X_S. Thus the set $\mathcal{V}$ is reduced to $\{n_i, \dots, n_k\}$.

Pelc [25] proposed a questioning strategy for adaptive games in which the questioner wins if he determines ω, or proves a lie took place. For a search space of size $|\Omega|$, and a maximum number of lies ℓ, the winning strategy requires $\lceil \log_2 |\Omega| \rceil + \ell$ questions. To find ω, the questioner first performs a binary search requiring $\lceil \log_2 |\Omega| \rceil$ questions to converge to a value ω'. It then asks the responder

ℓ times if $\omega \leq \omega'$. Since the responder is limited in lies, the questioner can determine if ω' is the secret value or the responder has lied.

Following the winning strategy proposed by Pelc, let the source win if either a misbehaving link is identified or the source can prove a lie has occurred. The source can converge to a single link by performing a binary search. The source initializes $\mathcal{V} = \{n_1, \ldots, n_k\}$ and selects node with index $i = \lceil \frac{|\mathcal{V}|}{2} \rceil$, for audit. As previously described, $\mathcal{V}$ is reduced to either $\{n_1, \ldots, n_i\}$ or $\{n_i, \ldots, n_k\}$. The source continues to audit nodes in $\mathcal{V}$ until $|\mathcal{V}| = 2$. In the case of a single misbehaving node, the source identifies the misbehaving link as shown in the following Proposition.

Proposition 2. *For a single misbehaving node, the source always converges to the misbehaving link in* $\log_2(|P_{SD}|)$ *audits.*

Proof. Let n_M denote the misbehaving node. Initially, $\mathcal{V} = P_{SD}$ and hence $n_M \in \mathcal{V}$. Let the source select a node n_i upstream of n_M for audit. Being upstream, n_i responds honestly that it forwarded packets to the next hop, reducing $\mathcal{V}$ to $\{n_i, \ldots, n_k\}$, with $n_M \in \mathcal{V}$. Similarly, if a node n_j downstream of n_M is audited, it will respond that no packets were forwarded, reducing $\mathcal{V}$ to $\{n_1, \ldots, n_j\}$. If n_M is audited, its response will indicate that misbehavior occurs either upstream of downstream. In either case $n_M \in \mathcal{V}$, since the audited node always remains in $\mathcal{V}$. The convergence of the binary search will end in a suspicious set $\mathcal{V} = \{n_{M-1}, n_M\}$ or $\mathcal{V} = \{n_M, n_{M+1}\}$, depending on whether n_M indicated that misbehavior occurs upstream of downstream. In any case, the identified link is a misbehaving one since per the definition, its two incident nodes provide conflicting audit replies. Since the binary search converges in $\log_2(|P_{SD}|)$, in case $|M| = 1$ the source will locate n_M in $\log_2(|P_{SD}|)$ steps.

If two or more nodes collude, the source may converge on a link in which both nodes are behaving, as shown in the following example. In Figure 2(a), $M = \{n_1, n_4\}$ with nodes n_1, n_4 colluding. Initially, n_4 drops all packets, while n_1 behaves. Let node n_2 be audited and report no misbehavior, thus $\mathcal{V} = \{n_2, n_3, n_4\}$. Assume now that nodes n_1, n_4 switch their behavior with node n_1 dropping packets while n_4 is behaving, as shown in Figure 2(b). If node n_3 is audited, it will report misbehavior upstream, reducing $\mathcal{V}$ to $\{n_2, n_3\}$ and thus removing n_4 from $\mathcal{V}$. Hence, link (n_2, n_3) is incorrectly identified as misbehaving.

Pelc solves this problem through the repetitive questioning of the result, thereby exhausting the responder's lies. In our case, a simultaneous audit on nodes n_i, n_{i+1} of an identified link $\mathcal{V} = \{n_i, n_{i+1}\}$ is sufficient to identify a misbehaving link or the occurrence of a lie. If $|X_i \cap X_{i+1}| \ll |X_i|$, a misbehaving link is identified. Else, the source concludes that a lie occurred. Returning to our previous example, in Figure 2(c), n_2 and n_3 are simultaneously audited. Since both nodes are honest, they return identical audit replies and no misbehaving link is identified. In this example, the responder has lied by changing the value of ω during the search, i.e., initially $\omega = n_4$, then $\omega = n_1$. However, S can identify that a lie occurred.

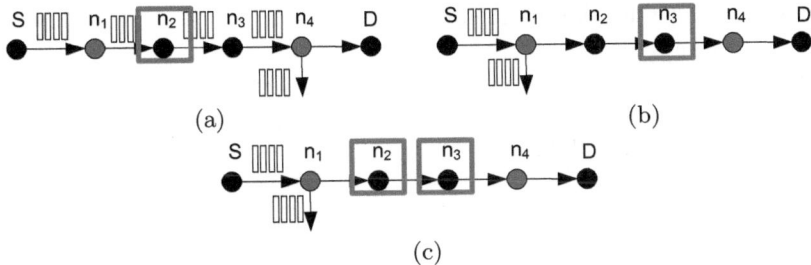

Fig. 2. (a) Nodes n_1, n_4 collude, with n_4 dropping all packets. Audited node n_2 claims misbehavior is downstream. (b) Nodes n_1, n_4 alter their behaviors, with n_1 dropping all packets. Audited node n_3 claims misbehavior is upstream. (c) Source simultaneously audits n_2, n_3 to verify if misbehaving link exists.

When the source identifies a lie occurred, is can also reach to the following conclusion: either (a) $n_M \in \mathcal{V}$ but lied during the simultaneous audit, or (b) $|M| \geq 2$ with at least one misbehaving node upstream of n_{i+1} and one downstream of n_i. Note that if $|M| = 1$ and the misbehaving node stops misbehaving (due to the fact that it is being audited) the destination alerts the source that misbehavior has stopped in P_{SD}. In such a case, the source will take two steps. First, any outstanding audits will be discarded. Second, the search will be suspended at the current state until misbehavior re-appears on P_{SD}. When misbehavior is resumed, the source continues the search from where it left off the last time misbehavior occurred.

If the destination does not alert the source that performance in P_{SD} has been restored, the source concludes that $|M| \geq 2$. This is evident in our example by the responses of n_2; on the first audit in Figure 2(a), it claims that misbehavior is downstream, while in Figure 2(c), it claims misbehavior is upstream. Let the audit process converge to link (n_i, n_{i+1}). Since the source knows that at least one misbehaving node is upstream of n_i and one is downstream, it attempts to isolate the effect of the misbehavior of each node by partitioning P_{SD} into $P_{Sn_i} = \{n_1, \ldots, n_i\}$ and $P_{n_{i+1}D} = \{n_{i+1}, \ldots, n_k\}$. The source repeats the audits recursively for each path partition $P_{Sn_i}, P_{n_{i+1}D}$. However, note that the destination can only determine if misbehavior occurs in P_{SD}, not which partition.

To treat each partition individually, the source considers n_i as a pseudo-destination and n_{i+1} as a pseudo-source. In P_{Sn_i}, node n_i is always audited simultaneously with any other node. Similarly node n_{i+1} is audited simultaneously with any other node in P_{n+1D}. Note that if n_i is the misbehaving node, it has only two strategies, (a) respond honestly, or (b) lie. If n_i lies, it immediately implicates itself in a misbehaving link, since both n_i, n_{i+1} are always audited. If n_i responds honestly, the search in P_{Sn_i} will converge to the misbehaving link (assuming one misbehaving node in P_{Sn_i}). For the realization of the cut questions, the source initializes $\mathcal{V}_{Sn_i} = \{n_1, \ldots, n_i\}$ and selects $n_j, j = \lceil \frac{|\mathcal{V}_{Sn_i}|}{2} \rceil$ for audit. The cut question "Is $n_M < n_j$?" is true if $|X_S \cap X_j| \ll |X_S|$ and $|X_S \cap X_i| \ll |X_S|$. The second condition verifies misbehavior on P_{Sn_i}.

Algorithm 1. Cut Questioning Algorithm

1: $n_i \leftarrow n_1, n_j \leftarrow n_{|P_{SD}|}, \mathcal{V} = \{n_i, \ldots, n_j\}$
2: **while** $|\mathcal{V}| > 2$ **do**
3:     $h = \lceil \frac{|\mathcal{V}|}{2} \rceil$, $Audit(n_h)$
4:     **if** $|X_S \cap X_h| \approx |X_S|$ **then**
5:         $n_i \leftarrow n_h$
6:     **else**
7:         $n_j \leftarrow n_h$
8:     **end if**
9: **end while**
10: $Audit(n_i, n_j)$
11: **if** $|X_i \cap X_j| \ll |X_i|$ **then**
12:     **return** X_i, X_j
13: **else**
14:     **return** $|M| \geq 2, Partition\ P_{SD}$
15: **end if**

Likewise on $P_{n_{i+1}D}$, the audit response of n_{i+1} acts as a verification if packets from X_S have reached this partition. Node n_{i+1} therefore acts as a pseudo-source for $P_{n_{i+1}D}$. Much like n_i, if n_{i+1} lies it immediately implicates itself in a misbehaving link since (n_i, n_{i+1}) is always audited. Thus the source can identify multiple misbehaving links using this adaptive auditing strategy. This strategy is presented in Algorithm 1.

Adaptive Audits with Membership Questions. Our scheme can also use an adaptive auditing strategy based on membership questions to identify the misbehaving nodes. Membership questions are constructed by combining two cut questions. To answer the question, "Is $n_M \in A = \{n_i, \ldots, n_j\}$?" the source audits n_i, n_j simultaneously and compares their audit replies. If $|X_i \cap X_j| \approx |X_i|$, then n_i, n_j claim $n_M \notin A$, since all packets forwarded by n_i are received by n_j. Else, they claim $n_M \in A$. Dhagat et al. [10] proposed an adaptive questioning strategy which proceeds in stages. During each stage, the questioner either believes the responder's answer and places it in a trusted set T, or discards it if it contradicts prior answers. Let $\mathcal{V}_j$ represent the set of possible values for ω at stage j, with $\mathcal{V}_1$ initialized to Ω.

Suppose that $\mathcal{V}_j$ is the current stage, with $|\mathcal{V}_j| > 1$, and let set $\{r_{j-1,a}, r_{j-1,b}\}$ represent the answers to round $j - 1$. The questioner divides $\mathcal{V}_j$ into two equal-sized subsets, A and B. The responder is asked "Is $\omega \in A$?" If the answer $r_{j,a}$ is "yes", the questioner adds $\{r_{j,a}\}$ to T and moves to the next stage with $\mathcal{V}_{j+1} = A$. Else, the questioner asks "Is $\omega \in B$?" If the answer $r_{j,b}$ is "yes," $\{r_{j,a}, r_{j,b}\}$ are added to T and the questioner moves to $\mathcal{V}_{j+1} = B$. If both $r_{j,a}, r_{j,b}$ are negative, the questioner removes $\{r_{j-1,a}, r_{j-1,b}\}$ from T, and returns to stage $\mathcal{V}_{j-1}$. The questioner then selects a different partition of $\mathcal{V}_{j-1}$ for stage j and repeats the questioning on each partition. Dhagat et. al. showed that the responder's secret value ω can be identified after $q = \lceil \frac{2 \log_2 |\Omega|}{1-3\beta} \rceil$ questions, when $\beta < \frac{1}{3}$, with β being the fraction of q than are lies [10]. To prevent repeated lies

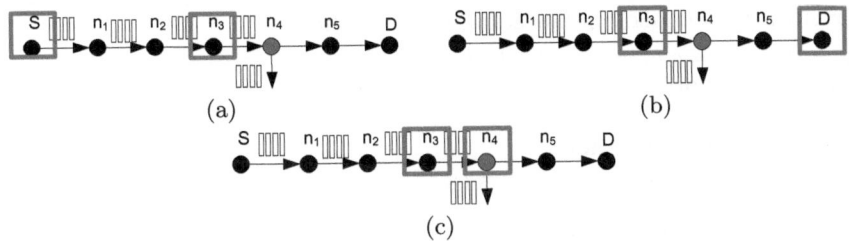

Fig. 3. (a) Let $\mathcal{V}_1 = \{S, n_1, \ldots, n_5, D\}$ with $A = \{S, n_1, n_2, n_3\}, B = \{n_3, n_4, n_5, D\}$ and $n_M = n_4$. The source audits A, concluding $n_M \notin A$. (b) The source then audits B, concluding $n_M \in B$. (c) The source proceeds to stage $\mathcal{V}_2 = \{n_3, n_4, n_5, D\}$ and continues the auditing strategy.

from the same misbehaving node, the source selects a new node and repeats the membership questions, until $|\mathcal{V}_j| = 2$.

Mapping Dhagat's questioning strategy to misbehavior identification, the source begins from stage $\mathcal{V}_1 = \{S, n_1, \ldots, n_k, D\}$. Set $\mathcal{V}_1$ is divided into two subsets, $A = \{S, \ldots, n_i\}$ and $B = \{n_i, \ldots, D\}$ with $i = \lceil \frac{|\mathcal{V}_1|}{2} \rceil$. The source first asks "Is $n_M \in A$?" by simultaneously auditing nodes S, n_i. If S and n_i return conflicting audit replies, the source knows that $n_M \in A$, adds $\{r_{1,a}\}$ to T, and proceeds to stage $\mathcal{V}_2 = \{S, \ldots, n_i\}$. Else, the source questions "Is $n_M \in B$?" by simultaneously auditing nodes n_i, D, whose audit replies define answer $r_{1,b}$. If n_i, D return conflicting audit replies, i.e., $|X_i \bigcap X_D| \ll |X_i|$, the source knows that $n_M \in B$, adds $\{r_{1,a}, r_{1,b}\}$ to T, and proceeds with $\mathcal{V}_2 = \{n_i, \ldots, D\}$. If both $r_{1,a}, r_{1,b}$ are negative, the source concludes a lie has occurred.

In Figure 3(a), $n_4 = n_M$. The source splits $\mathcal{V}_1 = \{S, n_1, \ldots, n_5, D\}$ to sets $A = \{S, n_1, n_2, n_3\}, B = \{n_3, n_4, n_5, D\}$, and audits S, n_3 to realize the membership question "Is $n_M \in A$?" Since n_3 is honest, the source asks "Is $n_M \in B$?" by simultaneously auditing n_3, D, as shown in Figure 3(b). Since n_3, D are honest, the source concludes $n_M \in B$. In Figure 3(c), the source moves to the next stage by dividing $\mathcal{V}_2 = B$ into two memberships sets. The process is repeated until $|\mathcal{V}_j| = 2$. In our example, the source converges to the misbehaving link (n_3, n_4). The source's auditing strategy is presented in Algorithm 2.

Proposition 3. *For a single misbehaving node, the source converges to the misbehaving link in less than* $4 \log_2(|P_{SD}|) + 2$ *audits.*

Proof. Let the source be at stage $\mathcal{V}_j = \{n_i, \ldots, n_k\}$ with $n_M \in \mathcal{V}_j$ and select node n_h for audit, creating membership sets $A = \{n_i, \ldots, n_h\}$ and $B = \{n_h, \ldots, n_k\}$. If $n_M \neq n_i, n_h, n_k$, then all audit responses will be honest and the source will conclude either $n_M \in A$ or $n_M \in B$, thus proceeding to the next stage with $\mathcal{V}_{j+1} = A$, $\mathcal{V}_{j+1} = B$ and $n_M \in \mathcal{V}_{j+1}$. As long as the source audits honest nodes, the set of suspicious nodes V_j will be reduced by half.

Now assume one of the n_i, n_h, n_k is n_M. When audited, n_M will either respond honestly, or lie. If n_M responds honestly, the search will proceed to state $\mathcal{V}_{j+1}$

Algorithm 2. Membership Questioning Algorithm

```
1:  V₁ = {nᵢ, ..., n_k}, nᵢ ← S, n_k ← D, T = r₁,ₐ
2:  while |V_j| > 2 do
3:      h = ⌈|V_j|/2⌉,  r_{j,a} = audit(nᵢ, n_h)
4:      if |Xᵢ ∩ X_h| ≪ |Xᵢ| then
5:          T ← {r_{j,a}},   j = j + 1,  V_j = {nᵢ, ..., n_h}
6:      else
7:          r_{j,b} = audit(n_h, n_k)
8:          if |X_h ∩ X_k| ≪ |X_h| then
9:              T ← {r_{j,a}, r_{j,b}},   j = j + 1,  V_j = {n_h, ..., n_k}
10:         else
11:             return  j = j − 1
12:         end if
13:     end if
14: end while
15: return  Xᵢ, X_k
```

with $n_M \in V_{j+1}$ and $|V_{j+1}| = \frac{|V_j|}{2}$. Thus the search continues to converge. If n_M lies, the source will obtain negative answers from both membership questions, unable to reduce V_j further, thus returning to stage V_{j-1} with $n_M \in V_{j-1}$. The source will then pick a different n_h, and repeat the set splitting, thus preventing the same lie from repeating.

In the absence of lies, the total number of membership questions needed for convergence to the misbehaving link is $2\log_2(|P_{SD}|)$. This is true, since at each stage we split the suspicious set in half similar to a binary search. To realize a membership question we need to simultaneously audit two nodes, requiring a total of $4\log_2(|P_{SD}|)$ audits in the worst case. If n_M is audited and lies, the search backtracks to the previous stage, resulting in the waste of two audits. For a single misbehaving node n_M and the fact that the source always selects a different node after a backtrack, n_M will be audited only once. Thus, in the worst case, the source requires $q \leq 4\log_2(|P_{SD}|) + 2$ audits.

Corollary 2. *The source never converges to a link with two behaving nodes.*

Proof. According to Algorithm 2, the source must receive conflicting reports from two simultaneously audited nodes to proceed from stage $j - 1$ to stage j. Hence, to terminate with $V_j = \{n_i, n_{i+1}\}$ the source must receive conflicting audit replies from n_i, n_{i+1} when simultaneously audited. However, via Corollary 1, this cannot occur if n_i, n_{i+1} are behaving nodes.

It is possible that multiple neighboring colluding nodes can delay the search indefinitely. Assume all nodes in V_j collude. Once in stage V_{j+1}, the replies to the audits from the colluding nodes yield membership questions on both partitions negative, thus forcing the source to return to stage V_j. Auditing any other node in V_j will yield the same results since nodes in V_j are colluding. If the source has audited all possible partitions of V_j, and thus all $n_i \in V$, with no progress to the next stage, it terminates the search and proceeds to the identification phase.

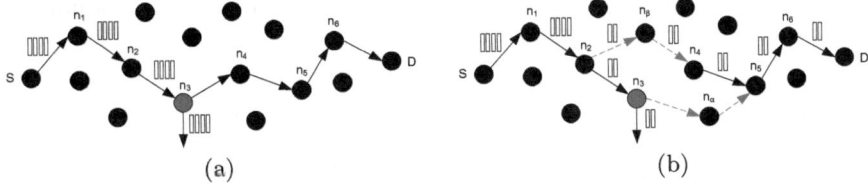

Fig. 4. (a) Node n_3 drops packets, with link (n_3, n_4) being the misbehaving link. (b) Slight alteration to routing path.

4.3 Misbehaving Node Identification

Once the source has converged to a misbehaving link (n_i, n_{i+1}), it can no longer proceed to identify the misbehaving node. The two conflicting audit responses from n_i, n_{i+1} indicate that either n_i or n_{i+1} is lying. From the routing point of view, identifying the misbehaving link is sufficient for restoring the performance in P_{SD} since the source can now avoid this link. However, we would like to identify and isolate the misbehaving node to prevent it from further affecting other paths. This is accomplished through the idea of path division. The path P_{SD} is divided in such a way that new independent observations can be made with respect to n_i and n_{i+1}. We first illustrate the idea of path division for a single misbehaving nodes and then generalize to many.

Single Misbehaving Node. Without loss of generality assume that the audit process converged to (n_M, n_{M+1}), where n_M is the misbehaving node. The source divides P_{SD} into two paths such that packets are routed through either n_M or n_{M+1}, and attempts to re-identify the misbehaving link. This can be achieved by bypassing each node in P_{SD} via an alternative path. Instead of performing the entire audit process, the source concentrates on the nodes around n_M, n_{M+1}, For example, in Figure 4(a), the source has identified link (n_3, n_4) as the misbehaving one. In Figure 4(b), the source splits the traffic between two paths that bypass n_3, n_4 in turn via nodes n_α, n_β. Path segment $\{n_2, n_3, n_4, n_5\}$ is replaced by the segments $\{n_2, n_\beta, n_4, n_5\}$ and $\{n_2, n_3, n_\alpha, n_5\}$, thus isolating n_3, n_4 from each other. The source simultaneously audits nodes n_β, n_4 and n_3, n_α to identify the misbehaving link. The source identifies link (n_3, n_α) as misbehaving, and hence identifies the misbehaving node n_3.

Multiple Misbehaving Nodes. Assume now the existence of multiple misbehaving nodes in P_{SD}. If the cut auditing strategy is employed, the source will split P_{SD} to smaller paths in order to isolate the effect of each misbehaving node. The source can then perform the path division in each subpath as in the case of a single misbehaving node. Note that, as in the case of a single misbehaving node, the newly added nodes must not be misbehaving in order to avoid framing honest nodes. If the membership questioning strategy is employed, the source will converge to a set $\mathcal{V}_j$ containing at most one honest node. To identify the misbehaving one, all nodes in $\mathcal{V}_j$ must be excluded in turn from P_{SD} according to the path division process. That is, the source constructs $|\mathcal{V}_j|$ individual paths with each node in $\mathcal{V}_j$ being present on only one path.

4.4 Mobility

We now relax our assumption that P_{SD} does not change during the identification process. Let a node n_i be removed from P_{SD}. If $n_i \notin \mathcal{V}$, then its removal has no effect on the search. The source identifies misbehaving links from the nodes in $\mathcal{V}$. Let $n_i \in \mathcal{V}$. There are two cases, either n_i is a behaving node, or n_i is misbehaving. If n_i is behaving, then removing it is analogous to reducing $\mathcal{V}$ to a smaller set that still contains the misbehaving node. If n_i is misbehaving, then the performance in P_{SD} is restored or one less misbehaving node is present.

Consider now adding a new node n_i to P_{SD}. If n_i is added between nodes in $\mathcal{V}$, then regardless of n_i's behavior, this is equivalent to n_i being in $\mathcal{V}$, in the first place and not yet been audited. Let n_i be added in P_{SD} outside $\mathcal{V}$. If n_i is an honest node, there is no effect on the audit process. If n_i is a misbehaving node, then this is equivalent to the situation in which $|M| \geq 2$ and one of the n_M has been removed from $\mathcal{V}$. However, we have shown that both auditing strategies can address the case of multiple misbehaving nodes. In the case of cut questions, the source splits P_{SD} into two paths while in the case of membership questions, the source converges on the misbehaving node in $\mathcal{V}$. Once this node is removed, the source will continue to identify the newly added misbehaving node.

5 The Audit Mechanism

We now describe how the source can perform audits in a resource-efficient manner. The audit mechanism is adopted from [18] and is based on the compact representation of a membership set via Bloom filters [4]. The goal of auditing a node $n_i \in P_{SD}$ is to force n_i to commit to the set of packets X_i that it received and forwarded to the next hop. Contradicting commitments are used to identify misbehaving links and eventually misbehaving nodes. To respond to an audit, the node n_i records the packets forwarded for a period of time, and reports them to the source. Based on this report, the source compares the packets in X_i with the packets in X_S originally sent to the destination. Buffering the packets themselves requires a large amount of storage and significant overhead for transmission back to the source. On the other hand, Bloom filters provide a storage-efficient way of performing membership testing [4]. The audit process occurs in three steps; sending an audit request, constructing the audit reply, and processing the audit reply. We now describe these steps in detail.

Sending an Audit Request: The source audits a node n_i according to the algorithms described in Section 4. The source selects the audit duration a_d, measured in number of packets, and the initial packet sequence number a_s from which the audit will begin. The value of a_d is a parameter that must be sufficiently large to differentiate misbehavior from normal packet loss. The audit request is routed to n_i via P_{SD}. Values a_s and a_d are randomized thereby preventing any misbehaving nodes from conjecturing the start and duration of audits, unless they are audited themselves. Note that an audit request may fail to reach the audited node n_i since a misbehaving node along P_{Sn_i} may drop it, or n_i is the misbehaving node and chooses not to respond. In this case, the source tries a threshold number of times to audit n_i. Failure to obtain a reply is interpreted as "Node n_i did not forward packets in X_S to the next hop." This is true since either n_i is the misbehaving node or a misbehaving node is upstream of n_i.

Constructing an Audit Reply: When a node n_i is audited, it constructs a Bloom filter of the set of packets it receives and forwards, from a_s to $a_s + a_d$, denoted by $X_i = \{x_{a_s}, x_{a_s+1}, \ldots, x_{a_d}\}$. By using a Bloom filter, packets in X_i can be compactly represented in an m-bit vector v_i with $m \ll |X_i|$ [4]. After a_d packets have been added to v_i, node n_i signs v_i, and sends it to S via the reverse path P_{n_iS}. The signed Bloom filter binds the audited node to the set of packets X_i that it claims to have forwarded to the next hop, in a publicly verifiable manner. Based on n_i's signature, any node can verify the authenticity and integrity of v_i. To assess the behavior of audited nodes, the source constructs its own Bloom filter v_S in the same manner as n_i. When S receives n_i's Bloom filter, it compares it against v_S and compute what fraction of packets in X_S was forwarded by n_i.

Processing the Audit Reply: When S receives v_i, it verifies its authenticity and discards v_i if the signature check fails. Otherwise, given the vector length m, the cardinalities of X_i, X_S, filters v_i, v_S, and the number z of hash functions used to generate the Bloom filters, S computes the metric [5],

$$|X_S \bigcap X_i| \approx |X_S| + |X_i| - \frac{\log_2\left(\frac{<v_S, v_i>}{m} + \left(1 - \frac{1}{m}\right)^{z|X_S|} + \left(1 - \frac{1}{m}\right)^{z|X_i|}\right)}{z \log_2\left(1 - \frac{1}{m}\right)} \quad (1)$$

6   Performance Evaluation

6.1   Simulation Setup

We randomly deployed 100 nodes within an 80×80 square and selected 10 source/destination pairs. For each pair, we constructed the shortest path and randomly selected the set of misbehaving nodes. We generated traffic from S to D according to the constant bit-rate (CBR) model. Each misbehaving node randomly selected a behavioral state of either *behave* or *misbehave*, with equal probability. It then randomly selected the duration of the state from the interval $[1, 400]$ packets. We focus on two metrics of interest: (a) the *communication overhead* defined as the number of messages transmitted/received by nodes in P_{SD}, weighed by $1/0.5$, respectively [12], and (b) the *identification delay* defined as the time elapsed from the occurrence of misbehavior until the misbehaving nodes are identified, normalized over the audit duration. Simulations were performed in a packet-level C simulator.

6.2   Auditing Strategy Comparison

We first compared the performance of the two auditing strategies; the strategy based on cut questions as described by Algorithm 1, which we will refer to as CUT, and the strategy based on membership questions as described by Algorithm 2, which we will refer to as MEM.

Communication Overhead. In Figure 5(a), we show the communication overhead required to identify one misbehaving node as a function of the path length. We observe that CUT requires less communication overhead than MEM. This is expected, as the realization of cut questions requires only one audit, whereas membership questions require two audits. Both auditing strategies audit in a binary fashion, thus resulting in logarithmic increase in communication overhead as a function of the path length. In Figure 5(b), we show the communication overhead required to identify two misbehaving nodes as a function of path length.

Identification Delay. In Figure 5(c), we show the delay required to identify one misbehaving node as a function of the path length. Both CUT and MEM incur approximately the same delay due to their binary search approach. In Figure 5(d), we show the delay required to identify two misbehaving nodes as a

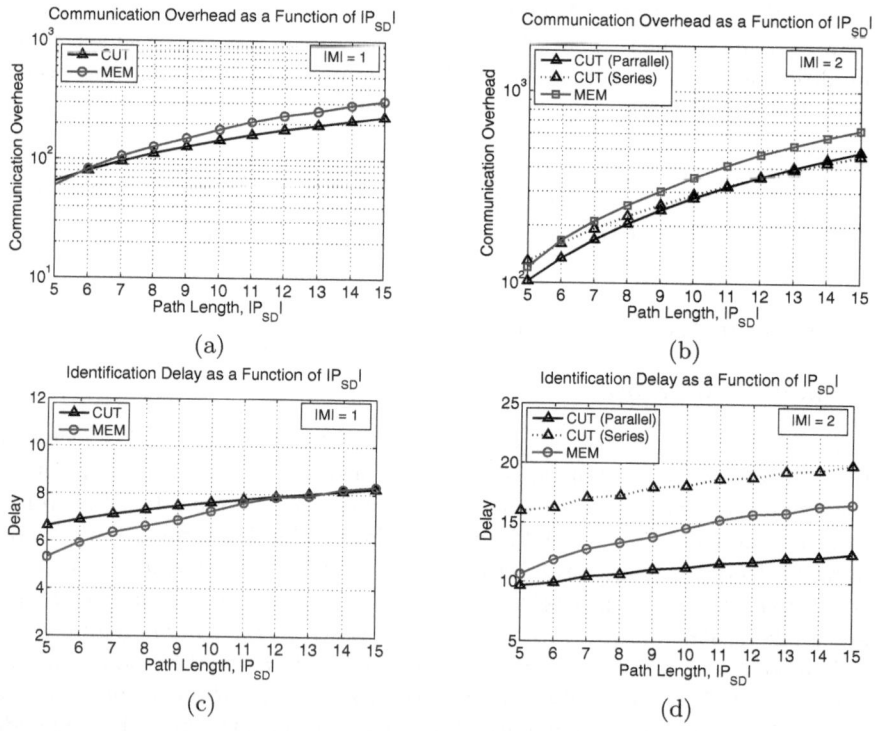

Fig. 5. Communication overhead for (a) one misbehaving node, (b) two misbehaving nodes. Identification delay for (c) one misbehaving node, (d) two misbehaving nodes.

function of the path length. In CUT, after the path is partitioned, the auditing of the two partitions is dependent on the misbehavior strategies of nodes in M. Assume that only one misbehaving node drops packets at a time. Thus the search will only audit the path partition which is reporting misbehavior. This causes the source to search the partitions in series, i.e., one at a time. If both misbehaving nodes drop packets, the source can audit the two path partitions in parallel, since each path partition contains a source (or pseudo-source) and a destination (or pseudo-destination). This parallel auditing decreases the delay.

For CUT, we plot both the case of search in series and parallel, giving an expected range for the delay. Note that the delay of MEM falls within this range; closer to the parallel CUT for smaller path sizes and closer to the series CUT as the path length increases. This is due to the nature of the auditing strategies employed. In CUT, the source cannot determine if a lie occurred until performing the simultaneous audit at the end of the auditing strategy. In MEM, the source determines if a lie occurred by looking for contradictions at every stage. Therefore, if a lie is found, the penalty is only the waste of two audits. This results in a tradeoff in which MEM incurs an additional overhead per stage compared to CUT by checking for contradictions at the expense of delay.

6.3   Comparison with Other Schemes

We now compare the performance of our algorithms to CONFIDANT [6], 2ACK [20], and AWERBUCH [1]. For CONFIDANT, every one-hop neighbor of a transmitting node was assumed to operate in promiscuous mode, thus overhearing transmitted messages. For 2ACK, a fraction p of the messages transmitted by each node was acknowledged two hops upstream of the receiving node. We set $p = \{1, 0.5, 0.1\}$ [20]. AWERBUCH identifies misbehaving links by requesting selected nodes in P_{SD} to acknowledge each packet back to the source. For comparison, we select the adaptive auditing strategy utilizing cut questions. The plots of Figure 5(a)-(d) can be used for comparisons with MEM. We first considered the overhead during a fixed duration of time, i.e., the time required to identify the misbehaving node using CUT.

Fixed Time Communication Overhead. In Figure 6(a), we show the communication overhead as a function of the path length. The Y axis is shown in logarithmic scale. The communication overhead for CUT is between 1-2 orders of magnitude less compared to other schemes. This gain is due to the fact that CUT does not expend energy on a per-packet basis to monitor the behavior of each node. The 2ACK scheme presents the highest communication overhead since every packet requires a 2-hop acknowledgment upstream per link traversed.

In Figure 6(b), we show the communication overhead as a function of the audit duration a_d for a path of eight nodes. Schemes 2ACK, CONFIDANT, and AWERBUCH all incur a linear increase in communication overhead, due to the per-packet behavior evaluation. On the other hand, the communication overhead for CUT and MEM is incurred on a per-audit basis, and is independent of audit duration. While our algorithms provide significant savings in communication

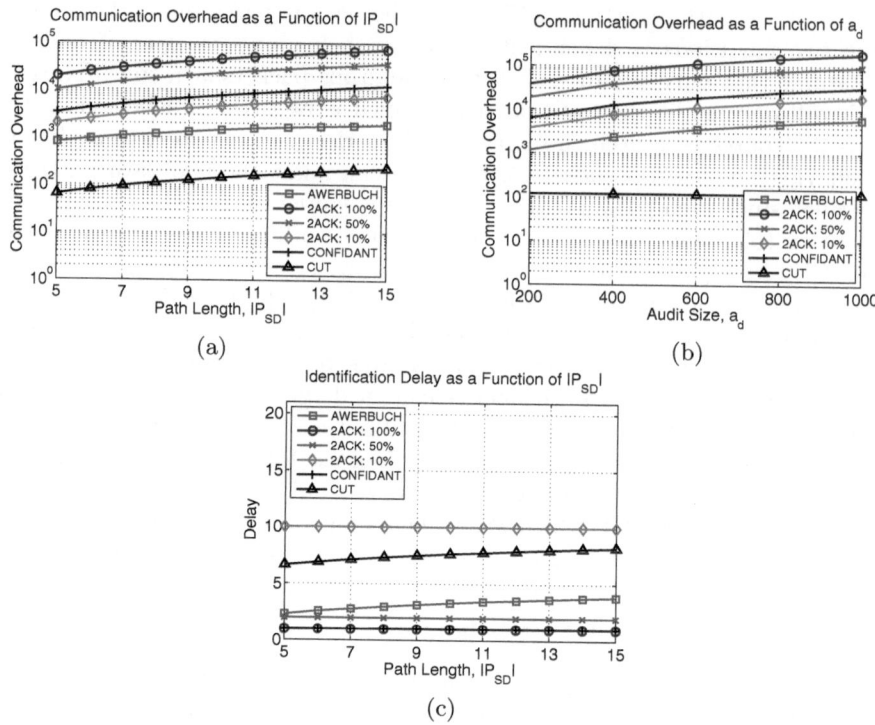

Fig. 6. (a) Communication overhead as a function of $|P_{SD}|$ for an audit size of 200 packets. The overhead is computed over time required by the CUT scheme to converge to the misbehaving node. (b) Communication overhead as a function of audit size for $|P_{SD}| = 8$. (c) Delay as a function of $|P_{SD}|$ in units of number of audits.

overhead, they require a longer time to identify the misbehaving nodes. On the other hand, the proactive schemes require only the duration of one audit to identify misbehavior. This is due to the fact that proactive protocols monitor all nodes in the path P_{SD} in parallel. Fortunately, for schemes CUT and MEM, the delay grows logarithmically with $|P_{SD}|$. Hence, the increase in identification delay is small compared to the savings in communication overhead.

In Figure 7(c), we show the identification delay as a function of path length. CONFIDANT requires a single audit duration to identify the misbehaving node since all nodes in P_{SD} are monitored in parallel. AWERBUCH performs a binary search, incurring a logarithmic increase in delay. The 2ACK scheme also requires a single audit duration for identification when all packets are acknowledged. However, the identification delay increases when only a fraction of the packets are acknowledged. For example, when only 10% of the packets are acknowledged, 2ACK and CUT incurr similar delay. However, as shown in Figure 6(b), CUT incurs an order of magnitude less in communication overhead.

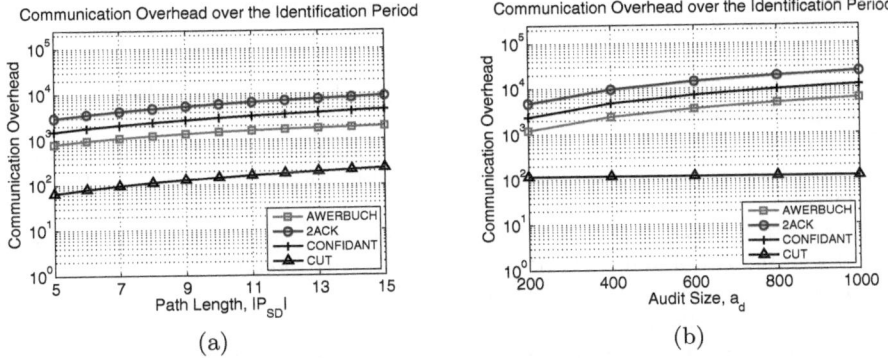

Fig. 7. (a) Communication overhead for an audit size of 200 packets. For each scheme, the overhead is computed for the time required to identify misbehavior, (b) communication overhead as a function of audit size for $|P_{SD}| = 8$.

Comparison Based on Identification Delay. We now evaluate the communication overhead incurred by each scheme until the misbehaving node is identified. In Figure 7(a), we show the communication overhead as a function of the path length, for an audit size of 200 packets. In Figure 7(b), we show the communication overhead as a function of the audit size, for a path of eight nodes. We observe that even in the case where the communication overhead is measured only during the identification delay, CUT significantly outperforms the other schemes. The CONFIDANT, 2ACK and AWERBUCH schemes are sensitive to path length and audit size. On the other hand, CUT illustrates a graceful tradeoff between communication overhead and delay.

7   Conclusion

We addressed the problem of identifying misbehaving nodes that refuse to forward packets to the destination in a wireless multi-hop network. We mapped this problem to the classic Rényi-Ulam game of 20 questions. From this mapping we employed communication efficient questioning strategies which allow the source to locate the set of misbehaving nodes. We showed that our scheme significantly reduces the communication overhead associated with misbehavior identification compared to previously proposed schemes. This reduction in resource expenditure comes at the expense of a logarithmic increase in the identification delay.

References

1. Awerbuch, B., Holmer, D., Rotaru, C.-N., Rubens, H.: An on-demand secure routing protocol resilient to byzantine failures. In: WiSe 2002 (2002)
2. Balakrishnan, K., Deng, J., Varshney, P.K.: Twoack: Preventing selfishness in mobile ad hoc networks. In: WCNC 2005 (2005)

3. Berlekamp, E.: Error Correcting Codes. Wiley, NY (1968)
4. Bloom, B.H.: Space/time trade-offs in hash coding with allowable errors. Communications of the ACM 13(7), 422–426 (1970)
5. Broder, A., Mitzenmacher, M.: Network applications of bloom filters: A survey. Internet Mathematics 1(4), 485–509 (2004)
6. Buchegger, S., Boudec, J.-Y.L.: Performance analysis of the confidant protocol (cooperation of nodes: Fairness in dynamic ad-hoc networks). In: MobiHOC 2002 (2002)
7. Buchegger, S., Boudec, J.-Y.L.: Self-policing mobile ad-hoc networks by reputation systems. IEEE Communications Magazine, 101–107 (2005)
8. Buttyan, L., Hubaux, J.-P.: Enforcing service availability in mobile ad-hoc wans. In: MobiHOC 2000, pp. 87–96 (2000)
9. Buttyan, L., Hubaux, J.-P.: Stimulating cooperation in self-organizing mobile ad hoc networks. ACM/Kluwer Mobile Networks and Applications 8(5) (2003)
10. Dhagat, A., Gács, P., Winkler, P.: On playing "twenty questions" with a liar. In: SODA 1992, pp. 16–22. Society for Industrial and Applied Mathematics (1992)
11. Dong, Y., Go, H., Sui, A., Li, V., Hui, L., Yiu, S.: Providing Distributed Certificate Authority Service in Mobile Ad Hoc Networks. In: SecureComm 2005 (2005)
12. Feeney, L.M., Nilsson, M.: Investigating the energy consumption of a wireless network interface in an ad hoc networking environment. In: INFOCOM 2001 (2001)
13. Ganeriwal, S., Srivastava, M.: Reputation-based framework for high integrity sensor networks. In: SASN 2004, pp. 66–77 (2004)
14. He, Q., Wu, D., Khosla, P.: Sori: A secure and objective reputation-based incentive scheme for ad hoc networks. In: WCNC 2004 (2004)
15. Hu, Y., Johnson, D., Perrig, A.: SEAD: secure efficient distance vector routing for mobile wireless ad hoc networks. Ad Hoc Networks 1(1), 175–192 (2003)
16. Jakobsson, M., Hubaux, J.-P., Buttyan, L.: A micropayment scheme encouraging collaboration in multi-hop cellular networks. In: Proc. of Financial Crypto (2003)
17. Johnson, D., Maltz, D., Hu, Y. C.: The dynamic source routing protocol for mobile ad hoc networks (dsr). draft-ietf-manet-dsr-09.txt (2003)
18. Kozma Jr., W., Lazos, L.: REAct: Resource-Efficient Accountability for Node Misbehavior in Ad Hoc Networks based on Random Audits. In: WiSec 2009 (2009)
19. Liu, A., Ning, P.: Tinyecc: A configurable library for elliptic curve cryptography in wireless sensor networks. In: IPSN 2008 (2008)
20. Liu, K., Deng, J., Varshney, P., Balakrishnan, K.: An acknowledgment-based approach for the detection of routing misbehavior in manets. IEEE Transactions on Mobile Computing 6(5), 536–550 (2006)
21. Marti, S., Giuli, T., Lai, K., Baker, M.: Mitigating routing misbehavior in mobile ad hoc networks. In: MobiCom 2000, pp. 255–265 (2000)
22. Michiardi, P., Molva, R.: Core: A collaborative reputation mechanism to enforce node cooperation in mobile ad hoc networks. In: CMS 2002 (2002)
23. Padmanabhan, V.-N., Simon, D.-R.: Secure traceroute to detect faulty or malicious routing. SIGCOMM Computer Communication Review 33(1) (2003)
24. Papadimitratos, P., Haas, Z.: Secure routing for mobile ad hoc networks. In: SCS Communication Networks and Distributed Systems Modeling and Simulation Conference (CNDS 2002), pp. 1–27 (2002)
25. Pelc, A.: Detecting errors in searching games. Journal of Combinatorial Theory Series A 51(1), 43–54 (1989)
26. Perkins, C., Royer, E., Das, S.: Ad hoc On-Demand Distance Vector (AODV) Routing (2003)

27. Perrig, A., Szewczyk, R., Tygar, J., Wen, V., Culler, D.: SPINS: Security Protocols for Sensor Networks. Wireless Networks 8(5), 521–534 (2002)
28. Raghani, S., Toshniwal, D., Joshi, R.: Dynamic Support for Distributed Certification Authority in Mobile Ad Hoc Networks. In: Proceedings of the 2006 International Conference on Hybrid Information Technology, vol. 1, pp. 424–432. IEEE Computer Society, Washington (2006)
29. Rényi, A.: A Diary on Information Theory. Wiley, New York (1984)
30. Rivest, R., Meyer, A., Kleitman, D., Winklmann, K., Spencer, J.: Coping with errors in binary search procedures. J. Comput. System Sci. 20, 396–404 (1980)
31. Spencer, J., Winkler, P.: Three thresholds for a liar. Combinatorics, Probability and Computing 1, 81–93 (1992)
32. Ulam, S.: Adventures of a Mathematician. Scribner, New York (1976)
33. Yi, S., Kravets, R.: MOCA: Mobile Certificate Authority for Wireless Ad Hoc Networks. In: 2nd Annual PKI Research Workshop Pre-Proceedings, vol. 51
34. Zhong, S., Chen, J., Yang, Y.R.: Sprite: A simple cheat-proof, credit-based system for mobile ad-hoc networks. In: INFOCOM 2003 (2003)

Multichannel Protocols for User-Friendly and Scalable Initialization of Sensor Networks

Toni Perković, Ivo Stančić, Luka Mališa, and Mario Čagalj

FESB, University of Split, Croatia
{toperkov,istancic,lmalisa,mcagalj}@fesb.hr

Abstract. We consider the classical problem of establishing *initial* security associations in wireless sensor networks. More specifically, we focus on pre-deployment phase in which sensor nodes have not yet been loaded with shared secrets or other forms of authentic information.

In this paper, we propose two novel *multichannel* protocols for initialization of large scale wireless sensor networks. The first protocol uses only secret key cryptography and is suitable for CPU-constrained sensor nodes. The second protocol is based on public key cryptography. Both protocols involve communication over a bidirectional radio channel and an unidirectional out-of-band *visible light channel*. A notable feature of the proposed "public key"-based key deployment protocol is that it is designed to be secure in a very strong attacker model, where an attacker can eavesdrop, jam and modify transmitted messages by adding his own message to both a radio and a visible light channel; the attacker however cannot disable the visible light communication channel. We show that many existing protocols that rely on the visible light channel are insecure in this strong adversary model.

We implemented the proposed protocols on the Meshnetics wireless sensor platform. The proposed protocols are cheap to implement, secure in the very strong attacker model, easy to use and scalable. We also designed and tested a simple random number generator suitable for sensor platforms.

1 Introduction

Deployment of cryptographic keys into individual sensor nodes is an imperative for secure operation of a sensor network. While there is a large body of work on key management in scenarios where cryptographic keys are already deployed into the nodes [9,11,12,21,27], very few studies exists on the equally important problem of establishing initial security associations in large wireless sensor networks.

Many existing systems consider the key pre-deployment to be a trivial matter. Thus, we can read that "the key distribution is relatively simple; nodes are loaded with a shared key before deployment". Long experience with WiFi networks have taught us that very often such "relatively simple" setup procedures render the security features useless (users easily give up and thus leave their networks unprotected), even when dealing with only a few network devices. Some

Y. Chen et al. (Eds.): SecureComm 2009, LNICST 19, pp. 228–247, 2009.

other solutions propose to send the key in the clear over the radio channel or alternatively, imprint the keys onto the nodes at production time (ZigBee [1]). The problem with this approach is that customers may not trust the keys deployed by the factory.

Solutions that require physical contact are not scalable, especially if the user is required to initialize a large number of nodes. More advanced solutions have been proposed in [8,5,3,24,24,19,30] some of which do not scale well and/or require specialized node hardware, and some are insecure in the realistic attacker model introduced in this paper.

When dealing with initialization of network nodes on a large scale, a secure, fast, cost effective and above all user-friendly solution is mandatory. In this paper, we propose two novel *multichannel* protocols for initialization of large scale wireless sensor networks. Similar to [30], our protocols involve communication over a radio channel and the out-of-band visible light channel (VLC). The first protocol uses only secret key cryptography and is suitable for CPU-constrained sensor nodes. The "secret key"-based initialization of sensor nodes is depicted in Figure 1(a). In this protocol, each sensor node establishes a unique secret key with a *base station (BS)*. The base station comprises a simple web camera and one sensor node all attached to an ordinary PC. In the first phase of the protocol, the sensor nodes transmit secret keys to the base station over a *protected* visible light channel (Figure 1(a)). In the second phase, each sensor node runs a key verification protocol with the base station over a bidirectional radio channel. Once the keys are verified, the base station can serve as a *trusted third party* and mediate establishment of security associations between any pair or any group of sensor nodes.

Our second protocol uses public key cryptography. The "public key"-based sensor node initialization process is summarized in Figure 1(b). As with the previous protocol, the ultimate goal is to establish security association between each sensor node and the base station. This protocol is based on the multichannel pairing protocol from [38,5]. Thus, each sensor node first exchanges its public key (through specially formed commitment/openning pairs) with the base station over a radio channel (Figure 1(b)). In turn, each sensor node transmits a *short authentication string* (SAS) using a visible light channel (Figure 1(b) - right). The proposed "public key"-based protocol is similar to [30], with the difference that our protocol is designed to be secure in a very strong attacker model, where an attacker can eavesdrop, jam and modify transmitted messages by adding his own message to both a radio and a visible light channel; the attacker however cannot disable the visible light communication channel[1].

The paper is organized as follows: in Section 2 we state the problem and assumptions. In Sections 3 and 4 we present the "secret key"- and "public key"-based protocols (including security analysis of both protocols). We describe the implementa-

---

[1] It was brought to our attention recently that a similar approach has been suggested in [31]. The initialization method in [31], however, is developed for a weaker attacker model than the one we consider here.

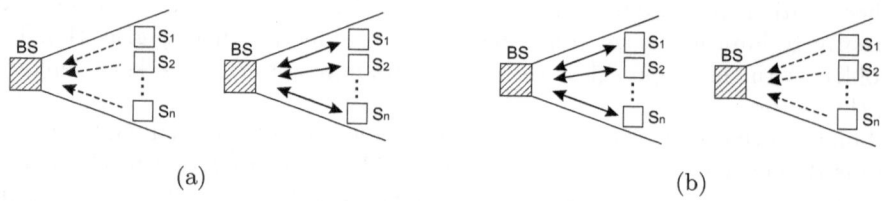

Fig. 1. Two phases of node initialization for (a) secret key and (b) public key deployment protocol. In (a) nodes transmit the key to the base station via VLC (dashed arrows) and perform authentication via a radio channel (full line arrows), while in (b) they exchange public keys over a radio channel and perform authentication via VLC.

tion of the protocols and a simple random number generator in Section 5. Related work is provided in Section 6. Finally, we conclude in Section 7.

2   Problem Statement and System Model

We consider the following problem: *How to securely initialize a large number of sensor nodes in a user-friendly fashion?* Since the initialization will be performed by potentially non-expert personnel, a solution has to be easy both to learn and use (*user-friendliness*). In addition, the hardware cost per node has to be minimized (*cost-efficiency*).

2.1   System Model

We assume that a user is equipped with a base station used for verification and monitoring as shown in Figure 2.

Base Station. The base station comprises a monitor, a simple web camera and one sensor node (a verification node) all attached to an ordinary PC. The verification node serves as a radio modem to the base station.

Uninitialized Sensor Nodes. Nodes may be equipped with a single LED (we used two LEDs in our implementation) used for key transmission via out-of-band VLC and with radio transceivers. In addition, each node has a "pushbutton" used to either restart or finalize the initialization process.

Cardboard box. A simple cardboard box is used to block the escape of light during the key transmission via VLC. The cardboard box is required only for the "secret key" - based key deployment protocol.

2.2   Attacker Model

An adversary has full control over the radio channel. He can eavesdrop, drop, delay, replay and modify messages sent via radio. Thus, he is able to initiate communication with any device (a node or the base station) and at any given

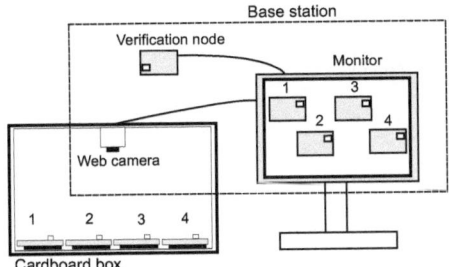

Fig. 2. Secret key deployment setup comprises a base station and a simple cardboard box

time during the key transmission. Furthermore, the adversary can install his own web camera in the same place where the initialization is taking place. We assume that devices involved in key deployment (PC and nodes themselves) are not compromised. Taking into account these constraints, we define: (a) *a passive adversary* who only observes the visual channel and can eventually record a secret key if the key transmission takes place in insecure conditions (outside the cardboard box), and (b) *an active adversary* who in addition can initiate communication with any device during the initialization phase.

In the case of "public key" - based initialization, we consider a stronger adversary model where an attacker can eavesdrop and modify messages sent over a light channel at all times (we elaborate this in Section 4).

3   Secret Key Deployment

In this section we propose secret key based key deployment protocol and provide initial security assessment of the proposed key deployment method.

3.1   Key Transmission and Verification

Prior to the start of node initialization, the user connects a web camera and a verification node to a PC. Next, the user places the web camera on top of the box from the inside, as shown in Figure 2. At this stage, the user turns the nodes ON and places them inside the box. Next, the user closes the box, runs the program on the PC and initiates the node initialization procedure. The box remains closed until the key transmission and verification is performed on all nodes which is subsequently indicated on the monitor.

Key transmission. Our "secret key"-based deployment is build upon ISO/IEV 9798-2 [4] three-pass key authentication protocol (Figure 3). We modify this protocol to include the communication over VLC (dashed arrow in Figure 3). The modified protocol evolves as follows.

The node S_i generates n-bit random key K_{S_iB} and k-bit random string N_{S_i}. The base station generates k-bit random string N_{B_i}. The node, equipped with

$$\text{Node } S_i \qquad\qquad\qquad\qquad \text{Base Station } BS$$
$$\text{Pick } K_{S_iB} \in_U \{0,1\}^n$$
$$\text{Pick } N_{S_i} \in_U \{0,1\}^k \qquad\qquad\qquad \text{Pick } N_{B_i} \in_U \{0,1\}^k$$

(1) $\qquad\qquad\qquad\qquad S_i \| K_{S_iB}$
$\qquad\qquad\qquad\qquad\qquad\qquad\qquad\qquad - - - - \rightarrow$

(2) $\qquad\qquad\qquad\qquad B \| N_{B_i}$
$\qquad\qquad\qquad\qquad\qquad\qquad\qquad\qquad \leftarrow$

(3) $\qquad\qquad S_i \| \{ N_{S_i} \| N_{B_i} \| B \}_{K_{S_iB}}$
$\qquad\qquad\qquad\qquad\qquad\qquad\qquad\qquad \longrightarrow$

$\qquad\qquad\qquad\qquad\qquad\qquad\qquad \text{Verify } N_{B_i}, K_{S_iB}$

(4) $\qquad\qquad B \| \{ N_{S_i} \| N_{B_i} \}_{K_{S_iB}}$
$\qquad\qquad\qquad\qquad\qquad\qquad\qquad\qquad \leftarrow$

(5) $\qquad$ Verify N_{S_i}, K_{S_iB}

Fig. 3. Modification of ISO/IEV 9798-2 three pass key authentication protocol. The dashed arrow represents key transmission over secure VLC.

minimally one LED, sends the key via VLC (step 1) to the base station (web camera), as shown in Figure 3. At the same time, the base station performs three tasks: (i) collects keys K_{S_iB} generated by the nodes (step 1), (ii) initiates key verification over a radio channel (steps 2-5), and finally, (iii) notifies the user which node has been successfully initialized via the monitor. Section 5 provides details of the key transfer over VLC.

Key verification. After the key is transmitted over VLC, the base station initiates the key verification protocol. All messages in the key verification are exchanged over the radio channel. The base station (using the verification node) sends random nonce N_{B_i} over the radio channel to node S_i (step 2). Next, S_i forms a packet by encrypting concatenations $N_{S_i} \| N_{B_i} \| B$ with the key K_{S_iB}. The node S_i sends this message (and its identity) to the base station (step 3). The base station extracts the random nonce N_{S_i}, verifies the key K_{S_iB} and the random nonce N_{B_i}. If the verification is successful, the BS encrypts concatenation $N_{B_i} \| N_{S_i}$ using K_{S_iB} and sends it back to node S_i. The node S_i receives and verifies both the key K_{S_iB} and the random nonce N_{S_i}. The whole procedure is considered as completed if all the nodes are successfully initialized, which is finally indicated by the GUI on the monitor. At the end, the user opens the box and completes the initialization with the short push on the node's button. This feature is used to ensure the "proof of presence" property to prevent an active attack (as described in Section 3.3).

3.2 Sensor Node State Diagram

Both user and base station need to know the status of the initialization process at any given time. For that reason, the current state of the node will be indicated with a LED according to the state diagram shown in Figure 4. During the initialization process, the node can take one of the four following states: Uninitialized,

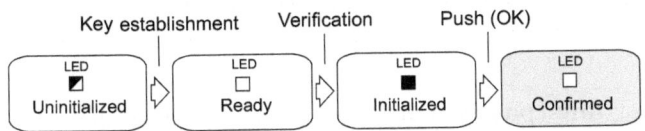

Fig. 4. Node's state diagram. A colored square indicates that the LED is ON, while a half colored that the LED is blinking.

Ready, Initialized and Confirmed. Next, we describe each of these states as well as the transitions between them.

Uninitialized state. Initially, when the user powers the node ON it is in the Uninitialized state. Prior to the start of key transmission the LED blinks with a predefined frequency. During this phase the node generates the key and, upon completion, sends it via VLC to the camera. After the key transmission is complete, the node advances to the Ready state (step 1 in Figure 3).

Ready state. The node remains in this state for a predefined period of time (e.g., a few seconds). In this state the node has sent the key and awaits the base station to initiate the key verification protocol over a radio channel. During this phase the node's LED is OFF (Figure 4). If the node does not receive any messages from the base station within the predefined period of time, it automatically restarts and returns back to the Uninitialized state. The node repeats the whole procedure which involves new key generation and transmission over VLC. Alternatively, the node receives a message from the base station and starts the key verification (steps 2-5 in Figure 2). If the key verification is successful, the node advances to the Initialized state.

Initialized state. In this state the node's LED is turned ON (Figure 4). At the same time, the base station notifies the user via the monitor about the node's position within the box as well as its current state. If the key verification succeeded on both sides (the node's and the base station's), the user is instructed to remove the nodes from the box and to shortly push the button on the node to finalize the initialization. The push of the button serves as "proof of presence", an aspect we describe in Section 3.3. However, if the node or the base station failed to verify the key, the user is instructed over the monitor to restart the initialization on selected nodes with a longer push on the button in order to repeat the node initialization. After the short push of the button, the node advances to the Confirmed state.

Confirmed state. In this state the node and the base station established a secret key and verified it, and the initialization process is finalized.

3.3   Initial Security Assessment

Our "secret key"-based protocol is build upon ISO/IEV 9798-2 [4] three-pass key authentication protocol that was proven to be secure when used over a radio channel. Therefore, we focus on possible attacks over the VLC, as we extended the ISO/IEV 9798-2 [4] protocol by including a transmission of a secret key via the VLC.

Camera recording (passive) attacker. A camera-recording attacker attempts to learn the secret key simply by recording the key sent from the node via VLC (step 1 in Figure 2). In this model the attacker does not interact in any way with the node initialization procedure.

Let us consider the case in which the node starts sending the key under insecure conditions (e.g., outside of the box). Thus, the attacker records the key, and the node advances from Uninitialized to Ready state (node's LED turns ON). In this state, the node waits a predefined period of time for the base station to initiate the key verification (Figure 4). After the predefined time period has passed during which the base station didn't initiate the key verification protocol, the node returns back to the Uninitialized state and repeats the whole procedure again (generates a new key and, again, sends it via VLC). The base station waits to receive a notification from the user that the system is ready (operates in *secure conditions*). Only then will the base station begin to process keys received over VLC and initiate the key verification protocol. Under secure conditions, the attacker does not have an access to the key transmitted by the node and therefore cannot successfully perform the key verification with the node.

Active attacker. In this attacker model, the attacker controls both the radio channel and communication over VLC when sensor node(s) are out of the cardboard box. Let us assume that the attacker captures the key sent by a node via VLC under insecure conditions (e.g., the node outside of the box). At this stage, the node is in Ready state and awaits the base station to initiate key verification (Figure 4). Next, the attacker initiates the key verification over the radio channel using the captured key. If the verification is successful on the node's side, the node advances to the Initialized state (the LED turns ON as shown in Figure 4). In this state the node waits for the user to confirm the initialization (push on a button). The user doesn't know that the attacker placed the node in the Initialized state so she picks the node up, and places it inside the box. Once the compromised node is placed inside the box, the base station recognizes a constantly powered ON LED on it and warns the user (via the monitor) to restart the initialization of that node. This is done by a longer push on the node's button. This form of active attacks does not work as the attacker does not have physical access to the node, therefore he cannot force the node to advance to Confirmed state. The user basically "proves her presence" through the push button.

4   Public Key Deployment

In this section we extend the attacker model to a more powerful adversary who can observe the electromagnetic radiation emanating from the LEDs. We assume the LEDs emanate radio signals which cannot be blocked by a simple cardboard box and we also assume that the attacker is able to easily eavesdrop on the leaked signals. This is a variant of an attack previously introduced in [18].

To establish keys between nodes and the base station by using a bidirectional radio channel and an unidirectional out-of-band VLC, we use SAS protocols [5,38]. The protocols make the key exchange process more usable, but at the cost of having to introduce public key cryptography. Recent work on elliptic curve cryptography has shown promising results regarding key distribution on resource constrained devices like our sensor nodes. In TinyPBC [26] and NanoECC [36] times less than 1 and 2 seconds, respectively, for point multiplication in binary fields were achieved.

Many prominent solutions that use LEDs and cameras [32,30] assume that the Visible Light Channel is authentic, which is not the case in our attacker model. To convey information via VLC they use on-off keying (switch the LED ON or OFF). An attacker equipped with a directional light source (e.g. a laser) has the capability to modify a message sent via VLC. In our model the attacker can modify messages by flipping $0 \rightarrow 1$, but not vice versa ($1 \rightarrow 0$) as the attacker cannot force a switched ON LED to turn OFF. In this case we speak of a *semi-authentic* visible light channel.

In the following sections we describe how to perpetrate such attacks and we also propose solutions on how to protect against them.

4.1   Attacks on Visible Light Channel

We consider prominent device pairing methods proposed in [32] and [30]. Both of the methods were developed for an authentic VLC (an attacker cannot modify messages sent via VLC). The proposed methods are secure within the authentic VLC model but, as we will show, are insecure in our semi-authentic VLC model (an attacker can flip $0 \rightarrow 1$).

Protocol [32] in the semi-authentic model. In [32] two devices (S_1 and S_2 as shown in Figure 5(a)) exchange public key values via a radio channel using the

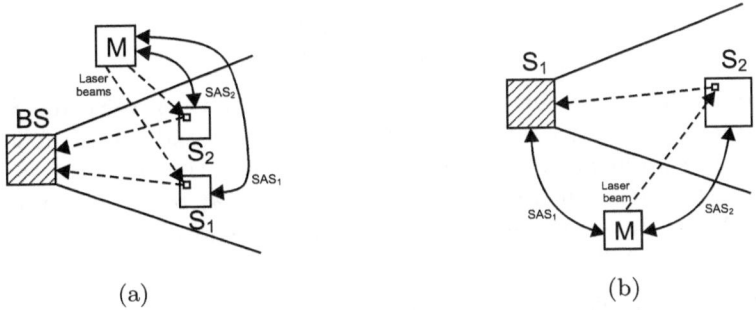

(a)   (b)

Fig. 5. Attacker M, with the aid of a laser, tries to modify short authentication strings exchanged over VLC (dashed arrows) between devices S_1, S_2 and BS (full-line arcs represent communication over a radio channel)

SAS protocol [5,38]. To authenticate these messages, each device simultaneously transmits short authentication strings (SAS) using visible light. The camera (BS in Figure 5(a)) captures both of these authentication strings and compares them. As BS does not know the SAS beforehand, the attacker can mount a MITM attack and modify these strings with a laser. Attacker M exchanges public keys with two devices S_1 and S_2 via a radio channel. When transmission via VLC occurs, the attacker points the lasers into the nodes' LEDs and appropriately modifies the bits (flips $0 \rightarrow 1$). The simplest attack is the one in which the attacker flips all bits $0 \rightarrow 1$. In this case, the base station will see all 1s and inform the user about the correct authentication. Please note that all 1s is a legitimate SAS.

Protocol [30] in the semi-authentic model. In an approach similar to [32], two devices (S_1 and S_2 in Figure 5(b)) exchange public key values over a radio channel. In this scheme, at least one device has an integrated web camera. In order to verify the exchanged public key values, device S_2 sends the SAS via VLC (using LEDs) to the device S_1. Here, an attacker tries to mount a MITM attack by exchanging different public keys with devices S_1 and S_2, (Figure 5(b)). To succeed, the attacker has to ensure that $SAS_1 = SAS_2$. Due to the property of the protocols [5] in which the probability for SAS_1 and SAS_2 to be equal is 2^{-k} (k is the length of the SAS) the attacker will establish two different authentication strings SAS_1 and SAS_2 with a high probability. However, in the semi-authentic model where the adversary can modify the bits (flip $0 \rightarrow 1$) this probability is significantly reduced. Indeed, if the ith bits of SAS_1 and SAS_2 are equal, an attacker will not need to modify them in any way. On the other hand, if the ith bits of SAS_1 and SAS_2 equal 1 and 0, respectively, an attacker could flip $0 \rightarrow 1$ by using the laser. Finally, if the ith bits of SAS_1 and SAS_2 are 0 and 1, the attacker will be unable to flip $1 \rightarrow 0$ for he cannot switch OFF an already powered ON LED. This is summarized below:

SAS_{1i}	SAS_{2i}	Attack
0	0	yes
0	1	no
1	0	yes
1	1	yes

Thus, we conclude that that 3 combinations of ith bits of SAS_1 and SAS_2 are beneficial for the attacker (all combinations but the second one). It follows that the probability for an attacker to modify the bits is $3/4$, therefore, the probability of a successful attack increases to $(3/4)^k$ as opposed to 2^{-k} (in the case of authentic VLC). If $k = 15$, the probability in a single attack increases from 2^{-15} to approximately 2^{-6}.

Virtual node attack. Let us assume the user wants to initialize one node (S_1) and the attacker (M) wants to inject his own virtual node (S_2) as shown in Figure 6. Attacker M simply exchanges public key values over a radio channel with BS and points his laser within the visible area of the base station's camera. The pointed laser is used to create a virtual node (device S_2 in Figure 6), and as

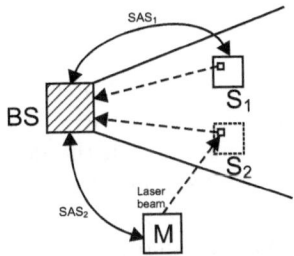

Fig. 6. An example of the virtual node attack; dashed arrows and full-line arcs represent communication over a semi-authentic VLC and a radio channel, respectively

such, to "blink" the correct short authentication string in such a way that the base station's camera detects it. The BS compares the SAS_2 it received from the attacker's laser with the one established over radio, sees that they match, and accepts the public key values from M as authentic.

4.2 "Public Key"–Based Deployment Protocol

We assume that each node S_i and the base station BS previously generated public key values pk_{S_i} and pk_B. In order to exchange authenticated public key values over a radio channel, we propose using the protocol introduced in [5,38], and shown in Figure 7. Please note, the base station performs this protocol individually with each node. The protocol evolves as follows:

(i) The user counts the number of nodes he/she wants to initialize and enters the number into the base station control software via a keyboard. We will show

$$
\begin{array}{ll}
\text{Node } S_i & \text{Base Station } BS \\
\text{Pick } N_{S_i} \in_U \{0,1\}^k & \text{Pick } N_{B_i} \in_U \{0,1\}^k \\
m_{S_i} \leftarrow 1\|S_i\|pk_{S_i}\|N_{S_i} & m_{B_i} \leftarrow 0\|B\|pk_B\|N_{B_i} \\
(c_{S_i}, d_{S_i}) \leftarrow \text{commit}(m_{S_i}) & (c_{B_i}, d_{B_i}) \leftarrow \text{commit}(m_{B_i})
\end{array}
$$

$$\xleftarrow{\quad c_{B_i} \quad}$$

$$\xrightarrow{\quad c_{S_i} \quad}$$

$$\widehat{m}_{B_i} \leftarrow \text{open}(\widehat{c}_{B_i}, \widehat{d}_{B_i}) \xleftarrow{\quad d_{B_i} \quad}$$

$$\mathsf{SAS}_{S_i} \leftarrow \widehat{N}_{B_i} \oplus N_{S_i} \xrightarrow{\quad d_{S_i} \quad} \widehat{m}_{S_i} \leftarrow \text{open}(\widehat{c}_{S_i}, \widehat{d}_{S_i})$$
$$\mathsf{SAS}_{B_i} \leftarrow N_{B_i} \oplus \widehat{N}_{S_i}$$

$$\xdashrightarrow{\quad \mathsf{SAS}_{S_i} \quad} \text{Verify } \mathsf{SAS}_{S_i} = \mathsf{SAS}_{B_i}$$

If $\mathsf{SAS}_{S_i} = \mathsf{SAS}_{B_i}$, the base station informs the user
to accept public key values as authentic.

Fig. 7. SAS protocol by [5,38]. The dashed arrow represents communication over a semi-authentic VLC.

later that by entering the number of nodes we can prevent the virtual node attack and make the size of the SAS invariant of the number of nodes to be initialized.

(ii) The user switches the nodes ON and places them in front of the camera, with the LEDs facing the camera.

(iii) The node's LED starts flashing with the delimiter 111000 to indicate to the BS they are ready to be initialized and to enable the BS to count them.

(iv) Next, the user instructs the base station to begin with the protocol shown in Figure 7. Having exchanged commit/open pairs with the BS, each node S_i first calculates the respective SAS_{S_i} (Figure 7), Manchester encodes SAS_{S_i} and begins transmitting it repetitively via a VLC (using on-off keying, switching LED OFF and ON). The Manchester encoded short authentication string, denoted $M(\mathsf{SAS}_{S_i})$, is separated with delimiter 111000. The usage of the delimiter and Manchester encoding was inspired by I-codes [6] and was used to prevent the flipping attacks (Section 4.3). Finally, the node transmits (blinks) the following repetitive sequence:

$$\underbrace{\cdots 1\,1\,1\,0\,0\,0}_{\text{delim.}}\ \underbrace{1\,0\,0\,1 \cdots 1\,0}_{M(\mathsf{SAS}_{s_i})}\ \underbrace{1\,1\,1\,0\,0\,0}_{\text{delim.}}\ \underbrace{1\,0\,0\,1 \cdots 1\,0}_{M(\mathsf{SAS}_{s_i})}\ \underbrace{1\,1\,1\,0\,0\,0 \cdots}_{\text{delim.}}$$

(v) If the SAS verification is successful for **all** the nodes, the user is instructed to finalize the initialization procedure by pushing a button on each of the nodes. If one or more nodes fail to initialize properly (e.g. due to errors in transmission, attacks etc.) the initialization procedure is aborted for all the participating nodes.

4.3   Short Security Analysis

Due to the lack of space, in this section we provide only a short security analysis of the public key deployment protocol.

Flipping attacks. In order to prevent flipping attacks we used Manchester encoded SAS_{S_i} for the transmission via VLC. Note that such a message contains an equal number of 0s and 1s. Due to the on-off keying modulation and the fact that an attacker is unable to switch OFF the LED (flip $1 \to 0$), any attempt of flipping will be detected by the BS as an excess of 1s. This construction is proved secure in [6].

Virtual node attack. According to the protocol, the base station knows exactly how many nodes it has to initialize (step (i) of the protocol). In addition, the BS counts itself the nodes by detecting respective delimiters (111000) transmitted over VLC. In order to successfully inject his own virtual nodes, the attacker has to block transmission of the delimiter 111000 over VLC for at least one of the nodes. However, the attacker cannot do this, for he is unable to turn OFF an already switched ON LED. In addition, any attempt of flipping $0 \to 1$ in the delimiter will be detected by the BS [6].

All or none. The design choice to abort the initialization procedure if at least one node fails to initialize properly makes the SAS invariant to changing the number of nodes to be initialized. Indeed, from the above analysis we know that an attacker can neither add new (virtual) nodes, remove existing (legal) ones, nor perform bit flipping attacks. It follows that the attacker can only try to perform a man-in-the-middle attack against one or more legal nodes. Now, if an attacker attempts to mount the attack against m nodes (out of n) and the respective short authentication strings are mutually independent, the probability of a successful attack against at least one sensor node, in a single attempt, will be at least $\min\{m \cdot 2^{-k}, 1\}$ [5]. For example, if the attacker attacks $m = 100$ nodes and $k = 15$, the probability for the attacker to succeed against at least one node is around 2^{-8}. However, by restricting the attacker to be successful against all the nodes, the probability for the attacker to succeed is reduced to $(2^{-15})^m = 2^{-15 \cdot m}$. Therefore, the best strategy for the attacker is to mount an attack against exactly one node (i.e., $m = 1$), which implies the probability of success (in a single attempt) to be bounded by $2^{-k} + \varepsilon$ (k being the size of SAS and ε a negligible probability) [5].

5   Implementation

We next describe the implementation of our secret-key deployment protocol. More specifically, we describe the implementation of a simple random number generator (RNG) and the key recognition software that enables communication over the light channel. We used Meshnetics ZigBee sensor nodes equipped with Green and Red LEDs, Atmel AT-mega1281V microcontrollers and AT86RF230 RF transceivers. Each sensor module features 128KB of flash memory and 8KB of RAM with data rate of 250 kbps in frequency band from $2.400 - 2.483$ GHz. For software developing and testing of the initialization procedure, a PC with the following configuration was used: Intel dual core processor clocked at 2.66GHz, 2GB of RAM, a Logitech notebook deluxe webcam with VGA resolution at 30fps interfaced via USB to the computer and Windows XP SP3 operating system.

5.1   Random Number Generator

The key feature for secure communication lies in a good random number generator. In this section, we describe our Random Number Generator (RNG). We first describe some related work on random number generators suitable for devices with limited processing capabilities.

TinyRNG [15] uses transmission bit errors as a source of randomness. These bits are randomly distributed as well as uncorrelated and may not be manipulated by an adversary. In [14] two oscillators are used, one oscillating much faster than the other. Generated bit stream's randomness is based on the frequency instability of a free running oscillator. The slow oscillator samples the higher frequency oscillator. They have shown that if the jitter in the slow oscillator signal is sufficient, the output of the RNG will have very little bit-to-bit correlation. Tkacik [37] also uses two free-running oscillators whose frequency vary with voltage and temperature. Random numbers are generated as exclusive-or of previously selected and permuted 32 bits of the LFSR (linear feedback shift register) and CASR (cellular automata shift register). Each shift register is clocked by these oscillators. However, an initial seed is required for each register.

Design of a Random Number Generator. In our implementation we used the approach from [14]. The generation of random numbers goes as follows: Meshnetics ZigBee nodes are equipped with two usable oscillators, an Internal Calibrated RC Oscillator (4 MHz) and a Watchdog Oscillator (128 kHz) [25]. The software running on the sensor nodes creates two timers; one timer is associated with the slower oscillator and the other timer with the faster one. The timers are configured with clock dividers in such a way that the slower timer fires once per second, while the faster one fires roughly 50000 times per second. On every tick of the slower timer, the number of ticks from the faster timer is logged. Figure 8(b) shows two traces of the number of ticks from the faster timer during the period of 512 ticks from the slower timer (roughly 512 seconds). As shown, the source of randomness comes from the instability (jitter) of the two used oscillators (Figure 8(a)).

Digital postprocessing. Table 5.1 shows the digital postprocessing and the random number generation process. As shown, on each successful low frequency timer tick the number of high frequency ticks is counted. Next, this value is

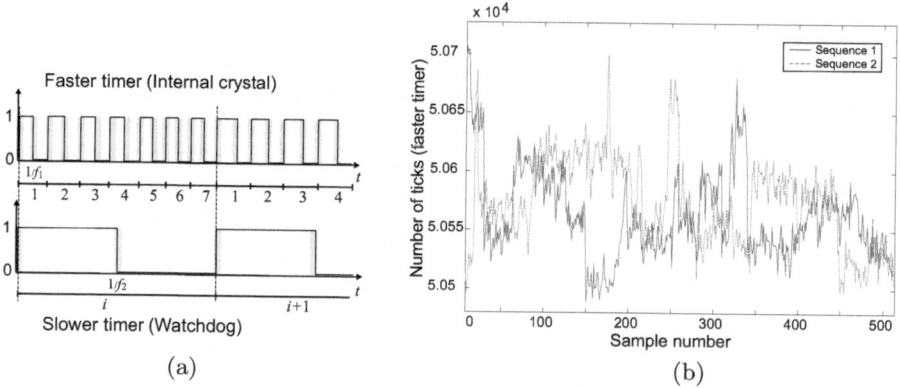

(a)   (b)

Fig. 8. (a)An example of oscillator frequency instability (jitter). (b) Two traces showing the number of ticks of a faster timer relating to one tick of the slower one.

Table 1. An example of digital postprocessing performed on the generated raw bitstream as well as the generation of random numbers. The generated bit stream is 110100.

Number of ticks (Watchdog)	Number of ticks (Internal RC)	Partial binary representation	Last two digits
1	50607	10101111	11
2	50605	10101101	01
3	50640	11010000	00

converted into a binary representation (last eight binary digits are presented in Table 5.1) from which the last two bits are taken. We could extract more than two bits at the expense of a more complex extractor. Since for our purpose the entropy is sufficient, we choose to use this simple extractor. The results of statistical tests are presented in the next section.

Statistical Tests. ENT [40] and NIST [13] statistical test suites were used to test the randomness of our generated bitstreams. Statistical tests were conducted on a 3×10^6 long bitstream which we obtained from 7 ZigBee nodes over the period of approximately 3 days.

ENT [40] is a pseudorandom number frequency test that performs a variety of tests such as Entropy, Arithmetic mean, Monte Carlo value for Pi, Serial correlation coefficient and Chi square distribution. Table 5.1 contains the results of the ENT test performed on a 3×10^6 long bitstream.

Only the last two bits were taken from the binary representation of the faster timer tick count. If more than two bits are taken, the bitstream fails the "Chi square" part of the ENT test suite. But, as we already mentioned, our RNG directly samples the number of faster timer ticks, without the requirement for other complex extractors.

NIST STS [13] contains 15 tests out of which only 8 were performed due to the minimum bitstream requirement (3×10^6 bits were produced) for each test. Each test is used to calculate the P-value which shows the strength of the null hypothesis. The hypothesis passes the test if the P-value is higher than 0.01 in which case the sequence is considered to be random. As shown in Table 5.1, the generated sequence passed the tests (P-value is higher than 0.01).

Table 2. ENT test results

Entropy = 0.999999 bits per bit.
Optimum compression would reduce the size of this 3×10^6 bit file by 0 percent.
Chi square distribution for 3267632 samples is 2.40 and randomly would exceed this value 12.14 percent of the times.
Arithmetic mean value of data bits is 0.4996 (0.5 = random).
Monte Carlo value for Pi is 3.155460889 (error 0.44 percent).
Serial correlation coefficient is 0.000264 (totally uncorrelated = 0.0).

Table 3. NIST test results

TEST	P-VALUE	PROPORTION	TEST	P-VALUE	PROPORTION
frequency	0.148094	0.9922	fft	0.468595	1.0000
block-frequency	0.500934	0.9922	aperiodic	all passed	all passed
cumulative-sums	0.311542	0.9922	apen	0.275709	0.9844
cumulative-sums	0.031497	0.9922	serial	0.671779	0.9844
runs	0.437274	0.9922	serial	0.637119	0.9922

These tests were performed over the raw bits. Since the output bitstream passes both NIST and ENT test suites, no additional randomness extractors (universal hash functions [41,7], von Neumann extractor [39], or simply applying a cryptographic hash function over the bitstream) are necesarry.

These results are preliminary; future work will include a more detailed study of factors which impact the work of RC oscillators (e.g. voltage and temperature), and therefore directly impact the quality of the generated random numbers.

5.2 Communication over a Visible Light Channel

After the key generation follows the key transmission via an out-of-band Visible Light Channel (VLC). The sensor nodes are programmed in such a way that generated key bits are Manchester encoded prior to transmission which ensures lower bit error rates during the transmission over VLC. The bits are transmitted in such a way that logical 0 and 1 of our bitstream are represented with LED ON and OFF states, respectively. The duration of each state (single LED's blink) is approximately 200 ms. In Figure 9 we give an example of a bitstream's "life-cycle"; from the bit generation to the bit transmission phase. As shown in Figure 9, the generated bits are separated in such a way that the first and the second LED (Green and Red LED) transmit odd and even bits, respectively, of Manchester encoded binary stream via VLC. In this way we achieve easier key recognition on the side of the base station, as described in the sequel.

Computer Vision. Once the user places sensor nodes inside of the box, we use our computer vision (CV) system to derive the secret key from the nodes' LED blinking sequence. We developed our CV system in MATLAB 2007 GUI [16], and achieved transmission speeds of 10 bits per second (5 b/s per each LED).

The image processing part of our CV system is CPU demanding. In order to achieve real-time performance, we process only certain parts of an webcam-obtained image - so called "Areas of Interest" (small rectangles encompassing LEDs of each node). The algorithm was designed to work with two LEDs on each node (Green and Red LED). To determine the Area of Interest (AoI) for each node, which is the first step, a few seconds of buffered frames is required. Once the areas are determined, the rest of the algorithm is performed in real-time. All of the following steps are performed only over Areas of Interest. The rest of the image does not contain any relevant information, and thus is excluded from future processing.

Image transformation. In the second step, the selected image parts (AoIs) are converted from RGB to HSV color space, known to be more reliable for detecting

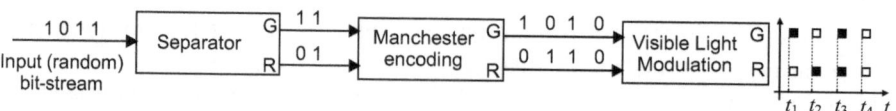

Fig. 9. An example of the bit stream sent via VLC using Manchester encoding. G and R stand for Green and Red LED, respectively.

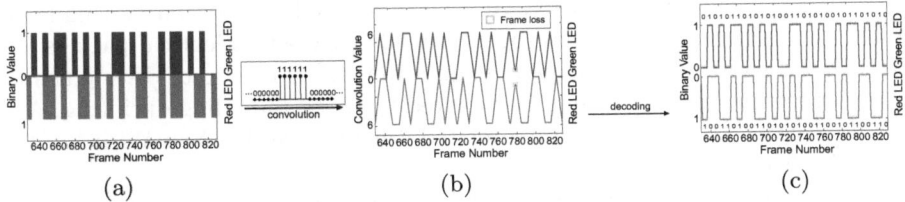

Fig. 10. A key recognition process: (a) detecting the status of LED indicators, (b) applying the convolution over the sampled area and (c) the bit identification process after the convolution

colors in low and changing light conditions [33]. Obtained images are tested for their levels of Hue, Saturation and Value, which enables us to detect the state (ON/OFF) of each LED. Color detector relies mainly on the level of Hue, while levels of Saturation and Value are just used in order to avoid false detection due to noise at low illumination conditions in the dark box.

Recognition of VLC signal. Due to a high frame loss and transmission error rates, during the transmission each bit is repeated in 6 consecutive frames (6 samples per bit). Decoding starts by detecting first 18 frames of the packet delimiter (3 binary ones in a sequence on both LEDs). Next, the key recognition algorithm performs the mathematical operation of convolution over the frames following the delimiter with a mask of six consecutive 1s (Figure 10(a)). As a result, data arrays containing values ranging from 0 to 6 are obtained (Figure 10(b)), where elements with extremes 0 and 6 are decoded as bits 0 and 1, respectively. Plateaus (areas with multiple, identical and consecutive elements) are decoded as double 0s or 1s, depending on their values (0 or 6). As Manchester encoding was used, only the convoluted signal's slope is analyzed, and not their values. This results in a method highly robust to de-synchronization effects. As shown by an example in Figure 10(b), frame loss during transmission via VLC does not affect correct bit recognition in any way.

6   Related Work

Recently, many key deployment schemes such as Zig Bee [1], SPINS [27], LEAP [42] and Transitory Master Key [10] have been proposed. Others [9,11,12,21,28] propose random key pre-distribution schemes. All of these schemes rely on an unspecified secure key deployment mechanism between devices.

In On-off Keying, the presence of an RF signal represents a binary 1, while its absence represents a binary 0 [5,6]. By using an unidirectional encoding scheme, On-off Keying ensures that an attacker is unable to modify a packet during transmission.

In Shake Them Up [8], user establishes a secret key between two nodes by holding and shaking the devices together while they send identical packets over the radio. This way, they assume that an adversary is unable to distinguish the source of the packets. This may be violated by using radio fingerprinting. Also, this does not scale well. The three related schemes are Are You With Me [20], Smart-Its Friends [17] and [22].

In Key Infection [2], two nodes establish a secret key by sending it in the clear over radio. They assume an attacker is unable to eavesdrop all the keys from all the nodes (e.g., 10.000 nodes) during key deployment. Based on simplicity and cost effectiveness, this scheme is insecure against a determined adversary. Moreover, an adversary is capable of injecting his own key, also violating key authentication.

In Resurrecting Duckling [35], a physical contact is required to securely establish a secret key. Based on the assumption that physical contact is secure, key authenticity and secrecy are ensured. But, since it requires specialized additional hardware, this scheme is not cost effective.

In Message In a Bottle [19], keys are sent in the clear to the nodes located inside a Faraday cage that ensures key secrecy and authenticity. However, the number of simultaneously initialized nodes determines the size of the Faraday cage. Moreover, a scale is used to determine the number of nodes within the Faraday cage based on total Faraday cage weight. In order to ensure key secrecy and authenticity for a large number of nodes, this scheme requires specialized setup hardware.

In HAPADEP [34] both data and verification information is sent over an audio channel. The pairing devices are both required to have speakers and microphones. In a related paper, Saxena and Uddin [30] present a device pairing method with an unidirectional channel based on devices equipped with LEDs and a video camera as the receiver. Their method is used for asymmetric pairing scenarios. Again, Saxena et. al. [32] use an auxiliary device (a laptop equipped with a web camera) to compare a short authentication string sent from the nodes to the laptop via unidirectional visible light channel. Both protocols are prone to laser attacks where an adversary may inject his/her malicious key by modifying the messages sent via the light channel with a directional light source (e.g. laser emitter).

Talking to strangers [3] requires specialized setup hardware (e.g. audio or infrared) in order to setup a public key. Seeing Is Believing uses an installation device with a camera or a bar code reader to create an out-of-band secure channel [24]. Key authenticity is achieved through certified public keys.

Mayrhofer and Welch [23] use an out-of-band laser channel constructed with off the shelf components for transmitting short authentication strings. According to [23], the proposed solution does not ensure complete authenticity of the the laser channel. Roman and Lopez [29] discuss general aspects of communication over a visible light channel.

7   Conclusion

We made several contributions in this paper. We proposed two novel *multichannel* protocols for initialization of large scale wireless sensor networks. The first protocol uses only secret key cryptography and is suitable for CPU-constrained sensor nodes. The second protocol is based on public key cryptography. Both protocols involve communication over a bidirectional radio channel and an unidirectional out-of-band *visible light channel*.

We demonstrated the importance of considering a very strong and realistic attacker model, where an attacker can eavesdrop, jam and modify transmitted messages in both a radio and a visible light channel; many existing protocols that rely on a visible light channel were shown to be insecure in this strong adversary model. Our "public key" - based protocol is designed to be secure in this very strong attacker model. Moreover, we showed that principle "all or none" keeps invariant the size of *short authentication strings* to changing the number of sensor nodes to be initialized.

The proposed protocols are implemented on the Meshnetics ZigBee sensor nodes. We showed that the proposed protocols are cheap to implement (a sensor node has to be equipped with one LED and a "push button") and scalable. We also designed and tested a simple random number generator suitable for CPU-constrained sensor nodes.

Acknowledgment

The authors would like to thank the anonymous reviewers for their thorough reviews and helpful suggestions.

References

1. ZigBee Alliance. ZigBee Specification (Document 053474r06, Version 1.0). Technical report (June 2005)
2. Anderson, R., Chan, H., Perrig, A.: Key Infection: Smart Trust for Smart Dust. In: IEEE International Conference on Network Protocols (2004)
3. Balfanz, D., Smetters, D.K., Stewart, P., Wong, H.C.: Talking to Strangers: Authentication in Ad-hoc Wireless Networks. In: Symposium on Network and Distributed Systems Security (2002)
4. Boyd, C., Mathuria, A.: Protocols for Authentication and Key Establishment. Springer, Heidelberg (2003)
5. Cagalj, M., Capkun, S., Hubaux, J.: Key Agreement in Peer-to-Peer Wireless Networks. In: Proceedings of the IEEE Special Issue on Cryptography and Security (2006)
6. Cagalj, M., Hubaux, J.P., Capkun, S., Rengaswamy, R., Tsigkogiannis, I., Srivastava, M.: Integrity (I) Codes: Message Integrity Protection and Authentication Over Insecure Channels. In: Proceedings of the IEEE Symposium on Security and Privacy (2006)
7. Carter, L., Wegman, M.N.: Universal Classes of Hash Functions. Journal of Computer and System Sciences 18(2) (1979)

8. Castelluccia, C., Mutaf, P.: Shake Them Up!: A Movement-based Pairing Protocol for CPU-constrained Devices. In: ACM MobiSys (2005)
9. Chan, H., Perrig, A., Song, D.: Random Key Predistribution Schemes for Sensor Networks. In: Proceedings of the IEEE Symposium on Security and Privacy (2003)
10. Deng, J., Hartung, C., Han, R., Mishra, S.: A Practical Study of Transitory Master Key Establishment For Wireless Sensor Networks. In: Proceedings of the First International Conference on Security and Privacy for Emerging Areas in Communications Networks (2005)
11. Du, W., Deng, J., Han, Y.S., Varshney, P.K.: A Pairwise Key Pre-Distribution Scheme for Wireless Sensor Networks. In: Proceedings of the 10th ACM conference on Computer and Communications Security, CCS (2003)
12. Eschenauer, L., Gligor, V.D.: A Key-Management Scheme for Distributed Sensor Networks. In: Proceedings of the 9th ACM conference on Computer and Communications Security (2002)
13. Rukhin, A., et al.: A Statistical Test Suite for Random and Pseudorandom Number Generators for Cryptographic Applications (2001), http://csrc.nist.gov/rng/
14. Fairfield, R.C., Mortenson, R.L., Coulthart, K.B.: An LSI random number generator (RNG). In: Blakely, G.R., Chaum, D. (eds.) CRYPTO 1984. LNCS, vol. 196, pp. 203–230. Springer, Heidelberg (1985)
15. Francillon, A., Castelluccia, C.: TinyRNG: A Cryptographic Random Number Generator for Wireless Sensors Network Nodes. In: Int. Symposium on Modeling and Optimization in Mobile, Ad Hoc and Wireless Networks (2007)
16. MATLAB Online Users Guide, http://www.mathworks.com (last access, September 2008)
17. Holmquist, L.E., Mattern, F., Schiele, B., Alahuhta, P., Beigl, M., Gellersen, H.W.: Smart-Its Friends: A Technique for Users to Easily Establish Connections between Smart Artefacts. In: International Proceedings of the 3rd international conference on Ubiquitous Computing (2001)
18. Kuhn, M.G.: Electromagnetic eavesdropping risks of flat-panel displays. In: Martin, D., Serjantov, A. (eds.) PET 2004. LNCS, vol. 3424, pp. 88–107. Springer, Heidelberg (2005)
19. Kuo, C., Luk, M., Negi, R., Perrig, A.: Message-In-a-Bottle: User-Friendly and Secure Key Deployment for Sensor Nodes. In: ACM SenSys (2007)
20. Lester, J., Hannaford, B., Borriello, G.: "Are you with me?" - using accelerometers to determine if two devices are carried by the same person. In: Ferscha, A., Mattern, F. (eds.) PERVASIVE 2004. LNCS, vol. 3001, pp. 33–50. Springer, Heidelberg (2004)
21. Liu, D., Ning, P., Du, W.: Group-Based Key Pre-Distribution in Wireless Sensor Networks. In: ACM Workshop on Wireless Security (2005)
22. Mayrhofer, R., Gellersen, H.: Shake Well Before Use: Two Implementations for Implicit Context Authentication. In: Ubicomp (2007)
23. Mayrhofer, R., Welch, M.: A Human-Verifiable Authentication Protocol Using Visible Laser Light. In: International Conference on Availability, Reliability and Security (2007)
24. McCune, J.M., Perrig, A., Reiter, M.K.: Seeing-Is-Believing: Using Camera Phones for Human-Verifiable Authentication. In: Proceedings of the IEEE Symposium on Security and Privacy (2005)
25. Murray, K.D.: 8-bit AVR Microcontroller with 64K/128K/256K Bytes In-System Programmable Flash, http://www.atmel.com (last access, March 2008)

26. Oliveira, L.B., Scott, M., Lopez, J., Dahab, R.: TinyPBC: Pairings for Authenticated Identity-Based Non-Interactive Key Distribution in Sensor Networks. In: 5th International Conference on Networked Sensing Systems, INSS (2008)
27. Perrig, A., Szewczyk, R., Tygar, J.D., Wen, V., Culler, D.E.: SPINS: Security Protocols for Sensor Networks. Wireless Networks 8(5) (2002)
28. Ramkumar, M., Memon, N.: An Efficient Key Predistribution Scheme for Ad-hoc Network Security. IEEE Journal on Selected Areas in Communications (2005)
29. Roman, R., Lopez, J.: KeyLED - Transmitting Sensitive Data Over Out-of-Band Channels in Wireless Sensor Networks. In: IEEE WSNS (2008)
30. Saxena, N., Uddin, M. B.: Automated Device Pairing for Asymmetric Pairing Scenarios. In: Chen, L., Ryan, M.D., Wang, G. (eds.) ICICS 2008. LNCS, vol. 5308, pp. 311–327. Springer, Heidelberg (2008)
31. Saxena, N., Uddin, M.B.: Bootstrapping Key Pre-Distribution: Secure, Scalable and User-Friendly Initialization of Sensor Nodes. In: ACNS (2009)
32. Saxena, N., Uddin, M.B., Voris, J.: Universal Device Pairing Using an Auxiliary Device. In: Proceedings of the 4th Symposium on Usable Privacy and Security, SOUPS (2008)
33. Shapiro, G., Stockman, G.C.: Computer Vision. Prentice-Hall, Englewood Cliffs (2001)
34. Soriente, C., Tsudik, G., Uzun, E.: HAPADEP: Human-Assisted Pure Audio Device Pairing. In: Proceedings of the 11th International Conference on Information Security, ISC (2008)
35. Stajano, F., Anderson, R.: The Resurrecting Duckling: Security Issues for Ad-hoc Wireless Networks. In: 7th International Workshop. Springer, Heidelberg (1999)
36. Szczechowiak, P., Oliveira, L.B., Scott, M., Collier, M., Dahab, R.: NanoECC: Testing the Limits of Elliptic Curve Cryptography in Sensor Networks. In: Verdone, R. (ed.) EWSN 2008. LNCS, vol. 4913, pp. 305–320. Springer, Heidelberg (2008)
37. Tkacik, T.E.: A Hardware Random Number Generator. Revised Papers from the 4th International Workshop on Cryptographic Hardware and Embedded Systems, CHES (2003)
38. Vaudenay, S.: Secure Communications Over Insecure Channels Based on Short Authenticated Strings. In: Shoup, V. (ed.) CRYPTO 2005. LNCS, vol. 3621, pp. 309–326. Springer, Heidelberg (2005)
39. von Neumann, J.: Various Techniques Used in Connection With Random Digits. Applied Math Series (1951)
40. Walker, J.: Hotbits, http://www.fourmilab.ch/random/ (last access, March 2009)
41. Yuksel, K., Kaps, J.P., Sunar, B.: Universal Hash Functions for Emerging Ultra-Lowpower Networks. In: Proceedings of the Communications Networks and Distributed Systems Modeling and Simulation Conference (2004)
42. Zhu, S., Setia, S., Jajodia, S.: LEAP: Efficient Security Mechanisms for Large-Scale Distributed Sensor Networks. In: Proceedings of the 10th ACM conference on Computer and Communications Security, CCS (2003)

Aggregated Authentication (AMAC) Using Universal Hash Functions

Wassim Znaidi[1], Marine Minier[1], and Cédric Lauradoux[2]

[1] CITI Laboratory, Lyon University
F-69621, France
firstname.lastname@insa-lyon.fr
[2] UCL / INGI / GSI, Place Saint Barbe, 2
Louvain-la-Neuve, Belgique
cedric.lauradoux@uclouvain.be

Abstract. Aggregation is a very important issue to reduce the energy consumption in Wireless Sensors Networks (WSNs). There is currently a lack of cryptographic primitives for authentication of aggregated data. The theoretical background for Aggregated Message Authentication Codes (AMACs) has been proposed by Chan and Castelluccia at ISIT 08.

In this paper, we propose a MAC design based on universal hash functions and more precisely on the Krawczyk's constructions. We show how those designs can be used for aggregation and how it can be easily adapted for WSNs. Our two AMAC constructions offer a small memory footprint and a signification speed to fit into a sensor. Moreover, when compared with scenarios without aggregation, the method proposed here induces a simulated energy gain between 3 and 9.

Keywords: Sensor networks, aggregation, authentication, MACs.

1 Introduction

The purpose of wireless sensors networks (WSNs) is to collect data and then transmit them at a gathering point. There are two classes of nodes in such a network. *Data nodes* have limited resources (CPU, memory and energy) and are on their own, *i.e.* the energy is the critical resource. *Gathering nodes* are considered more powerful (base stations) and they have an access to a power supply. We consider in this paper, an *hop-by-hop* scheme for the data forwarded by the *data nodes* to a *gathering node* considering a fixed topology. In this model, the most expensive operation is the data transmission. It is then highly valuable to reduce the size and the number of the transmitted messages. The messages aggregation has been used for this purpose. We apply a function over all the data produced by the collection nodes instead of concatenating them. This function is evaluated successively by each data gathering node resulting in a communication scheme with a constant message size.

If aggregation is a very powerful technique to save energy, it has to be used carefully. Sensors can be deployed at a large scale and over a large area. It is very

Y. Chen et al. (Eds.): SecureComm 2009, LNICST 19, pp. 248–264, 2009.
© Institute for Computer Science, Social-Informatics and Telecommunications Engineering 2009

likely that they get compromised or attacked and thus an attacker can influence the result of the aggregation function [22]. Security is therefore a critical issue. The confidentiality, the integrity and the origin of aggregated data must be preserved. There exists several works concerning aggregated encryption [10,6,8] or aggregated authentication [19,23,15]. Recently, several works [7,16,3] have established the foundation for aggregated MACs. This work aims to fill the gap between the theoretical results and practical design for aggregated MACs.

A MAC algorithm could be seen as a signature only valid between two users that share the same secret key: a MAC allows to guarantee the integrity of the transmitted message and to verify the identity of the sender for the user sharing the symmetric key. A MAC is thus an algorithm that takes as input a message m and a key K and that produces a fingerprint $tag = MAC_K(m)$. The receiver of the message $m'||tag(m)$ verifies if $tag' = MAC_K(m')$ is equal or not to tag. In the case of a WSN, this verification must be performed at each stage of the aggregation to establish a complete trust chain over all the results.

In this paper, we propose two AMAC constructions based on the well-known universal hash functions, *i.e.* *CRC Hash* and *LFSR hash* proposed by Krawczyk in [17].

Contributions of the paper are as follows:

- We show how to use existing universal hash functions to design an aggregated MAC scheme and which level of security can be achieved.
- We identify the parameters of the functions suitable for the constraints of sensors.
- We present a comparison of the performances of several aggregation scenarios.

The simulations performed for different scenarios show that the gain in terms of energy between our method and methods without aggregation varies between 3 and 9.

In Section 2, we give a reminder on aggregation and message authentication codes. We particularly focus on universal hash functions based MACs. We present in Section 3 our new designs for aggregate MACs and we discuss the security issues. The performance of our schemes are evaluated in Section 4. Then, we conclude.

2 Preliminaries

In this section, we first introduce a formal definition of message aggregation and of AMAC as done in [7] and describe the relative constructions proposed in the literature. Then, we introduce three particular MAC designs which are linear and based upon universal hash functions.

2.1 Formal Definition of Aggregation and Related Work

The basic communication model in an hop-by-hop WSNs is the concatenation. Let consider a WSN with n nodes sending message of ℓ bits. Each node i concatenates its contribution x_i to the result of the previous step. This model is easily implemented but it consumes a significant amount of bandwith and energy: the last node in the protocol has for instance to transmit $x_1, x_2, \cdots, x_i, \cdots, x_{n-1}, x_n$. This overhead can be reduced by using the aggregation: given n messages $(x_1, x_2, \cdots, x_n)$ of length ℓ sent by the n different nodes, the aggregated result m of length ℓ is defined by a function f:

$$f : m = f(x_1, x_2, \cdots, x_n).$$

Some examples of aggregation functions usually used are the median or the mean as explained in [22]. The security of an aggregation scheme relies on encryption and authentication. The properties of those two mechanisms are very specific in the context of aggregation.

Encrypted and Aggregated Data. The confidentiality of the messages sent in an aggregation scheme requires to perform the aggregation over the ciphertexts rather than the plaintexts. This problem is usually solved with homomorphic ciphers. Many public-key cryptosystems, *e.g.* RSA or ElGamal, can be used for this purpose but they are not generally suitable for sensors. A method of homomorphic ciphers based upon stream ciphers and suitable for WSNs applications has been developed in [6]. Let consider a node i receiving an ℓ-bit message p_{i-1}. The node i aggregates its contribution x_i to the message p_{i-1} in the following way:

$$p_i = p_{i-1} + c_i \quad \mod q$$
$$= p_{i-1} + x_i + keystream_i \quad \mod q$$

where q is a well chosen prime number and where $keystream_i$ is the keystream produced by the node i with its secret key k_i and a stream cipher.

Aggregated Authentication. The aim of aggregated authentication is to provide a way to verify the aggregated result, *i.e.* $f(x_1, x_2, \cdots, x_i, \cdots)$, rather than each message x_i individually. Different schemes have been proposed for the aggregation of authentication. They used Merkle tree [19] or MACs algorithms [23,3,15,7]. We especially focus on the solution proposed by A. Chan and C. Castelluccia [7]. They have proposed a formalization of aggregate message authentication code (AMAC) and they study its security. More formally, an AMAC algorithm is defined as follows:

- Key Generation (KG). Let $KG(1^\lambda, n) \rightarrow (k_1, k_2, ..., k_n)$ be a probabilistic algorithm. Then, k_i (with $1 \leq i \leq n$) is the secret key used to generate a verification tag by node i. The gathering node also called the sink possesses all k_i's used for tag verification.
- Tag Generation (MAC). $MAC_{k_i}(x_i) \rightarrow tag_i$ takes a secret key k_i and a message x_i as input to generate a verification tag tag_i for x_i. The message sent out from node i is a 3-tuple $(\{i\}, x_i, tag_i)$.
- Tag Verification (Ver). Let m be an f-aggregate of messages $x_1, x_2, \cdots, x_i, \cdots$ and hdr be the set of all contributing identities. Then

$$Ver_{k_1, k_2, \cdots, k_i, \cdots}(m, tag_1, tag_2, \cdots, tag_i, \cdots) \rightarrow 0/1$$

takes the aggregate m and the tag tag_i and secret key k_i for each $i \in hdr$ and outputs 1 if m is a correct aggregate (i.e. $m = f(x_1, x_2, \cdots, x_i, \cdots)$) and 0 otherwise.

Note that no aggregation algorithm is specified in AMAC; the aggregation is done in plaintexts. When an aggregating node with identity k receives two measurement values and their tags from downstream, say, $(\{i\}, x_i, tag_i)$ and $(\{j\}, x_j, tag_j)$, it would pass

$$(\{i, j, k\}, f(x_i, x_j, x_k), tag_i, tag_j, tag_k)$$

as the aggregation result to its parent where x_k is its own measurement.

Note also that aggregation of verification tags is not considered here. So all the tags are needed in the verification: let $m = f(x_1, ..., x_i, ...)$, then the correctness requirement of AMAC is as follows:

$$\text{Ver}_{k_1, \cdots, k_i, \cdots}(m, MAC_{k_1}(x_1), ..., MAC_{k_i}(x_i), ...) = 1.$$

here, the tags are not aggregated.

At this time, no instance of this scheme has been proposed. We propose to use the universal hash functions defined by Carter and Wegman in [4] to design a MAC corresponding to the requirements of [7] and where the tags could also be aggregated leading to an AMAC scheme defined as follows: let $m = f(x_1, \cdots, x_i, \cdots)$ and tag be the value $g(MAC_{k_1}(x_1), ..., MAC_{k_i}(x_i), ...)$ be the tag aggregation considering that g is an aggregation functions that could be (or not) equal to f, then the corresponding verification is thus:

$$\text{Ver}_{k_1, \cdots, k_i, \cdots}(m, tag) = 1.$$

2.2   MACs Based Upon Universal Hash Functions

In this section, we will first introduce the definition of an universal hash function and the original MAC schemes proposed by Krawczyk based upon universal hash functions and the one proposed by Sarkar.

Universal Hash Functions. A universal hash function is a family of functions indexed by a parameter called the key and it must verify that the probability over all keys that all distinct inputs collide is small. This notion was introduced by Carter and Wegman in [4].

Definition 1. *Let f_k be a function of an (ℓ, n)-family H from an ℓ-bit set to an n-bit set with the parameter k taken in a set $\mathcal{K}$. The family H is ϵ-almost universal if the probability of collisions for a random distribution of the value k over the set $\mathcal{K}$ (i.e. $Pr_k(f_k(M) = f_k(M')), \forall k \in_R \mathcal{K}$) is smaller than ϵ. We also say that H is ϵ-almost XOR universal (ϵ-AXU) if the associated differential probability for a random distribution of the value k over the set $\mathcal{K}$ is bounded by ϵ, i.e. $\forall(M, M', a), Pr_k(f_k(M) - f_k(M') = a) \leq \epsilon$.*

Definition 2. *We also say that a family of functions H is $\oplus$-linear if for all M, M', we have $f_k(M \oplus M') = f_k(M) \oplus f_k(M')$ for all instance f_k in H.*

The Definition 2 is particularly important for aggregated authentication.

These functions can be used for message authentication if the output is processed with another function. A MAC designs using the Definition 1 assumed the following scenario: the parties have already exchanged their secret key k, then to exchange a message M of length ℓ, the sender sends M and the tag $tag = f_k(M) \oplus r$. The shared secret key k is thus composed of a particular f_k function drawn randomly from an (ℓ, n)-family of hash functions and a random pad r. At reception, the receiver verifies the "tag" tag, corresponding with the MAC will be recomputed and checked for consistency. In practice, the fingerprint $f_k(M)$ will be encrypted with a stream cipher that will produce r.

Krawczyk has shown in [17] that the design of MAC of the above kind (i.e. combined with a one-time pad) requires to have a family of functions that is ϵ-almost XOR universal. Moreover, the family of functions can also be $\oplus$-linear.

Many universal hash families have been proposed in the literature to build MACs. One of the first examples was the evaluation of a particular polynomial in a particular point k as done in [2]. Let consider an ℓ-bit message m split into t blocks m_i such that $\ell = p^t$. In this case, the universal hash function f_k is defined as:

$$f_k(m_1, \cdots, m_t) = \sum_{i=1}^{t} m_i \cdot k^i \quad \mod p.$$

This function is multi-linear and the base field could be $\mathbb{F}_p$ or $\mathbb{F}_{2^n}$. In [21], V. Shoup gave a classification of the universal hash functions that could be used for MACs constructions in 3 categories: The first one is composed of the polynomial evaluations over a prime field or a finite field; the second one is composed of polynomial divisions over $\mathbb{F}_2$ described by Krawczyk in [17] and known as cryptographic CRC and as "LFSR (Linear Feedback Shift Register) hash"; the third category is composed of polynomial division over $\mathbb{F}_{2^k}$. V. Shoup particularly studied in his article the last class. The reader can find more details on MAC algorithms based on universal hash functions in [18] and on their security in [13].

Cryptographic CRC. As described in [21], the first scheme proposed by H. Krawczyk in [17] is based upon modular division using an irreducible polynomial over the field $\mathbb{F}_2$. It is a cryptographic variant of the well-known Cyclic Redundancy Codes (CRC), standards for errors detection in networks. More precisely, each message M is seen in its equivalent polynomial representation $M(x)$ over the field $\mathbb{F}_2$, the coefficients being the bits of M. Thus, for each irreducible polynomial $q(x)$ of degree n over $\mathbb{F}_2$, the associated family of universal hash functions is $h_q(M) = M(x) \cdot x^n \mod q(x)$. Notice here that it is necessary to multiply $M(x)$ by x^n to ensure the security of the scheme for the notion of ϵ-AXU.

The (ℓ, n) family of hash functions h_q is the set of irreducible polynomials of degree n and of the messages of size ℓ. This family is $\oplus$-linear, ϵ-almost universal (with $\epsilon \leq \frac{n+\ell}{2^n-1}$) and ϵ-almost XOR universal (with $\epsilon \leq \frac{n+\ell}{2^n-1}$).

The hardware and software implementation of such mechanisms is really efficient because the modular division for polynomials in $\mathbb{F}_2$ could be performed using a simple LFSR. The corresponding extension proposed by Shoup and proved secure is the extension of this construction to the case where the base field is $\mathbb{F}_{2^k}$. In this case, the corresponding ϵ value is about $\frac{\ell}{2^{kn}}$.

Linear Feedback Shift Register (LFSR) Hashing. In the same article, Krawczyk introduced a second construction based upon random matrices. More precisely, given A a boolean Toeplitz matrix of size $n \times \ell$ (i.e. each lower diagonal is fixed, i.e. if $k-i = l-j$ for all indices then $A_{i,j} = A_{k,l}$) and given a message M of size ℓ. The universal hash function $h_A(M)$ is then the binary multiplication of the matrix A by the column vector composed of the bits of the message M: $h_A(M) = A \cdot M$.

A simple method to build such matrices is the LFSR use: given $q(x)$ an irreducible polynomial of degree n over $\mathbb{F}_2$; given $s_0, s_1, \cdots$ the output sequence of the bits generated by the LFSR defined according to $q(x)$ and the initial state of the LFSR $s = (s_0, s_1, \cdots, s_{n-1})$. For each irreducible polynomial $q(x)$ and for each non-zero initial state of the LFSR, we associate the hash function $h_{q,s}(M)$ defined as the linear combination $\bigoplus_{j=0}^{\ell-1} M_j \cdot (s_j, s_{j+1}, \cdots, s_{j+n-1})$ where M_j is the bit number j of M. In other words, at each clock, the LFSR updates its internal state taking into account each message bit. This hash functions family is $\oplus$-linear, ϵ-almost universal (with $\epsilon \leq \frac{n+\ell}{2^n-1}$ and ϵ-almost XOR universal (with $\epsilon \leq \frac{\ell}{2^n-1}$).

Multi-linear Universal Hash Functions. In [20], P. Sarkar proposed the following evaluation: given a field $\mathbb{F}_p$ and an extension of this field $\mathbb{F}_{p^n}$ with $n \geq 1$; given ϕ a linear transformation from $\mathbb{F}_{p^n}$ into itself such as the minimal polynomial of ϕ in $\mathbb{F}_p[x]$ be of degree n and be irreducible over $\mathbb{F}_p$; the message to cipher M is cut into $l \leq n$ elements $(M_1, \cdots M_l)$ over $\mathbb{F}_p$. The hash functions family is:

$$G_K(M) = M \cdot (K, \phi(K), \cdots, \phi^{l-1}(K))$$
$$= M_1 K + M_2 \phi(K) + \cdots + M_l \phi^{k-l}(K)$$

where K belongs to $\mathbb{F}_{p^n}$.

The family G_K is thus a linear combination of $(M_1, \cdots M_l)$ and of $(K, \phi(K), \cdots, \phi^{l-1}(K))$, it is multi-linear (i.e. linear in each of its component), ϵ-almost universal with $\epsilon \leq 1/q^n$ and also ϵ-almost XOR universal with the same ϵ value.

Sarkar also noted that ϕ could be easily implemented because it can be seen as a LFSR over $\mathbb{F}_p$. The author studied the particular p values allowing a fast implementation in hardware and in software. The examples given are $q = 2$, $n = 128$; $q = 2^8 + 1$, $n = 16$; $q = 2^{16} + 1$, $n = 8$; $q = 2^{32} + 15$, $n = 4$. He also gave some examples of extensions over the field $\mathbb{F}_2$ that we will not detail here.

3   New Designs

We are going to present in this section the possible applications of the previous functions in the case of a WSN. We simplify the study case for a better understanding and reduce the number of nodes to two nodes i and j depending on

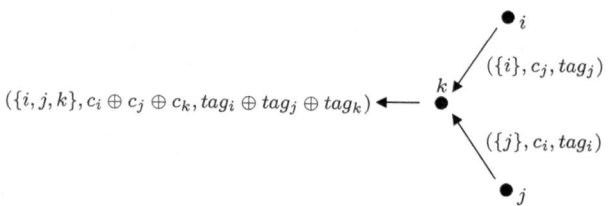

Fig. 1. XOR aggregation with three nodes: i, j and k

one aggregator node k as shown in Fig. 1. This last one is directly connected to the sink. This simple scheme could be easily generalized.

3.1 XOR Aggregation: How to Adapt the Krawczyk's Approaches for WSNs

The first construction described by Krawczyk could be directly applied to the MAC aggregation if the XOR operation is used. This approach could be directly combined with the XOR data aggregation proposed in [6].

Suppose that a WSN is composed of N nodes i. Each node receives during an initialization phase a cipher key K_{Ei} shared between the node and the base station (the sink), an authentication key K_{Ai} also shared between the node and the base station and a polynomial $q(x)$ shared by all the nodes and the base station. Suppose now that the simple network described in Fig. 1 describes a tree with three nodes i, j and k directly rooted at the sink.

When the node i wants to send (on demand or at regular intervals) a message m_i, it ciphers this message using a pseudo-random stream generated using a stream cipher algorithm E (for example RC4 or SNOW v2 [11]). The common cipher key K_{Ei} and a common initial value IV_i used once must be shared between the node i and the base station for the correct use of the E stream cipher. Thus, node i ciphers $C_i = m_i \oplus r_i$ where r_i is the pseudo-random stream produced by $E(K_{Ei}, IV_i)$. Note that as mentioned in [6] the node i must also transmit to the base station its Id i and the unique value IV_i that also plays the role of a counter to discard replay attacks. We first propose that IV_i be the only transmitted value writing $IV_i = i||CTR_i$, i.e. as a concatenation between the node Id and a counter CTR_i incremented by 1 at each new sending. (Note also that this value must be transmitted each time in case of the non-transmission of a particular message).

The node i also produces the corresponding MAC of the message m_i using the Krawczyk's construction: it computes $tag_i = (m_i(x) \cdot x^n \mod q(x)) \oplus r_i'$ where r_i' is the pseudo-random stream produced using E initialized with K_{Ai} and IV_i. Thus, node i transmits to its parent k the value: $\{hdr, data, tag\}$ with $hdr = IV_i$, $data = C_i$, $tag = tag_i$.

Suppose now that the node j wants to transmit the message m_j to the aggregator node k, then it sends to k $\{IV_j, C_j, tag_j\}$. The node k transmits to the base station (considering that it sends itself m_k):

$$\{IV_i, IV_j, IV_k, C_i \oplus C_j \oplus C_k, tag_i \oplus tag_j \oplus tag_k\}.$$

The base station deciphers

$$C_i \oplus C_j \oplus C_k \oplus r_i \oplus r_j \oplus r_k = m_i \oplus m_j \oplus m_k = M$$

using the knowledge of the different keys and of the different IVs. It verifies:

$$tag_i \oplus tag_j \oplus tag_k \oplus r_i' \oplus r_j' \oplus r_k'$$
$$= m_i \cdot x^n \oplus m_j \cdot x^n \oplus m_k \cdot x^n \mod q(x)$$
$$= (m_i \oplus m_j \oplus m_k) \cdot x^n \mod q(x)$$
$$= M \cdot x^n \mod q(x).$$

Thus, the base station could verify the aggregated tags according to the received sum.

Examples of values sizes Concerning the key sizes, the minimal size is 128 bits. The polynomial $q(x)$ could be a primitive polynomial of size 64 bits. the IV values could be of 48 bits length (24 bits for example for the node Id and 24 bits reserved for the counter[1]. The messages must have a 64 bits length (this value could not be smaller than the degree of the polynomial q). In this case, the final messages size is defined according the number of nodes N transmitting an information: $48N + 64 + 64$. Considering a 96 bits polynomial and a 96 bits message to transmit, the total size of transmitted information becomes $48N + 96 + 96$. The security bounds of the underlying universal hash functions are equal to 2^{-56} (resp. 2^{-87}). The induced overhead on the network clearly depends on the header size and thus on the nodes number sending back an information.

The following on demand mechanism could reduce this overhead: when the base station wants to receive values from the network, it broadcasts a unique value IV of 24 bits. The nodes receiving it use the previous method to cipher and to authenticate their data using the IV value $IV_i = i\|IV$. The header has only to be constituted of the responding Id nodes and could thus be replaced by a ciphered Bloom filter of size m as proposed in [1]. The cipher key K of the filter is shared between all the nodes and the base station. In this case, each node ciphers k times its Id by putting at 1 the bit corresponding to this position in the Bloom filter. This improvement seems to be efficient even if the probability to obtain a false positive in the Bloom filter is about $0,6185^{m/n}$ where n is the number of elements to insert in considering that k is about $0,7\frac{m}{n}$. For a network with 200 nodes, a Bloom filter of size 2048 bits with $k = 7$ has a false positive probability less than 1%. This probability could be reduced if instead of bits, the Bloom filter is composed of 4 bits word or of bytes. If a Bloom filter is used, the required computations performed by each sensor are increased by k additional

---

[1] Once all the IVs values used, the cipher keys must be changed to discard WEP like attacks.

hash computations and the base station must test if all the Id nodes are or are not in the Bloom filter leading to kN additional hash computations.

We have proposed a direct use of the Krawczyk constructions in the case where we try to obtain the XOR of the messages and not the sum. We have presented examples using the first construction, the same reasoning could be applied using the second construction. In this case, each node must be initialized with the same matrix A to conserve the $\oplus$-linearity, the other parameters being the same than the one previously described. A deduced MAC size of 64 or 96 bits gives reasonable security bounds.

3.2   Aggregation over $\mathbb{F}_p$

In this section, we extend the Krawczyk's MAC constructions over $\mathbb{F}_p$ with p prime to transform the proposed MAC from $\oplus$-linear to +-linear as the one proposed by Sarkar.

Extension of Krawczyk's Over $\mathbb{F}_p$. We first introduce the following notations: we denote by $\mathbb{F}_p^l$ the vectorial space of size l over $\mathbb{F}_p$ and $\mathbb{F}_p^n$ the one of size n. The message M we want to compute the MAC is written $M = (M_0, \cdots, M_{l-1})$ where each M_i belongs to $\mathbb{F}_p$. In the same way, the output of the hash function is considered as an element of $\mathbb{F}_p^n$, a vector of size n over $\mathbb{F}_p$.

In this case, the first Krawczyk's construction becomes: given a message $M = (M_0, \cdots, M_{l-1})$ in $\mathbb{F}_p^l$ seen as a polynomial with coefficients in $\mathbb{F}_p$: $M(x) = M_0 x^{l-1} + M_1 x^{l-2} + \cdots + M_{l-1}$; given an irreducible polynomial $q(x)$ of degree n with coefficients in $\mathbb{F}_p$ with its leading coefficient equal to 1; the hash function is defined as $h_q(M) = M(x) \cdot x^n \mod q(x)$ seen as a vector of size n of elements of $\mathbb{F}_p$. The obtained tag is thus $tag = h_q(M) + r$ where r is a vector of size n of random numbers taken in $\mathbb{F}_p$.

We need to compute $Pr_h(h_q(M) - h_q(M') = c(x))$ to prove that this function is ϵ-almost universal. This function is trivially +-linear. Thus, directly using the results of [21] and of [17], if $h_q(M) - h_q(M') = c(x)$, by linearity, we have $h_q(M - M') = c(x)$, i.e. $q(x)$ divides $(M - M') \cdot x^n - c(x)$. This polynomial has a maximal degree equal to $l + n$ whereas $q(x)$ has a degree of n, the number of factors of degree n of this polynomial is $\frac{n+l}{n}$. The maximal number of functions h_q that map $M - M'$ in $c(x)$ is $\frac{n+l}{n}$. The number of all possible functions is the set of irreducible polynomials of degree n over $\mathbb{F}_p[x]$; there are $\frac{p^{n-1}}{n}$ such functions. We thus directly deduce that $Pr_h(h_q(M) - h_q(M') = c(x)) \leq \frac{n+l}{p^{n-1}}$ and that the family of proposed functions is ϵ-almost + universal with $\epsilon = \frac{n+l}{p^{n-1}}$ and it is multi-linear.

We have generalized the first Krawczyk's construction for a prime field $\mathbb{F}_p$ keeping the willing linear properties. Notice that the second scheme proposed in [17] could be generalized in the same way: in this case, this last generalization is very closed to the one proposed by Sarkar except the definition of the final set seen in the Krawczyk's generalization as a vectorial space ans seen in the Sarkar's construction as a field extension. Let us now explain how to use those generalizations for WSNs.

Applications for WSNs. In this section, we will describe how to use those new constructions for MAC aggregation in WSNs. We suppose here that the messages (for example temperatures) sensed by the nodes will be sent by packets, one packet being constituted of l single messages of size p. Each node sensing l different values stores those values $M_0, \cdots, M_{l-1}$ before sending them together to the base station at each given time interval or on demand, l being known and fixed in advance. In the case where a sensor has not collected l values but a smaller number, it replaces in the sent message the missing values by some 0s. Let us illustrate the proposed method using the example of Fig. 1.

Using the same assumptions than the one of Section 3.1, we suppose that each node i shares with the base station a cipher key K_{Ei}, an authentication key K_{Ai}, a stream cipher E and an irreducible polynomial $q(x)$ of degree n with coefficients in $\mathbb{F}_p$ common with all the other nodes and the base station.

The node i stores l messages and sends them to the base station using the proposed method. For ciphering messages, it directly uses the method described in [6], i.e. it computes for each message M_j ($j \in [0, .., l-1]$), $M_j + r_j \mod p$ where r_j is a pseudo-random number smaller than p and where p is a prime number defined as $p \geq 2^{\lceil \log_2 M*N \rceil}$ (where M is the maximal size that can take a single message) and where N is always the total number of nodes. So, first, node i ciphers its l messages $M^i = (M_0^i, \cdots, M_{l-1}^i)$:

$$C^i = M^i + r^i$$
$$= (C_0^i = M_0^i + r_0^i \mod p, \cdots,$$
$$C_{l-1}^i = M_{l-1}^i + r_{l-1}^i \mod p)$$

where each r_j^i is a random number in $[0, p-1]$. Those values are obtained using algorithm E initialized with the key K_{Ei} and an IV value IV_i. From this point, node i computes the MAC of all the l messages M^i seen as a polynomial with coefficients in $\mathbb{F}_p$: it first computes a vector of size n: $h_q(M^i) = (h_0^i, \cdots, h_{n-1}^i) = (M^i \cdot x^n \mod q(x))$. Finally, we have:

$$tag^i + r'^i = (tag_0^i, \cdots, tag_{n-1}^i)$$
$$= (h_0^i + r_0'^i \mod p, \cdots,$$
$$h_{n-1}^i + r_{n-1}'^i \mod p)$$

where r_j' are random numbers belonging to $[0, p-1]$ obtained using E initialized with K_{Ai} and IV_i. Node i transmits to its parent node k: $\{hdr, data, tag\}$ with $hdr = IV_i$, $data = C^i$ and $tag = tag^i$.

Following the same previous example, the node j depending on the same parent k transmits its own l messages and the associated MAC: $\{IV_j, C^j, tag^j\}$. The node k transmits to the base station (considering that it has also l messages to transmit):

$$\{IV_i, IV_j, IV_k, C^i + C^j + C^k, tag^i + tag^j + tag^k\}.$$

The base station deciphers $C^i + C^j + C^k - r^i - r^j - r^k = M^i + M^j + M^k = M$ using its knowledge of each K_{Ei} and of IV_i. M is a vector of size l. More precisely: $M = (\sum_i M_0^i \mod p, \cdots \sum_i M_{l-1}^i \mod p)$. It then verifies:

$$tag^i + tag^j + tag^k + r'^i + r'^j + r'^k$$
$$= M^i \cdot x^n + M^j \cdot x^n + M^k \cdot x^n \mod q(x)$$
$$= (M^i + M^j + M^k) \cdot x^n \mod q(x)$$
$$= (\sum_i M_0^i \mod p, \cdots ,$$
$$\sum_i M_{l-1}^i \mod p) \cdot x^n \mod q(x)$$
$$= M \cdot x^n \mod q(x)$$

Thus, the base station is able to verify the value of the aggregated tags according the received sum value.

Examples of values sizes. As previously mentioned, for key and IV sizes we keep the sames as the ones defined in Section 3.1. So, let us define the p value. It depends on the size of the network and of the maximal size of the message to send. With the temperature example and a network composed of 200 nodes, we consider that the temperature is smaller than 5000 Celsius degree. we directly deduce that the p size is about 20 bits. We thus need to choose a prime number easy to implement such as $2^{20} + 7$. For the particular values closed to the powers of 2, we could choose the prime numbers given in [20]. So if $p = 2^{20} + 7$, an irreducible polynomial of degree 6 gives an admissible security bounds and we can send the messages by packets of size 10. Thus the generated MAC for each node will be of size 126 bits, the concatenation of 10 messages will be of length 300 bits. If N nodes transmit their values, the packets size will be bounded over by $48N + 126 + 300$.

As explained in Section 3.1, a ciphered Bloom filter (with the same properties) could be used for overhead reductions.

3.3   Security Analysis in the AMAC Model

In [7], A. Chan and C. Castelluccia proposed two security models for aggregation schemes in WSNs. The first one is called Concealed Data Aggregation (CDA) and concerns a security model for data aggregation whereas the second one called AMAC is dedicated to the security of Aggregated Message Authentication Codes.

The security model for CDA defines a security notion against adaptive chosen ciphertexts attacks and the indistinguishability notion in this model (IND-CCA2). The particular game used here authorizes the usual challenges in this model (cipher oracle and decipher oracle). The authors noticed that the construction defined in [6] does not verify the IND-CCA2 property. This comes directly from the intrinsic nature of the construction: we could distinguish two particular ciphertexts and their sum.

In the same article, the authors defined the AMAC security notion using a generation oracle and a verification oracle. In this model, they demonstrated that an adversary is able to win the following game: given two messages, ($hdr = \{i\}, m_i, tag_i$) sent by node i and ($hdr = \{j\}, m_j, tag_j$) sent by node j that the adversary knows; if the adversary sends to the verification oracle the aggregated of those two messages ($hdr = \{i, j\}, m_i + m_j, tag_i + tag_j$) then the adversary is able to forge a tag for a valid message.

The schemes proposed here have this security weakness intrinsically linked with the wishing linear properties. However, we could legitimately question the practical implications of a such attack especially concerning our MAC schemes. The discussion below only concerns the AMAC case of study, the data aggregation scheme used here being the one studied in [7]. Indeed, if we suppose that the base station needs the complete value of the header to correctly decipher the messages sum and the aggregated value of the tags, this implies that the adversary (that can not replay old packets due to the presence of the IV value into the header) could only send to the base station information that it already knows. If the base station does not possess those information, this implies that the two nodes i and j were not able to transmit their values to the base station (because for example the parent node is dead). In this last case, the adversary helps for good operations in the network.

Moreover, in the AMAC security model proposed in [7], the header is not taken into account. We can imagine to redefine a security model where the header is included. In this case, the security of the scheme could rely in part at least on the header itself by ciphering it for example or by using a ciphered Bloom filter as explained in Section 3.1. One of the simplest methods consists in always ciphering the header using the AES and a unique key shared by all the nodes in the network and the base station. This method is not robust against node compromise and do not allow to verify if the header has been modified during process or not. To discard this last problem, we could add to this ciphered header a MAC chain for which each node contributes by over-ciphering data as done in [9]. The only deduced constraint is that the use of an additive homomorphic cipher is prohibited to cipher and authenticate the header.

4   Performance Comparison

In this section, we present the performance evaluation of different aggregation models. We have considered four cases:

- **Scenario 1**: the communication scheme uses no aggregation, *i.e.* the concatenation, nor for the data neither for the authentication.
- **Scenario 2**: the data are aggregated using a stream cipher as proposed in [6]. The authentication is not aggregated.
- **Scenario 3**: the data and the authentication are respectively aggregated with a stream cipher over $\mathbb{F}_2$ and with the AMAC proposed in Section 3.1.
- **Scenario 4**: the data and the authentication are respectively aggregated with a stream cipher [6] and with the AMAC proposed in Section 3.2.

We test those four schemes using the LEACH [14] election mechanism and the WSnet simulator [12]. First, we briefly describe LEACH, the simulation parameters and we discuss the different results.

4.1   LEACH: Low-Energy Adaptative Clustering Hierarchy

LEACH [14] is a clustering-based protocol which minimizes energy dissipation in WSN's. LEACH selects randomly nodes as cluster heads (special aggregator nodes), so the energy dissipation in the communications with the sink is spread to all nodes in the network. LEACH is composed of two steps: the set-up phase and the steady phase. During the set-up phase, a sensor node is elected as cluster head if it generates a random number (between 0 and 1) greater than a given threshold T defined as:

$$T = \begin{cases} \frac{P}{1-P*[r \mod 1/P]}, & \text{if } n \in G \\ 0, & \text{Otherwise} \end{cases}$$

where P is the desired percentage of cluster heads, r is the current round of the protocol and G is the set of nodes that have not been selected as a cluster head in the previous rounds. Using this threshold, each node will be a cluster-head at some point after $1/P$ rounds. After an advertisement information, each node selects its cluster head. During the steady phase, each node sends their sensing values to its cluster head which aggregates the received data before sending them to the base station. After the steady phase, the network goes into the set-up phase again for a new round and a new cluster heads election.

4.2   Different Scenarios and Evaluation Parameters

We have implemented the LEACH protocol on the WSnet simulator. we have set $P = 0.2$ and we have tested our approach on a random nodes distribution. Each simulation is run with n sensor nodes and $n \in [100; 600]$ distributed randomly over a square field of 400m by 400m. Our simulations use the IEEE 802.11 physical and MAC layers which are fully simulated in the WSnet environment. We have also used the RC4 stream cipher but any other stream cipher can be used. In this study, we have simulated four different scenarios: the first scenario consists in no aggregation at all, nor for the data neither for the MACs. The second scenario simulates the data aggregation technique presented in [5] which we add the concatenation of all tags generated by sensor nodes. The two last cases simulate our own proposals described in Section 3.1 and in Section 3.2. Note that the operations performed in Scenario 3 are based upon XOR operations whereas the usual + is used only in Scenario 4.

4.3   Simulation and Results

We have simulated the 4 scenarios described above based on aggregator nodes elected using the LEACH protocol. We consider in all simulations that each node senses a value at each second and sends it with a given probability equal to 90 %.

We have tested the average delay time for a packet to travel until reaching the sink and the average energy consumption per nodes for the four scenarios. For each test, we have repeated the tests over 100 simulations for the four scenarios.

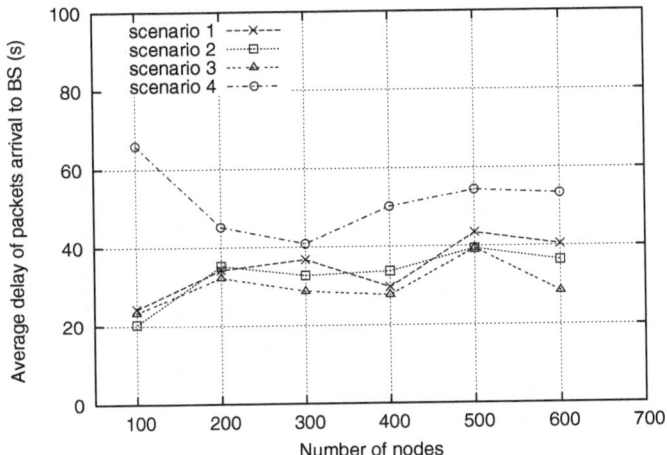

Fig. 2. Energy consumption for different aggregation schemes

Fig. 2 presents the average delay for a packet to reach the sink. Clearly, Scenario 4 is the slowest one and has the maximum delay for the average packet transfer as every node must wait l time intervals before sending the l aggregated values to first its cluster head that aggregates them and forwards to the sink. Moreover, the cluster head must wait values from other members before sending all the aggregated values to the base station. So, as expected, the Scenario 4 has the highest latency compared with the other scenarios. This proposal is really adapted to applications that require simple data gathering or environmental monitoring without any emergency needs. Among the three other scenarios, the Scenario 3 is the one that presents the best average delay compared with the two first scenarios. This comes from the fact that this scenario minimizes the size of the sent packets leading to a better transmission time for every cluster head.

Fig. 3 presents the gain on energy consumption between Scenario 4 and each of the other scenarios. As one could see on the figure, this gain is between 2.5 and 9 which is really significant. Moreover, Scenario 3 has the best energy consumption after Scenario 4. Those real improvements in terms of energy keeping is directly linked with the size of the sent packets because in the three first scenarios the number of sent packets is about the same whereas in the last case, the number of sent packets is divided by l here equal to 10.

More formally, considering the WSN as a tree of depth d and of width t, considering that one bit is sent by each nodes, the number of sent bits when no aggregation is performed is equal to $\sum_{i=0}^{d}(d-i)t^{d-i}$, whereas when we consider an aggregated scheme, the total number of sent bits is $\sum_{i=0}^{d} t^{d-i}$. In those cases, considering that each node send a message of m bits and a tag of ℓ bits, the number of bits sent if Scenario 1 is used is $(\ell+m)\sum_{i=0}^{d}(d-i)t^{d-i}$, for Scenario 2 it is $\ell\sum_{i=0}^{d}(d-i)t^{d-i}+m\sum_{i=0}^{d}t^{d-i}$, for Scenario 3 it is $\ell\sum_{i=0}^{d}t^{d-i}+m\sum_{i=0}^{d}t^{d-i}$ and for Scenario 4, considering that a message of length m and a MAC of length

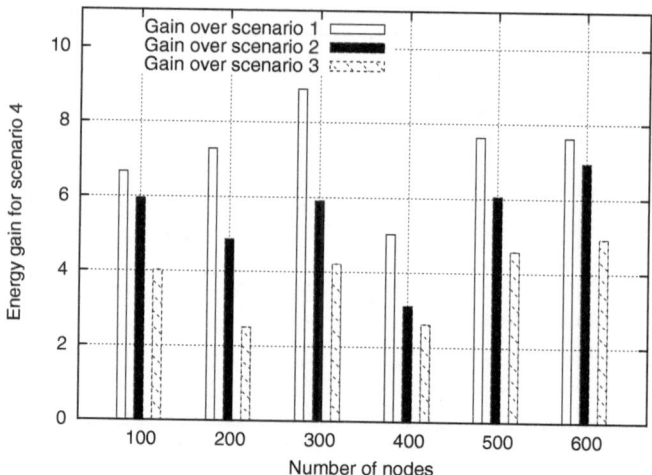

Fig. 3. Gains on energy consumption when comparing Scenario 4 with the three first scenarios

ℓ is sent one time during l periods, it is about

$$\frac{\ell \sum_{i=0}^{d} t^{d-i} + m \sum_{i=0}^{d} t^{d-i}}{l}.$$

Thus, performance evaluations confirm theoretical evaluations: in the theoretical approach, the gain would be better than as shown by evaluations but this fact is directly linked with the perfect structure of the network supposed in the theoretical model; the number of aggregator nodes is also greater in the theoretical approach.

5   Conclusion

In this paper, we have presented a simple method based upon universal hash function to aggregate MACs in a Wireless Sensors Network. To reach this aim, we extended the two schemes originally proposed by Krawczyk in [17] to simplify the data treatment. We have also discussed the security of our schemes in the model proposed in [7]. We have validated our approaches by intensive simulations that show the pertinence of our schemes and a significant gain in terms of energy when our last proposal is used.

Due to the small sizes of the sent messages in a WSN, it seems judicious to send several messages in a same time to be sure to generate a correct (and sufficiently long) MAC. If only a single message is sent, the required functions for this operation look more like expansion functions than compression functions. In our future work, we will particularly focus on this expansion aspect and on the implementation of universal hash functions in MSP430 sensors in order to evaluate their software performances on small devices.

References

1. Aad, I., Castelluccia, C., Hubaux, J.P.: Packet coding for strong anonymity in ad hoc networks. In: IEEE Securecomm (August 2006)
2. Bernstein, D.J.: The poly1305-aes message-authentication code. In: Gilbert, H., Handschuh, H. (eds.) FSE 2005. LNCS, vol. 3557, pp. 32–49. Springer, Heidelberg (2005)
3. Bhaskar, R., Herranz, J., Laguillaumie, F.: Efficient authentication for reactive routing protocols. In: AINA (2), pp. 57–61. IEEE Computer Society, Los Alamitos (2006)
4. Carter, L., Wegman, M.N.: Universal Classes of Hash Functions. Journal of Computer and System Sciences - JCSS 18(2), 143–154 (1979)
5. Castelluccia, C.: Securing very dynamic groups and data aggregation in wireless sensor networks. In: IEEE MASS - The Fourth IEEE International Conference on Mobile Ad-hoc and Sensor Systems, Pisa, Italy, October 2007, pp. 1–9 (2007)
6. Castelluccia, C., Mykletun, E., Tsudik, G.: Efficient aggregation of encrypted data in wireless sensor networks. In: Mobile and Ubiquitous Systems: Networking and Services - MobiQuitous 2005, pp. 1–9 (2005)
7. Chan, A.C.-F., Castelluccia, C.: On the (Im)possibility of aggregate message authentication codes. In: IEEE International Symposium on Information Theory - ISIT 2008, pp. 235–239. IEEE, Los Alamitos (2008)
8. Chan, A.C.-F., Castelluccia, C.: On the Privacy of Concealed Data Aggregation. In: Biskup, J., López, J. (eds.) ESORICS 2007. LNCS, vol. 4734, pp. 390–405. Springer, Heidelberg (2007)
9. Chan, H., Perrig, A., Song, D.: Secure hierarchical in-network aggregation in sensor networks. In: CCS 2006: Proceedings of the 13th ACM conference on Computer and communications security, pp. 278–287. ACM, New York (2006)
10. Domingo-Ferrer, J.: A Provably Secure Additive and Multiplicative Privacy Homomorphism. In: Chan, A.H., Gligor, V.D. (eds.) ISC 2002. LNCS, vol. 2433, pp. 471–483. Springer, Heidelberg (2002)
11. Ekdahl, P., Johansson, T.: A new version of the stream cipher SNOW. In: Nyberg, K., Heys, H.M. (eds.) SAC 2002. LNCS, vol. 2595, pp. 47–61. Springer, Heidelberg (2003)
12. Ben Hamida, E., Chelius, G., Gorce, J.-M.: Scalability versus accuracy in physical layer modeling for wireless network simulations. In: 22nd ACM/IEEE/SCS Workshop on Principles of Advanced and Distributed Simulation (PADS 2008), Rome, Italy (June 2008)
13. Handschuh, H., Preneel, B.: Key-Recovery Attacks on Universal Hash Function Based MAC Algorithms. In: Wagner, D. (ed.) CRYPTO 2008. LNCS, vol. 5157, pp. 144–161. Springer, Heidelberg (2008)
14. Heinzelman, W., Chandrakasan, A., Balakrishnan, H.: Energy-efficient communication protocol for wireless microsensor networks. In: Proceedings of the Hawaii Conference on System Sciences (January 2000)
15. Hu, L., Evans, D.: Secure aggregation for wireless networks. In: Workshop on Security and Assurance in Ad hoc Networks, pp. 384–394 (2003)
16. Katz, J., Lindell, A.Y.: Aggregate message authentication codes. In: Malkin, T.G. (ed.) CT-RSA 2008. LNCS, vol. 4964, pp. 155–169. Springer, Heidelberg (2008)
17. Krawczyk, H.: LFSR-based hashing and authentication. In: Desmedt, Y.G. (ed.) CRYPTO 1994. LNCS, vol. 839, pp. 129–139. Springer, Heidelberg (1994)

18. Nevelsteen, W., Preneel, B.: Software performance of universal hash functions. In: Stern, J. (ed.) EUROCRYPT 1999. LNCS, vol. 1592, pp. 24–41. Springer, Heidelberg (1999)
19. Przydatek, B., Song, D., Perrig, A.: SIA: Secure information aggregation in sensor networks. In: ACM SenSys 2003 (November 2003)
20. Sarkar, P.: A New Universal Hash Function and Other Cryptographic Algorithms Suitable for Resource Constrained Devices. Cryptology ePrint Archive, Report 2008/216 (2008), http://eprint.iacr.org/
21. Shoup, V.: On Fast and Provably Secure Message Authentication Based on Universal Hashing. In: Koblitz, N. (ed.) CRYPTO 1996. LNCS, vol. 1109, pp. 313–328. Springer, Heidelberg (1996)
22. Wagner, D.: Resilient aggregation in sensor networks. In: 2nd ACM workshop on Security of ad hoc and sensor networks - SASN 2004, pp. 78–87. ACM, New York (2004)
23. Yang, Y., Wang, X., Zhu, S., Cao, G.: Sdap: a secure hop-by-hop data aggregation protocol for sensor networks. In: MobiHoc 2006: Proceedings of the 7th ACM international symposium on Mobile ad hoc networking and computing, pp. 356–367. ACM, New York (2006)

Sec-TMP: A Secure Topology Maintenance Protocol for Event Delivery Enforcement in WSN[*]

Andrea Gabrielli[1], Mauro Conti[2], Roberto Di Pietro[3], and Luigi V. Mancini[1]

[1] Dipartimento di Informatica, Università di Roma "La Sapienza", Roma, IT
[2] Department of Computer Science, Vrije Universiteit Amsterdam, Amsterdam, NL
[3] Dipartimento di Matematica, Università di Roma "Tre", Roma, IT

Abstract. Topology Maintenance in Wireless Sensor Networks (WSNs), that is, alternating duty cycles with sleep cycles while having an adequate number of nodes monitoring the environment, is a necessary requirement to allow the WSNs to move from niche applications to widespread adoption; topology maintenance is even mandatory when the WSNs are used in a security sensitive context.

In this work, we present the first scalable Secure Topology Maintenance Protocol (Sec-TMP) for Wireless Sensor Networks that does not require pair-wise node confidentiality. The aim of Sec-TMP is to enforce event delivery to the BS while providing a standard topology maintenance service to the WSN. Sec-TMP enjoys the following features: it does not require pair-wise node confidentiality; it does not need any underlying routing—just one-hop communications are used; and, it is highly scalable. Sec-TMP reaches its goal being also resilient to the known attacks on TMPs: snooze attack; sleep deprivation attack; and, network substitution attack. Furthermore, Sec-TMP confines node replication attack: once a node is captured, the protocol limits the possible usage of the corresponding node's ID to a single neighbourhood. Finally, extensive simulations support our findings.

Keywords: Sensor Network Security, Topology Maintenance Protocol, Attack-Resilient.

1 Introduction

Wireless Sensor Networks (WSNs) are designed to fulfil a variety of tasks; law enforcement, disaster recovery, search-and-rescue, to cite a few [1]. WSNs are often unattended and operate in harsh environment. Furthermore, they operate without relying on existing infrastructure; for example, nodes are scattered by an airplane and once on the ground they start communicating to each other. The communication radius of a node determines its neighbourhood.

One of the most challenging research problem of WSNs is topology maintenance [6,5]: if more than the required number of nodes are present in a given

---

[*] This work is partially supported by Caspur under grant HPC-2007.

Y. Chen et al. (Eds.): SecureComm 2009, LNICST 19, pp. 265–284, 2009.

area, some of the nodes could switch from working state to sleeping state to save energy and to avoid communication congestion. Nodes in sleeping state could be activated in a further moment, if required. Different Topology Maintenance Protocols (TMPs), that assume a trusted environment, have been already proposed in the literature [6,5,13,23]. While being efficient and effective, if a malicious node is inserted in the above cited solutions, the WSN functionalities can be subverted. Indeed, just few preliminary works consider the security of TMPs, such as [18,15].

The main goal of our Sec-TMP protocol is to enforce the delivery of events intended to be received by the BS, or to detect that such a delivery failed, while providing the functionalities of a standard topology maintenance protocol. Assume a specific event (e.g. a fire alarm) is sensed in a given neighbourhood. The adversary goal could be to prevent the BS from learning such an event. To this aim, it could exploit a non-secure TMP so that only malicious nodes will be working in that neighbourhood when the BS will come to collect data—nodes that will not signal the firing event.

As an example, we can think to any data gathering scenario in a hostile environment such as a battlefield.

We assume WSNs do not necessary have an underlying routing protocol: each node is just programmed to sense data and pass-it on to the Base Station (BS). In order to increase the network lifetime, and to be resilient to attacks, nodes run our Sec-TMP protocol. In particular, assume node b is driven —by our TMP protocol— to a sleeping state while b has some data to communicate to the BS. Before setting itself in sleeping state, b sends its sensed data to its neighbour, say a, that has committed itself to be in working state (see Figure 1). Node a will eventually deliver b's data to the BS.

Contributions. In this paper we propose, to the best of our knowledge, the first Secure Topology Maintenance Protocol (Sec-TMP) for Wireless Sensor Networks (WSN) that: (i) does not require any pair-wise node confidentiality; (ii) is scalable (newly deployed nodes would be involved in the topology maintenance protocol by pre-existing nodes in the network); (iii) it is resilient to standard attacks TMP are subject to [18,15]: snooze attack; sleep deprivation attack; and, network substitution attack; and, (iv) tames the effect of node replication attack.

Organization. In Section 2, we describe the related work in the area. In sections 3 and 4 we describe the system assumptions and the threat model, respectively. In Section 5, we give an overview of the proposed protocol. In Section 6, we describe the Sec-TMP protocol in detail. In Section 7, we analyze the security of

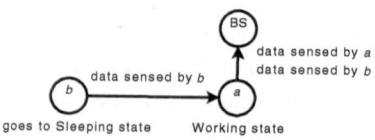

Fig. 1. Collaboration instance

our proposal and we discuss the obtained simulation results in Section 8. Finally, we conclude in Section 9.

2   Related Work

One of the main issue in WSNs is energy consumption since power is often provided to a node by a small battery, and it is often impossible to replace nodes battery. For instance, this is the case when WSNs are deployed in a hostile or not easy reachable environment.

Many different approaches have been proposed in the literature to prolong the lifetime of a WSN. A classification of these techniques is presented in [2]. There are protocols, such as [24], that are based either on changing and adjusting the transmission power of each node, or on geometrical structure-based methods to select next-hop neighbours. There is a class of so called Power Management protocols [2], such as [20], that aim to save the nodes' energy by switching off the radio in the active nodes when they do not need to communicate.

Between the proposed solutions to extend network lifetime, there are the Topology Maintenance Protocols (TMPs). The TMPs protocols leverage the node redundancy to schedule working periods between nodes. Only a subset of the nodes are in a working state, while the others goes in a sleeping state (stand-by) to save energy. Several TMPs have been proposed in the literature (e.g. SPAN [6], ASCENT [5], GAF [26], PEAS [13], CCP [23], AFECA [25]). These TMPs differ not only in the approach to schedule sleeping periods, but they differ also in their objectives. For example, SPAN [6] and ASCENT [5] aim to maintain only network connectivity. Others, such as PEAS [13] and CCP [23], aim to maintain both connectivity and sensing coverage. Some of them, as for example GAF [26], rely upon nodes location information; they require nodes with a GPS or some other location system. The CCP approach is similar to the one of SPAN, and they share the same state activity diagram and the same communication pattern.

The secure TMP protocol that we propose in this paper belongs to the Topology Maintenance Protocols that leverages sleep-wake period management. So, the comparison of our solution and the focus of this paper is within this category. In particular, all the cited protocols in the considered classification are vulnerable to attacks, as described in [15].

The first work that addressed the security issues on TMPs, [18], described the snooze attack against some previous protocols (e.g. GAF [26], SPAN [6], and AFECA [25]). However, [18] does not discuss the use of the snooze attack to reduce the sensing coverage. Moreover, [18] does not take the sleep deprivation and the network substitution attacks into consideration, nor do they discuss any possible countermeasure. In [21] the sleep deprivation attack is introduced in a context different than that of TMPs (and no countermeasures are described).

In [15], the security vulnerabilities of topology maintenance protocols for wireless sensor networks are analyzed. In particular, two new attacks in the context of topology maintenance protocol, namely the sleep deprivation attack and the network substitution attack are described. In [15], authors describe how these attacks can be launched against PEAS, ASCENT, and CCP, and suggest some countermeasures to make these cited protocols robust against the exposed attacks.

We observe that the solutions proposed in [15] require authenticated node pair-wise communication, e.g. a pair-wise scheme such as [12,10,7] must be used in conjunction with the TMP protocol. Furthermore, [15] requires the used pair-wise key establishment scheme to be resilient to node replication attacks [8]. In particular, in the proposed countermeasures it is required that, even if the adversary captures a node w, the identity of the compromised node cannot be successfully impersonated outside the neighbourhood of w. One possible protocol for achieving this goal in a sensor network is LEAP [27]. However, any scheme that meets these requirements can be used.

In [14], the authors propose the idea of applying mechanism inspired by biological systems and processes in order to increase the security and fault-tolerance of TMP protocols. However, no practical implementation is analysed.

The motivation of the importance of events detection and data survival in unattended wireless sensor networks has recently been highlighted in [11].

Finally, also the mobility of the BS, or the mobility of the nodes, has been considered for different purposes. As an example, in [3] the mobility of the BS is leveraged to balance the power consumption of static nodes in the network. In [16], the authors proposed a protocol to move nodes along the deployment area to maintain the coverage.

3   System Assumptions and Notation

In the remainder of this work we assume the system model described in this section. In particular, we assume a static network—each node has an initial location that does not change as the time goes by. However, nodes do not need to be deployed all at the same time—newly deployed nodes will cooperate to the aim of the topology maintenance with the nodes already present in the network area.

Our Secure Topology Maintenance Protocol (Sec-TMP) does not require any underlying routing protocol—it only resorts to one hop messages. Furthermore, each node, say a, can only contact the Base Station (BS) directly (via one-hop), when the BS is within the a's transmission radius. The BS is not always reachable by every nodes. The BS has a GPS, and it moves over the network in an unpredictable way. It can also be absent from the network for a while (i.e. no node is able to reach the BS). As a result, from the single node point of view, the BS can be reachable (appear/disappear) in an unpredictable way.

In Sec-TMP, time synchronization between nodes is not necessary—nodes are not required to share a common time. However, we assume that the clock drift (difference between nodes' clock speed) is negligible.

Sec-TMP does not require any pair-wise key to be shared among nodes. Node a only shares a symmetric key, K_a, with the BS. So, we remark that the BS (and not the other nodes) is the only one that checks the authenticity of the messages originated (not just forwarded) by the nodes. Furthermore, the nodes do not know the set of the legitimate node IDs. To ease exposition, we assume that the following hypothesis holds: for each neighbourhood, the number of nodes physically present is greater or equal than the number of nodes, (d), desired to be in a working state. Otherwise, the security properties stated in this paper cannot be guaranteed.

We define the neighbourhood of node a as the circular area having (i) center corresponding to a's location; and, (ii) radius R_t—the Transmission range—. The nodes in the neighbourhood of a are its one-hop neighbours (also called just neighbours).

Table 1 summarizes the notation used in the paper.

Table 1. Summary of Notation

Symbol	Meaning
N	Number of network nodes
d	Desired number of node in Working state for each neighbourhood
R_t	Transmission range (radius)
T_s	Sleeping time
K_a	Symmetric key shared between node a and the BS
λ	The desired probing rate towards nodes in Working state
ρ	Average of the received neighbours densities
τ	Time interval between consecutive BS contact
p	Probability that a node starts in Working state
$Dens_a$	Neighbour density of the node a
t_{BS}	Time information used by the BS
$counter_b$	A counter used by the node b

4   Threat Model

In this section, we describe the behavior and the capabilities of the adversary. In particular, we assume the adversary can compromise nodes and make them colluding, or playing an even harmful attack: having laptop-class devices storing and using information retrieved from compromised nodes. We observe that this type of attacks are particularly challenging. In fact, they cannot be prevented by authentication mechanisms since the adversary knows all the crypto material possessed by the compromised nodes.

The aim of the adversary is to avoid a sensed data to reach the Base Station. Under our assumption of data forwarding, the adversary can directly reach its target if it is able to do the following: leverage the proposed TMP protocol to avoid non-compromised (honest) nodes from being in Working state. However, the adversary could try to achieve its goal also by using standard TMP attacks [15,18]:

– Sleep Deprivation Attack. The adversary tries to induce a node to remain active. This attack has two effects. First, by increasing the energy expenditure of sensor nodes, it reduces the lifetime of the node and of the network as well. Second, in the case of a densely populated area, it can lead to increased energy consumption due to congestion and contention at the data link layer.

- Snooze Attack. The adversary forces the nodes to remain in sleeping state. The adversary can launch this attack to reduce the sensing coverage in a region of the network. This kind of attack can be applied to the whole network or to a subset of nodes.
- Network Substitution Attack. The adversary takes control of the entire network or of a portion of it by using a set of colluding malicious nodes.

As an example, if the adversary is able to let all the node exhaust their batteries through sleep deprivation attack, there would be no sensed data at all —so, the adversary reaches its goal.

We assume that, for a neighbourhood composed of d nodes, the adversary is able to uses at most $d-1$ compromised nodes identity. We stress that it is not necessary for the adversary to capture the node IDs in the same neighbourhood where they are intended to be used by the adversary. Furthermore, we assume that the adversary can eavesdrop on the communications of other nodes and inject data packets into the network. Nodes are not considered to be tamper-proof. As a consequence, all the information (including cryptographic keys) stored in a node the adversary compromised with are considered leaked to the adversary. As for the BS, we assume that it is trusted.

Finally, we assume the adversary is also able to perform the following type of attack (that Sec-TMP is implicitly able to defend from): Node replication attack. That is, the adversary cloning the identity (and the cryptographic material) of a captured node in other malicious nodes.

5   Protocol Overview

In Sec-TMP, each node has three operating states: Working, Sleeping, and Probing. The state diagram is described in Figure 2.

- Sleeping: the node does nothing but saving energy (i.e. turning off the radio) and waiting for the time out T_s—indicating the sleeping time—to expire. When T_s expires the node moves in Probing state.
- Probing: the node probes its neighbours to determine whether to go either in Sleeping or in Working state. The node sends a PROBE message within its transmission range R_t, and it waits for P-REPLY messages in response.
- Working: the node executes the regular node operations such as sensing and communicating to the BS as required by our protocol. When the BS claims its presence, and the node off-loads the data it stores to the BS, it also sends

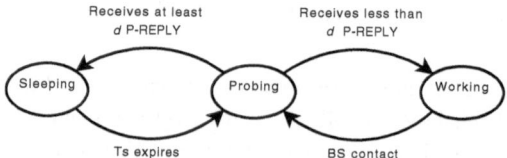

Fig. 2. State transition diagram

a request to the BS to receive a proof of the upload activity. Then, it goes in Probing state. Moreover, if the node receives a PROBE message from one of its neighbours, it replies to the neighbour sending a P-REPLY. Via the P-REPLY, the node informs the neighbours that it is in Working state.

Once a node is deployed, it starts in Sleeping state having an initial sleeping time-out T_s. The initial T_s is randomly picked over a given time interval. This choice is motivated to avoid (probabilistically) that all the nodes go in Probing state all together.

We remind that we do not assume any confidentiality layer in the node pairwise communications. The main idea underlying our proposal is to have a mechanism that, if a node a states to be in a Working state (e.g. replying to a PROBE originated by node b), then node a is actually forced to be in Working state until the next b's probe. Otherwise, a will be considered failed or malicious. In the proposed protocol implementation, we actually extend this mechanism from a single node, to a set of at least d nodes: i.e. a node a can actually move from the Working state to the Sleeping state if there is some other node that, in turn, moved from the Sleeping state to the Working state.

Enforcement is realized using the evidences issued by the BS to the node in Working state. Let us assume node b sends a PROBE to node a, and a replies being in Working state. If the BS enters the communication range of a before the next b's probe, a has to prove to b it was in Working state when the BS come —that is, it uploaded b's data to the BS. To this aim, we require a to ask the BS for a specific proof (provided by an authenticated token[1]). Node a sends back the token to b, allowing b to check that a was in Working state and interacted with the BS. This is required for any node to which a replied to as a consequence of a PROBE. However, it is not necessary that the BS releases a new token before the next b's probe. If this is the case, node a replies to b using the last token received from the BS. As we describe in Section 7.1, this behaviour does not introduce a security issue. Note that there is a specific transient case to deal with: before the first BS arrival, no node can provide a token to a probing node. We describe this situation (protocol Start-up), in Section 6.1.

6   Protocol Description

In this section, we give a detailed description of our protocol. In particular, we first present the protocol Start-up in Section 6.1. Then, we describe the behaviour of a node while it is in the different protocol states (Figure 2): Section 6.2 describes the Probing state, Section 6.3 describes the Working state and, finally, Section 6.4 describes the Sleeping state.

6.1   Protocol Start-Up

Before the first BS arrival, no node can provide a token to a probing node. As a consequence, every node could be in Working state after the first PROBE,

---

[1] Note that the token authenticity is enforced using symmetric keys.

until the first BS arrival. Note that this solution would be energy-consuming and could cause transmission congestion.

Assuming no adversary is present in the network before the first BS arrival, we can easily start-up our protocol as follows: instead of considering d tokens, a probing node b considers d P-REPLYs without requiring any token. In particular, the probing node b will accept P-REPLYs without requiring tokens, until it receives a P-REPLY with a token (i.e. the BS has contacted a neighbour of b). As soon as the BS arrives the first time, every nodes in Working will receive the correct tokens.

However, we do assume that an adversary can be present from the time of the network deployment (also, be there during the start-up). So, one might think to the following attack. If a single adversary node is present in every neighbourhood (this is not against our assumptions), it can send as many probe replies (P-REPLY) as it wants. In this way, a single adversary node can put every other node in the neighbourhood in Sleeping state.

To avoid this attack, we use a probabilistic approach. At the beginning, every node decides, with probability p, to go in Working state. Otherwise, with probability $1-p$, the node goes in Probing state and accepts P-REPLYs without requiring tokens. These latter nodes accept P-REPLYs without requiring tokens, until they receive a P-REPLY with a token.

As a result, even in the presence of an adversary, a honest node will be in Working state in a neighbourhood with a given probability p. Then, at the first BS arrival, it will receive the token from the BS.

Note that node a ends the start-up phase when it receives a token (either from the BS, if a is in Working state, or from a node in Working state, if a sends a PROBE). Thus, the end of the start-up propagates in the network thanks to the nodes in Working state.

In the following, we describe the operations performed by nodes when the start-up phase has come to end.

6.2   Probing State

In this section, we describe the operations that a node performs while it is in Probing state (Algorithm 1).

The aim of the Probing state is to determine the next state (Sleeping or Working) of the executing node. To do so, a node, say b, broadcasts a PROBE (line 5). The PROBE is authenticated by the node b with its key K_b, and it contains a counter value, $counter_b$. The $counter_b$ is different for each following PROBE message. As explained in Section 7.1, the $counter_b$ is necessary to avoid replay attacks. Once the PROBE has been broadcasted (line 5), the nodes waits for a time δ (set in line 4) to receive the associated reply P-REPLY from its neighbours in Working state.

In each P-REPLY message, there is a token generated by the BS. The token contains: a time information t_{BS}, the identity of the replying node a, and the neighbour density of the replying node, that is $Dens_a$.

```
input  :
          b, ID of the executing node
          a, ID of the replying node
          D_reply, Desired number of P-REPLY (i.e. d)
output:
          the next node state (Working/Sleeping)
          T_s, Sleeping time
begin
     C_reply = 0
     // counter of P-REPLY
     Set TimeOut δ
     Broadcast ⟨PROBE, ID_b, counter_b, MAC_{K_b}(counter_b)⟩

     while δ not elapsed do
         for any received P-REPLY message MSG do
             if ⟨P − REPLY, b, a, t_BS, Dens_a, RMAC⟩ ← MSG AND
                MAC_{K_b}(t_BS, Dens_a, a) = RMAC then
                  C_reply ++

     if C_reply < D_reply then
          move to Working state
     else
          T_s = sleepTime
          move to Sleeping state
end
```

Algorithm 1. Sec-TMP: Probing

The t_{BS} is used to inform the probing node about the time when the token has been created. This value is used in the following way: if b receives a P-REPLY with a given BS time, say t'_{BS}, then b will not consider the tokens having $t''_{BS} < t'_{BS}$ for the computation of d. In fact, let us assume node c stated to be in Working state; if it is not able to show to b a BS token with time t'_{BS}, that means c did not have the chance to communicate with the BS during the BS arrival time t'_{BS}: c was sleeping or not following the protocol. In either case, the P-REPLY of c will be ignored. Observe that the BS needs time to move and talk with the different nodes in the neighborhood. As a result, the t_{BS} values released to two neighbour nodes can actually be different by a small amount of time. For easy of exposition, in the following we do not explicitly consider this problem. However, we note that this could be solved either (i) implementing t_{BS} as just a counter managed by the BS, so not changed while talking with different nodes in the same BS passage; (ii) comparing the different t_{BS} values taking into consideration this small amount of time as negligible.

Once a node collected all the P-REPLYs, it has to take a decision about its next state. In particular, if less than the desired P-REPLYs are received (check line 10), the executing node will move to Working state (line 11). Otherwise, if the number of P-REPLYs is enough for the node, it will set the Sleeping time (line 13) and it will go in Sleeping state (line 14).

The node sets the Sleeping time according to an exponential distribution [13]:

$$f(T_s) = (\lambda/\rho)e^{-(\lambda/\rho)T_s} \qquad (1)$$

where λ is the desired probing rate towards working nodes (λ is the same for each node), and ρ is the estimated neighbours density. The idea behind this is to leverage the network density; we decrease the probing rate of a sleeping node together with the increase of its neighbours density. The ρ value is computed by each node independently as the average of the received densities ($Dens_a$ of line 8). The $Dens_a$ is the neighbour density of the replying node a. Note that the value $Dens_a$ is computed by the BS. In particular, the BS computes $Dens_a$ by counting the number of TOKEN requested by a node a in Working state. Then, the BS inserts $Dens_a$, into the TOKENs replied to a (Algorithm 2, line 6). This value represents the neighbours density of node a —node a requests a TOKEN for each of its neighbours.

6.3   Working State

In this section we describe the operations performed by a node that is contacted by the BS. When node a is aware of the presence of the BS within its communication range, it runs Algorithm 2. In particular, node a sends to the BS a request (REQ type message) containing the last PROBEs a received from its neighbours (line 2). The REQ message is authenticated using K_a. If the BS fails authenticate the message, then the BS just discards the request. However, if the authentication succeeds, with this message node a implicitly requires the BS to produce a token (that is different for each neighbour of a) to prove that a was in Working state and interacted with the BS. In fact, for each received token (TOKEN message, line 6) that passes the authentication check (line 7), a updates the token available for the corresponding neighbour b (line 8). That is, this token will be used as a P-REPLY for the next PROBE of node b. Furthermore, at the end of every BS contact, a moves to Probing state (line 10). This latter operation is required to recover from a situation that tends to put in Working state all the nodes in a neighbourhood. The latter situation is rooted in failure of a working node. Assume that d nodes in the neighbourhood are in Working state, as required by the protocol. Assume that one of these nodes, say a, exhausts its battery. As a results, less that d nodes are in Working state. In this setting, as soon as the first node in this neighbourhood, say b, sends out a PROBE, it will receive less than d P-REPLYs with a token. As a consequence, node b will go in Working state. Any other further nodes in this neighbourhood, say c, executing a probe, would receive less than d P-REPLYs with a token (also if $\geq d$ node are now in Working state), because less than d nodes would be able to send a P-REPLY with a token. So, if a working node fails, the result is that every node executing a probe in the neighbourhood of the failed node would go in Working state, having $D > d$ nodes in Working state.

However, when the BS arrives, all the nodes in Working state are required to: ask tokens for their neighbours, and go in Probing state. As a result, only the last d out of the D nodes executing the probe will go back in Working state. In general, this procedure could be leveraged to select the d out of the D nodes in some optimal way. As an example, it could be desirable that the d nodes out of D are the ones with more available energy —to do so, we can distribute the time of the probe execution of the D nodes such that the higher is the battery power, the later is the performed the probe.

Note that, before releasing a TOKEN for a node b, the BS verifies the b's PROBE message. The BS discards the request without releasing the TOKEN, if either the authentication of the PROBE using K_b fails, or the BS has already seen the $counter_b$ value. The $counter_b$ value avoids replay attacks, as described later in Section 7.1.

Finally, we remind that the BS has a GPS: this allows the BS to build a map of the network topology as follows. The first time it receives a message from a node a, the BS bounds the location of a into a region compliant with the BS location at the moment of the contact (depending on the transmission radius of nodes). In particular, after the first contact with the BS, node a location is bounded into a circular area of radius R_t. For each following contact with node a, the BS refines the boundary region. The BS stores the network topology map for security reasons described later in Section 7.1.

---

input :
  a, ID of the executing node
  NP_a, PROBEs of all the neighbours node
output:
  CT_a, Executing node's table of tokens

begin
  $\langle REQ, a, NP_a, MAC_{K_a}(NP_a) \rangle \rightarrow BS$
  Set TimeOut δ
  while δ *not elapsed* **do**
    for *any received TOKEN message MSG* **do**
      $\langle TOKEN, a, Dens_a, b, t_{BS},\ MAC_{K_b}(t_{BS}, Dens_a, a), RMAC \rangle \leftarrow MSG$
      if $MAC_{K_a}(MAC_{K_b}(t_{BS}, Dens_a, a)) - RMAC$ **then**
        Update $CT_a(b, \langle TOKEN, Dens_a, t_{BS}, MAC_{K_b}(t_{BS}, Dens_a, a) \rangle$

    if *received a valid TOKEN message* **then**
      move to Probing state

end

Algorithm 2. Sec-TMP: BS-contact

6.4   Sleeping State

A node in Sleeping state does nothing but waiting for the T_s timer expiration. In this state, the node saves energy having its radio turned off. When the sleeping timer T_s expires, the node turns the radio on and enters the Probing state.

7   Security Analysis

We remind that the protocol goal is to enforce the delivery of a sensed event from the generator node to the BS, while providing a standard topology management service. That is, assuming at most $d - 1$ compromised nodes in the neighbourhood, there is at least a non compromised node in Working state for each neighbourhood. The idea is that if there is a sensed event that should be reported to the Base Station, there will be at least one (honest) node doing that.

As outlined in Section 4, the adversary could try to reach its goal in a direct way —i.e. leveraging the specific behaviour of our protocol— or through standard attacks on TMPs. In Section 7.1, we describe why our protocol features cannot be leveraged by the adversary. In Section 7.2, we discuss the resilience of our protocol to the standard attacks on TMPs [15,18] (to the best of our knowledge, all the known attacks on TMPs): Sleep Deprivation attack; snooze attack; and, network substitution attack. Finally, in Section 7.3 we discuss how Sec-TMP tames the node replication attack.

7.1 Sec-TMP Security Property

In the following we first revise the possible attack, and later explain how Sec-TMP thwarts it.

Using a node's ID in more than one neighbourhood. Once an adversary captured a node, it can know the ID (and the symmetric key) of that node (we remind from Section 3 that we do not assume to have tamper proof sensors). The adversary can plug all the known IDs in a single device (e.g. a laptop class node) that pretends to be in every neighbourhood. In this way, compromising a total of d nodes, one might think that the adversary could be able to use the d IDs in every neighbourhood. So, taking over the network.

In Section 6.3 we described how the BS can estimate the network topology. We now observe that this allows the BS to prevent the described attack. In fact, assume the adversary already used a node's ID, say a, in a given location, say loc_a, to ask for tokens. Then, the BS will link the node's ID a to the claimed location loc_a. When a further asks other tokens, BS will check whether the current a's location is coherent with loc_a. If this is not the case, the BS will not give any token to a —and possibly take further actions. Furthermore, the neighbours declared by node a (for which it ask the BS appropriate tokens) should also be coherent with the other information the BS collected about the network. As an example, a node b cannot be a's neighbour if b appeared in a location not coherent with the neighbourhood area of a.

Using the same TOKENs to reply to PROBE messages with multiple identities (Sybil Attack [19]). An adversary can eavesdrop the communications between honest nodes and the BS. Assume that the honest node a requests a TOKEN for one of its neighbours, say b. The adversary eavesdrops and stores this TOKEN received by node a. Assume that the number of working neighbours of b is less than d. Then, after the following probe, node b will go in Working state. One might think that the adversary could send multiple P-REPLYs to b, using the stored TOKENs and identities different from a (Sybil attack [19]). In this way, the adversary could induce b to figure out it has more than d working neighbours —that is, the goals of the adversary is to force node b to go in Sleeping state.

However, the BS includes into each TOKEN released, the identity of the requesting node, in this case a. Note that, the TOKEN message is authenticated with the symmetric key K_b (Algorithm 2, line 6). As a consequence, the authentication check of the TOKEN (received from the adversary) performed by probing node b fails, because the sender identity does not match (check done in line 8, Algorithm 1). Thus, the attack fails because the honest node b discards the P-REPLYs sent by the adversary.

Using old BS's TOKENs to reply to PROBE messages (Replay Attack). As for the previous attack, assume that an adversary eavesdrops the communication between honest nodes and the BS. Assume that a honest node a requests a TO-KEN for one its neighbours, say b. The adversary eavesdrops and stores the TO-KENs received by node a. After node a fails (for instance, because its battery depleted), the adversary could send the stored TOKENs to node b. In this way, the adversary tries to induce b to believe that a is still alive and in Working state.

Such an attack would fail for the following reason. The BS includes into the released TOKENs, a time t_{BS} (and the identity of the TOKEN requesting node, say a). The t_{BS} value changes (increases) for each BS arrival. Because of the mechanism described in 6.3, every node that ends the sleeping time before the next BS arrival will remain in Working state. Since there are at most $d - 1$ compromised nodes in the neighbourhood, after a while a honest node, say c, will be in Working state (remind that, by the protocol, d nodes must be in Working state). The next time the BS will come over the neighbourhood, c will ask the BS for a TOKEN for node b. Being c a honest node, it will take the correct TOKEN with the updated t'_{BS}. When b executes the probe, if it receives a TOKEN with the updated t'_{BS}, b will ignore any other TOKEN with $t''_{BS} < t'_{BS}$.

Using old PROBE messages (Replay Attack). Assume that an adversary replies old PROBE messages of a node b to its neighbour a that is in Working state. The scope of such an attack can be to induce the BS to consider b alive even if b failed. As a result, the BS would compute a false neighbours density. Indeed, the number of TOKENs requested from a node a is used by the BS to compute the neighbours density of node a. The density estimation is then included into the TOKEN that the BS sends back to the node a. Eventually, the density is used by the neighbour of a that are in Sleeping state to set the next sleeping timer (see Section 6.2). If the adversary could successfully use old b's PROBE messages, it could maliciously falsify the BS estimation. In this way, it could increase the length of the nodes sleeping periods. Furthermore, the adversary could increase the nodes energy depletion. In fact, the BS continues to release the TOKEN for the already failed node, b, to the node a. As a consequence, node a receives the TOKEN, stores it, and then sends it to b in the following P-REPLY.

We observe that the adversary cannot succeed because of the $counter_b$ included into each PROBE. In particular, each time a node probes the neighbourhood, it includes into the PROBE message the value of the node $counter_b$. If the attacker uses old $counter_b$ values, the node a discards the corresponding PROBE requests—and possibly take further actions. Observe that, if the attacker pretends to use forged $counter_b$ values, the node a cannot further identify the $counter_b$ as a forged one (a does not know the keys that b shares with the BS). However, this can be done by the BS that know the key that b should use to generate the MAC part of the b's message.

Node impersonation. The PROBE and the REQ messages are authenticated by the senders through keyed MAC (line 5 Algorithm 1, and line 2 Algorithm 2, respectively). Thus, an adversary can send valid PROBE and REQ messages on behalf of an identity a if, and only if, the adversary has compromised the node a (i.e. insider adversary).

Maintaining the network in start-up. Assume the adversary is present from the moment of the initial network deployment. As described in Section 6.1, the adversary can send P-REPLYs without any token, using d different identities, to maintain a node in Sleeping state. However, the adversary cannot force the nodes that start in Working state to go in Sleeping state. In fact, to do so, the adversary would need P-REPLYs with a token that can be only obtained from the BS. A node a ends the start-up when it receives a token (either from the BS, if a is in Working state, or from a node in Working state, if a sends a PROBE). Thus, the end of the start-up propagates in the network thanks to the nodes in Working state. When the node a ended the start-up, it pretends P-REPLYs with a token. As a consequence, the adversary can increase the time needed by the network to end the protocol start-up (i.e. all the nodes have received at least a P-REPLY with a token). However, it cannot indefinitely keep the network in the start-up. In Section 8.2, we report the simulation results of the time needed by the network to complete the start-up, assuming an adversary is present from the moment of the network deployment.

Event hiding. Assume the start-up completed and an adversary wants to hide to the BS an event generated by node b. To do so, the adversary could leverage the TMP protocol to have: (i) b in Sleeping state; and, (ii) just malicious nodes in Working state in the b's neighborhood—a malicious node would not send the target event to the BS.

As for (i), if the adversary wants b to move to Sleeping state, it must be able to provide d P-REPLYs with token. In fact, node b ended the start-up and it goes in Sleeping state only if it receives at least d P-REPLYs with token. The same condition is required for (ii). In fact, any other honest node should be forced to switch to Sleeping state by the adversary. Otherwise, an honest node in Working state will report the event to the BS.

As we assume that the adversary uses at most $d-1$ identity in a neighborhood, it would not be able to provide d malicious P-REPLYs.

7.2   Sec-TMP Resilience to Standard TMPs Attacks

In this section, we describe how Sec-TMP faces standard TMPs known attacks [18,15].

Resilience to Sleep Deprivation Attack. In this attack, the adversary wants to induce a node, say a, to remain in Working state, even if the node a already has d neighbours in Working state. Note that the node a, regardless if it is in Sleeping or Working state, periodically goes in Probing state. In Probing state, a sends out a PROBE. If it receives at least d P-REPLYs, it goes in Sleeping state. Thus, avoiding the reception of the P-REPLYs by node a is the only way the adversary has to successfully launch a Sleep Deprivation Attack against a. In other words, the adversary, to reach its goal, has to jam the node a to prevent the reception of the P-REPLY messages. However, due to the periodically and asynchronously transitions of the nodes to Probing state, this operation must be performed quite often, making the attack not affordable for node-class adversary, and resource consuming even for a laptop-class adversary. Moreover, note that a denial-of-service attack involving continuous jamming (e.g. constant or deceptive jamming) can be performed in any sensor network, regardless of the topology maintenance protocol being used. Hence, we do not consider this as an attack that is specific to topology maintenance protocols. It is worth noticing that the selective jamming can also be applied during BS tokens releasing phase. However, this behaviour can be detect by the BS, that can possibly react.

Resilience to Snooze Attack. In the snooze attack, the adversary wants to induce a node, say a, to remain in Sleeping state, even if the node a has less than d neighbours in Working state, as required by the Sec-TMP protocol. As previously described in 7.1, the adversary has to compromise d node identities that are neighbours of a. Thus, Sec-TMP is resilient to an adversary that compromises up to $d - 1$ nodes within a neighbourhood.

Resilience to Network Substitution Attack. The adversary substitutes legitimate nodes with malicious nodes in a portion of the network. To apply this attack, the adversary has to induce all the legitimate nodes in that portion of the network to go in Sleeping state. Thus, the resilience of Sec-TMP to this attack is the same of the resilience to the snooze attack (Section 7.2).

7.3   Sec-TMP to Thwart Node Replication Attack

We observe that our Sec-TMP protocol, while designed as a TMP for event delivery enforcement, it has also some ability to detect the node replication attack [4]: the adversary captures a node, and clone the identity (and the cryptographic material) of the captured node in other malicious nodes. In fact, as described in Section 6.3, the BS estimates the network topology. This allows the BS to detect if the same node ID, say a is used in two different locations, e.g. loc_a and loc'_a. Remind that the BS can estimate the location of (i) the node that directly asks for tokens —a in Figure 1; and, (ii) the nodes tokens are asked for —b in Figure 1. Once detected a cloned ID, the BS can take the appropriate actions, such as revoking the node.

8   Simulations and Discussion

In this section, we describe the simulation results we obtained for the Sec-TMP protocol.

We implemented a simulator of our protocol. We assumed nodes uniformly distributed in a $50 \times 50 m^2$ area (nodes remain stationary after deployment). We considered deployments with N=250, 500, 750, 1000, 2000, and 4000 nodes. In particular, for each network size, the shown results are the average of 100 different random network deployment.

Table 2. Sensor parameters for simulations

Parameter	Value
R_t	10 meters
R_s	10 meters
Tx consumption	0.0074mW per bit
Rx consumption	0.003575mW per bit
Idle consumption	13.8mW per second
Sleeping consumption	0.075mW per second
Signature consumption	2% of the packet transmission cost
Initial energy of a node	60 Jules

The parameters that represent the node characteristics are reported in Table 2. The values are similar to the hardware characteristics of the Berkeley Motes [9] sensors. In particular, we use the energy model proposed in [22], and the Tiny-Sec [17] model for the power consumption of symmetric cryptography operations.

In Section 8.1, we describe the simulation results related to the network coverage lifetime. Finally, in Section 8.2, we study the start-up completion time while considering the presence of an adversary.

8.1   Network Lifetime and Area Coverage

In this subsection, we evaluate the ability of Sec-TMP to increase the coverage lifetime of the network together with the increase of the number of deployed nodes.

To measure the coverage, we logically divide the entire sensing region into adjacent $5x5m^2$ patches. The coverage of the deployment area is approximated by monitoring the coverage of the top left corner of each patch, excluding those points that are on the border of the deployment area. A similar approach is used in [23]. The coverage lifetime is defined as the time interval from the activation of the network until the time of the following event: the percentage of the total area being monitored (by at least K nodes in Working state, K-coverage) drops below a specified threshold. The coverage lifetime characterizes how long the system ensures that the events are monitored with a probability of success higher than the specified threshold. In particular, we considered the threshold to be 80%. The coverage degree K is set to 1. With the minimum number of sensors in the network being equal to 250, we are quite sure we have enough nodes for 1-coverage in the area.

Figure 3 reports the coverage lifetime (y-axis) for networks with 250, 500, 750, 1000, 2000, and 4000 nodes (x-axis). Results are reported for different values of d (1,5, and 10), and for different values of BS arrival interval, τ (600 and 900 seconds). From Figure 3, we can conclude that Sec-TMP achieves the main goal of a TMP, that is the network lifetime grows almost linearly with the number of nodes. Note that, the curves trend are similar. However, while d increases, the network lifetime gain is smaller. This is because of the greater number of simultaneously active nodes. As discussed in the Section 7, the value d is related to the resilience of Sec-TMP to the adversaries. More precisely, an adversary has to compromise at least d nodes within the transmission range of a to successfully attack node a. We can say, the greater the required resilience to adversaries,

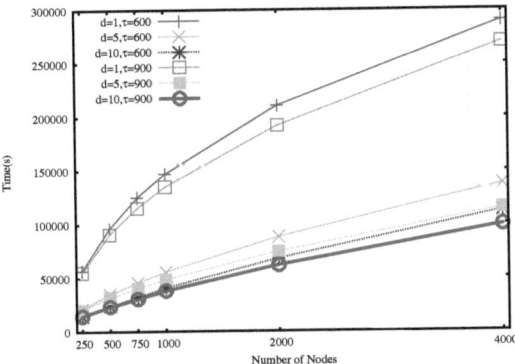

Fig. 3. Network coverage lifetime

the smaller the performance of Sec-TMP. For simulations with τ equal to 600 seconds, the gain in network lifetime is bigger than for simulations with τ equal to 900 seconds. This is expected because, as described in Section 6.3, the more nodes fails, the more nodes go in Working state; however, the number of nodes in Working state start decreasing with the BS arrival. Thus, when $\tau = 600$ the number of nodes in Working state is reduced quickly by the BS, compared to the case of $\tau = 900$. As a consequence, the increase in the network lifetime is higher with smaller τ.

8.2   Start-Up Completion Time

In this section, we study the time needed by the network to end the protocol start-up when an outsider adversary is present. In particular, we assume the adversary is present from the moment of the initial network deployment. As described in Section 6.1, the adversary can send P-REPLYs without any token to maintain a node in Sleeping state. However, the adversary cannot force the nodes that start in Working state to go in Sleeping state. In fact, to do so, the adversary would need P-REPLYs with token that can be obtained from the BS only. Without loss of generality, for the following simulations, we assume a single adversary node is present in every neighbourhood: the adversary responds to each PROBE with d P-REPLYs, using d different identities. The probability p that a node starts in Working state is equal to 0.1.

In Figure 4, we plot the network start-up time (y-axis), that is the time from the deployment of the network until all the nodes completed the start-up. Note that, a node a ends the start-up when it receives a token (either from the BS, if a is in Working state, or from a node in Working state, if a sends a PROBE). Thus, the end of the start-up propagates in the network thanks to the nodes in Working state. In Figure 4, we can see that the time to end the start-up decreases with the increase of the network density. This is due to the fact that, the higher is the network density, the higher is the probability to have a neighbour in Working state. In fact, both the number of nodes that start in Working state and the

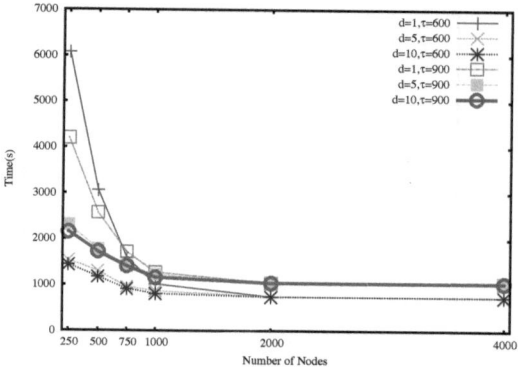

Fig. 4. Time to end the protocol start-up, when an adversary is present

number of neighbours is higher with a higher nodes density. The number of nodes that are simultaneously in Working state grows together with values of d. This is the reason why, when τ is fixed (i.e. τ=600), the time to end the start-up is higher for higher values of d. Finally, we can see from Figure 4 that the time to end the start-up increases tighter with τ. Indeed, if it takes longer for the BS to join the WSN, it takes longer to release the tokens.

9   Concluding Remarks

In this paper, we presented the first Secure Topology Maintenance Protocol (Sec-TMP) for Wireless Sensor Networks that enjoys the following features: it does not require any pair-wise node confidentiality; it does not need any underlying routing—just one-hop communications are used; and, it is highly scalable. The aim of Sec-TMP is to enforce event delivery to the BS while providing the functionalities of a standard topology management protocol. Sec-TMP reaches its goal being also resilient to the known attacks on TMPs: snooze attack; sleep deprivation attack; and, network substitution attack. Furthermore, Sec-TMP confines node replication attack: once a node is compromised, the protocol limits the possible usage of the corresponding node's ID to a single neighbourhood. Finally, simulation results support our findings.

References

1. Akyildiz, I.F., Su, W., Sankarasubramaniam, Y., Cayirci, E.: Wireless sensor networks: a survey. International Journal of Computer and Telecommunications Networking - Elsevier 38(4), 393–422 (2002)
2. Anastasi, G., Conti, M., Francesco, M.D., Passarella, A.: Mobile Ad Hoc and Pervasive Communications. In: How to Prolong the Lifetime of Wireless Sensor Networks, ch. 6. American Scientific Publishers (2007)

3. Basagni, S., Carosi, A., Melachrinoudis, E., Petrioli, C., Wang, Z.M.: Controlled sink mobility for prolonging wireless sensor networks lifetime. ACM/Springer Journal on Wireless Networks (WINET) 14(6), 831–858 (2008)

4. Bryan, P., Perrig, A., Gligor, V.: Distributed detection of node replication attacks in sensor networks. In: Proc. of the 26th IEEE International Symposium on Security and Privacy (S&P 2005), pp. 49–63 (2005)

5. Cerpa, A., Estrin, D.: Ascent: Adaptive self-configuring sensor networks topologies. IEEE Transactions on Mobile Computing 3(3), 272–285 (2004)

6. Chen, B., Jamieson, K., Balakrishnan, H., Morris, R.: Span: An energy-efficient coordination algorithm for topology maintenance in ad hoc wireless networks. ACM Wireless Networks Journal 8(5), 481–494 (2002)

7. Conti, M., Di Pietro, R., Mancini, L.V.: Ecce: Enhanced cooperative channel establishment for secure pair-wise communication in wireless sensor networks. Ad Hoc Networks (Elsevier) 5(1), 49–62 (2007)

8. Conti, M., Di Pietro, R., Mancini, L.V., Mei, A.: A randomized, efficient, and distributed protocol for the detection of node replication attacks in wireless sensor networks. In: Proc. of the 8th ACM International Symposium on Mobile Ad Hoc Networking and Computing (MobiHoc 2007), pp. 80–89 (2007)

9. Crossbow Technology Inc. MICA Sensor Node, http://www.xbow.com

10. Di Pietro, R., Mancini, L.V., Mei, A.: Energy efficient node-to-node authentication and communication confidentiality in wireless sensor networks. Wireless Sensor Networks 12(6), 709–721 (2006)

11. Di Pietro, R., Mancini, L.V., Soriente, C., Spognardi, A., Tsudik, G.: Catch me (if you can): Data survival in unattended sensor networks. In: Proc. of the 6th IEEE International Conference on Pervasive Computing and Communications (PERCOM 2008), pp. 185–194 (2008)

12. Eschenauer, L., Gligor, V.D.: A key-management scheme for distributed sensor networks. In: Proc. of the 9th ACM International Conference on Computer and Communications Security (CCS 2002), pp. 41–47 (2002)

13. Ye, F., Zhong, G., Lu, S., Zhang, L.: Peas: A robust energy conserving protocol for long-lived sensor networks. In: Proc. of the 23rd IEEE International Conference on Distributed Computing System (ICDCS 2003), pp. 28–37 (2003)

14. Gabrielli, A., Mancini, L.V.: Bio-inspired topology maintenance protocols for secure wireless sensor networks. In: Liò, P., Yoneki, E., Crowcroft, J., Verma, D.C. (eds.) BIOWIRE 2007. LNCS, vol. 5151, pp. 399–410. Springer, Heidelberg (2008)

15. Gabrielli, A., Mancini, L.V., Setia, S., Jajodia, S.: Securing topology maintenance protocols for sensor networks: Attacks and countermeasures. In: Proc. of the 1st IEEE/CreateNet International Conference on Security and Privacy for Emerging Areas in Communication Networks (SecureComm 2005), pp. 101–112 (2005)

16. Jiang, Z., Wu, J., Agah, A., Lu, B.: Topology control for secured coverage in wireless sensor networks. In: Proc. of the 4th IEEE International Conference on Mobile Adhoc and Sensor Systems (MASS 2007), pp. 1–6 (2007)

17. Karlof, C., Sastry, N., Wagner, D.: Tinysec: A link layer security architecture for wireless sensor networks. In: SenSys 2004, pp. 162–175 (2004)

18. Karlof, C., Wagner, D.: Secure routing in wireless sensor networks: attacks and countermeasures. Ad Hoc Networks 1(2-3), 293–315 (2003)

19. Newsome, J., Shi, E., Song, D., Perrig, A.: The sybil attack in sensor networks: Analysis & defenses. In: Proc. of the 3rd IEEE and ACM International Symposium on Information Processing in Sensor Networks (IPSN 2004), pp. 259–268 (2004)

20. Rhee, I., Warrier, A., Aia, M., Min, J., Sichitiu, M.L.: Z-mac: a hybrid mac for wireless sensor networks. IEEE/ACM Transactions on Networking 16(3), 511–524 (2008)
21. Stajano, F., Anderson, R.: The resurrecting duckling: Security issues for ad-hoc wireless networks. In: Proc. of the 7th International Workshop on Security Protocols, pp. 172–182 (1999)
22. Wander, A., Gura, N., Eberle, H., Gupta, V., Shantz, S.C.: Energy analysis of public-key cryptography for wireless sensor networks. In: PERCOM 2005, pp. 324–328 (2005)
23. Wang, X., Xing, G., Zhang, Y., Lu, C., Pless, R., Gill, C.: Integrated coverage and connectivity configuration in wireless sensor networks. In: Proc. of the 1st ACM International Conference on Embedded Networked Sensor Systems (SenSys 2003), pp. 28–39 (2003)
24. Wang, Y., Li, F., Dahlberg, T.A.: Energy-efficient topology control for three-dimensional sensor networks. International Journal of Sensor Networks 4(1/2), 68–78 (2008)
25. Xu, Y., Heidemann, J., Estrin, D.: Adaptive Energy-Conserving Routing for Multihop Ad Hoc Networks. Research Report 527, USC/Information Sciences Institute (2000)
26. Xu, Y., Heidemann, J., Estrin, D.: Geography-informed energy conservation for ad hoc routing. In: Proc. of the 7th ACM International Conference on Mobile Computing and Networking (MobiCom 2001), pp. 70–84 (2001)
27. Zhu, S., Setia, S., Jajodia, S.: Leap: Efficient security mechanisms for large-scale distributed sensor networks. In: Proc. of the 10th ACM International Conference on Computer and Communications Security (CCS 2003), pp. 62–72 (2003)

Hierarchical Self-healing Key Distribution for Heterogeneous Wireless Sensor Networks

Yanjiang Yang[1], Jianying Zhou[1], Robert H. Deng[2], and Feng Bao[1]

[1] Institute for Infocomm Research, Singapore
{yyang,jyzhou,baofeng}@i2r.a-star.edu.sg
[2] School of Information Systems, Singapore Management University
robertdeng@smu.edu.sg

Abstract. Self-healing group key distribution aims to achieve robust key distribution over lossy channels in wireless sensor networks (WSNs). However, all existing self-healing group key distribution schemes in the literature consider homogenous WSNs which are known to be unscalable. Heterogeneous WSNs have better scalability and performance than homogenous ones. We are thus motivated to study *hierarchial* self-healing group key distribution, tailored to the heterogeneous WSN architecture. In particular, we revisit and adapt Dutta *et al.*'s model to the setting of hierarchical self-healing group key distribution, and propose a concrete scheme that achieves computational security and high efficiency.

Keywords: Wireless sensor network, self-healing group key distribution, wireless sensor network security.

1 Introduction

A wireless sensor network (WSN) consists of a large number of sensor nodes collecting and reporting the environmental data to a base station. A sensor node is a small sensing device capable of wireless communications through radio signals. Due to the low cost requirement, sensor nodes are extremely constrained in hardware, having limited computation capability, storage capacity, and radio transmission range. Worse yet, sensor nodes are usually powered by batteries, hence restricted power supply is yet another major limitation of WSNs.

WSNs are easily susceptible to adversaries who can intercept or interrupt the wireless communications. It is thus crucial to ensure secure communication when WSNs are deployed for mission-critical applications. A fundamental service to achieve secure communication is key distribution, whereby sensor nodes establish (secret) keys. Unfortunately, it is commonly acknowledged that key distribution in WSNs is not trivial, considering the resource-constrained nature of sensor nodes. Hence lots of efforts have been dedicated to the study of key management and distribution in WSNs [3,4,6,5,7,8,9,10,11,12,13,15]. These methods are categorized into group key distribution [3,6,8,10,11] and pairwise key distribution [4,5,7,9,12,15].

Y. Chen et al. (Eds.): SecureComm 2009, LNICST 19, pp. 285–295, 2009.

Among the existing group key distribution schemes, self-healing group key distribution [6,11,13] particularly suits WSNs. A prominent property of this type of group key distribution is *self-healing*, which allows group members to recover lost group keys of previous sessions based solely on the key update message of the current session. This makes group key distribution resilient to lossy wireless communication of WSNs.

All the self-healing group key distribution schemes in the literature considered homogeneous WSNs where all sensor nodes are assumed to be the same. However, homogeneous WSNs are not scalable. We are thus motivated to study self-healing group key distribution in *heterogenous* WSNs. A heterogenous WSN is composed of not only resource constrained sensor nodes, but also a number of more powerful high-end devices. Specifically, a WSN is partitioned into a number of *groups*, and a high-end device is placed into each group, acting as the *group manager*. A group manager is more powerful, and thus does not suffer from the resource scarceness problem as much as a sensor node does.

Our Contributions. Tailored to the heterogeneous WSN architecture, we propose the concept of *hierarchical* self-healing group key distribution. In particular, we formulate a security model for hierarchical self-healing group key distribution by revisiting and adapting Dutta *et al.*'s model [6]. We then propose a concrete scheme, proven secure under the model. Our scheme is "authenticated", compared to Dutta *et al.*'s schemes, in the sense that every non-revoked sensor node can ascertain the validity of the group keys it generated from the key update messages, without involving any extra communication overhead. As communication is more energy consuming than computation in WSNs, this property is important to prevent sensor nodes communicating using invalid group keys.

2   Related Work

Public key cryptosystems are in general too expensive for WSNs, so symmetric key primitives such as secret key encryption or cryptographic hash function are often preferred. As such, key management and distribution in WSNs boils down to sharing of secret keys among sensor nodes. To achieve this objective, a commonly used approach is to pre-load a set of secrets inside sensor nodes before their deployment. These pre-loaded secrets are then used either directly as pair-wise keys between a pair of neighboring sensor nodes, i.e., pair-wise key distribution [4,5,7,9,10,12,15], or as a basis to establish new common keys shared by a group of sensor nodes, i.e., group key distribution [3,6,8,10,11].

Among the existing group key distribution schemes, self-healing group key distribution is particularly suitable for WSNs, because of its self-healing and membership revocation properties. Staddon *et al.* [13] first proposed the concept and a concrete construction of self-healing group key distribution based on secret sharing of two dimensional polynomials. Their construction, however, is not efficient, suffering from high communication and storage overhead. Liu *et al.* [11] then generalized the security notions in [13], and presented a new scheme with better efficiency by combining personal secret distribution with the self-healing technique of [13]. Blundo *et al.* [1] analyzed the security definitions in [11,13] and concluded that it is impossible for any scheme to achieve all of the security requirements formulated in [11,13]. They then formulated a new definition for self-healing group key distribution and came up with a new scheme [2].

All the above self-healing group key distribution schemes are intended to achieve information theoretic security. In [6], Dutta *et al.* proposed a novel computationally secure scheme, based on a combination of a reverse one-way hash chain and a forward one-way hash chain. Their idea in achieving self-healing is that along the reverse hash chain, the hash value of $h^j(.)$ (associated with an earlier session) can be computed from any pre-image $h^i(.)$ (associated with a later session), where $i < j$ and $h^i(.) = \underbrace{h(h(\cdots h(.)))}_{i \ times}$. While Dutta *et al.*'s model is weaker, their schemes tremendously improve the efficiency of the information theoretically secure schemes. Our proposed scheme is based on Dutta *et al.*'s idea of a combination of reverse and forward one-way hash chains, but we rectify the vulnerability of their construction (it can be shown that their schemes cannot achieve t-Revocation). The main differences between our scheme and Dutta *et al.*'s schemes are twofold. First, our scheme is hierarchical, tailored to the heterogeneous WSNs. Second, our scheme achieves authenticated group key distribution, allowing every non-revoked sensor node to verify whether or not its generated group keys are valid, without requiring any extra communications.

3   Heterogeneous WSN Architecture

We partition a WSN into a number of *groups*. A high-end device is placed into each group, acting as the *group manager*. In contrast to sensor nodes, the high-end group managers have relatively higher computation capability, larger storage size, and longer radio range. They also have longer power supply, and can even be line-powered in some circumstances, e.g., when a WSN is deployed to monitor a building, the group managers can easily tap on the electricity lines to get power supply. Therefore unlike sensor nodes, group managers do not suffer too much from the resource scarceness problem. The introduction of high-end group managers into a WSN makes the once homogeneous network *heterogeneous*, as depicted in Figure 1.

In this architecture, downlink messages broadcast by the base station directly reach sensor nodes, whereas uplink messages sent by a sensor node to the base stattion is forwarded via its group manager, which acts as an intermediary between the base station and the sensor nodes within its jurisdiction. A sensor node may reach the group manager directly, or by traversing a *short* multi-hop path. Since group managers are not severely constrained by resources, communication at the level of group managers (including the base station) does not suffer from the limits upon sensor nodes.

Intuitively, the inclusion of powerful group managers provides shortcuts for data delivered from the sensor nodes to the base station, so the overall system performance and in turn the lifetime of the network are expected to be

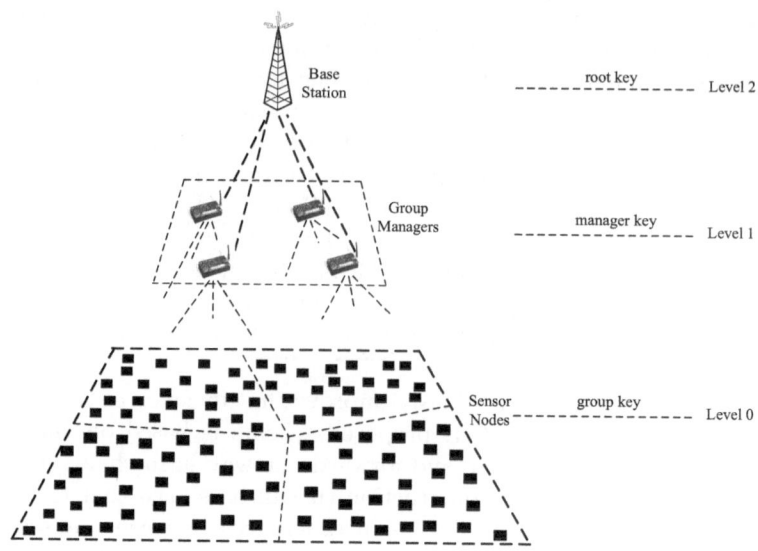

Fig. 1. Heterogeneous Wireless Sensor Network

greatly improved. Indeed, the effect of adding powerful nodes to WSNs has been analyzed in [14]: only a modest number of reliable, long-range backhaul links and line-powered nodes are required to have a significant effect, and if properly deployed, heterogeneity can triple the average delivery rate and yield a 5-fold increase in the lifetime of a large battery-powered sensor networks.

4   Model and Definition

System Model. Three types of entities are involved in our hierarchical group key distribution system: the base station, group managers, and a large number of sensor nodes. The sensor nodes are partitioned into a number of n_G groups, and each group has a group manager. A group has a unique group ID, and we use G_ν to denote the ID of group $\nu \in \{1, \cdots, n_G\}$. Each sensor node in a group is uniquely identified by an ID number i, where $i \in I \subseteq \{1, \cdots, n\}$, and I is the set of all node ID numbers of that group and n is the largest possible ID# in the system.

In correspondence to the heterogenous architecture, the keys held by the entities form a hierarchy, as shown in Figure 1: the base station holds a *root key* at level 2, each group manager has a distinct *manager key* at level 1, and sensor nodes in every group hold a common *group key* during each session at level 0. Traffic generated at lower level can be decrypted or authenticated using the keys at higher levels, but not the other way around. This key hierarchy helps to implement "separation of duty" within the system, e.g., it is not necessary for the sensor nodes to process the control messages broadcast by the base station to the group managers.

A group manager takes charge of distribution of group keys within its group. A group key is associated with each session. To distribute a group key for a new session, the group manager broadcasts a *key update message* to all its sensor nodes. The group key is then computed by a sensor node based on the received key update message and its preloaded *personal secret*. Denote the personal secret of sensor node i as S_i, which is a vector of m elements where m is the maximum number of sessions supported by the WSN. Each element in S_i corresponds to a session and we use $S_i[j]$ to denote the element corresponding to the jth session, $j \in \{1, \cdots, m\}$. $S_i[j]$ becomes *obsolete* once the group key for the jth session is established; otherwise $S_i[j]$ is *fresh*. A sensor node can be revoked or non-revoked. Only non-revoked sensor nodes are able to compute the group keys. To be resilient to the lossy channel of WSNs, the generation of group keys is self-healing in the sense that a non-revoked sensor node can recover group keys of all previous sessions as long as it successfully receives the key update message of the current session.

Adversary Model. We assume that the base station and the group managers are trusted, as we mainly concern with the distribution of group keys among sensor nodes. An adversary is able to passively eavesdrop on, or actively intercept, modify, insert, or drop key update messages from a group manager to all its sensor nodes. We also allow the adversary to compromise up to t sensor nodes in a group, where t is a system parameter.

Definition. We formally define the concept and security requirements of hierarchical self-healing group key distribution, by revisiting and extending the definition in [6].

Definition 1. (*Hierarchical Self-healing Group Key Distribution with t-Revocation*) *Let n, m, t be system parameters. $\mathcal{D}$ is hierarchical self-healing group key distribution with t-revocation, if the following holds:*

a. (*Key Hierarchy*) *The manager keys held by the group managers are derived from the root key of the base station, but it is computationally infeasible to compute the root key from the manager keys. The same relationship should hold between group keys and the corresponding manager key.*

b. (*Secrecy of Personal Secret*) *For any $U \subset \{1, \cdots, n\}, |U| \leq t$, it is computationally infeasible for the nodes in U to collectively determine the fresh elements of S_i for any $i \notin U$.*

c. (*Authenticated Generation of Group Key*) *Let gK_j be the group key for session j, and B_j be the broadcast key update message from the group manager, where $j \in \{1, \cdots, m\}$. For any non-revoked sensor node in the group, gK_j is efficiently computed from B_j and $S_i[j]$ in an authenticated manner. On the contrary, it is computationally infeasible to compute the group session key from the key update message or a personal secret alone.*

d. (*t-Revocation*) *For any session j, let R_j be the set of revoked nodes at the start of session j, where $|R| \leq t$, it is computationally infeasible to compute gK_j from the broadcast message B_j and $\{S_i\}_{i \in R_j}$.*

e. (*t-wise Forward Secrecy*) Let $U \subseteq \{1, \cdots, n\}$ denote the sensor nodes which joined the group after session j. Given that $|U| \leq t$, it is computationally infeasible for all members in U to collectively compute $gK_1, \cdots, gK_j$, even with the knowledge of $gK_{j+1}, \cdots, gK_m$.

f. (*Self-healing*) A non-revoked sensor node between sessions j_1 and j_2, $1 \leq j_1 < j_2 \leq m$, can efficiently compute any gK_j, $j_1 \leq j \leq j_2$, from B_{j_2} and its personal secret.

5 Our Construction

5.1 Scheme Details

We suppose that the set of revoked users is monotonic, i.e., a revoked user never rejoins the network. Let F_q be a finite field, where q is a large prime number. All arithmetic operations are performed in F_q. Let $h, h_R, h_F : \{0,1\}^* \to F_q$ be cryptographic hash functions, and n_G, n, m, t be system parameters.

– **System Initialization.** The base station chooses a root key $rK = [rk_1, rk_2]$, where rk_1 and rk_2 are random numbers of appropriate length. For each group $G_\nu, \nu = 1, 2, \cdots, n_G$, the base station computes a manager key as $mK_{G_\nu} = [mk_1, mk_2]$, where $mk_1 = h(G_\nu, rk_1)$ and $mk_2 = h(G_\nu, rk_2)$. Clearly, it is computationally infeasible to compute one manager key from another without knowing the root key. Then the base station securely passes the manager keys to the corresponding group managers. We do not specify how this can be done, but it often suffices by using some out-of-band channel. Upon receipt of the manager keys, the group managers begin the preparation for setting up group keys. Without loss of generality, let's consider a particular group G_ν whose manager key is mK_{G_ν}. The group manager sets mk_1 to be the seed s_R for a one-way hash chain of length $m + 1$, i.e.,

$$k_R^j = h_R(k_R^{j-1})$$
$$= h_R(h_R(k_R^{j-2})) = \cdots = h_R^j(s_R), 1 \leq j \leq m+1 \qquad (1)$$

and sets mk_2 to be the seed s_F for another hash chain of length m, i.e.,

$$k_F^j = h_F(k_F^{j-1})$$
$$= h_F(h_F(k_F^{j-2})) = \cdots = h_F^j(s_F), 1 \leq j \leq m \qquad (2)$$

The group key gK_j for session $j \in \{1, \cdots, m\}$ is defined to be:

$$gK_j = k_R^{m-j+1} + k_F^j$$
$$= h_R^{m-j+1}(s_R) + h_F^j(s_F) \qquad (3)$$

We can see that the hash chain associating with $h_R()$ is used in the *reverse* order thus called the reverse hash chain, and that associated with

$h_F()$ called the forward hash chain. The group manager then selects m random t-degree polynomials $f_1(x), \cdots, f_m(x) \in F_q[x]$, each corresponding to a session. The personal secret for the member sensor node i is defined to be $S_i = [f_1(i), \cdots, f_m(i)]$. The group manager sends S_i together with k_R^{m+1} and s_F to each node i in a secure manner. Note that k_R^{m+1} will be used as the initial *authenticator* (denoted as $Auth$) in the process of group key generation.

– **Broadcast.** At the start of each session, the group manager broadcasts a key update message to enable sensor nodes to generate a new group key. Let $R_j = \{i_1, ..., i_w\}$ be the set of revoked sensor nodes upon the start of session $j \in \{1, \cdots, m\}$ and $|R_j| = w \leq t$. The group manager chooses a random set $R_j' = \{i_t', \cdots, i_{w+1}'\} \subset \{1, \cdots, n\} \setminus I$, where I is the set of all node IDs of that group. That is, the group manager chooses $t - w$ random IDs that are not in that group. Next, the group manager computes k_R^{m-j+1} from s_R by Equation (1), and then computes the following polynomials:

$$r_j(x) = (x - i_1) \cdots (x - i_w)(x - i_{w+1}') \cdots (x - i_t')$$
$$b_j(x) = k_R^{m-j+1}.r_j(x) + f_j(x)$$

We call $r_j(x)$ the revocation polynomial and $f_j(x)$ the masking polynomial. Finally, the group manager broadcasts the key update message B_j to the sensor nodes in its group, where

$$B_j = R_j \cup R_j' \cup \{b_j(x)\}$$

– **Session Key Generation.** Upon receipt of B_j, if node i is not revoked, it is able to compute $k_R^{m-j+1} = \frac{b_j(i) - f_j(i)}{r_j(i)}$. Then it can validate k_R^{m-j+1} using the authenticator $Auth$. For example, if $Auth = k_R^{m+1}$, then the validation is to test $Auth \stackrel{?}{=} h_R^j(k_R^{m-j+1})$. If the validation fails, the node aborts the key generation. Otherwise, it continues to compute $k_F^j = h_F^j(s_F)$ using s_F (Equation (2)), and in turn the group key $gK_j = k_R^{m-j+1} + k_F^j$. The node also updates $Auth$ by setting $Auth = k_R^{m-j+1}$. For efficiency reason, the node can also choose to keep k_F^j instead of s_F for future sessions.

– **Addition of New Group Member.** A newly added member in session j is not allowed to compute group keys of previous sessions. To add a new member with ID $\alpha \in \{1, \cdots, n\}$ starting from session j, the group manager computes and gives $S_\alpha = \{f_j(\alpha), f_{j+1}(\alpha), \cdots, f_m(\alpha)\}$ and $k_F^j = h_F^j(s_F)$ to the node.

5.2 Efficiency

Our scheme is highly efficient in terms of storage, communication, and computation overhead. For storage, the personal secret together with the authenticator accounts for $(m + 1) \log q$ bits storage in each sensor node (compared to Dutta *et al.*'s scheme, ours only needs $\log q$-bit more storage for the authenticator).

For communications, our scheme generates $t(\log q + \log n) \approx t \log q$ bits key update message (since $n \ll q$), which is almost the same as the bit length of the key update message in Dutta et $al.$'s scheme. For computation, no costly public key primitive is involved in our scheme, and the computation overhead inflicted upon sensor nodes includes only cryptographic hash function and polynomial operations.

5.3   Security Analysis

Theorem 1. *The above construction is a hierarchical self-healing group key distribution scheme with respect to Definition 1.*

Proof. It is not difficult to check that our scheme meets the properties of key hierarchy, authenticated computation of group key, and self-healing. We thus, in what follows, focus on showing that our scheme satisfies other security requirements.

$\diamond$ Secrecy of Personal Secret. Personal secrets are computed from t-degree polynomials. For any t-degree polynomial corresponding to a session, a set U of sensor nodes, where $|U| = \tau \leq t$, contributes τ points over the polynomial. It is thus impossible (in an information theoretic sense) for τ nodes to determine the polynomial **solely** from the personal secrets they have, and in turn any other value of the polynomial. It remains to check whether the broadcast key update messages reveal information on personal secrets. Let us consider a particular non-revoked node i in session j. From the broadcast message B_j, node i calculates $k_R^{m-j+1} = \frac{b_j(i) - f_j(i)}{r_j(i)}$. Then with k_R^{m-j+1}, node i can actually compute any $f(i'), i' \neq i$, as $f(i') = b_j(i') - k_R^{m-j+1}.r_j(i')$. This suggests that once a group session key is established, the element of a sensor node's personal secret corresponding to that session is revealed to all other non-revoked nodes. This is exactly the reason why we distinguish between obsolete and fresh elements within a personal secret. We stress that the fresh elements of a personal secret remain secret, since they are computed from different polynomials.

$\diamond$ t-Revocation. Without loss of generality, let's consider the last session m, and assume the set of t revoked nodes $R_m = \{1, 2, \cdots, t\}$, the maximum number of allowed revoked nodes, at the start of the session. Our goal is to show that these t revoked nodes cannot compute gK_m from the broadcast key update message B_m and their personal secrets. We model the coalition of the t revoked nodes as a polynomial-time algorithm $\mathcal{A}$, which takes View of the protocol as input and outputs a guessed group key gK'_m for session m. We say $\mathcal{A}$ breaks t-revocation if gK'_m is authenticated with respect to the group keys of previous sessions (or the authenticator). We prove, by contradiction, that if $\mathcal{A}$ breaks t-revocation, then we can construct a polynomial-time algorithm $\mathcal{B}$ for inverting one-way hash function $h_R(.)$, using $\mathcal{A}$. In particular, given $y = h_R(x)$, $\mathcal{B}$ computes x by invoking $\mathcal{A}$ as follows.

$\mathcal{B}$ sets $k_R^2 = y$ while leaves k_R^1 undefined (k_R^1 should be x by definition), and then computes $k_R^3 = h_R(y), k_R^4 = h_R(k_R^3) = h_R^2(y), \cdots, k_R^{m+1} = h_R(k_R^m)$.

$\mathcal{B}$ continues to select a random s_F and compute the forward hash chain $k_F^j = h_F(k_F^{j-1}) = h_F(h_F(k_F^{j-2})) = \cdots = h_F^j(s_F), 1 \le j \le m$. $\mathcal{B}$ sets the j-th group session key as $gK_j = k_R^{m-j+1} + k_F^j, 1 \le j \le m-1$, and leaves gK_m undefined (gK_m should be $x + h_F^m(s_F)$ by definition). $\mathcal{B}$ selects m random polynomials $f_1(x), \cdots, f_m(x) \in F_q[x]$, each of degree t. For each node $i \in \{1, 2, \cdots, t\}$, $\mathcal{B}$ computes the personal secret $S_i = \{f_1(i), \cdots, f_m(i)\}$. For each session $1 \le j \le m-1$, $\mathcal{B}$ computes the broadcast key update message $B_j = R_j \cup R_j' \cup \{b_j(x)\}$, where $R_j \subseteq R_m$, and $b_j(x) = k_R^{m-j+1}.r_j(x) + f_j(x)$, with R_j' and $r_j(x)$ being constructed in exactly the same manner as in our scheme. To compute the broadcast key update message for session m, $\mathcal{B}$ selects a random $k_R^1 \in F_q$, and computes $B_m = R_m \cup \{b_m(x)\}$, where $b_m(x) = k_R^1.r_m(x) + f_m(x)$, with $r_m(x) = (x-1)(x-2) \cdots (x-t)$. Then $\mathcal{B}$ sets View of the protocol as

$$
\text{View} = \left\{ \begin{array}{c} k_R^{m+1} \\ s_F \\ \{f_j(1), f_j(2), \cdots, f_j(t)\}, j = 1, \cdots, m \\ B_j, j = 1, \cdots, m \\ gK_1, \cdots, gK_{m-1} \end{array} \right\}.
$$

Finally, $\mathcal{B}$ gives View to $\mathcal{A}$, which in turn outputs gK_m', its guess for the actual group session key gK_m. $\mathcal{B}$ outputs $gK_m' - h_F^m(s_F)$. It is easy to see that for any session $j, 1 \le j \le m-1$, the simulation by $\mathcal{B}$ in constructing View is perfect with respect to the original scheme. We next show that the simulation for session m (where a random k_R^1 is used) is also perfect to $\mathcal{A}$. To see this, $\mathcal{A}$ has $b_m(x)$ (from B_m) and $\{f_m(1), \cdots, f_m(t)\}$ at its disposal. First, from $f_m(1), \cdots, f_m(t)$, $\mathcal{A}$ cannot determine $f_m(.)$ as it only has t points over the t-degree polynomial which has $t+1$ unknown coefficients, thus $\mathcal{A}$ cannot compute any $f_m(i), i \notin \{1, \cdots, t\}$. Here, we need to stress that $b_m(x)$ does not help $\mathcal{A}$ to determine $f_m(x)$. The reason is that $b_m(x)$ at the points of $\{1, \cdots, t\}$ equals $f_m(1), \cdots, f_m(t)$, respectively, thereby revealing no more information on $f_m(x)$. Second, on $b_m(x)$, there are two cases to be considered.

1. $i \in \{1, \cdots t\}$: from $b_m(x)$, $\mathcal{A}$ can evaluate $b_m(i)$ which equates $f_m(i)$ regardless of k_R^1. $\mathcal{A}$ thus can check against $f_m(i)$ which is already at its disposal. The simulation is perfect to $\mathcal{A}$.
2. $i \notin \{1, \cdots, t\}$: from $\mathcal{A}$'s point of view, $b_m(i) = k_R^1.r_m(i) + f_m(i)$ is random, because in each $b_m(i)$ there are two unknown variables k_R^1 and $f_m(i)$; for every value of k_R^1, there is a corresponding value $f_m(i)$. Hence the adversary $\mathcal{A}$ who cannot break the polynomial $f_m(.)$ has no way to distinguish whether or not the genuine k_R^1 is used in $b_m(i)$. The simulation is thus again perfect to $\mathcal{A}$.

Combining together the above arguments, we conclude that $\mathcal{B}$ inverts h_R with the same advantage as $\mathcal{A}$ breaks t-revocation.

$\diamond$ t-wise Forward Secrecy. An intuition on t-wise forward secrecy is that if a set of nodes U join the system in session j, they are given $k_F^j = h_F^j(s_F)$. To compute the group key for an earlier session $j' < j$, they need $k_F^{j'}$. In our scheme, the only information relates to $k_F^{j'}$ are $k_F^j, k_F^{j+1}, \cdots, k_F^m$. Computing $k_F^{j'}$ from these group keys of later sessions clearly involves inverting the one-way cryptographic hash function $h_F(.)$. As it is straightforward to construct an adversary $\mathcal{B}$ for inverting $h_F(.)$, based on an adversary $\mathcal{A}$ that breaks t-wise forward secrecy (similar to the above proof), we omit the details of the proof. ☐

6   Conclusion

We studied hierarchical self-healing group key distribution for heterogenous WSNs. In particular, we formulated a model for hierarchical self-healing group key distribution, and proposed concrete schemes that achieve provably computational security and high efficiency.

Acknowledgement

This work is supported by A*STAR project SEDS-0721330047.

References

1. Blundo, C., D'Arco, P., Santis, A., Listo, M.: Definitions and Bounds for Self-healing Key Distribution. In: Díaz, J., Karhumäki, J., Lepistö, A., Sannella, D. (eds.) ICALP 2004. LNCS, vol. 3142, pp. 234–245. Springer, Heidelberg (2004)
2. Blundo, C., D'Arco, P., Santis, A., Listo, M.: Design of Self-healing Key Distribution Schemes. Designs, Codes and Cryptography 32(1-3), 15–44 (2004)
3. Blundo, C., Santis, A., Herzberg, A., Kutten, S., Vaccaro, U., Yung, M.: Perfectly-secure key distribution for dynamic conferences. In: Brickell, E.F. (ed.) CRYPTO 1992. LNCS, vol. 740, pp. 471–486. Springer, Heidelberg (1993)
4. Chan, H., Perrig, A., Song, D.: Random Key Pre-distribution Schemes for Sensor Networks. In: IEEE Symposium on Security and Privacy, pp. 197–213 (2003)
5. Du, W.L., Deng, J., Han, Y.S., Varshney, P.K.: A Pairwise Key Pre-distribution Scheme for Wireless Sensor Networks. In: ACM Conference on Computer and Communication Security, CCS 2003, pp. 42–51 (2003)
6. Dutta, R., Change, E.C., Mukhopadhyay, S.: Efficient Self-healing Key Distribution with Revocation for Wireless Sensor Networks Using One Way Key Chains. In: Katz, J., Yung, M. (eds.) ACNS 2007. LNCS, vol. 4521, pp. 385–400. Springer, Heidelberg (2007)
7. Eschenauer, L., Gligor, V.D.: A Key-Management Scheme for Distributed Sensor Networks. In: ACM Conference on Computer and Communication Security, CCS 2002 (2002)
8. Huang, D., Mehta, M., Medhi, D., Harn, L.: Location-aware key management scheme for wireless sensor networks. In: 2nd ACM workshop on Security of Ad Hoc and Sensor Networks
9. Liu, D., Ning, P.: Improving Key Pre-distribution wih Deployment Knowledge in Static Sensor Networks. ACM Transactions on Sensor Networks (2005)
10. Liu, D., Ning, P., Du, W.L.: Group-based Key Pre-distribution in Wireless Sensor Networks. In: ACM Workshop on Wireless Security (2005)

11. Liu, D., Ning, P., Sun, K.: Efficient Self-Healing Group Key Distribution with revocation Capability. In: ACM Conference on Computer and Communication Security, CCS 2003 (2003)
12. Perrig, A., Szewczyk, R., Wen, V., Culler, D., Tygar, J.D.: SPINS: Security Protocols for Sensor Networks. Wireless Networks Journal (WINE) (September 2002)
13. Staddon, J., Miner, S., Franklin, M., Balfanz, D., Malkin, M., Dean, D.: Self-healing Key Distribution with Revocation. In: IEEE Symposium on Security and Privacy, S&P 2002, pp. 241–257 (2002)
14. Yarvis, M., et al.: Exploiting Heterogeneity in Sensor Networks. In: IEEE INFOCOM 2005 (2005)
15. Zhu, S., Setia, S., Jajodia, S.: LEAP: Efficient Security Mechanisms for Large-scale Distributed Sensor Networks. In: ACM Conferenc on Computer and Communication Security, CCS 2003, pp. 62–72 (2003)

User–Centric Identity Using *e*Passports

Martijn Oostdijk, Dirk-Jan van Dijk, and Maarten Wegdam

Novay, P.O. Box 589, 7500AN Enschede, The Netherlands
{martijn.oostdijk,dirk-jan.vandijk,maarten.wegdam}@novay.nl

Abstract. The worldwide introduction of *e*Passports presents a unique opportunity for the online identity community to implement trustworthy identity providers. The *e*Passport provides citizens with a strong authentication token within a global Public Key Infrastructure backed by government administrations. This paper studies the possibilities for leveraging the *e*Passport for user-centric identity and reports on an experiment in which *e*Passports are combined with the user-centric identity management framework Information Card. Note that no changes to already deployed *e*Passports are needed for our solution to work.

1 Introduction

Most online services (*e*Commerce, *e*Government) are only meaningful if (aspects of) the identity of users can be established in a trustworthy manner by whoever offers the service. At the same time users will only use a service when they feel that they are in control over who they share identity information with. This paper investigates, through a practical experiment, how strong authentication means (i.e. *e*Passports) can be combined with user-centric identity management.

Over the last couple of years electronically readable travel documents (*e*Passports) have been introduced in most countries of the world. An *e*Passport contains an embedded chip with card holder data which allows an automated inspection system (typically operated by border control officials) to read out data from the chip and, more interestingly, to verify the integrity of the data and the authenticity of the chip. The embedded chip communicates in a contactless manner based on standardized communication protocols, ensuring that the chip can be contacted when it is in the proximity of an inspection system.

While there are concerns about the privacy consequences of the introduction of *e*Passports [5,8,10,16,17], primarily caused by the combination of contactless communication with privacy sensitive biometric data, it also presents a unique opportunity for creating trustworthy online identities as it potentially provides citizens with a strong authentication token within a global Public Key Infrastructure (PKI) backed by government administrations [9]. Moreover, the technical standards which describe how the inspection system verifies the authenticity of *e*Passports are open and publicly available from the International Civil Aviation Organization (ICAO[1]) [6]. Although originally not

---

[1] See http://icao.int/

Y. Chen et al. (Eds.): SecureComm 2009, LNICST 19, pp. 296–310, 2009.
Institute for Computer Science, Social-Informatics and Telecommunications Engineering 2009

intended as such by ICAO, *ePassports*, as they are being deployed presently, seem ideal for authenticating users of third-party online services such as web stores.

At the same time an entirely different revolution is taking place in the online identity community which places the end-user at the center by relaying all communication between identity providers and service providers (also called relying parties) through the user's client. Web 2.0 services are driving this revolution, which is therefore sometimes dubbed Identity 2.0 or user-centric identity, and it is being enabled by identity management systems such as OpenID [13] and Information Card [12].

The objective of the research described in this paper is to study the possibilities for leveraging the *ePassport* for user-centric identity. This would establish online identities backed by government issued hardware tokens in a user-centric manner. This paper describes a prototype user-centric online identity solution based on Information Cards which uses *ePassports* for authentication. The implementation makes it possible for users to show aspects of their *ePassport* (aspects such as information stored in the chip's memory but also proof of authenticity of the chip) to relying parties (online service providers) with the help of an identity provider. The identity provider only needs to store minimal information about the user's passport and can be seen as a privacy filter from the user's perspective.

The solution is built using open standards and is published as open source software.

The remainder of this paper is organized as follows: Section 2 presents a brief introduction to the ICAO standards and describes the various features of the *ePassport*. Section 3 introduces user-centric identity management and in particular the Information Card framework. Section 4 explains how these standards are combined into a working prototype information card identity provider that uses *ePassports* for authentication and lists some of the results. Section 5 discusses the lessons learned. Section 6 presents concluding remarks and pointers for future work.

2   The ICAO *ePassport*

This section provides a short introduction to the ICAO standard for *ePassports*, for more details see [5,9,16].

The ICAO standard for *ePassports* is described in Doc 9309 [6], which itself is based on many other standards and specifications. The specification consists of two parts:The first part describes the format of the contents of the *ePassport*, the so-called Logical Data Structure. The second part describes a variety of mandatory and optional security controls which are implemented by the *ePassport* to protect the information in the Logical Data Structure against various forms of attack.

2.1   Logical Data Structure

The contents of an *ePassport* are structured in terms of so-called data groups. Together with an index file (COM) and a signature file (SOd) these form the Logical Data Structure. Table 1 lists the data groups found in a typical Dutch *ePassport*.

Table 1. Contents of the Logical Data Structure

COM	An index of which DGs are present
DG1	The contents of the MRZ (name, date of birth, ...)
DG2	JPEG or JPEG2000 image of face
DG11	Optional passport holder's full name if too long for MRZ
DG15	Public key for Active Authentication
SOd	Security document with signature over Logical Data Structure hashes

Two files are always present: COM contains an index which indicates which of the 16 possible data groups are present. SOd is the security document which contains the issuing country's signature over the contents of the data groups. It contains hashes for each of the data groups present in the Logical Data Structure and a signature over these hashes. This allows Passive Authentication as described in Section 2.2.

The first data group, DG1, contains the textual information about the passport holder that is also (optically) printed in the Machine Readable Zone (MRZ) on the data page of the ePassport. The information in DG1 contains the passport holder's name, date of birth, gender, as well as the document's number and date of expiry. In the Dutch case DG1 also contains the passport holder's citizen number (the Dutch equivalent of a social security number).

The public key in DG15 is used for a security mechanism called Active Authentication which is described in Section 2.2.

For the purposes of this paper the Logical Data Structure elements of interest are: DG1 which contains textual information about the passport holder and the document itself, DG15 which contains a public key, and the SOd which contains a signature over the different data groups.

2.2 Security Controls

To protect against attacks such as skimming, altering, unauthorized access and cloning the ePassport contains a number of security controls.

Basic Access Control: When attempting to read the Logical Data Structure, the ePassport requires the inspection system to first show knowledge of an access key comprised of three items in the MRZ: the passport document number, the date of birth of the passport holder, and the date of expiry of the passport. By requiring the inspection system to prove knowledge of these items, the passport is convinced of the fact that the inspection system has seen the data page of the physical passport booklet, which means that whoever is operating the inspection system has access to the booklet and has the passport holder's consent to read it. BAC prevents skimming in which an attacker gets access to an ePassport without the holder's knowledge or consent.

Extended Access Control: Some data groups contain information of a highly sensitive privacy nature, such as biometric templates. To protect against unauthorized parties reading such files an additional access control mechanism may be implemented on top of BAC. Whether an inspection system can complete the EAC protocol when presented with an *e*Passport depends on whether it has acquired a document verifier certificate (DVC) from the *e*Passport's issuing state.

Passive Authentication (PA): The security document (SOd) attached to the Logical Data Structure contains hashes of all data groups and a signature over these hashes. The signature is set using a Document Signing Private Key and can be checked using the Document Signing Public Key Certificate (DSC), which in European Union passports is included inside the SOd. The DSC, in turn, is signed using the Country Signing Private Key and can be checked using the Country Signing Public Key Certificate (CSC). This latter certificate, at least in the Dutch case, can be downloaded from a government website. PA prevents altering the data in the Logical Data Structure (either by changing or replacing the chip or by intervening with the communications between chip and inspection system).

Active Authentication (AA): The inspection system can challenge the chip to prove authenticity by signing on request a random nonce using a document specific private key. The corresponding public key can be read from DG15 (which is part of the Logical Data Structure, and therefore part of the signed data in the SOd) so that the inspection system can check the resulting signature. AA prevents cloning, as the private key cannot be extracted from the *e*Passport by an attacker. The verification algorithm for AA is specified as an ISO standard [7].

For European Union passports BAC is mandatory. EAC is not widely used presently, but will be as soon as fingerprints are included in *e*Passports across Europe (expected mid 2009). PA is mandatory for all ICAO *e*Passports, however not all European Union member states allow third party access to their CSC. AA is optional and only a few countries implement it.

2.3 Software for Accessing *e*Passports

The ICAO standards have been implemented by various countries and manufacturers of identity products since 2005/2006. Open source initiatives to read *e*Passports soon followed, mostly with the purpose to test the various official implementations. Shortly before the introduction of the Dutch *e*Passport in 2006, software was developed at Radboud University to test the Dutch implementation of the *e*Passport [5]. Some of the results were later disseminated as an open source project, JMRTD (http://jmrtd.org), which we used in our prototype[2]. The software consists of a framework for reading and verifying passports using off-the-shelf hardware as well as a reference implementation in Java Card of the *e*Passport itself.

The JMRTD API offers data structures for the information stored in the Logical Data Structure, making it possible to interpret the data. The API also implements the various security controls used by the passport such as BAC, PA, and AA.

---

[2] Other open source implementations are the OpenMRTD project (http://openmrtd.org) and the RFIDIOt project (http://rfidiot.org)

3 User-Centric Identity

This section introduces user-centric identity management and focuses in particular on the Information Card specification [12].

Online identity management is a game for three: the user, the relying party, and the identity provider. The user wants to use a service provided by the relying party. At the same time the relying party wants to have some assurance about the client's identity. The identity provider helps the user and the relying party in providing this assurance. Sections 3.1 and 3.2 explain in more detail how this game is played.

Whether the user (with the help of the identity provider) succeeds in convincing the relying party that the claimed identity is correct depends on the level of trust that the relying party has in the identity provider. At the same time, the user also needs to trust the identity provider to only use information rendered for the purpose of acquiring the service from relying party. This privacy problem is discussed in Section 5.2.

User-centric identity management approaches place the user (contrary to e.g. the identity provider) in the center of the solution, which includes among others that the flow of information goes via the user. We use here the Laws of Identity created by Kim Cameron [4][3] to further define user-centric identity management. For brevity we only list them:

1. User control and consent
2. Minimal disclosure for a constrained user
3. Justifiable parties
4. Directed identity
5. Pluralism of operators and technologies
6. Human integration
7. Consistent experience across contexts

There are two prominent specifications for user-centric identity management: OpenId and Information Card. OpenID is a lightweight approach to user-centric identity that has its origin in 2005 for preventing spam through blog post comments. It is specified by the OpenID foundation[4]. Several open source initiatives exist[5]. For this paper however we used Information Card since it enforces privacy sensitivity through user-centricity, and is in the process of becoming a more formal standard.

The original Information Card specification is by Microsoft, and is called the Identity Selector Interoperability Profile. It was drafted with the above laws in mind, and the Information Card adheres to them. This specification was used as input to the Organization for the Advancement of Structured Information Standards (OASIS), which is now in the process of standardizing Information Card. The main factors of the Information Card specification that contribute to adherence to the Laws of Identity are the use of a card metaphor, and the routing of all identity claims through the user's client, as we explain below.

---

[3] An interesting side-note about Cameron's paper: One of the laws (the law of directed identity) is illustrated with a self-service passport reader example.
[4] See http://openid.net
[5] See http://wiki.openid.net/Libraries

The user interacts with the system through a so-called *Card Selector* (or Identity Selector). The selector presents the user with a number of visual cards containing claims (called fields) about the user's identity. The selector is a metaphor for a wallet containing all sorts of plastic cards.

All traffic between identity provider and relying party is routed through the user's client, literally putting the user in the center. This keeps the user informed and moreover gives the user the option of aborting the transaction at different stages of the authentication process.

Microsoft's CardSpace is a closed source card selector embedded into the Windows operating system. Microsoft's .NET framework offers building blocks for constructing identity providers and relying parties. Open source alternatives for card selectors, identity providers, and relying parties exist[6] as well. Some of the Information Card implementation details in this paper are CardSpace specific, and may be handled differently by other implementations.

3.1 Enrolling at the Identity Provider

Enrollment is the process of registering a user's identity with the identity provider. In Information Card enrollment results in a so-called managed card which (in the user's experience) is retrieved and made accessible in the Card Selector. Information Card allows two different types of cards: Self-issued cards contain claims made by the user about the user. Managed cards, on the other hand, contain claims by an online identity provider about the user. Retrieving a managed card is done by selecting some authentication mechanism which is used for proving possession of the managed card in the future. The managed card is said to be backed by that authentication mechanism. Typically a self-issued card is used as backing for a managed card but other options, like traditional username and password or an X.509 certificate corresponding to some private key (possibly on a hardware token), are also possible.

3.2 Using a Managed Card to Authenticate at the Relying Party

After the user has retrieved a managed card from the identity provider he or she can start visiting relying parties. Upon such a visit the relying party sends a policy back to the user's client which, amongst others, contains field names (such as "Last name" or "Date of Birth") deemed necessary by the relying party before the service can be acquired and specifies which identity provider the relying party trusts. The user is now typically presented with the Card Selector which shows only those cards which comply with the relying party's policy. The user selects one of those cards and in case it is the managed card the identity provider is called upon to fill in the values of those fields specified in the relying party's policy. The result is a so-called *security token* (generated by Security Token Service, a component of the identity provider) which is first sent back to the client.

Obviously, the token contains sensitive information and its confidentiality needs to be protected. To do this, it may be encrypted with the relying party's public key.

---

[6] See, for instance, the Higgins project at http://www.eclipse.org/higgins/ and the DigitalMe selector at http://www.bandit-project.org/index.php/Digital_Me

Furthermore the token is signed by the identity provider providing proof of authenticity of the originator (the identity provider) and integrity of the token itself which can
be verified by the relying party.

Since the user has to concur with the identity provider, if the security token is encrypted for the relying party or in a token format unknown to the card selector, a so-
called *display token* is also sent to the client. The display token contains the same
claims as the security token, except that the contents of the display token are encrypted with the user's public key rather than the relying party's public key, so that
the user can inspect the values filled in for each field. If the user concurs with the
identity provider that the supplied claims are correct the user's client forwards the
security token to the relying party.

4   Combining ePassports and User-Centric Identity

This section describes how the scenarios in Section 3 change when Information Card
is combined with *e*Passports.

In the altered scenarios for enrollment and authentication we have the traditional
three parties involved in user-centric identity management as described in Section 3,
namely the relying party, identity provider and the user's client. These parties are
depicted in Figure 1.

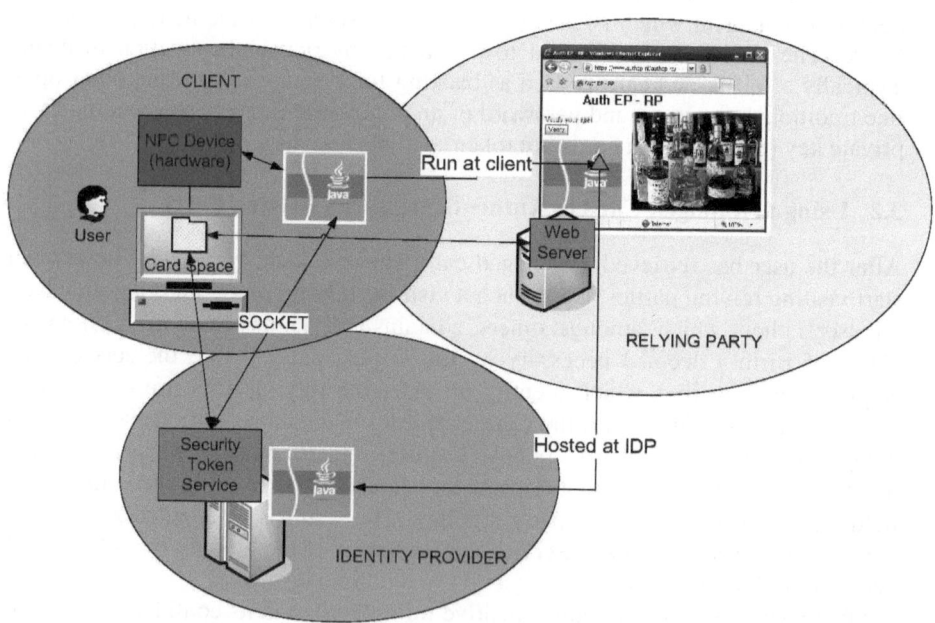

Fig. 1. The three parties in the *e*Passport Information Card scenario

4.1   Enrolling the ePassport at the Identity Provider

Before the user can use his *e*Passport at a relying party, he needs to enroll it with the identity provider. The user visits the identity provider's website and requests a managed card. The managed card will be tied to the user's *e*Passport. The user also supplies the BAC keys to the identity provider at this point. The identity provider needs the BAC keys in order to communicate with the *e*Passport chip during the authentication scenario as described in 4.2. Remember that the BAC keys are based on the user's date of birth, the *e*Passport's date of expiry, and the *e*Passport's document number.

At enrollment time the user sends an empty self-issued card to the identity provider which is used to back the managed card. The user also enters the date of birth, date of expiry, and the document number at the identity provider's website. The identity provider stores this information and sends a managed card (whose picture resembles a passport) to the user's card selector. The managed card contains no information apart from an id which the identity provider can use to later resolve the user's BAC keys (which it needs to communicate with the *e*Passport) and authenticate the user.

At enrollment time the user also needs to install a so-called Java policy file, allowing signed mobile code coming from the identity provider's web server to access the contactless card reader hardware. The role of the policy file is explained in Section 4.2. As an alternative to using this Java mobile code approach the user could be asked to install some local software, which might be perceived as being more transparent from the average user's perspective.

4.2   Using the ePassport to Authenticate at a Relying Party

Figure 2 shows the different entities involved in the authenticate-with-*e*Passport scenario and the traffic that is exchanged between them.

A TCP connection from the identity provider to the user's contactless card reader is created as soon as the user loads the relying party's login page. In the current implementation this is accomplished by placing a Java applet owned by the identity provider on the relying party's web page (to be more precise, what is placed on the relying party's web page is an HTML applet tag linking to applet code on the identity provider's web server). The applet is signed by the identity provider and also loaded from the identity provider's web server so that the Java Runtime Environment (JRE) at the user's client trusts this piece of mobile code enough to allow it to set up a connection back to the identity provider's server. The JRE was given permission to connect to the contactless card reader in a Java policy file which was installed during enrollment. The TCP connection is used for subsequent communication between the identity provider and the user's *e*Passport.

Using the managed card acquired during enrollment the user can attempt to login at the relying party. An information card policy is sent to the identity provider via the Card Selector just like in the normal Information Card scenario. One extra step is taken by the identity provider after receiving a token request from the client. In this extra step the identity provider checks if the user has a valid passport and it reads the user's details from the passport. As soon as the client actually requests a token at the identity provider, the identity provider will look at the provided managed card and

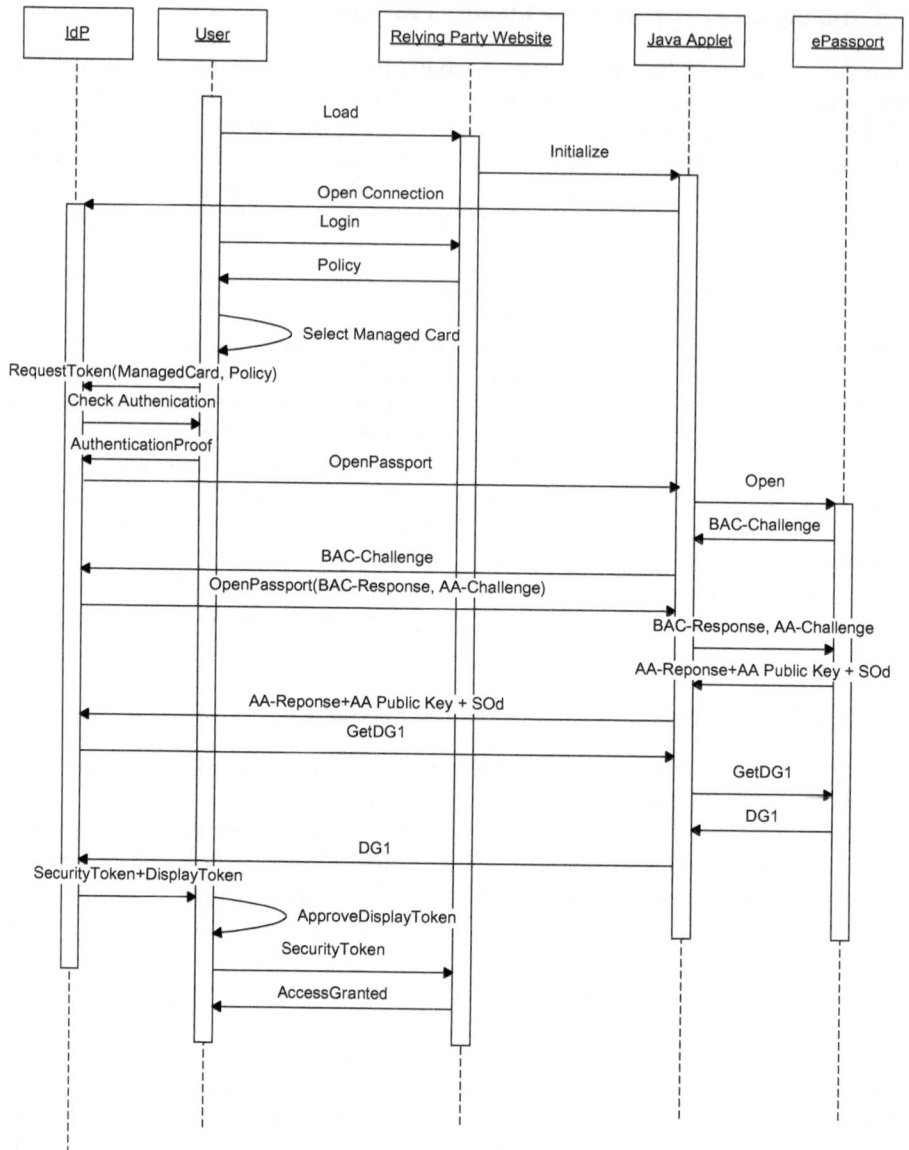

Fig. 2. Message sequence chart of the authenticate-with-*e*Passport scenario

send the appropriate BAC data to the passport authenticating the identity provider at the passport. The identity provider will request the *e*Passport's AA public key and SOd. With the SOd it can check if the public key has been signed by the issuing country. It can then send a random challenge to the *e*Passport which encrypts it using the AA private key. This proves that the passport is authentic and not a simple clone. The identity provider will request the minimal needed information from the *e*Passport to confirm to the token request. The token is sent back to the client and from here on the normal Information Card scenario continues.

To summarize, the identity provider uses BAC, AA, and PA and then reads DG1. Based on the results of the security protocols the identity provider knows that the information in DG1 correctly identifies a citizen of the issuing country (for as far as the identity provider trusts the country's CSC, of course). Remember that DG1 contains basic textual card holder information (name, date of birth, date of expiry of document, document number, gender, nationality, and in the Dutch case even the citizen number). The information in this data group is used in the token created by the identity provider and only the required fields (as requested by the relying party's policy) are sent to the relying party (via the user's client). No other information is sent to the relying party and the relying party needs to trust the identity provider that it has done its job in checking the validity of the user's ePassport.

5 Discussion of Lessons Learned

The previous sections of this paper report on an experiment in which ePassports are combined with the user-centric identity management framework Information Card. It validates that it is in principle possible to create an identity solution combining real smart cards and 'virtual' information cards. This section discusses the lessons learned.

5.1 The Need for an Online Identity Provider

A surprising aspect of the described solution is that the role of the identity provider is somewhat different from traditional Information Card identity providers. A typical identity provider has knowledge about identities of users. The identity provider in our experiments, on the other hand, stores no information about the user except for the BAC access keys. The token sent to the relying party is freshly constructed based on information read 'live' from the ePassport, not on information stored at the identity provider. In fact, one could argue that the government issuing the ePassport should be considered as the real provider of identity here and that no separate identity provider should be needed.

Disadvantages of having a separate identity provider lie in the trust that the other parties need to have on the identity provider's integrity. These are discussed in Section 5.2. Still, this identity provider is necessary to deal with limitations in Information Card and ICAO specifications:

1. The Information Card specification simply requires an external online identity provider. One could, of course, always choose to implement it as a client-side or relying party-side service.

2. The ICAO specifications were never explicitly designed to allow limited disclosure of data in the sense of Cameron's second law. The information in the ePassport is structured in data groups and the complete data group needs to be sent to an inspection system before it can determine whether its hash in the SOd checks out. The online identity provider is needed to act as a privacy filter.

3. The identity provider stores the BAC keys and uses these to open a secure connection to the *e*Passport. Other parties - on the Internet but also client-side malware - cannot interfere with the communications between identity provider and *e*Passport[7]. This means that the client is not a part of the Trusted Computing Base. The BAC keys should not be distributed to untrusted parties for security and privacy reasons, providing them to each relying parties is therefore not an option.

A smartcard solution that would be designed specifically to facilitate privacy-sensitive assertions on identity claims, would need to be able to sign not only specific parts of the identity data, but also derived data from these identity data. For example, sign an 'above 18' claim, derived from the birth date.

5.2 The Need to Trust the Identity Provider

Since user-centric identity management introduces an online identity provider, a question that will need to be answered is to what degree both the user and the relying party must *trust* the identity provider:

- The user needs to trust the identity provider, which has full access to his *e*Passport, with respect to the privacy sensitive data stored in the chip and with respect to the authentication functionality provided by the chip. Obviously, our implementation of the identity provider is well-behaved with respect to privacy: it will only read relevant data groups and only use this to construct a token which is sent (via the user's card selector, i.e. with the user's consent) to the relying party. Still, the user needs to trust that our identity provider is implemented as advertised.
- The relying party needs to trust the identity provider with respect to the correctness of information issued about the user. The relying party needs to trust that the identity provider has done its job in inspecting that the *e*Passport is authentic (using PA) and is present during the transaction (using AA). The identity provider is not able to convey evidence (such as a signature) that proves to the relying party that the information in the token originates from an *e*Passport without sending the complete data group DG1.

A particularly frightening threat to the user's privacy would be an evil identity provider that keeps track of all relying parties visited by a user. Such a big-brother identity provider could even use the *e*Passport's active authentication functionality to construct undeniable proof-objects of the user's involvement in transactions with relying parties (i.e. have the user's *e*Passport 'sign' the transaction so as to later confront the user with such proof). Note, however, that there is no need for the identity provider to know the relying party's identity, it only needs a public key from the relying party to encrypt the security token.

Another threat is formed by the simple challenge-response nature of the AA protocol. A rogue identity provider can relay challenges and responses to another identity provider and pretend to have direct access to an *e*Passport.

---

[7] Unfortunately, client-side malware can communicate with the *e*Passport if it manages to get hold of the user's date of birth and the document's number and date of expiry through some other (totally unrelated) means.

How an identity provider can build up enough trust so that both users and relying parties trust it is a general trust problem. The following three suggestions are merely recommendations:

- The server should be run by an independent party. Although the government can be trusted in terms of correctness of information (i.e. the relying party would be happy with the government as online identity provider), the general sentiment amongst end-users may be that governments should not be present during each and every online transaction of their citizens. Governments may not be too eager to become online identity providers for their citizens either.
- The implementation of the identity provider should be open and transparent. Open source helps. However, it remains impossible to validate online whether a server is actually running the source code that is published.
- Trust in the identity provider could perhaps be established using *reputation*. Reputation based Trust (occasionally referred to as web-of-trust) is a mechanism that is sometimes part of Web 2.0 applications. *CACert.org* is an example of a real-world web-of-trust: it establishes a distributed network of trust by having people meet *in real life* and present their (non-electronic, old fashioned) passport in order to gain trust points. Persons with enough trust points can get their SSL site certificate signed by the CACert certificate authority. Web users can add the CACert certificate authority's root certificate in their browser's list of trusted certificates if they feel that the described procedure warrants that the CACert certificate authority only signs site certificates when the site's identity has been adequately checked.

The privacy problems with the identity provider are beyond the scope of the feasibility study described in this paper. A solution lies either in building up enough trust in the identity provider or in changing the *e*Passport standards so that it can be used to generate trustworthy proof of authenticity and at the same time conform to Cameron's second law. The latter option makes an online identity provider redundant. Recent interest in integrating different hardware electronic identity cards[8] (*e*ID), such as described in [3], indicates that we may be heading towards a world where online identity providers are no longer necessary or at least play a different role.

5.3 Not a Global PKI for Online User Authentication

Unfortunately our solution does not provide a global PKI for online user authentication. Apart from practical problems (not every citizen has an *e*Passport, contactless card readers are not widely available yet), the ICAO standards leave countries plenty of options in *not* implementing the various features which are essential to the proposed identity solution. The features that have to be present to allow online verification of an *e*Passport are:

- The *e*Passport must support passive authentication so that the authenticity and integrity of the LDS can be verified.

---

[8] See also the Stork European project at http://www.eid-stork.eu/

- The *e*Passport must support active authentication so that the authenticity of the chip can be verified.
- The data groups with identifying information must not be protected by extended access control (unless relevant keys are know to the identity provider).
- The document signing public key certificate must be available to the Identity Provider and its validity should be checked.

Passive authentication is present in all passports. Active authentication, on the other hand, is optional and recent interoperability tests[9] seem to indicate that less than thirty percent of countries currently chose to implement active authentication.

The data groups of interest for this paper, DG1 and DG15, are not protected by extended access control: they are readable to an inspection system once the basic access control protocol is successfully completed. Future versions of the *e*Passport may contain interesting identifying information in other data groups that are protected by extended access control. In that case the identity provider needs to be trusted by the ICAO Public Key Directory (PKD) to get access keys for performing the extended access control protocol.

Some countries have their country signing certificate available on a public government website. An overview list appeared in an ICAO published report [2]. As an alternative to publishing this certificate on a website the PKD can be used. The PKD shares document signing certificates rather than country signing certificates. An additional advantage of the PKD is that it provides a central online service for obtaining validity information about certificates in the form of so-called certificate revocations lists.

Our solution was tested using the Dutch *e*Passport which supports all features necessary for our purposes. The ICAO standards for *e*Passports are definitely not designed with online verification and limited disclosure of contents in mind.

6 Concluding Remarks

The main conclusion from this research is that with the help of an online identity provider it is possible to leverage *e*Passports for user-centric identity management. Our prototype validates this for Information Card, and we expect a similar outcome for e.g. OpenID, or for a more identity provider-centric approach to identity management such as SAML.

The online identity provider needs to be trusted by both user and relying party, but due to specifics of the Information Card specification and the *e*Passport specification, cannot be done without.

From a privacy perspective, and to enhance availability, it may be preferred to have an identity solution which does not depend on an online identity provider at all. After all, the *e*Passport can be considered as a smartcard that embeds several important claims about a person (such as name, birthday, gender), which are already signed by a trusted government agency. To be useful in a privacy sensitive manner, the *e*Passport or another similar hardware token with embedded claims would have to be able to sign individual claims, including derived claims such as 'above 18".

---

[9] See http://www.e-passports2008.org/

Apart from further exploring the above, there are some technical loose ends that require our attention for future work:

- OpenID is the other popular upcoming user-centric identity management standard. It would certainly be interesting to investigate how our solution integrates with OpenID.
- Currently the managed information card associated with an *e*Passport is backed by a self-issued information card of the user. A much more elegant alternative would be to use the smart card framework supported by the card selector (the Crypto API for Microsoft Windows or PKCS 11 for certain other alternatives) instead. In essence this creates a "soft-token" interface for the *e*Passport's active authentication signature which can be used to back a managed card. Such an approach of integrating smart cards with Information Card would also be more in line with [1] and commercial smart card based identity solutions such as TrustBearer's OpenID product[10].
- The introduction of Near Field Communication (NFC) technology in all sorts of mobile devices may solve the practical problem of (lack of) contactless card reader availability. While porting *e*Passport software to J2ME may be an interesting challenge, the eCL0WN tool[11] for Nokia NFC phones proves that it is feasible to read out *e*Passports using NFC. Similarly, IBM is working on using contactless cards to strengthen online authentication [12]. Combined with user-centric identity management systems for mobile phones (IBM has demonstrated a card selector for the Android platform [11]) this leads the way to *mobile-centric identity management*, which combines user-centric with mobile phones as the most ubiquitous and personal device people have. The *e*Passport could, for example, be helpful in the provisioning process in such a mobile-centric identity management system.

Acknowledgments. This research was funded by the NLnet (http://nlnet.nl/) foundation.

References

1. Aussel, J.-D.: Smart Cards and Digital Identity. Teletronikk 3/4, 66–78 (2007) ISSN 0085-7130
2. Broekhaar, S., Verschuren, J.: How to Obtain CSCA Certificates – The CSCA Overview List, MRTD report, 2, ICAO, 32–35 (2007)
3. Bruegger, B.P., Hühnlein, D., Kreutzer, M.: Towards global eID-Interoperability. In: BIOSIG 2007. LNI, vol. 108, pp. 127–140 (2007)
4. Cameron, K.: The Laws of Identity – as of 5/12/2005, Microsoft Corporation (2005)
5. Hoepman, J.-H., Hubbers, E., Jacobs, B., Oostdijk, M., Schreur, R.W.: Crossing Borders: Security and Privacy Issues of the European e-Passport. In: Yoshiura, H., Sakurai, K., Rannenberg, K., Murayama, Y., Kawamura, S.-i. (eds.) IWSEC 2006. LNCS, vol. 4266, pp. 152–167. Springer, Heidelberg (2006)

---

[10] See http://www.trustbearer.com/
[11] See http://seclists.org/fulldisclosure/2008/Dec/0567.html, Jeroen van Beek.

6. ICAO: Machine Readable Travel Documents, ICAO Doc 9303, part 1: Specifications for Electronically Enabled Passports with Biometric Identification Capability, 6th edn., vol. 2 (2006)
7. ISO: Information technology — Security techniques — Digital signature schemes giving message recovery — Part 2: Integer factorization based mechanisms, ISO/IEC 9796-2, 2nd edn. (2002)
8. Juels, A., Molnar, D., Wagner, D.: Security and Privacy Issues in E-passports. In: Proc. SecureComm 2005, pp. 74–88. IEEE Computer Society, Los Alamitos (2005)
9. Lekkas, D., Gritzalis, D.: e-Passports as a means towards the first world-wide Public Key Infrastructure. In: López, J., Samarati, P., Ferrer, J.L. (eds.) EuroPKI 2007. LNCS, vol. 4582, pp. 34–48. Springer, Heidelberg (2007)
10. Liu, Y., Kasper, T., Lemke-Rust, K., Paar, C.: E-Passport - Cracking Basic Access Control Keys. In: Meersman, R., Tari, Z. (eds.) OTM 2007, Part II. LNCS, vol. 4804, pp. 1531–1547. Springer, Heidelberg (2007)
11. Nadalin, A.J.: Mobile Identity. In: The European e-Identity Conference, The Hague (2008), http://www.eema.org/downloads/annual08/nadalin2c.pdf
12. Nanda, A.: Identity Selector Interoperability Profile, V1.0, Microsoft Corporation (2007)
13. OpenID: OpenID Authentication 2.0 – Final (2007),
http://openid.net/specs/openid-authentication-2_0.html
14. Ortiz-Yepes, D.A.: Enhancing Authentication in eBanking with NFC-Enabled Mobile Phones. ERCIM News 76, 63–64 (2009)
15. SAML, OASIS specification (2005),
http://saml.xml.org/saml-specifications
16. Vaudenay, S., Monnerat, J., Vuagnoux, M.: About Machine-Readable Travel Documents. In: Proc. International Conference on RFID Security 2007, pp. 15–28 (2007)
17. Vaudenay, S.: E-Passport Threats. IEEE Security & Privacy, 72–75 (November/December 2007)

Defending against Key Abuse Attacks in KP-ABE Enabled Broadcast Systems

Shucheng Yu[1], Kui Ren[2], Wenjing Lou[1], and Jin Li[2]

[1] Department of ECE, Worcester Polytechnic Institute, MA 01609
yscheng@wpi.edu, wjlou@ece.wpi.edu
[2] Department of ECE, Illinois Institute of Technology, IL 60616
{kren,jin.li}@ece.iit.edu

Abstract. Key-Policy Attribute-Based Encryption (KP-ABE) is a promising cryptographic primitive which enables fine-grained access control over sensitive data. However, key abuse attacks in KP-ABE may impede its wide application especially in copyright-sensitive systems. To defend against this kind of attacks, this paper proposes a novel KP-ABE scheme which is able to disclose any illegal key distributor's ID when key abuse is detected. In our scheme, each bit of user ID is defined as an attribute and the user secret key is associated with his unique ID. The tracing algorithm fulfills its task by tricking the pirate device into decrypting the ciphertext associated with the corresponding bits of his ID. Our proposed scheme has the salient property of *black box* tracing, i.e., it traces back to the illegal key distributor's ID only by observing the pirate device's outputs on certain inputs. In addition, it does not require the pirate device's secret keys to be *well-formed* as compared to some previous work. Our proposed scheme is provably secure under the Decisional Bilinear Diffie-Hellman (DBDH) assumption and the Decisional Linear (DL) assumption.

1 Introduction

There is a trend that more data are stored or delivered across third parties over Internet for either reliable storage or ease of sharing. For example, individuals would store their personal information on portal web sites such as Google, and commercial content providers may deliver their product through content delivery networks (CDNs) such as Akamai. Such a trend raises the concern that sensitive data stored or cached by these third-party sites will be compromised. Moreover, in some critical or copyright-sensitive application scenarios, it requires differentiated service in the way that, data are defined with sets of attributes and each user is limited to access data of some particular set of attributes or their combinations. In this kind of applications, each user's access privilege is assigned by the user's role or the price that this user paid. One example of this kind of applications is *targeted broadcast* system, e.g., a digital video recorder (DVR) system. In such a system, the content provider might broadcast episodes of TV shows and each of the shows may be assigned a set of attributes such as *name*,

Y. Chen et al. (Eds.): SecureComm 2009, LNICST 19, pp. 311–329, 2009.
© Institute for Computer Science, Social-Informatics and Telecommunications Engineering 2009

season number, genre, so on and so forth. Users will obtain the access privilege to contents of some particular combination of these attributes by paying the corresponding price to the content provider. The user's access privilege can be encoded as a policy such as (*"name=heros"* AND (*"season 2"* OR *"season 3"*)). As content providers might provide their services across third party CDNs, for the purpose of access control it is desirable to encrypt the media products using certain cryptographic primitive since traditional centralized access control methods such as the reference monitor approach might not be suitable in this scenario.

Key-policy attribute-based encryption (KP-ABE)[1] is such a cryptographic primitive that was proposed to resolve the exact problem of fine-grained data access control in one-to-many communications. In KP-ABE, a ciphertext is associated with a set of attributes, and each user secret key is embedded with an access structure which is the logic combination of certain set of attributes. Users can decrypt a ciphertext if and only if the set of attributes associated with the ciphertext satisfy the access structures embedded in their secret keys. Beside this property, KP-ABE also has nice properties of collusion resistance and provable security under standard difficulty assumptions. All these properties seem to make KP-ABE a perfect tool to enforce access control in the above copyright-sensitive applications.

However, the following issue may impede its direct application in targeted broadcast systems of our interests: In the current KP-ABE construction [1], a user secret key is defined over an access structure and does not have the one-to-one correspondence with any particular user. This results in the fact that a paid user is able to "share" his secret key and abuse his access privilege without being identified. More seriously, pirates may take this advantage to make profits by abusing the access privilege. We call this kind of misbehavior by *key abuse attacks*. As a matter of fact, key abuse attacks are extremely harmful for copyright-sensitive application scenarios. Imagine that in a DVR system protected by KP-ABE, key abusers can easily distribute content decryption keys to others by ways such as sending via email. Due to the cost of this is extremely low, it is more destructive than directly distributing the content itself. Therefore, before KP-ABE can be safely applied to aforementioned applications, key abuse attacks should be well addressed. The ideal way for defending against key abuse attacks is to technically prevent illegal users from using others' decryption keys. However, it is difficult to realize since it may require on-line servers to monitor the usage of user decryption keys, or the user secret key to be physically associated with the user. In conventional broadcast encryption, the issue of key abuse is addressed by using a technique called *traitor tracing* which has been well studied by previous works [2,3,4,5]. The key idea of traitor tracing is to enable the content provider to trace any suspicious pirate device and thus discover illegal key distributor's identitie(s) and collect evidences of key abuse. Then the content provider can sue the illegal key distributors by presenting these evidences to law authorities. Specifically, the content provider would choose particular types of ciphertexts and trick pirate devices into decrypting them. Success of decryption will provide the evidence of pirating. At a high level view, we can play the same trick in KP-ABE to defend against key abuse attacks. However, underlying techniques adopted by existing traitor tracing systems can not be directly applied to KP-ABE because receivers are represented individually in conventional broadcast encryption while not in KP-ABE. Therefore, it is desirable to propose a novel solution for defending against key abuse attacks in KP-ABE.

1.1 Our Contribution

In this paper, we resolve the issue and provide an abuse free KP-ABE (AFKP-ABE) scheme based on the Decisional Bilinear Diffie-Hellman (DBDH) assumption and the Decision Linear (D-Linear) assumption. AFKP-ABE has properties of partially collusion resistance and black box tracing according to the definition in [5]. In addition, AFKP-ABE is efficient since both the secret key size and the ciphertext size are $\mathcal{O}(logN)$, where N is the total number of users. The main technical challenge of our construction of AFKP-ABE is to realize black box tracing, i.e., tracing the pirate device only by observing its outputs on some inputs. To achieve this goal, one frequently used method is to trick the pirate device into decrypting tracing ciphertexts and success of decryption will provide the evidence of pirating as mentioned before. In the context of KP-ABE, however, this implies that an unsuspected user may not be able to correctly decrypt a tracing ciphertext even if the attributes embedded in the ciphertext satisfy his access structure, and thus has the chance to detect the ongoing tracing activity. A pirate can take advantage of this and collude with other pirates to detect tracing activities.

The main idea of our construction is as follows. Each user is assigned a unique ID which is chosen from the identity space. Then, we define bits of user identities as attributes and embed them in user secret key. We call these attributes by *identity-related attributes* and other attributes by *normal attributes*. Normal (non-tracing) encryption algorithm associates the identity-related attributes to the ciphertext in the way that all the bits of the identity space are set to "don't care". The tracing algorithm just associates the suspicious identity corresponding identity-related attributes to the ciphertext. This turns out that only the user with the suspicious identity is able to correctly decrypt the tracing ciphertext. Note that in this construction the only difference between a normal encryption algorithm and the tracing algorithm is on the input set of identity-related attributes. To make the tracing algorithm indistinguishable from the regular encryption algorithm, we hide these identity-related attributes when encrypting so that pirate devices are not able to tell which and how many identity-related attributes are used. In addition we also hide some of the normal attributes. The intuition behind this is to prevent the pirate device from being able to check if normal attributes of the ciphertext satisfy his access structure and thus detect the tracing activity. We achieve the goal of hiding attributes using the technique from anonymous ciphertext-policy attribute-based encryption [6] in which the ciphertext policy is hidden to receivers. Our definition of the KP-ABE tracing system is based on the definition of the traitor tracing system by Boneh et al. [5].

1.2   Related Work

Attribute-Based Encryption. Sahai and Waters[7] first introduced attribute-based enctyption (ABE) for encrypted access control. In an ABE system, both the user' private key and the ciphertext are associated with a set of attributes. If only at least k attributes overlap between the ciphertext and his private key, can the user decrypt the ciphertext. Based on ABE, Goyal et al. [1] proposed a key-policy attribute-based encryption (KP-ABE) scheme and introduced the concept of ciphertext-policy attribute-based encryption (CP-ABE). The first CP-ABE construction was proposed by Bethencourt et al. [8]. Cheung et al. [9] proposed the first provably secure CP-ABE. In CP-ABE, the user secret key is associated with a set of attributes and ciphertexts are embedded with an access structure. A user is able to decrypt the ciphertext only if the attributes associated with his secret keys satisfy the access structure of the ciphertext. KP-ABE is defined in the reverse way than CP-ABE. User secret keys in KP-ABE are embedded with an access structure and ciphertexts are associated with a set of attributes. Successful decryption of the ciphertext requires a match between the user's access structure and the ciphertext attribute set.

Anonymous CP-ABE. In conventional CP-ABE schemes [8,9], ciphertext policies should be revealed in the ciphertext so that receivers are able to combine correct secret keys for decryption. To better protect user privacy, some application scenarios may require ciphertext policies to be hidden to receivers. We call this branch of CP-ABE schemes by *anonymous CP-ABE*. The first anonymous CP-ABE scheme was proposed by Kapadia et al. [10]. However, this scheme is not collusion-resistant and it needs an online semi-trusted server to participate in data encryption. Yu et al. proposed two collusion-resistant anonymous CP-ABE schemes [11,12] based on [9]. But the security of these schemes is based on strong assumptions. Nishide et al. [6] proposed the first provably secure anonymous CP-ABE based on the DBDH assumption and the D-Linear assumption. In their proposed scheme, each attribute could have several values. A public key component is defined over each value of an attribute. User secret key is associated with exactly one value of each attribute. The ciphertext has a *well-formed* ciphertext component for each intended attribute value and *mal-formed* ciphertext components for unintended attribute values. It sets an attribute as "don't care" by presenting well-formed ciphertext components for all the values of this attribute. If there is one ciphertext component corresponding to the user attributes is mal-formed, this user will not be able to decrypt the ciphertext. Because the scheme is designed in such a way that it is hard to distinguish well-formed ciphertext components from mal-formed ones, receivers are not able to tell which or how many attributes appear in the ciphertext policy. Our construction is partially based on this scheme. We refer to [6] for more details on this scheme. Anonymous CP-ABE can also be realized by using a recently invented cryptographic primitive called predicate encryption by Katz et al.[13]. However, their construction requires the bilinear group to be of the order of product of three large primes. Moreover, their security proof is based on new complexity assumptions. Recently, Li et. al proposed two accountable attribute-based schemes

[14,15] which solve the similar issue of key forgery in the setting of CP-ABE. We claim that our work is proposed in parallel with these schemes and in different models.

Traitor Tracing. Traitor tracing systems were proposed for use in broadcast environments to help content providers trace back to the original source of pirates. In a traitor tracing system, each user (with a decoder) is assigned a personal decryption key. The content provider encrypts the content such that only authorized users are able to decrypt. Suppose a group of colluding users P contribute their personal keys to build a pirate decoder. The tracing scheme should be able to trace back to each member of P. The first traitor tracing system is proposed by Chor et al[2]. Since that, many traitor tracing schemes [3,4,5] have been proposed. These scheme can be categorized by the following properties[5]: public key/private key broadcast encryption, public/private traceability, collusion resistance, black box tracing, stateful/stateless decoder. For example, [5] is a traitor tracing system for public key broadcast and enables private black box tracing against arbitrary colluding stateless decoders. Other important properties of a traitor tracing system include secret key size and ciphertext size.

The rest of this paper is organized as follows. Section 2 reviews some technique preliminaries pertaining to our construction. Section 3 presents formal definitions and models of our proposed abuse free key-policy attribute-based encryption scheme. In section 4 we give our construction of such a scheme as well as our security proof to it. In section 5, we discuss potential application scenarios in which our scheme would be applicable. We conclude this paper in Section 6.

2   Preliminaries

2.1   Bilinear Maps

Our design is based on some facts about groups with efficiently computable bilinear maps.

Let $\mathbb{G}_0$ and $\mathbb{G}_1$ be two multiplicative cyclic groups of prime order p. Let g be a generator of $\mathbb{G}_0$. A bilinear map is is an injective function $e : \mathbb{G}_0 \times \mathbb{G}_0 \to \mathbb{G}_1$ with the following properties:

1. *Bilinearity*: for all $u, v \in \mathbb{G}_0$ and $a, b \in \mathbb{Z}_p$, we have $e(u^a, v^b) = e(u, v)^{ab}$.
2. *Non-degeneracy*: $e(g, g) \neq 1$.
3. *Computability*: There is an efficient algorithm to compute $e(u, v)$ for all $u, v \in \mathbb{G}_0$.

2.2   Complexity Assumptions

Decisional Bilinear Diffie-Hellman (DBDH) Assumption. Let $a, b, c, z \in \mathbb{Z}_p$ be chosen at random and g be a generator of $\mathbb{G}_0$. The DBDH assumption [16] states that no probabilistic polynomial-time algorithm $\mathcal{B}$ can distinguish the tuples $(A = g^a, B = g^b, C = g^c, e(g, g)^{abc})$ from the tuple $(A = g^a, B = g^b, C = g^c, e(g, g)^z)$ with non-negligible advantage.

The Decision Linear (D-Linear) Assumption. Let $z_1, z_2, z_3, z_4, z \in \mathbb{Z}_p$ be chosen at random and g be a generator of $\mathbb{G}_0$. The D-Linear assumption [17] states that that no probabilistic polynomial-time algorithm $\mathcal{B}$ can distinguish the tuple $(g, g^{z_1}, g^{z_2}, g^{z_1 z_3}, g^{z_2 z_4}, g^{z_3+z_4})$ from the tuple $(g, g^{z_1}, g^{z_2}, g^{z_1 z_3}, g^{z_2 z_4}, g^z)$ with non-negligible advantage.

3   Definitions and Models

In this section, we present the definition of our abuse-free KP-ABE (AFKP-ABE) scheme as well as its security definition. The security definition of our scheme is consistent to traitor tracing schemes [5].

3.1   Description of AFKP-ABE

The AFKP-ABE scheme has the following five algorithms:

Setup$(1^\lambda$, n). The setup algorithm is a randomized algorithm. It takes as input the security parameter 1^λ and n, the length of a user identity. It outputs a master key MK and public parameters PK.

Enc(M, γ, PK). The encryption algorithm is a randomized algorithm. It takes as input a message M, a set of attributes γ, and the public parameters PK. It outputs a ciphertext E. On different input γ, this algorithm can be used either for normal (non-tracing) operations of content distribution, or for the purpose of tracing.

KeyGen(T, MK, PK). The key generation algorithm is a randomized algorithm. It takes as input an access structure T, the master secret key MK, and the public parameters PK. It outputs a user secret key SK.

Dec(E, SK, PK). The decryption algorithm is a deterministic algorithm. It takes as input the ciphertext E for a set of attributes γ, a user secret key SK for an access structure T, and the public parameters PK. If $\gamma \models T$, i.e., γ satisfies T, it outputs the message M. Otherwise it outputs $\perp$ with overwhelming probability.

Trace$^{\mathcal{D}}(\varepsilon)$. This algorithm takes input a parameter ε (which should be polynomially related to λ), and has black-box access to an ε-useful decoder box $\mathcal{D}$ which is constructed by the adversary. It outputs a set of guilty colluders in polynomial time.

3.2   Security Definition

The security of ABKP-ABE is defined by the following two security games.

Game 1. The first game captures the idea of *Semantic Security*. In our scheme we follow the definition of the standard game used by KP-ABE [1] which proceeds with the following steps.

- *Init.* The adversary declares the set of attributes, γ, that he wishes to be challenged upon.
- *Setup.* The challenger runs the Setup algorithm of AFKP-ABE and gives the public parameters to the adversary.
- *Phase 1.* The adversary is allowed to issue queries for private keys for many access structures T_i, where $\gamma \nvDash T_i$ for all i.
- *Challenge.* The adversary submits two equal length messages M_0 and M_1. The challenger flips a random coin b, and encrypts M_b with γ. The ciphertext is passed to the adversary.
- *Phase 2.* Phase 1 is repeated.
- *Guess.* The adversary outputs a guess b_0 of b.

The advantage of an adversary $\mathcal{A}$ winning this game is defined as $Adv_{SS} = Pr[b_0 = b] - \frac{1}{2}$.

Game 2. The second game captures the notion of *Traceability against partial collusion.* Our definition of the traceability game is based on that of [5]. Given λ, n, and ε, the game proceeds with the following steps.

- *Setup.* The adversary $\mathcal{A}$ outputs a set $U = \{u_1, u_2, \ldots, u_t\}$ of colluding users with the only restriction that no pair of users have exactly the same access privilege. The access structure associated with user $u_i \in U$ is denoted by T_i.
- *Key Generation.* The challenger runs the key generation algorithm $KeyGen$ to provide the user secret key for each user in U.
- The adversary $\mathcal{A}$ outputs a pirate device $\mathcal{D}$.
- The challenger runs $Trace^{\mathcal{D}}(\varepsilon)$ to obtain a set S.

We say that the adversary $\mathcal{A}$ wins the game if the following two conditions hold:

1. The decoder $\mathcal{D}$ is ε-useful, i.e., for a randomly chosen M in the finite message space, we have that $Pr[\mathcal{D}(Enc(M, \gamma, PK)) = M] \geq \varepsilon$ if there exists a user $u_i \in U$ with $\gamma \models T_i$, where γ is chosen in the way that makes Enc run under normal (non-tracing) operation.
2. The set S is either empty, or is not a subset of U.

We denote the probability that the adversary $\mathcal{A}$ wins this game by Adv_{TR}. If U contains exactly one user, this game captures the notion of *Traceability against single pirate.*

Definition 1. *We say that AFKP-ABE is secure if Adv_{SS} and Adv_{TR} are negligible (in λ) for any polynomial time adversary $\mathcal{A}$ and any constant $\varepsilon > 0$.*

To prove the security of AFKP-ABE in Game 2, another required security game is the *Indistinguishability Game* which captures the notion that, it is hard to distinguish ciphertexts generated by normal (non-tracing) operations from those generated by tracing operations. Its concrete security definition is given in Appendix(see Section 6.2).

4   Our Construction

In this section, we present our construction of the secure AFKP-ABE scheme.

4.1   Main Idea

The intuition of our construction can be summarized as the follows. We define a n-bit user identity space and each bit of them is defined as an attribute with two occurrences, one for bit value 0 and the other for bit value 1. Each user is then assigned a unique ID from the identity space. The encryption algorithm will associate these identity-related attributes to the ciphertext in the following way: for normal (non-tracing) operations, all these n attributes are set as "don't care"; for tracing operations, they are set to represent the suspicious identity. In tracing operations, a user is able to decrypt the ciphertext only if his identity equals the suspicious one. To make tracing ciphertexts indistinguishable from normal ciphertexts, we hide these identity-related attributes in the way that any user is not able to tell which and how many of them are set as "interested" (i.e., not "don't care"). In addition, we also hide some normal attributes so that upon a fail decryption the user can not tell if it is caused by the mismatch of his ID or by his access privilege (without considering his ID). Thus, he is not able to distinguish a tracing activity from a normal (non-tracing) one. The security goal of our construction is to build such a KP-ABE scheme in which 1) any user without the correct decryption key is not able to tell a single bit of the message, and 2) given a pirate device, the authority is able to trick it into decrypting tracing ciphertexts and thus discover the identity of the original owner of the decryption key held by this device.

Definition of Attributes. We define three set of attributes: *public normal attributes, hidden normal attributes* and *hidden identity-related attributes*. We denote the universe of each of them by $\mathcal{U}_{PN}, \mathcal{U}_{HN}$, and $\mathcal{U}_{HID}$ respectively. The letter P in the subscription denotes the word "public", H means "hidden", N represents "normal", and ID is the abbreviation of "identity". $\mathcal{U}_{PN}$ and $\mathcal{U}_{HN}$ contain attributes to be used by normal encryptions. $\mathcal{U}_{HID}$ contains identity-related attributes for describing the suspected user's identity and is particularly used for tracing. In ciphertexts, the associated attributes from $\mathcal{U}_{HN}$ and $\mathcal{U}_{HID}$ have to be hidden such that any receiver is not able to tell which and how many of them are used, while attributes from $\mathcal{U}_{PN}$ are public. Each attribute in $\mathcal{U}_{HID}$ has two occurrences, one for bit value 0 and the other for bit value 1. Similarly, we assume that attributes in $\mathcal{U}_{HN}$ also have binary values like those in $\mathcal{U}_{HID}$. This assumption is just for concise presentation of our scheme. Extending our scheme to support the non-binary case is trivial. From now on we will call the union of $\mathcal{U}_{HID}$ and $\mathcal{U}_{HN}$ as *hidden attributes* by capturing their common property of "hidden". We denote the universe of hidden attributes as $\mathcal{U}_H$, and thus $\mathcal{U}_H = \mathcal{U}_{HN} \cup \mathcal{U}_{HID}$. We denote the number of attributes in $\mathcal{U}_{HN}$ by m and that in $\mathcal{U}_{PN}$ by k. Therefore, the total number of hidden attributes is $m + n$.

   According to the above discussion, it is clear that in a ciphertext there could be three types of attributes: attributes from $\mathcal{U}_{PN}$, attributes from $\mathcal{U}_{HN}$, and those

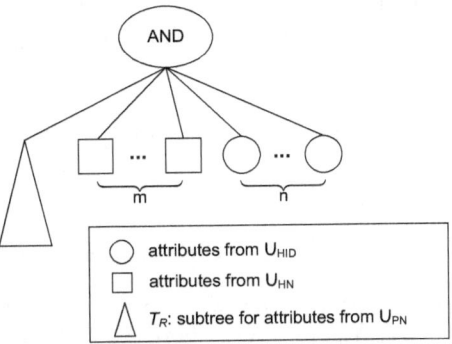

Fig. 1. Illustration of the construction of our access structure

from $\mathcal{U}_{HID}$. We denote the set of these three type of attributes in a ciphertext by γ_{PN}, γ_{HN}, and γ_{HID} respectively. Therefore, we have $\gamma = \gamma_{PN} \cup \gamma_{HN} \cup \gamma_{HID}$, where γ represents the set of all the attributes interested by the encryptor.

Access Structure. Our definition of the access structure (implemented using an access tree) is the same as KP-ABE [1], i.e., each interior node of the tree is a threshold gate and the leaves are associated with attributes. However, our construction has the following restrictions on the access structure: (1) each access structure should deal with all the hidden attributes and all of them should appear on the second layer of the tree; (2) the root node has to be an AND gate; (3) all the attributes from $\mathcal{U}_{PN}$ should appear in a subtree which we denote by T_R. Interior nodes of the subtree T_R could be any kind of threshold gates. The structure of the access tree in our construction is illustrated by Fig. 1. In addition, each non-root node has a unique index given by its parent. For the convenience of representation, we will denote a node x's parent by x_{pa} and x's index by $idx(x)$.

4.2   AFKP-ABE Scheme

In the description, $\mathbb{G}_0$ is a bilinear group of prime order p and g is a generator of $\mathbb{G}_0$. We use $e : \mathbb{G}_0 \times \mathbb{G}_0 \rightarrow \mathbb{G}_1$ to represent a bilinear map. The Lagrange coefficient $\Delta_{i,S}(x)$ is defined as follows, where $i \in \mathbb{Z}_p$, $x \in \mathbb{Z}_p$ are variables, and $S \subset \mathbb{Z}_p$ is some set.

$$\Delta_{i,S}(x) := \prod_{j \in S \setminus \{i\}} \frac{x-j}{i-j}.$$

We use strings of length n to represent user IDs. "don't care" bit of an ID is represented by a "$*$".

Setup$(1^\lambda, n)$ Define $\mathcal{U}_H = \{1, \cdots n, n+1, \cdots m+n\}$, where the first n elements are for $\mathcal{U}_{HID}$ and the last m for $\mathcal{U}_{HN}$, and $\mathcal{U}_{PN} = \{1, 2, \cdots k\}$. For each attribute $i \in \mathcal{U}_{PN}$, choose a random number t_i from $\mathbb{Z}_p$. Then for each hidden attribute

$j \in \mathcal{U}_H$, choose random numbers $\{a_{j,t}, b_{j,t}\}_{t=0,1}$ from $\mathbb{Z}_p$ and random points $\{A_{j,t}\}_{t=0,1}$ from $\mathbb{G}_0$. Finally, choose a random number y from $\mathbb{Z}_p$. The public parameters PK' are published as

$$PK = (Y = e(g,g)^y, \{T_i = g^{t_i}\}_{i \in \mathcal{U}_{PN}}, \{A_{j,t}^{a_{j,t}}, A_{j,t}^{b_{j,t}}\}_{j \in \mathcal{U}_H, t=0,1})$$

and the master key MK is

$$MK = (y, \{t_i\}_{i \in \mathcal{U}_{PN}}, \{a_{j,t}, b_{j,t}\}_{j \in \mathcal{U}_H, t=0,1})$$

Enc(M, γ, PK) Define $\gamma = \gamma_{PN} \cup \gamma_{HN} \cup \gamma_{HID}$ as mentioned before. Let the ID represented by γ_{HID} be $X_n X_{n-1} \cdots X_1$, where $X_i = 0, 1$ or $*$ for each $1 \leq i \leq n$. The encryptor generates ciphertext components for γ_{HID} as follows. First choose a random number s from $\mathbb{Z}_p$. Then for each $1 \leq i \leq n$, pick random numbers $r_{i,0}$ and $r_{i,1}$ from $\mathbb{Z}_p$, and compute tuples $\{[\hat{E}_{i,t}, \check{E}_{i,t}]\}_{t=0,1}$ as follows.

(1) If $X_i = b$, where $b = 0|1$, the encryptor sets $[\hat{E}_{i,1-b}, \check{E}_{i,1-b}]$ as random (*mal-formed*), and $[\hat{E}_{i,b}, \check{E}_{i,b}] = [(A_{i,b}^{b_{i,b}})^{r_{i,b}}, (A_{i,b}^{a_{i,b}})^{s-r_{i,b}}]$ (*well-formed*).

(2) If $X_i = *$, for $t = 0, 1$ the encryptor sets $[\hat{E}_{i,t}, \check{E}_{i,t}] = [(A_{i,t}^{b_{i,t}})^{r_{i,t}}, (A_{i,t}^{a_{i,t}})^{s-r_{i,t}}]$ (*well-formed*).

Ciphertext components for γ_{HN} are generated in the same way as γ_{HID}. The encryptor generates ciphertext components for γ_{PN} as follows. For each $i \in \gamma_{PN}$, compute $E_i = T_i^s$. Finally, the ciphertext is output as follows.

$$E = (\gamma_{PN}, \tilde{E} = MY^s, E_0 = g^s,$$
$$\{E_i\}_{i \in \gamma_{PN}}, \{\{\hat{E}_{i,t}, \check{E}_{i,t}\}_{t=0,1}\}_{i \in \gamma_{HN} \cup \gamma_{HID}})$$

KeyGen(T, MK, PK) The access structure T is defined as mentioned before: the root node of the tree is an AND gate, all the hidden attributes appear on the second layer of the tree, and all the public normal attributes are in the subtree T_R. The trusted authority generates the user secret key as follows.

(1) For the subtree T_R, choose a polynomial q_x for each node x, including all the leaf nodes, of the tree in the top-down manner as follows. Starting from the root node r of T_R (with the threshold value k_r), choose a random number u from $\mathbb{Z}_p$ and set $q_r(0) = u$. Then randomly choose $k_r - 1$ other points to define the $(k_r - 1)$-degree polynomial q_r completely. For any other node x, q_x is generated in the same way and $q_x(0) = q_{x_{pa}}(idx(x))$.

After having defined the polynomials, the following secret key component is generated for each leaf node x in T_R:

$$D_x = g^{\frac{q_x(0)}{t_i}}$$

where i denotes the attribute in $\mathcal{U}_{PN}$ associated with node x. We use $\mathcal{L}_{T_R}$ to represent the set of all the leaf nodes in T_R.

(2) Secret key components for attributes from $\mathcal{U}_{HID}$ are generated as follows. Assume the user is assigned a unique identity $ID = X_n X_{n-1} \cdots X_1$, where $X_i = 0|1$ for each $1 \leq i \leq n$. Then for each attribute i in $\mathcal{U}_{HID}$, the authority

chooses random numbers v_i and λ_i from $\mathbb{Z}_p$ and outputs a triple $[\tilde{D}_i, \hat{D}_i, \check{D}_i]$ as follows.

$$\tilde{D}_i = g^{v_i}(A_{i,X_i})^{a_i,x_i b_i,x_i \lambda_i}, \quad \hat{D}_i = g^{a_i,x_i \lambda_i}, \quad \check{D}_i = g^{b_i,x_i \lambda_i}.$$

(3) Secret key components for attributes from $\mathcal{U}_{HN}$ are generated in the same way as $\mathcal{U}_{HID}$.

(4) The authority sets $v = \sum_{i \in \mathcal{U}_H} v_i$ and generates a secret key component $D_0 = g^{y-u-v}$.

Finally, the authority outputs the following as the user secret key (SK):

$$SK = (D_0, \{D_i\}_{i \in \mathcal{L}_{TR}}, \{\tilde{D}_i, \hat{D}_i, \check{D}_i\}_{i \in \mathcal{U}_H})$$

Dec(E, SK, PK) The receiver decrypts the ciphertext E by applying his secret key components to the ciphertext as follows.

(1) Apply secret key components for public normal attributes to the ciphertext. For each leaf node x of T_R, assuming x is associated with attribute $i \in \mathcal{U}_{PN}$, calculate the following (the result is denoted by F_x):

$$F_x = \begin{cases} e(D_i, E_i) = e(g,g)^{sq_x(0)}, & \text{if } x \in \gamma_{PN} ; \\ \perp, & \text{otherwise.} \end{cases} \qquad (1)$$

Then execute recursively for each non-leaf node z of T_R in the bottom-up manner as follows. For each child node x of z, if $F_x \neq \perp$ add x into a set S_z until S_z has k_z elements, where the set S_z is initialized to empty. If not able to construct such a k_z-sized set S_z, let $F_z = \perp$. Otherwise, calculate F_z as follows.

$$\begin{aligned} F_z &= \prod_{x \in S_z} F_x^{\Delta_{x,S_z}(0)} \\ &= \prod_{x \in S_z} (e(g,g)^{sq_x(0)})^{\Delta_{x,S_z}(0)} \\ &= e(g,g)^{sq_z(0)} \end{aligned}$$

where derivation of the last two steps holds because $q_x(0) = q_z(idx(x))$ and $q_z(0) = \sum_{x \in S_z}(q_z(idx(x)) \cdot \Delta_{x,S_z}(0))$.

This recursion ends up with outputting $F_r = e(g,g)^{sq_r(0)}$ if $\gamma_{PN} \models T_R$. Since $q_r(0) = u$, we have $F_r = e(g,g)^{su}$.

(2) Apply secret key components for hidden attributes to the ciphertext. If the set of hidden attributes in the access structure contains all the attributes in γ_{HN} and γ_{HID}, output the result F_H as follows.

$$\begin{aligned} F_H &= \prod_{i \in \mathcal{U}_H} \frac{e(E_0, \tilde{D}_i)}{e(\hat{E}_i, \hat{D}_i)e(\check{E}_i, \check{D}_i)} \\ &= e(g,g)^{sv} \end{aligned}$$

The message can be output as follows

$$M = \frac{\tilde{E}}{e(E_0, D_0)F_r F_H}$$

$$= \frac{Me(g,g)^{ys}}{e(g^s, g^{y-u-v})e(g,g)^{su}e(g,g)^{sv}}$$

Trace$^{\mathcal{D}}(\varepsilon)$ This algorithm takes as input ε and a ε-useful pirate device $\mathcal{D}$. We first show how to trace $\mathcal{D}$ which just holds one decryption key as follows. The tracing algorithm repeats the following steps $\frac{1}{\varepsilon}$ times for each identity ID_i in the system identity list:

- Step 1. Choose a set of attributes $\gamma = \gamma_{PN} \cup \gamma_{HN} \cup \gamma_{HID}$ such that γ satisfies the access structure of ID_i and γ_{HID} just contains the attributes corresponding to bits of ID_i.
- Step 2. Choose a random message M from the finite message space. Let $E \leftarrow Enc(M, \gamma, PK)$.
- Step 3. Test if $\mathcal{D}$ correctly decrypts E. If it does, stop and return with ID_i. Otherwise continue.

If at the end of these repetitions the algorithm does not return with any identity, return FAIL and stop the experiment. Tracing $\mathcal{D}$ which holds more than one decryption keys is similar with the exception that, in step 3 add ID_i into the guilty user set S instead of returning immediately, where S is initialized as empty. If at the end of these repetitions S is empty, return FAIL and stop the experiment.

4.3 Security Proof

We show the security of our scheme as follows.

Lemma 1. *If a polynomial-time adversary $\mathcal{A}$ can win Game 1 with non-negligible advantage Adv_{SS}, then we can build a simulator $\mathcal{B}$ that is able to solve the DBDH problem with advantage $\frac{1}{2}Adv_{SS}$.*

Proof. Security proof of this Lemma is presented in Appendix(see Section 6.1).

Lemma 2. *If a polynomial-time adversary $\mathcal{A}$ can win our Indistinguishability Game (see Appendix B) with advantage Adv_{IND}, then we can build a simulator $\mathcal{B}$ that is able to solve the D-Linear problem with advantage $\frac{1}{2}Adv_{IND}$.*

Proof. A sketch of the security proof for this Lemma is presented in Appendix(see Section 6.2).

Lemma 3. *If Adv_{IND} and Adv_{SS} are negligible, Adv_{TR} is negligible.*

Proof. Given a pirate device $\mathcal{D}$, our tracing algorithm $Trace^{\mathcal{D}}(\varepsilon)$ will try with each identity ID_i in the system identity list. We denote the attribute set chosen for testing ID_i by $\gamma^i = \gamma_{PN}^i \cup \gamma_{HN}^i \cup \gamma_{HID}^i$. We define the corresponding attribute set used for normal (non-tracing) encryption as $\bar{\gamma}^i = \gamma_{PN}^i \cup \gamma_{HN}^i \cup \bar{\gamma}_{HID}^i$. The only difference between the two sets of attributes is that, in γ_{HID}^i all the attributes corresponding to bits of ID_i are set as "interested", but in $\bar{\gamma}_{HID}^i$ all the identity-related attributes are set as "don't care". Based on this definition, we define the following two probabilities:

$$p_i = \Pr[\mathcal{D}(Enc(M, \gamma^i, PK)) = M]$$
$$p = \Pr[\mathcal{D}(Enc(M, \bar{\gamma}^i, PK)) = M]$$

where M is a random message picked from the message space. We distinguish between the following three types of ε-useful pirate devices that the pirate can generate, where ε is some fixed constant:

1. Pirate device $\mathcal{D}$ for which $|p - p_i|$ is non-negligible for some identity ID_i.
2. Pirate device $\mathcal{D}$ for which $|p - p_i|$ is negligible for each identity ID_i, but the tracing algorithm $Trace^{\mathcal{D}}(\varepsilon)$ outputs an empty set.
3. Pirate device $\mathcal{D}$ for which $|p - p_i|$ is negligible for each identity ID_i, but the tracing algorithm $Trace^{\mathcal{D}}(\varepsilon)$ outputs a set which is not contained in the set of colluding users U.

It is obvious that we can use any pirate producing type 1) devices to win the *Indistinguishability Game* with non-negligible advantage. We now show the rough idea of how we can use any pirate producing type 2) devices to win the *Indistinguishability Game* with non-negligible advantage. Assume the set of colluding users that the pirate claims to be able to collect is $U = \{u_1, u_2, \cdots, u_t\}$. Now denote the challenger of the *Indistinguishability Game* as $\mathcal{C}$, the simulator we want to build is $\mathcal{B}$, and the pirate is $\mathcal{A}$. Then the simulator we build executes as follows.

- Init. $\mathcal{B}$ presents $\mathcal{C}$ two attribute sets $\gamma_0 = \gamma^i$ and $\gamma_1 = \bar{\gamma}^i$ to be challenged upon, where γ^i is the attribute set that can be used to test user $u_i \in U$ by our tracing algorithm.
- Setup. $\mathcal{C}$ generates public parameters and give them to $\mathcal{B}$.
- Phase 1. $\mathcal{B}$ asks $\mathcal{C}$ to give him secret keys for all the users in U. Then $\mathcal{B}$ gives all these keys to $\mathcal{A}$ to answer key queries in the key generation phase of Game 2.
- Challenge. $\mathcal{B}$ submits two equal length messages M_0 and M_1 to $\mathcal{C}$. $\mathcal{C}$ flips a coin and encrypts M_b with γ_b. Then the ciphertext is given to $\mathcal{B}$.
- Phase 2. $\mathcal{B}$ submits more secret key queries.
- Guess. $\mathcal{B}$ asks $\mathcal{A}$ to decrypt the ciphertext given by $\mathcal{C}$. If the message returned by $\mathcal{A}$ is one of M_1 and M_0, $\mathcal{B}$ answers $b_0 = 1$. Otherwise, $\mathcal{B}$ answers $b_0 = 0$.

The advantage for our simulator $\mathcal{B}$ to win the *Indistinguishability Game* is $\frac{1}{\varepsilon}$ times the advantage that the type 2) devices, which are generated by $\mathcal{A}$, output the empty set.

It is easy to show that type 3) devices can be used to win Game 1 (the semantic security game). The intuition is that, type 3) devices can correctly decrypt a message which is encrypted for users whose secret keys are not known to type 3) devices with non-negligible advantage.

4.4 Efficiency Analysis

In AFKP-ABE, both the ciphertext size and the secret key size are linear to n, where n is the number of bits in the identity space. As the maximum number of users it can represent is $N = 2^n$, the complexity can be written as $\mathcal{O}(logN)$, where N is the total number of users. To trace a pirate, AFKP-ABE needs to try with every user's identity in the system list. When the number of users in a system is large, the tracing algorithm would be inefficient. To resolve this issue, we can first test with some normal ciphertexts using combinations of normal attributes. For example, we can use different combinations of attributes like location, age, etc. In practice, this process will hopefully rule out a significant portion of users. Our tracing algorithm can just test over the remaining set of users.

5 Application Scenarios of Our Scheme

In general, our proposed scheme is applicable to systems where 1) data can be categorized by their attributes and a user access privilege should be defined in the way that just allows the user to access certain intended subset of resources; 2) abuse of the access privilege should be prohibited. As we mentioned before, one important application scenario of our abuse free KP-ABE scheme is the area of copyright-sensitive targeted broadcast, especially commercial media broadcast systems. In these systems, contents usually have their commercial values and abuse of the access privilege usually causes legal concerns. Another important application scenario of our proposed scheme would be audit log systems. As these systems would be widely used in applications such as network management, audit logs may contain sensitive information and disclose of them to unauthorized parties would cause security concerns or privacy violations. Recently, we also witnessed application of KP-ABE in wireless networks environment. In [18], Yu et al. proposed a fine-grained data access control scheme for wireless sensor networks for mission-critical applications. In this paper, data access control is well resolved by combining KP-ABE with some other cryptographic primitives. However, the issue of access privilege abuse is not addressed since it is yet another serious issue if we consider the application of mission-critical scenarios such as battle fields. We believe our AFKP-ABE can serve to enhance their proposed scheme as the complexity of AFKP-ABE in terms of ciphertext size and secret key size is just $\mathcal{O}(logN)$, where N is the total number of users.

6 Conclusion and Future Work

In this paper, we focus on the key abuse attacks in KP-ABE enabled broadcast systems and proposed an abuse free KP-ABE (AFKP-ABE) scheme. To defend against the key abuse attacks, we introduce hidden attributes in the system such that the tracing algorithm can use them to identify any single pirate or partial colluding users. Our design enables black boxing tracing and does not require the well-formness of the user secret key. The complexity of AFKP-ABE in terms of ciphertext size and user secret keys size is just $\mathcal{O}(logN)$, where N is the total number of users. Our scheme is provably secure under DBDH assumption and D-Linear assumption. As a future work, we may focus on designing a tracing system against arbitrary colluders.

Acknowledgment

This work was supported in part by the US National Science Foundation under grants CNS-0716306, CNS-0626601, CNS-0746977, CNS-0831628, and CNS-0831963.

References

1. Goyal, V., Pandey, O., Sahai, A., Waters, B.: Attribute-based encryption for fine-grained access control of encrypted data. In: CCS 2006, pp. 89–98 (2006)
2. Chor, B., Fiat, A., Naor, M.: Tracing traitors. In: Desmedt, Y.G. (ed.) CRYPTO 1994. LNCS, vol. 839, pp. 257–270. Springer, Heidelberg (1994)
3. Boneh, D., Franklin, M.K.: An efficient public key traitor tracing scheme. In: Wiener, M. (ed.) CRYPTO 1999. LNCS, vol. 1666, pp. 338–353. Springer, Heidelberg (1999)
4. Kiayias, A., Yung, M.: Traitor tracing with constant transmission rate. In: Knudsen, L.R. (ed.) EUROCRYPT 2002. LNCS, vol. 2332, pp. 450–465. Springer, Heidelberg (2002)
5. Boneh, D., Sahai, A., Waters, B.: Fully collusion resistant traitor tracing with short ciphertexts and private keys. In: Vaudenay, S. (ed.) EUROCRYPT 2006. LNCS, vol. 4004, pp. 573–592. Springer, Heidelberg (2006)
6. Nishide, T., Yoneyama, K., Ohta, K.: Attribute-based encryption with partially hidden encryptor-specified access structures. In: Bellovin, S.M., Gennaro, R., Keromytis, A.D., Yung, M. (eds.) ACNS 2008. LNCS, vol. 5037, pp. 111–129. Springer, Heidelberg (2008)
7. Sahai, A., Waters, B.: Fuzzy identity-based encryption. In: Cramer, R. (ed.) EUROCRYPT 2005. LNCS, vol. 3494, pp. 457–473. Springer, Heidelberg (2005)
8. Bethencourt, J., Sahai, A., Waters, B.: Ciphertext-policy attribute-based encryption. In: SP 2007, Washington, DC, USA, pp. 321–334. IEEE Computer Society, Los Alamitos (2007)
9. Cheung, L., Newport, C.: Provably secure ciphertext policy abe. In: CCS 2007, pp. 456–465. ACM, New York (2007)
10. Kapadia, A., Tsang, P., Smith, S.: Attribute-based publishing with hidden credentials and hidden policies. In: NDSS 2007. LNCS, vol. 5037, pp. 179–192. Springer, Heidelberg (2007)
11. Yu, S., Ren, K., Lou, W.: Attribute-based on-demand multicast group setup with membership anonymity. In: Securecomm (2008)
12. Yu, S., Ren, K., Lou, W.: Attribute-based content distribution with hidden policy. In: NPSEC (2008)
13. Katz, J., Sahai, A., Waters, B.: Predicate encryption supporting disjunctions, polynomial equations, and inner products. In: Smart, N.P. (ed.) EUROCRYPT 2008. LNCS, vol. 4965, pp. 146–162. Springer, Heidelberg (2008)
14. Li, J., Ren, K., Kim, K.: A2be: Accountable attribute-based encryption for abuse free access control. Cryptology ePrint Archive, Report 2009/118 (2009), http://eprint.iacr.org/

15. Li, J., Ren, K., Zhu, B., Wan, Z.: Privacy-aware attribute-based encryption with user accountability. In: ISC (2009)
16. Boneh, D., Franklin, M.: Identity-based encryption from the weil pairing. In: Kilian, J. (ed.) CRYPTO 2001. LNCS, vol. 2139, pp. 213–229. Springer, Heidelberg (2001)
17. Boneh, D., Boyen, X., Shacham, H.: Short group signatures. In: Franklin, M. (ed.) CRYPTO 2004. LNCS, vol. 3152, pp. 41–55. Springer, Heidelberg (2004)
18. Yu, S., Ren, K., Lou, W.: FDAC: Toward fine-grained distributed data access control in wireless sensor networks. In: IEEE INFOCOM (2009)

Appendix

6.1 Security Proof for Lemma 1

Proof. In the DBDH game, the challenger chooses random numbers a, b, c from $\mathbb{Z}_p$ and flips a fair coin μ. If $\mu = 0$, set $z = abc$; If $\mu = 1$, set z as a random value in $\mathbb{Z}_p$. $\mathcal{B}$ is given $(A, B, C, Z) = (g^a, g^b, g^c, e(g, g)^z)$ and asked to output μ. To answer this challenge, $\mathcal{B}$ then simulates Game 1 as follows.

Init $\mathcal{B}$ runs $\mathcal{A}$. $\mathcal{A}$ chooses the set of attributes $\gamma = \gamma_{PN} \cup \gamma_{HN} \cup \gamma_{HID}$ it wants to be challenged upon. We denote the identity represented by γ_{HID} by $X_n X_{n-1} \cdots X_0$, where $X_i = 0, 1$ or $*$, for $1 \leq i \leq n$. We denote the set $\gamma_{HN} \cup \gamma_{HID}$ by γ_H.

Setup $\mathcal{B}$ creates public parameters as follows. First, set $Y = e(A, B) = e(g, g)^{ab}$. Then, for each attribute $i \in \mathcal{U}_{PN}$, generate T_i by the following steps:

– choose a random number $t_i \in \mathbb{Z}_p$.
– if $i \in \gamma_{PN}$, sets $T_i = g^{t_i}$; otherwise, set $T_i = g^{bt_i} = B^{t_i}$.

For each attribute $i \in \mathcal{U}_{HID}$, choose two random numbers $h_{i,0}$ and $h_{i,1}$ from $\mathbb{Z}_p$. Then proceed as follows.

– if $X_i = *$, $A_{i,t} = g^{h_{i,t}}$, $t = 0, 1$; otherwise, $A_{i,X_i} = g^{h_{i,X_i}}$ and $A_{i,1-X_i} = g^{bh_{i,1-X_i}} = B^{h_{i,1-X_i}}$.
– choose random numbers $\{a_{i,t}, b_{i,t}\}_{t=0,1}$ from $\mathbb{Z}_p$.

Attributes in $\mathcal{U}_{HN}$ are processed in the same way as $\mathcal{U}_{HID}$. Finally, output PK as in the real scheme.

Phase I. $\mathcal{A}$ submits a query for secret key of access structure T, where $\gamma \nvDash T$. Note that T has the structure of 1. $\mathcal{B}$ differentiates the following two cases and answers the query accordingly:

Case 1: In this case, $\gamma_{PN} \nvDash T_R$. $\mathcal{B}$ generates secret key components for hidden attributes as in the real scheme. To generate secret key components for attributes attached to T_R, $\mathcal{B}$ defines a recursive function $PolyDef(x)$ and runs it over the root node r of T_R. For each node x in T_R, use k_x and p_x to represent the node's threshold value and the number of its satisfied children respectively (the satisfied child is a child node of x that returns true over γ_{PN}).

PolyDef(x): It is defined by the following steps:

- Define q_x as follows.
  - If x is not r, set $q_x(0) = q_{x_{pa}}(idx(x))$; otherwise, set $q_x(0) = ab + br'$, r' is randomly chosen from $\mathbb{Z}_p$.
  - Select d ($= k_x - 1$) children of x. For each selected child i, choose a random number r'_i from $\mathbb{Z}_p$ and let $q_x(idx(i)) = br'_i$. This completes the construction of polynomial q_x. Note that, if $p_x \leq d$, the set of selected children should include all the p_x satisfied ones; otherwise, all the d selected children should be satisfied ones. We denote the set of these selected children of x plus x itself by X_s.
- For each remaining child j (not selected by the above step), calculate $q_x(j) = \sum_{i \in X_s} q_x(idx(i)) \Delta_{i,S_x}(j)$.
- For each child i of x, run $PolyDef(i)$.

When $PolyDef(r)$ terminates, $\mathcal{B}$ completes the construction of the polynomials for all the nodes in T_R. In particular, $p_r(0) = ab + br'$. Note that, in our construction of polynomials, for each node x, the polynomial values have the following properties:

(1) If $q_x(0)$ has the form of $R_x b$, then for each of its children i, $q_i(0)$ ($= q_x(idx(i))$) has the form of $R_i b$.

(2) If $q_x(0)$ has the form of $C_x ab + R_x b$, then for each of its children i, (i) if $i \in X_s$ (selected), $q_i(0)$ has the form of $R_i b$; otherwise, (ii) $q_i(0)$ has the form of $C_i ab + R_i b$.

(3) In (1) and (2), C_x, R_x, C_i, and R_i are functions of Lagrange coefficients and random numbers (i.e., r'_j's), and independent of a and b.

From these properties, we may categorize a leaf nodes x into one of the following three types:

(1) Type A: $x \in \gamma_{PN}$, i.e., x is a satisfied node. $q_x(0)$ has the form of $R_x b$.

(2) Type B: $x \notin \gamma_{PN}$ but one of x's ancestors (including x itself) is selected by its parent. $q_x(0)$ has the form of $R_x b$.

(3) Type C: all the other leaf nodes, $q_x(0)$ has the form of $C_x ab + R_x b$.

Therefore, the secret key component corresponding to each leaf node x of T_R is given as follows

$$D_x = \begin{cases} g^{\frac{R_x b}{t_x}} = B^{\frac{R_x}{t_x}}, & x \text{ in Type A.} \\ g^{\frac{R_x b}{t_x b}} = g^{\frac{R_x}{t_x}}, & x \text{ in Type B.} \\ g^{\frac{C_x ab + R_x b}{t_x b}} = A^{\frac{C_x}{t_x}} g^{\frac{R_x}{t_x}}, & x \text{ in Type C.} \end{cases} \tag{2}$$

The secret key component D_0 of SK is output as follows

$$g^{y-u-v} = g^{ab - q_r(0) - v} = g^{-br'} g^{-v} = B^{-r'} g^{-v}$$

where v is generated when constructing secret key components for hidden attributes. All the other components are generated as in the real scheme.

Case 2: In this case, $\gamma_{PN} \models T_R$, but the hidden attributes of T do not match with γ_H. Let a hidden attribute j that is not intended by γ_{HID} be the witness.

$\mathcal{B}$ generates secret key components corresponding to T_R as in the real scheme. $\mathcal{B}$ generates secret key components for hidden attributes as follows.

- For hidden attributes $1 \leq i \leq m + n$, pick v'_i randomly from $\mathbb{Z}_p$. Set $v_j = ab + v'_j$ and $v_i = v'_i$ for every $i \neq j$. Finally set $v = \sum_{i=1}^{m+n} v_i = ab + \sum_{i=1}^{m+n} v'_i$.
- compute the secret key components $[\tilde{D}_j, \hat{D}_j, \check{D}_j]$ of attribute j as follows.

$$
\begin{aligned}
\tilde{D}_j &= g^{v_j}(A_{j,X_j})^{a_{j,X_j} b_{j,X_j} \lambda_j} \\
&= g^{ab+v'_j}(A_{j,X_j})^{a_{j,X_j} b_{j,X_j} \lambda_j} \\
&= g^{ab+v'_j}(g^{bh_{j,X_j}})^{a_{j,X_j} b_{j,X_j} \lambda_j} \\
&= g^{v'_j}(g^{bh_{j,X_j}})^{a_{j,X_j} b_{j,X_j} \lambda'_j}
\end{aligned}
$$

where λ'_j is chosen by $\mathcal{B}$ and $\lambda_j = \frac{a}{h_{j,X_j} a_{j,X_j} b_{j,X_j}} + \lambda'_j$. $\mathcal{B}$ calculates $[\hat{D}_j, \check{D}_j]$ and $[\tilde{D}_i, \hat{D}_i, \check{D}_i]$ for $i \neq j$ as in the real scheme.
- Output D_0 of SK as: $D_0 = g^{ab-u-v} = g^{-u-\sum_{i=1}^{m+n} v'_i}$, where u is generated when constructing secret key components for T_R.

All the other components are generated as in the real scheme.

From the above description, we can see that $\mathcal{B}$ is able to construct a secret key of T in both cases. Furthermore, the distribution of the secret key of T is the same as that in the original scheme. The adversary $\mathcal{A}$ can repeat this step for polynomial times.

Challenge. The adversary $\mathcal{A}$ submits two equal length challenge messages m_0 and m_1 to $\mathcal{B}$. $\mathcal{B}$ flips a fair binary coin v and picks out m_v. The ciphertext of m_v is output as: $E = (\gamma_{PN}, \tilde{E} = m_v Z, E_0 = C, \{E_i = C^{t_i}\}_{i \in \gamma_{PN}}, \{\{\hat{E}_{i,t}, \check{E}_{i,t}\}_{t=0,1}\}_{i \in \gamma_H})$. Note that $\mathcal{B}$ can construct $\{\{\hat{E}_{i,t}, \check{E}_{i,t}\}_{t=0,1}\}_{i \in \gamma_H}$ because if the occurrence t of attribute i is in γ_H, $A_{i,t}$ does not contain the unknown value b, and if the occurrence t of i is not in γ_H, $\{\hat{E}_{i,t}, \check{E}_{i,t}\}$ are just chosen at random. If $\mu = 0$ it is easy to show that the ciphertext is a valid random encryption of message m_v. Otherwise, if $\mu = 1$, then $Z = e(g,g)^z$ and $\tilde{E} = m_v e(g,g)^z$. Since z is random, $\tilde{E}$ is just a random element of $\mathbb{G}_1$ from the adversary's view and contains no information about m_v.

Phase II. The simulator acts exactly as it did in Phase I.

Guess. The adversary $\mathcal{A}$ submits a guess v' of v. If $v' = v$, $\mathcal{B}$ outputs $\mu' = 0$, indicating that the given DBDH-tuple is a valid one. Otherwise it outputs $\mu' = 1$, indicating that the given DBDH-tuple is just a random quadruple. In the case of $\mu = 1$, the ciphertext E contains no information about m_v. Therefore, v' is just a random guess of v, and thus μ' is just a random guess of μ. Thus, we have $Pr[\mu' = \mu | \mu = 1] = \frac{1}{2}$. If $\mu = 0$, the ciphertext E is a valid encryption of m_v. Since by definition $\mathcal{A}$ has the advantage of Adv_{SS} to output a correct guess, i.e., $v' = v$, $\mathcal{B}$ outputs $\mu' = 0$ with the probability of $\frac{1}{2} + Adv_{SS}$, i.e., $Pr[\mu' = \mu | \mu = 0] = \frac{1}{2} + Adv_{SS}$. Therefore, the overall advantage of $\mathcal{B}$ in the DBDH game is $\frac{1}{2} Pr[\mu' = \mu | \mu = 0] + \frac{1}{2} Pr[\mu' = \mu | \mu = 1] - \frac{1}{2} = \frac{1}{2}(\frac{1}{2} + Adv_{SS}) + \frac{1}{2} \frac{1}{2} - \frac{1}{2} = \frac{1}{2} Adv_{SS}$.

6.2 Indistinguishability Game

This game captures the idea that ciphetexts generated by tracing operations are indistinguishable from those generated by normal (non-tracing) operations. In AFKP-ABE, these two types of ciphertexts are generated by running our encryption algorithm over different sets of attributes. To differentiate these two types of ciphertexts is actually equal to telling which set of attributes are used in a given data encryption operation. As we discussed, the attribute set γ used for an encryption operation is composed of three disjunctive subsets, i.e., $\gamma = \gamma_{PN} \cup \gamma_{HN} \cup \gamma_{HID}$. In a tracing operation, we set γ_{HID} to represent the suspicious identity, while in a normal (non-tracing) operation we set γ_{HID} to represent the identity of "$* * \cdots *$", i.e., each bit if ID is set as "don't care". We define the *Indistinguishability Game* by the following steps:

Init. The adversary $\mathcal{A}$ selects two sets of attributes to be challenged upon: $\gamma_0 = \gamma_{PN} \cup \gamma_{HN} \cup \gamma_{HID}$ and $\gamma_1 = \gamma_{PN} \cup \gamma_{HN} \cup \gamma^*_{HID}$, where γ_{HID} represents a certain identity ID_i, and γ^*_{HID} denotes the identity of "$* * \cdots *$", i.e., each bit if ID is set as "don't care". $\mathcal{A}$ submits these two sets of attributes to the challenger $\mathcal{C}$.

Setup. The challenger $\mathcal{B}$ runs the setup algorithm of AFKP-ABE and give public parameters PK to $\mathcal{A}$.

Phase 1. $\mathcal{A}$ asks for the secret key of access structure T. If $(\gamma_0 \models T \wedge \gamma_1 \models T)$ or $(\gamma_0 \nvDash T \wedge \gamma_1 \nvDash T)$, the challenger $\mathcal{B}$ answers the query and gives $\mathcal{A}$ the corresponding secret key SK_T. The adversary $\mathcal{A}$ can repeat this step polynomially many times.

Challenge. $\mathcal{A}$ submits two equal length messages M_0 and M_1 to $\mathcal{B}$. If $\mathcal{A}$ a secret key SK_T for which $(\gamma_0 \models T \wedge \gamma_1 \models T)$, it is required that $M_0 = M_1$. $\mathcal{B}$ flips a binary fair coin b and encrypts M_b using attribute set γ_b. The ciphertext is given to $\mathcal{A}$.

Phase 2. Repeat Phase 1. If $M_0 \neq M_1$, $\mathcal{A}$ can not submit secret key query for access structure T for which $(\gamma_0 \models T \wedge \gamma_1 \models T)$.

Guess. The adversary $\mathcal{A}$ outputs a guess b' of b.

Proof. We use a series of games to prove the security of this game as [6]. *Game* Ind_1 is defined in the same way as the original game except that in γ_0, γ_{HID} represents the identity of "$* * \cdots * X_1$", i.e., the upper $n - 1$ bits are set as "don't care" but keep the first bit the same as in the original game. *Game* Ind_2 is defined in the same way that γ_{HID} represents the identity of "$* * \cdots * X_2 X_1$", i.e., the upper $n - 2$ bits are set as "don't care" but keep the first bit the same as in the original game, so on and so forth. Our original game is thus *Game* Ind_n. To prove the security of our scheme, it is enough to prove that it is indistinguishable between *Game* Ind_i and *Game* Ind_{i+1}. We can use the similar technique used by [6] to prove this. For the space limit, we will present the complete proof of our Indistinguishability Game in the full version.

Breaking and Building of Group Inside Signature

S. Sree Vivek*, S. Sharmila Deva Selvi, S. Gopi Nath, and C. Pandu Rangan*

Indian Institute of Technology Madras,
Theoretical Computer Science Laboratory,
Department of Computer Science and Engineering,
Chennai, India
{svivek,sharmila,gopinath,prangan}@cse.iitm.ac.in

Abstract. Group Inside Signature (GIS) is a signature scheme that allows the signer to designate his signature to be verified by a group of people. Members other than the designated group cannot verify the signature generated by the signer. In Broadcast Group Oriented Signature (BGOS), a user from one group can designate his signature to be verified by members of another group. An Adaptable Designated Group Signature (ADGS), is one in which an user can designate his signature to be verified by a selected set of members who are from different groups. The two GIS schemes [5], [6] and the BGOS scheme [7], we consider are certificateless schemes and the ADGS scheme [8] which we consider here is an identity based scheme. In this paper, we present the cryptanalysis of all the four schemes that appeared in [5], [6], [7] and [8]. We also present a new identity based ADGS (N-ADGS) scheme and prove its security in the random oracle model. The existing model described in [8] for ADGS did not consider unlinkability which is one of the key properties required for ADGS. We provide the security model for unlinkability and also prove our scheme is unlinkable.

Keywords: Cryptanalysis, Group Inside Signature, Broadcast Group Oriented Signature, Adaptable Designated Group Signature, Provable Security, Random Oracle model.

1 Introduction

In general, digital signatures are publicly verifiable. Jackbson et.al (1996) [4] proposed the concept of Designated Verifier Signatures (DVS) and strong DVS (SDVS). In DVS, only a designated person can verify the signature, which is signed by a signer. DVS achieves this property by providing an ability called *Simulatability* to the designated verifier, which allows him to simulate the actual signers signature. In SDVS, any third party cannot verify the validity of the signature unless the private key of the designated verifier or the actual signer is exposed.

* Work supported by Project No. CSE/05-06/076/DITX/CPAN on Protocols for Secure Communication and Computation sponsored by Department of Information Technology, Government of India.

Y. Chen et al. (Eds.): SecureComm 2009, LNICST 19, pp. 330–339, 2009.

Extending a single party verification scheme to a designated group verification scheme is a challenging problem. In practice, there may be different group models. First, in networks like Local Area Networks, all group members reside in a single network and no member of the group may hang outside network. Certificateless GIS schemes [5] and [6] provide solutions for designating a signature to be verified inside such a group. Secondly, in distributed networks, the users of different companies or institutions naturally come under different work groups. If a member of one group wants to send a signed document to members of another group, BGOS [7] can be used. Moreover the signer wants to prevent the members outside the designated group from verifying the signature. The scheme in [7] focuses on this problem. Finally, in distributed networks, a signer may want several members to verify his signature, no matter whether those members are in same or different groups. The signer wants to prevent the members outside the defined group from verifying the signature. This model can be visualized as a more generalized version of the previous two models. ADGS scheme in [8] focuses on this problem. In fact even if a designated verifier v_i belongs to a group say G, while v_i can verify the signature of the sender, other members of G cannot verify the signature.

Suppose that a organization initiates a call for tender, asking for quotations to some companies for a set of instruments and tasks to be accomplished. Here, the requirement is that, the competing companies should not be able to verify the quotations quoted by their counter parts. So each company will encrypt and sign the quotation and send it to the organization. But nothing prevents the organization from revealing the quoted values once decrypted, since the organizations goal is to obtain quotations with low price. In this situation the organization could show the signed offers to some other companies and influence them to make better quotations. Here, we can use the ADGS scheme, because the company which proposes the quotation can designate the signature to the organization who has called for the tender and other companies can not verify the validity unless the verifier uses the private key of the organization.

Simulatability vs Unlinkability. The notion "Simulatability" in the context of DVS ensures that the designated verifier has the ability to simulate the transcript as if it is generated by the actual signer i.e., we can say that the designated verifier is also capable of generating the signature of the signer. Where as the notion of "Unlinkability" in the context of ADGS ensures that only the designated group members can verify the signature designated to them, members other than the designated group can not verify the signature. Thus, Unlinkability is different from Simulatability and should not be confused with each other.

Our Contribution. In this paper, we show that GIS in [5] and BGOS in [7] are not secure against both Type-I and Type-II adversaries, and the GIS in [6] is not secure against Type-I adversary. We also show that the basic ADGS scheme [8] is universally forgeable. We also propose a new Adaptable Designated Group Signature scheme (New-ADGS) and prove its security formally in the random oracle model. Due to page limitation, we omit the reviews of the broken schemes and the security proofs of the newly proposed ADGS scheme and is given in the full version of this paper [10].

2   Preliminaries

2.1   Bilinear Pairing

Let $\mathbb{G}_1$ be an additive cyclic group generated by P, with prime order q, and $\mathbb{G}_2$ be a multiplicative cyclic group of the same order q. A bilinear pairing is a map $\hat{e} : \mathbb{G}_1 \times \mathbb{G}_1 \to \mathbb{G}_2$ with the following properties.

- **Bilinearity.** For all $P, Q, R \in_R \mathbb{G}_1$ and $a, b \in_R \mathbb{Z}_q^*$, $\hat{e}(P + Q, R) = \hat{e}(P, R)$ $\hat{e}(Q, R)$, $\hat{e}(P, Q + R) = \hat{e}(P, Q)\hat{e}(P, R)$ and $\hat{e}(aP, bQ) = \hat{e}(P, Q)^{ab}$
- **Non-Degeneracy.** There exist $P, Q \in \mathbb{G}_1$ such that $\hat{e}(P, Q) \neq I_{\mathbb{G}_2}$, where $I_{\mathbb{G}_2}$ is the identity element of $\mathbb{G}_2$.
- **Computability.** There exists an efficient algorithm to compute $\hat{e}(P, Q)$ for all $P, Q \in \mathbb{G}_1$.

3   Cryptanalysis of Certificateless GIS and BGOS Schemes

In this section we show the weaknesses in two certificateless GIS schemes [5], [6] and a certificateless BGOS scheme [7].

3.1   Cryptanalysis of Certificateless GIS Scheme [5]

GIS scheme given in [5] allows the signer to designate his signature to be verified by a group of people who belong to the signer's group. Members other than the designated group should not be able to verify the signature generated by him. The scheme in [5] is not secure against Type-I and Type-II attacks.

Type-I Attack. On seeing a valid signature by an user on some message, anyone can commit a forgery on any message. During the unforgeability game between the challenger $\mathcal{C}$ and adversary $\mathcal{A}_I$, $\mathcal{C}$ gives $\mathcal{A}_I$ the public parameters *params* and $\mathcal{A}_I$ gives to $\mathcal{C}$ a target identity ID^*. $\mathcal{A}_I$ is supposed to generate a valid forgery for the target identity ID^* on some message and $\mathcal{A}_I$ is not allowed to query partial private key for the target identity ID^*. $\mathcal{A}_I$ interacts with $\mathcal{C}$ and access all the oracles with the restrictions given in the model. $\mathcal{A}_I$ can query signature on any message and user identity pair $\langle m, ID \rangle$. $\mathcal{A}_I$ can replace the public keys of suppose any user including user with identity ID^*. During the training-phase $\mathcal{A}_I$ receives a valid signature $\sigma = \langle m, U, V \rangle$ on a message m with target identity ID^* using the **Sign** oracle. Now we show how $\mathcal{A}_I$ can generate a valid signature σ^* on an arbitrary message m^* for the target identity ID^*, such that σ^* is not the output of previous queries to **Sign** oracle. This can be shown by the following computation done by $\mathcal{A}_I$

- Computes $U^* = U + hP_{i1} - H_0(ID^*)$, where $h = H_1(m, U)$ computed from σ.
- Computes $h^* = H_1(m^*, U^*)$
- Replaces public keys of ID^* as $P_{i1}^* = \frac{1}{h^*} H_0(ID^*)$ and $P_{i2}^* = \frac{1}{h^*} P$.
- $V^* = V$.

Now we claim that $\sigma^* = \langle m^*, U^*, V^* \rangle$ is a valid signature on the message m^* by the user with identity ID^* (with respect to its newly replaced public key). $\mathcal{C}$ can check the validity of the forged signature σ^* as follows.

Correctness of public keys. It is clear that $\langle P_{i1}^*, P_{i2}^* \rangle$ satisfies the verification

$$e(P_{i1}^*, P) \stackrel{?}{=} e(P_{i2}^*, H_0(ID_i)).$$

Correctness of forged signature. Note that $\mathcal{C}$ will use the current public key of ID^* that was set by $\mathcal{A}_I$.

- $\mathcal{C}$ has to check whether $e(V^*, P_{j1}) \stackrel{?}{=} e(U^*, D_{j1}) \, e(h^* P_{i1}^*, D_{j1})$. In fact

$$
\begin{aligned}
R.H.S &= e(U^*, D_{j1}) \, e(h^* P_{i1}^*, D_{j1}) \\
&= e(U + hP_{i1} - H_0(ID^*), D_{j1}) \, e(h^* P_{i1}^*, D_{j1}) \\
&= e(U + hP_{i1} - H_0(ID^*), D_{j1}) \, e(H_0(ID^*), D_{j1}) \\
&= e(U, D_{j1}) \, e(hP_{i1}, D_{j1}). \\
&= e(V, P_{j1}). \\
&= e(V^*, P_{j1}) \\
&= L.H.S
\end{aligned}
$$

Thus the forged signature σ^* passes the verification successfully.

Type-II Attack. Type-II attack is also possible on the same scheme. During the unforgeability game between the challenger $\mathcal{C}$ and adversary $\mathcal{A}_{II}$, $\mathcal{A}_{II}$ can interacts with $\mathcal{C}$ and access the **Sign** oracle with the restrictions given in the model. $\mathcal{A}_{II}$ can ask signature on any message and identity pair $\langle m, ID \rangle$. $\mathcal{A}_{II}$ has access to the master private key. So it can compute the private key of any user from its public keys $\langle P_{i1} P_{i2} \rangle$ as $D_i = kP_{i1}$. Since the public key $P_{i1} = x_i H_0(ID_i)$, so $\mathcal{A}_{II}$ can generate signature on behalf of any user and $\mathcal{A}_{II}$ can verify the signature of any user. Here, we can visualize $\mathcal{A}_{II}$ as the KGC because it knows the master private key in the scheme.

3.2   Cryptanalysis of Another Certificateless GIS Scheme [6]

Chunbo Ma et al. have proposed another GIS [6] scheme. In this section, we present *Type-I* forgery on the scheme [6]. Here adversary $\mathcal{A}_I$ who considered to be inside the group can sign on behalf of any user on any message. During the unforgeability game between the challenger $\mathcal{C}$ and adversary $\mathcal{A}_I$, $\mathcal{C}$ gives $\mathcal{A}_I$ the public parameters *params* and a target identity ID_A. $\mathcal{A}_I$ is supposed to generate a valid forgery for the target identity ID_A on some message and it is not allowed to query partial private key for the target identity ID_A. $\mathcal{A}_I$ interacts with $\mathcal{C}$ and access all the oracles with the restrictions given in the model. $\mathcal{A}_I$

can query signature on any message and user identity pair $\langle m, ID \rangle$. $\mathcal{A}_I$ can replace the public keys of any user including user with identity ID_A. During the training-phase $\mathcal{A}_I$ receives a valid signature $\sigma = \langle m, U, V \rangle$ on a message m with target identity ID_A as the signer from the **Sign** oracle and also obtains the private key of some other user say ID_B from the **Key Extract** oracle. Now $\mathcal{A}_I$ can generate a valid signature σ^* on a message m^* for the target identity ID_A by using the private key of ID_B, such that σ^* is not the output of previous queries to **Sign** oracle. This can be shown by the following computation done by $\mathcal{A}_I$. First $\mathcal{A}_I$ computes the value $e(g, g^k)$ even though $\mathcal{A}_I$ may not know the value $e(g, g^k)$ directly, it can compute $e(g, g^k)$ as follows.

$$
\begin{aligned}
e(D_B, P_{pub2})e(D_B, (P_{pub1})^{H_1(ID_B)}) &= e(g^{\frac{k^2}{k+H_1(ID_B)}}, g)e(g^{\frac{kH_1(ID_B)}{k+H_1(ID_B)}}, g)\\
&= e(g^{\frac{k^2}{k+H_1(ID_B)}} g^{\frac{kH_1(ID_B)}{k+H_1(ID_B)}}, g)\\
&= e(g, g^k)
\end{aligned}
$$

Hence, $e(g, g^k)$ can be computed by $\mathcal{A}_I$ and subsequently $\mathcal{A}_I$ generates the forgery by performing the following:

- Computes $r^* = e(g, g^k)^{a^*}$.
- Computes $V^* = H_0(m^*||r^*)$.
- Computes $U^* = SK_B^{(a^*+v^*)}$.
- Replaces ID_A's public keys $X_A^* = X_A$ and $Y_A^* = X_A^{(-H_1(ID_A))} X_B^{H_1(ID_B)} Y_B$.
- Broadcasts the signature σ^* (m^*, U^*, V^*, ID_A).

Now challenger $\mathcal{C}$ can verify the validity of the signature using the private key of any group member say C as follows:

Computes r' as

$$
\begin{aligned}
e(U^*, (X_A^*)^{H_1(ID_A)}.Y_A^*)e(S_C, X_C^{H_1(ID_C)}Y_C)^{-V^*} &=\\
=e(U^*, X_A^{H_1(ID_A)}.X_A^{-H_1(ID_A)} X_B^{H_1(ID_B)}Y_B) \, e(S_C, X_C^{H_1(ID_C)}.Y_C)^{-V^*}\\
=e(g, g)^{k(a^*+V^*)}e(g, g)^{-V^*k}.\\
=e(g, g)^{ka^*}\\
=r'
\end{aligned}
$$

Checks $V^* \stackrel{?}{=} H_0(m^*||r')$ if it holds σ^* is a valid forgery other wise not.

Since σ^* is a valid forgery which we showed now, we can claim that the scheme given in [6] is having *Type-I* forgery.

3.3   Cryptanalysis of Broadcast Group Oriented Signature [7]

In BGOS, an user from one group can designate its signature to be verifiable by members of other group. In this section we present the cryptanalysis of BGOS scheme, which too has both Type-I and Type-II attacks.

Type-I Attack on BGOS Scheme [7]. On seeing a valid signature by an user on some message, anyone can commit a forgery on any message. During the unforgeability game between the challenger $\mathcal{C}$ and adversary $\mathcal{A}_I$, $\mathcal{C}$ gives $\mathcal{A}_I$ the public parameters *params* and $\mathcal{A}_I$ gives to $\mathcal{C}$ a target identity ID_{bi}^*. $\mathcal{A}_I$ is supposed to generate a valid forgery for the target identity ID_{bi}^* on some message and it is not allowed to query partial private key for target identity ID_{bi}^*. $\mathcal{A}_I$ interacts with $\mathcal{C}$ and access all the oracles with the restrictions given in the model. $\mathcal{A}_I$ can query signature on any message and user identity pair $\langle m, ID \rangle$. $\mathcal{A}_I$ can replace the public keys of suppose any user including user with identity ID_{bi}^*. During the training-phase $\mathcal{A}_I$ receive a valid signature $\sigma = \langle m, U_1, U_2, V \rangle$ on a message m with target identity ID_{bi}^* using the **Sign** oracle. Now we show how $\mathcal{A}_I$ can generate a valid signature σ^* on an arbitrary message m^* for the target identity ID_{bi}^*, such that σ^* is not the output of previous queries to **Sign** oracle. This can be shown by the following computation done by $\mathcal{A}_I$

- Computes $U_1^* = U_1 + hP_{bi}$-$H_0(ID_{bi}^*)$ and $U_2^* = U_2 + hP_A^{(2)}$ - P.
- Computes $h^* = H_1(m^*, U_1^*)$.
- Replaces ID_{bi}^*'s public keys as $P_{bi}^* = \frac{1}{h^*}H_0(ID_{bi}^*)$ and $Q_{bi}^* = \frac{1}{h^*}P$.
- Replaces group $\mathcal{A}$'s public keys as $P_A^{(2)*} = \frac{1}{h^*}P$ and $Q_A^{(2)*} = \frac{1}{h^*}H_0(ID_A)$.
- $V^* = V$.

Now we claim that $\sigma^* = \langle m^*, U_1^*, U_2^*, V^* \rangle$ is a valid signature on the message m^* by the user with identity ID^*. $\mathcal{C}$ can check the validity of the forged signature σ^* as follows.

Correctness of Public Keys: The replaced public keys of group $\mathbb{A}$ $\langle P_A^{(2)*}, Q_A^{(2)*} \rangle$ passes the verification

$$e(P_A^{(2)*}, H_0(ID_A)) \overset{?}{=} e(P, Q_A^{(2)*})$$

The replaced public keys of user b_i $\langle P_{bi}^*, Q_{bi}^* \rangle$ also passes the following verification:

$$e(P_{bi}^*, P) \overset{?}{=} e(Q_{bi}^*, H_0(ID_{bi}))$$

Correctness of forged signature: Note that $\mathcal{C}$ will use the current public key of ID^* that was set by $\mathcal{A}_I$. $\mathbb{C}$ has to check $e(V^*, P_{ai}) \overset{?}{=} e(h^*P_{bi}^* + U_1^*, D_{aj}^{(2)})$ $e(h^*P_A^{(2)*} + U_2^*, D_{aj}^{(1)})$. Now,

R $.H.S =$
$$= e(h^*P_{bi}^* + U_1^*, D_{aj}^{(2)})e(h^*P_A^{(2)*} + U_2^*, D_{aj}^{(1)})$$
$$= e(h^*P_{bi}^* + U_1 + hP_{bi} - H_0(ID_{bi}^*), D_{aj}^{(2)})e(h^*P_A^{2*} + U_2 + hP_A^{(2)} - P, D_{aj}^{(1)})$$
$$= e(hP_{bi} + U_1, D_{aj}^{(2)})e(hP_A^{(2)} + U_2, D_{aj}^{(1)})$$
$$= e(V, P_{ai})$$
$$= e(V^*, P_{ai})$$
$$= L.H.S$$

Thus the forged signature σ^* passes the verification successfully.

Type-II Attack on BGOS Scheme [7]. Type-II attack is also possible on BGOS [7] scheme. During the Unforgeability game between the challenger $\mathcal{C}$ and adversary $\mathcal{A}_{II}$, $\mathcal{A}_{II}$ can interact with $\mathcal{C}$ and can access **Sign** oracle with the restrictions given in the model. $\mathcal{A}_{II}$ can ask signature on any message and identity pair $\langle m, ID \rangle$. The adversary $\mathcal{A}_{II}$ can access the master private key. So, $\mathcal{A}_{II}$ can compute the full private key of any user from group $\mathbb{A}$ using the public keys $\langle P_{ai}, Q_{ai} \rangle$ as $\langle \{ D_{ai}^{(1)}, D_{ai}^{(2)} \} \rangle = \langle sP_{ai}, tP_{ai} \rangle$ and any user from group B with public keys $\langle P_{bi}, Q_{bi} \rangle$ as $\langle \{ D_{bi}^{(1)}, D_{bi}^{(2)} \} \rangle = \langle sP_{bi}, tP_{bi} \rangle$. As a result the KGC can generate signature on behalf of any user and also verify the signature of any user in any group, which contradicts the statement of the authors.

4   Cryptanalysis of Identity Based ADGS Scheme [8]

In this section, we present the cryptanalysis of an identity based ADGS scheme [8]. We show that the ADGS scheme in [8], is universally forgeable by demonstrating two different ways to proceed with the attack.

Universal Forgery Without Having Access to Any Previous Signature. The scheme ADGS described above is universally forgeable. The adversary $\mathcal{A}$ can forge the signature of any user without seeing any valid signature previously signed by any user. $\mathcal{A}$ selects $r^*, k^*, t^* \in_R \mathbb{Z}_q^*$, computes $T_i^* = k^* Q_i \; for (i = 1$ to n $a_i \in \mathbb{U})$. and then computes the following values.

- $V_0^* = t^* s^* P.$
- $V_1^* = t^* k^* P.$
- $V_2^* = r^* k^* P.$
- $h^* = H_1(m^*).$
- $T_0^* = \frac{1}{h^*} k^* P.$
- $V^* = r^* P + P_{pub}.$

  $\mathcal{A}$ produces $\sigma^* = (m^*, V^*, V_0^*, V_1^*, V_2^*, T_0^*, ..., T_n^*)$ as a valid signature on message m^*.

Now the correctness of the forged signature σ^* can be shown as follows:

Correctness: The *L.H.S* is

$$
\begin{aligned}
e(V^*, T_i^*) &= e(r^* P + P_{pub}, k^* Q_i) \\
&= e(r^* P, k^* Q_i) e(P_{pub}, k^* Q_i) \\
&= e(r^* k^* P, Q_i) e(k^* P, s Q_i) \\
&= e(V_2^*, Q_i) e(\frac{1}{h^*} T_0^*, D_i) \\
&= R.H.S
\end{aligned}
$$

Thus, we show that $\mathcal{A}$ is capable of generating a valid ADGS on behalf of user with out knowing users secret key.

Universal Forgery on Seeing a Signature of an User. On seeing a valid signature by an user on some message, anyone can commit a forgery on any message. During the unforgeability game between the challenger $\mathcal{C}$ and adversary $\mathcal{A}$, $\mathcal{C}$ gives $\mathcal{A}$ the public parameters *params* and a target identity ID^*. $\mathcal{A}$ is supposed to generate a valid forgery for the target identity ID^* on some message and it is restricted to query private key for the target identity ID^*. $\mathcal{A}$ interacts with $\mathcal{C}$ and accesses all the oracles with the restrictions given in the model. $\mathcal{A}$ can query signature on any message and user identity pair $\langle m, ID \rangle$. $\mathcal{A}$ can replace the public keys of any user including user with identity ID^*. During the training-phase on receiving a valid signature $\sigma = \langle m, V, V_0, V_1, V_2, T_0, ..., T_n \rangle$ on a message m with target identity ID^* from the **Sign** oracle, $\mathcal{A}$ can generate a valid signature σ^* on a message m^* for the target identity ID^*, such that σ^* is not the output of previous queries to **Sign** oracle. This can be shown by the following computation done by $\mathcal{A}$

- Dividing V by h. $\frac{1}{h}V = (\frac{r}{h} + 1)D_0$ where $h = H_1(m)$.
- Computes $h^* = H_1(m^*)$.
- $V_0^* = V_0$ and $V_1^* = V_1$.
- $V_2^* = \frac{h^*}{h}V_2$.
- The remaining parameters $T_0, ..., T_n$, V_0 and V_1 are same as that of original signature.
- $V^* = h^*\frac{V}{h}$.

Now $\sigma^* = \sigma^* (m^*, V^*, V_0, V_1, V_2^*, T_0, ..., T_n)$ is a valid signature on the message by the user with identity ID^*. $\mathcal{C}$ can check the validity of the forged signature σ^* as follows.

Correctness: The *L.H.S* is

$$
\begin{aligned}
e(V^*, T_i) &= e((\frac{h^*}{h}r + h^*)D_0, k^*Q_i) \\
&= e(\frac{h^*}{h}rD_0, kQ_i)e(h^*D_0, kQ_i) \\
&= e(\frac{h^*}{h}rkD_0, Q_i)e(h^*kQ_0, D_i) \\
&= e(V_2^*, Q_i)e(\frac{1}{h^*}T_0^*, D_i) \\
&= R.H.S
\end{aligned}
$$

Now, it is clear that the forged signature σ^* passes the verification successfully.

5   New ADGS Scheme(N-ADGS)

In this section we present a new identity based ADGS scheme. Assume that a signer a_0 has to designate his signature to be verified by n users namely $\{a_1, ..., a_n\}$. All the n users may be from different groups and are selected by a_0. The signer a_0 forms the set $\mathbb{U} = \{a_1, ..., a_n\}$ to generate the signature. In our scheme designated members of the group cannot simulate the signers signature.

- **N-ADGS Initialize:**
  The PKG initializes the system by executing this algorithm. This algorithm takes the security parameter 1^k as input and produces two groups $\mathbb{G}_1$ and

$\mathbb{G}_2$ of prime order q, where $|q| = k$, a generator P of $\mathbb{G}_1$, a bilinear map $e :$
$\mathbb{G}_1 \times \mathbb{G}_1 \to \mathbb{G}_2$ and two cryptographic hash functions $H_1 : \{0,1\}^* \times \mathbb{G}_2 \times \mathbb{G}_1$
$\times \mathbb{G}_1 \times \mathbb{G}_1 \to \mathbb{Z}_q^*$ and $H_2 : \{0,1\}^* \to \mathbb{G}_1$. The master private key is $s \in_R \mathbb{Z}_q^*$
and the master public key is set to be $P_{pub} = sP$. Sets $\theta = e(P_{pub}, R)$ where
$R \in_R \mathbb{G}_1$. The public parameters are $\langle \mathbb{G}_1, \mathbb{G}_2, e, P, P_{pub}, P_{pub}, H_1, H_2, \theta, R \rangle$.

- **N-ADGS Key Generation/Extract:** This algorithm is executed by the
  PKG and on input of identity ID_i, PKG computes $Q_i = H_2(ID_i)$ and sets
  the private key as $D_i = sQ_i$. Now, D_i is sent to the user in a secure way.
- **N-ADGS Sign:** To sign a message m for a designated group of users $\mathbb{U} =$
  $(a_1, ..., a_n)$ with identities $(ID_1, ..., ID_n)$ the user with identity ID_0, private
  key D_0 and public key Q_0 performs the following steps:
  - Chooses r,k,t $\in_R \mathbb{Z}_q^*$ and computes $T_i = \langle T_{i1}, T_{i2} \rangle$ as $\langle t(Q_i + R), kQ_i \rangle$
    for$(i = 1$ to $n)$.
  - Computes $U_1 = rQ_0$, $U_2 = rkP$ and $U_3 = tP$.
  - Computes $\omega = e(D_0, U_3)$ and Computes $W = \theta^t \omega$.
  - Computes $h = H_1(m, \omega, U_1, U_2, U_3)$ and $V = rP_{pub} + hD_0$.
    Now $\sigma = (m, V, W, U_1, U_2, U_3, T_1, ..., T_n, \mathbb{U})$ is a valid signature on mes-
    sage m by ID_0, with the user group $\mathbb{U}$ as designated verifiers.
- **N-ADGS Verify:** Verification is a two step process. First step is to verify
  whether the verifier belongs to the group $\mathbb{U}$ and second step is to verify the
  validity of the signature.
  - *Judge Verifier:* Using the value $T_{i2} = kQ_i$, the verifier checks whether
    $e(T_{i2}, Q_0) \stackrel{?}{=} e(Q_i, U_1)$. If the verification holds then user with public key
    Q_i will do the next step in verification.
  - *Verify Signature:* Each designated verifier $a_i \in \mathbb{U}$ can verify the signature
    by performing the following steps.
    * Computes $\omega^{'} = We(D_i, U_3)e(P_{pub}, T_{i1})^{-1}$.
    * Computes $h^{'} = H_1(m, \omega^{'}, U_1, U_2, U_3)$.
    * Checks whether $e(V, T_{i2}) \stackrel{?}{=} e(h^{'}U_1, D_i)e(U_2, D_i)$.
    If the above check hold then the signature is valid. Otherwise the signa-
    ture is invalid.

5.1   Security Proof for N-ADGS

Unforgeability Proof

Theorem 1. *Our N-ADGS scheme is existentially unforgeable under chosen
message and identity attack if* **CDHP** *(Computational Diffie Hellman Problem)
is hard in* $\mathbb{G}_1$.

This proof appears in the full version of the paper [10].

Unlinkability Proof

Theorem 2. *Our N-ADGS scheme is unlinkable in the sense that members outside the group cannot verify the signature if* **DBDHP** *(Decisional Bilinear Diffie
Hellman Problem) is hard in* $(\mathbb{G}_1, \mathbb{G}_1, \hat{e})$.

This proof appears in the full version of the paper [10].

6   Conclusion

In this paper, we have presented attacks on two certificateless GIS schemes [5], [6], a certificateless BGOS scheme [7] and an identity based ADGS [8] scheme. We have proposed a new identity-based ADGS scheme. We leave as an open problem to construct efficient identity based ADGS with constant size signature independent of the number of designated verifiers. Our scheme is secure against existential forgery on adaptively chosen message and ID attack under the CDH assumption in the random oracle model and is unlinkable under the DBDH assumption.

References

1. Al-Riyami, S.S., Paterson, K.G.: Certificateless public key cryptography. In: Laih, C.-S. (ed.) ASIACRYPT 2003. LNCS, vol. 2894, pp. 452–473. Springer, Heidelberg (2003)
2. Boneh, D., Franklin, M.K.: Identity-based encryption from the weil pairing. SIAM J. Comput. 32(3), 586–615 (2003)
3. Hu, B.C., Wong, D.S., Zhang, Z., Deng, X.: Key replacement attack against a generic construction of certificateless signature. In: Batten, L.M., Safavi-Naini, R. (eds.) ACISP 2006. LNCS, vol. 4058, pp. 235–246. Springer, Heidelberg (2006)
4. Jakobsson, M., Sako, K., Impagliazzo, R.: Designated verifier proofs and their applications. In: Maurer, U.M. (ed.) EUROCRYPT 1996. LNCS, vol. 1070, pp. 143–154. Springer, Heidelberg (1996)
5. Ma, C., Ao, F., He, D.: Certificateless group inside signature. In: Proceedings, April 2005, pp. 194–200 (2005)
6. Ma, C., Ao, J.: Certificateless group oriented signature secure against key replacement attack. Cryptology ePrint Archive, Report 2009/139 (2009), http://eprint.iacr.org/
7. Ma, C., He, D., Ao, J.: Broadcast group oriented signature. In: 2005 Fifth International Conference on Information, Communications and Signal Processing, pp. 454–458 (2005)
8. Ma, C., Li, J.: Adaptable designated group signature. In: Huang, D.-S., Li, K., Irwin, G.W. (eds.) ICIC 2006. LNCS, vol. 4113, pp. 1053–1061. Springer, Heidelberg (2006)
9. Shamir, A.: Identity-based cryptosystems and signature schemes. In: Blakely, G.R., Chaum, D. (eds.) CRYPTO 1984. LNCS, vol. 196, pp. 47–53. Springer, Heidelberg (1985)
10. Sree Vivek, S., Sharmila Deva Selvi, S., Gopinath, S., Pandu Rangan, C.: Breaking and building of group inside signature. Cryptology ePrint Archive, Report 2009/188 (2009), http://eprint.iacr.org/

Use of ID-Based Cryptography for the Efficient Verification of the Integrity and Authenticity of Web Resources

Thanassis Tiropanis[1] and Tassos Dimitriou[2]

[1] University of Southampton, UK
tt2@ecs.soton.ac.uk
[2] Athens Information Technology, Greece
tdim@ait.edu.gr

Abstract. As the amount of information resources on the Web keeps increasing so are the concerns for information integrity, confidentiality and authenticity. In Web 2.0 users are producers as well as consumers of content and metadata, which makes guaranteeing the authenticity and integrity of information critical. The scale of the Web requires that any proposals in this direction require minimal (if any) infrastructural or administrative changes. This paper proposes the use of ID-based cryptography (IBC) to address requirements for integrity and authenticity of Web resources using either the URL/URI of a resource or the DNS name part of if. This approach presents certain challenges, which are discussed along with the pros and cons of different designs and implementations.

Keywords: Identity Based Cryptography, Integrity, Authenticity, Web 2.0.

1 Introduction

The number of Internet and Web users has been increasing at very high rates over the last decade and, considering that Internet penetration in developing countries is still relatively low there is space for ongoing increase in the coming years. The amount of content that is exchanged is constantly growing and a number of content distribution and delivery infrastructures (peer-to-peer, content repositories) are proving particularly popular. Web 2.0 applications allow users to be not just content consumers but also producers. The importance of content in this emerging paradigm of Web-service deployment and use has already been identified; according to O'Reilly [10] "data is the next Intel inside".

Currently there is research in progress on a new content-centric communication paradigm that aspires to transform networking by focusing not on enabling the communication between network end-points but on identifying content to be obtained from networks using client-server, peer-to-peer or other types of exchanges. There are expectations that this effort will lead to more efficient networks in terms of content distribution and more efficient services ([9], [11]).

Y. Chen et al. (Eds.): SecureComm 2009, LNICST 19, pp. 340–349, 2009.

This vision of content-centric communication however, is based on the assumption that a network infrastructure will be able to (i) identify each Web resource uniquely and (ii) provide guarantees on the integrity and authenticity of Web resources since they can be obtained not exclusively from their source network end-point but over a peer-to-peer or other type of content distribution network. The requirements of integrity and authenticity are increasingly critical as the content currently produced by an already large user base is set to keep growing. To this end we propose the use of Identity-Based Cryptography (IBC) as an efficient and scalable way of guaranteeing the authenticity and integrity of Web resources using an IBC-based system for efficiently signing and verifying Web based content. Our proposal involves the use of the URL/URIs (or the DNS name part of URL/URIs) of resources as IBC identifiers of every resource to be disseminated over the Web, Peer-to-Peer or other content dissemination networks combined. We also propose the use of identity-based digital signatures.

The proposed approach does not require infrastructural changes and we believe it can therefore be seamlessly introduced, making use of the existing XML Signature Syntax and Processing Standard of the W3C [14].

The rest of the paper is organized as follows: Section 2 reviews the existing literature on Identify-Based Cryptography (IBC), on Web 2.0 applications and on content dissemination infrastructure. Section 3 describes in detail how IBC can provide an efficient and scalable way to address integrity and authenticity concerns in content-centric communication. Section 4 discusses different deployment scenarios over the existing Web infrastructure, while Section 5 provides a discussion on the proposed approach and identifies further work items and related research directions.

2   Background

2.1   Supporting Identity Based Cryptography (IBC) on the Internet

Identity-based cryptography (IBC) was first introduced by Shamir back in 1984 [13]. While the original scheme of Shamir supported only signature operations, recently, there has been an increased interest in the use of IBC which was due to the discovery of a secure Identity Based Encryption scheme based on pairings over elliptic curves by Boneh and Franklin [1].

In an identity-based cryptosystem, public keys can be derived from arbitrary strings while the corresponding private keys are generated and distributed by an associated Trusted Authority (TA). Thus an identity-based cryptosystem enjoys most of the benefits of public key cryptography without the need for certificates and the problems they present. This in turn leads to a more lightweight approach to deploying public key cryptography [12]. In the sequel we review some basic IBC systems that have been proposed to date.

Boneh et al. [2] proposed a new approach to certificate revocation centered around the concept of an on-line SEmi-trusted Mediator called the SEM. The use of the SEM in conjunction with a simple threshold variant of the RSA cryptosystem enables the quick revocation of all security capabilities of a user. The proposal of mediated RSA was then used in an identity based setting in [3] as a simple solution and alternative to the Weil pairing scheme of [1].

One of the advantages of PKIs is that they can be organized into hierarchies which reflect the internal structure of a large organization or group. Recent work, however, has demonstrated the ability to implement similar hierarchies in an IBC context ([6,7]). In [17], this is taken one step further and an attempt was made to integrate this approach into existing standards and software, so as to ease deployment.

Finally, Crampton et al. [4] discuss how various Identity-based cryptographic techniques can be used to provide web services security. In particular, the authors compare Identity-based with traditional, certificate based techniques and they show how the first type can be used to secure XML messages in a more lightweight way compared to the second one.

2.2   Digital Signatures for Web Resources

The need to provide digital signatures for Web resources has been identified by the W3C and the IETF which engaged in common standardization activity (www.w3.org/Signature) and compiled requirements for XML Signatures in terms of a data model, syntax, format and processing (RFC2807 [15]). The core standard to emerge from this activity is the "XML Signature Syntax and Processing" [14] standard, which provides for digital signatures as XML documents.

XMLDsig can be used to sign not only XML documents but also resources in other formats. To verify the authenticity and integrity of a resource, one needs to have or obtain the key to be used; in a PKI setting, this is the public key of the entity that signed the resource. Considering that a large number of Web resources are specified in XML and that the use of the XML-compatible version of HTML (xHTML) is widely used today, it seems that XMLDsig provides a number of ways to package signatures into a large number of Web resources without changes to existing Web infrastructure.

3   IBC for Web Resources

Our proposal is based on mediated RSA that can be used to guarantee the integrity and authenticity of Web resources. The main idea behind this scheme (and any other IBC solution) is to generate and use public keys based on publicly available information that can be used to identify users or resources. On the other hand, private keys are generated by the Trusted Authority (TA) who possesses a master secret key (Section 3.1). Then, in Section 3.2, we explain what modifications (and simplifications) will be made in order to apply it for verifying Web resources.

3.1   Mediated RSA (mRSA)

One of the mRSA advantages is its *transparency*: in signature mode, mRSA yields standard RSA signatures which are much easier to incorporate with existing protocols. Mediated RSA involves a special entity called the SEM (SEcurity Mediator) which is a *partially* trusted server. To sign or decrypt a message, user Alice (one of the characters featuring in most cryptography scenarios) must obtain a message-specific token from the SEM.

The main idea behind mRSA is the splitting of an RSA private key into two parts using threshold cryptography. One part is given to the user while the other is given to the SEM. When the user and the SEM cooperate, the system is functionally equivalent to standard RSA. The fact that the private key is not held entirely by any one party is transparent to the outside, i.e. to those who use the corresponding public key to verify the signature (for more details see [3]).

3.2   Creating an Identity Based Infrastructure for Resource Authenticity

Our proposal uses mRSA to address the problem of authenticity and integrity of Web resources. Consider a set of services offered by some organization and a set of recourses associated with each such service. Ideally, we would like any third party to be able to authenticate these resources without the use of public key infrastructures or complicated protocols.

The basic idea behind mRSA is the use of a single *common* RSA modulus N among all users of a system. In our case, however, the "users" are the services offered by the organization with resources tied to these services. These resources must be integrity protected and authenticated by anyone interacting with a particular service. Thus, these resources correspond to the "messages" that need to be signed and must bear the signature of the corresponding service.

Using the same modulus by multiple entities in a normal RSA setting is totally insecure since anyone, using its own knowledge of a single key-pair, can factor the modulus and compute the other entities' private keys. However, this does not apply in our setting since the private key is shared between the entity and the SEM. Thus an attacker must compromise *both* to undermine the security of the system.

In the following, we use the full name of a service as the unique identifier (public key) for that service. We use the notation $ID_{Service}$ to denote the identity that will be used to compute the public RSA exponent. During initialization, a trusted authority (TA) sets up the RSA modulus N for all the services of the organization. N is equal to the product of two large safe primes p and q. The public exponent $e_{Service}$ is the result of a hash function such as SHA1 on $ID_{Service}$, with the rightmost bit set to one so that with high probability $e_{Service}$ is relatively prime to $\phi(N)$. This process is shown below:

Generate Public Key for $ID_{Service}$

Let k be the security parameter (say $k = 2048$)

- Generate random $k/2$-bit primes r and s such that $p = 2r + 1$ and $q = 2s + 1$ are also primes.
- Set $N = pq$
- For a particular service identified by $ID_{Service}$
  1. Set $e_{Service} = \text{hash}(ID_{Service}) \parallel 1$
  2. Set $d = 1/e_{Service} \bmod \phi(N)$
  3. Set $d_{Service}$ equal to a random number in $Z_N - \{0\}$
  4. Set $d_{SEM} = d - d_{Service} \bmod \phi(N)$

Once the private key is generated for a particular service, it can be used to sign a resource R through collaboration with the SEM. In what follows, we assume the existence of an appropriate encoding scheme that can be used to break the multiplicative properties of RSA. Typically, one can use the Probabilistic Signature Scheme (PSS) for RSA that can be found in then Public-Key Cryptography family of Standards PKCS#1. RSA-PSS incorporates processing schemes designed to provide additional security for RSA signatures. This encoding scheme, although not shown in detail, should be used and is denoted by Hash-PSS in the following description:

Sign resource R

- Set h equal to Hash-PSS(R)
- Compute partial signatures PS_{SEM} and $PS_{Service}$ as follows:
  1. $PS_{SEM} = h^{d_{SEM}} \bmod N$
  2. $PS_{Service} = h^{d_{Service}} \bmod N$
- Set $S = PS_{SEM} \cdot PS_{Service} \bmod N$
- Return signature S

Finally, any interested party that wants to ensure the authenticity of the resource R, it can do so by first computing the public key $e_{Service}$ from available information and then verifying the signature S.

Verify Signature S on resource R

- Retrieve domain modulus N
- Set $e_{Service} = \text{hash}(ID_{Service}) \parallel 1$
- Compute $h = S^{e_{service}} \bmod N$
- Verify whether h is equal to Hash-PSS(R)

Security Issues

The security of this scheme depends on whether someone can break into the SEM and the server and retrieve the corresponding private keys. In general, this is a safe assumption to make since the SEM can reside in hardened server that is more resistant to break-ins than usual machines.

This also solves the problem of the common modulus since even if a server is compromised, no attacker can use the key $d_{Service}$ to sign resources without the collaboration of the SEM. Additionally, knowing $d_{Service}$ for a particular service does not leak any information about either the primes that constitute the modulus or the private keys of other services. This is because both keys d_{SEM} and $d_{Service}$ are random quantities. Using a simulation argument, one can show that any attack that takes advantage of one of the two keys could be turned into an attack to standard RSA [3]. Thus knowledge of one of these keys does enable the attacker to sign fake resources (details omitted due to space restrictions).

4   IBC over the Existing Web Infrastructure

4.1   IBC over the Existing Web Protocols

In our proposal, IBC is to be used to verify Web resources that can be identified by their original URL or URI. This is achieved by applying the IBC scheme of Section 3.2, using the URL/URI (or the DNS name portion of it) as part of the key and by using the XMLDsig standard format and processes. In this way, checking the authenticity and integrity of a resource can be more efficient in comparison to PKI-based schemes as the URL/URI of the obtained resource is well known and the *modulus N* for the domain of the resource can be known or promptly obtained from a secure server or DNSSEC [5].

Although the role of the URL is to provide the location of resources instead of identification, we assume that when a resource is not identified by a URI, its URL serves as its identifier. We make a distinction between an *administrative domain* and a *DNS name*. An administrative domain can manage one or more DNS names. The *DNS name portion of a URL/URI* is the *host* field of the *authority* component of a URL or URI [16].

The digital signature for a resource can indicate whether the signature was produced using the whole of the URL/URI as key or just the DNS name in it, depending on the policy of the domain from which the resource originates. Effectively, our proposal is for two different modes of IBC-based resource validation:

- **MODE 1:** The 'modulus N' of the domain is used in combination with the *DNS name* part of the resource URL/URI to sign it. This means that the same private key can be used to sign any document in a specific domain regardless of its URL/URI. This can be flexible in terms of private key and digital signature management. On the other hand, there is a higher risk in using a single private key to sign all domain resources.

– **MODE 2:** The 'modulus N' of the domain is used in combination with the whole URL/URI to sign it. This requires a different private key for each URL/URI in a domain. This makes the management of private keys and digital signatures in a domain more complex but is ideal in cases when URIs represent user identities, such as OpenIDs.

Our proposal requires no changes to existing protocols and infrastructure, only some extra functionality on the client (e.g. Web browser) side, which can be implemented as a client plug-in. The *KeyInfo* element of a XMLDsig signature can be used to indicate the mode of validation the client is expected to use (Mode 1 or Mode 2).

Figure 1 shows an example of an administrative domain *myserver.com*, which runs three different servers with three different DNS names (*www.myserver.com*, *betamyserver.com* and *other.myserver.com*). The TA of the administrative domain *myserver.com* will provide the *modulus N* for all three DNS names and the SEM of *myserver.com* will sign each resource as detailed in Section 3.2. A client (Web browser, P2P client or other) can obtain and cache the 'modulus N' for any of the three DNS domains that can appear in URL/URIs, using a trusted server or secure DNS. When a client obtains a signed resource from these domains it will be able to verify its authenticity and integrity.

In order to support authenticity and integrity checks for non URL-based resources, the use of IBC for URI-identified resources can be implemented. Unlike a URL, a URI does not necessarily correspond to a communication end-point from which a resource can be obtained – it can be just an identifier. However, both URLs and URIs are expected to be maintained by the domain they belong

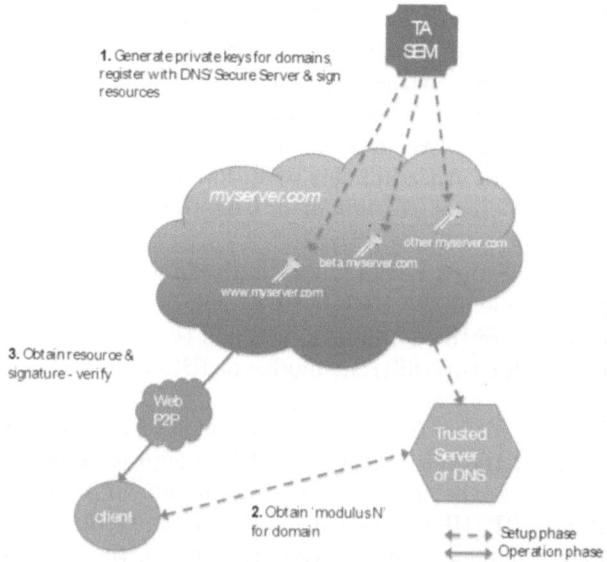

Fig. 1. Domain-wide IBC deployment for verifying Web resources

to. For this reason, resources that claim to be identified by a URI can be checked for their authenticity and integrity by a browser, program or user by using the 'modulus N' of the domain name portion of the URI. In this way, it would be possible to obtain any resource identifiable by a URI/URL via any type of content dissemination infrastructure and still be able to verify its authenticity and integrity. The 'modulus N' can be obtained in a number of ways as discussed next in Section 4.2.

The proposed two modes of resource validation can cater for different requirements of authenticity and integrity checks and support both domain-signed resources (Mode 1) or resources signed by individuals (Mode 2 for OpenID).

4.2 Scenarios for IBC Deployment on the Web

The following scenarios are envisaged for the deployment of IBC based authentication and integrity of Web resources that can be identified by a URL or URI. In these scenarios, the content identified by a URL or URI can be obtained over a number of different content delivery channels, not necessarily the Web. However, the Web infrastructure is used to obtain the 'modulus N' for the domains of each URL/URI.

Scenario 1: IBC for URL-Based or URI-Identified Resources ('Modulus N' Maintained Per DNS Domain)

In this scenario the 'modulus N' for the domain can be obtained by the client:

- From a Web server on the domain of the specific URL/URI. This requires the client to issue an HTTP GET request for a *standard relative URL* on the domain. For example, 'modulus N' for URL/URI of domain *www. myserver.com* could be obtained by issuing a GET request for the reserved relative URL '*modulusN.xml*' to the server of the domain, with absolute URL: *http://www.myserver.com/modulusN.xml*.

- If infrastructural changes for DNSSEC are adopted, 'modulus N' for a domain could be obtained by the DNSSEC [5] or by alternative directory services [8].

In all the above approaches, a client will obtain the 'modulus N' for the corresponding URL/URI domain and, when the URL-based resource is retrieved by the specific URL, will isolate the XMLDsig and proceed as detailed in Section 4.1. This approach is scalable and has the advantage that it can be easily deployed without necessarily making infrastructural changes. On the other hand, this scheme may have to rely on using a *reserved relative URL* on every domain (e.g. 'modulusN') and an agreed XML schema for 'modulus N' distribution (e.g. for file '*modulusN.xml*' in the example above).

Scenario 2: IBC for URL-Based or URI Identified Resources ('Modulus N' Obtained from Dedicated Secure Server)

This scenario applies when multiple DNS domains are shared within an organization or a virtual community. In this case, we assume that a dedicated secure server can be employed for the distribution of the 'modulus N' for URLs and URIs available for the participating domains. The client software (or browser plug-ins) can be configured to contact the designated secure server to obtain 'modulus N' when necessary. This approach has the benefit that it does not require a separate 'modulus N' for each DNS domain but it is not scalable and may require manual configuration by the user.

5   Conclusions and Further Work

In this work we described a new approach for authenticity of web resources. Our proposal has a number of advantages ranging from transparency, ease of use, and implicit authentication of resources to seamless introduction and support of context-centric networking and collaboration in virtual communities. Rather than using certificate based Public Key Cryptography (PKI), our proposal is based on the use of the more intuitive Identity Based Cryptography. Once resources are signed using a version of RSA called mediated RSA, anybody can verify the authenticity of these resources simply by using the name (URL/URI) of the resource as the verification key. In our proposal, we still need to use a type of domain certificate that includes the common RSA modulus N, but we should stress that this "certificate" is not like a normal public key certificate but rather a *long lived attribute* certificate for the entire domain that can be retrieved either by a dedicated server or by using DNSSEC if infrastructural changes are adopted (Section 4.2).

One other difference with traditional PKI systems is that the private key is generated by the trusted authority (TA). This enforcement, in general, may raise concerns related to key escrow and privacy surrounding the management of private keys. The first concern is not really an issue in our case since we are only dealing with signature (correspondingly client verification) of resources, so no encryption takes place. For the second concern which may lead to compromise of private keys and signature non-repudiation once a server has been compromised, one could use multiple servers and threshold cryptography. Furthermore, the use of the SEM ensures that an attacker must compromise *both* to undermine the security of the system.

Our approach opens some interesting directions for research. We plan to investigate performance issues when signing content of different types and sizes. This is an issue that needs to be addressed since most of the web traffic increase over the last few years has been attributed to the exchange of large volume multimedia content. We also plan to identify further requirements to support collaborative authoring of resources in virtual communities and investigate the use of OpenID to let authors sign portions of collaboratively produced documents, which, in turn, could be double signed by the domain of the community. The management of resources, private keys and signatures to support both modes can be a challenging task that we aim to explore further.

References

1. Boneh, D., Franklin, M.: Identity-based encryption from the Weil pairing. In: Kilian, J. (ed.) CRYPTO 2001. LNCS, vol. 2139, pp. 213–229. Springer, Heidelberg (2001)
2. Boneh, D., Ding, X., Tsudik, G., Wong, M., Wong, M.: Method for Fast Revocation of Public Key Certificates and Security Capabilities. In: 10th USENIX Security Symposium, pp. 297–308 (2001)
3. Boneh, D., Ding, X., Tsudik: Identity-Based Mediated RSA. In: Proceedings of 3rd International Workshop on Information and Security Applications, WISA 2002 (2002)
4. Crampton, J., Lim, H.W., Paterson, K.G.: What can identity-based cryptography offer to web services? In: Proceedings of the 2007 ACM Workshop on Secure Web Services, pp. 26–36. ACM, New York (2007)
5. Arends, R., Austein, R., Larson, M., Massey, D., Rose, S.: DNS Security Introduction and Requirements. IETF draft: draft-ietf-dnsext-dnssec-intro-13, October 10 (2004)
6. Gentry, C., Silverberg, A.: Hierarchical ID-based cryptography. In: Zheng, Y. (ed.) ASIACRYPT 2002. LNCS, vol. 2501, pp. 548–566. Springer, Heidelberg (2002)
7. Horwitz, J., Lynn, B.: Toward hierarchical identity-based encryption. In: Knudsen, L.R. (ed.) EUROCRYPT 2002. LNCS, vol. 2332, pp. 466–481. Springer, Heidelberg (2002)
8. Jones, J.P., Berger, D.F., Ravishankar, C.V.: Layering public key distribution over secure DNS using authenticated delegation. In: 21st Annual Computer Security Applications Conference (2005)
9. Metz, C., Bsales, J.: Five Ideas That Will Reinvent Modern Computing. PC Magazine (2007)
10. O'Reilly, T.: What Is Web 2.0. Design Patterns and Business Models for the Next Generation of Software. O'Reilly, Sebastopol (2005)
11. PARC (Palo Alto Research Center). Content-Centric Networking: PARC's Strategy for Pioneering a Self-Organizing Network That Meets Information Needs. Media Backgrounder (2006),
    http://www.parc.com/content/newsroom/CCN_backgrounder.pdf
12. Paterson, K.G., Price, G.: A comparison between traditional public key infrastructures and identity-based cryptography. Information Security Technical Report, vol. 8(3), pp. 57–72 (July 2003)
13. Shamir, A.: Identity-based cryptosystems and signature schemes. In: Blakely, G.R., Chaum, D. (eds.) CRYPTO 1984. LNCS, vol. 196, pp. 47–53. Springer, Heidelberg (1985)
14. Eastlake, D., Reagle, J., Solo, D., Hirsch, F., Roessler, T.: XMLDsig - XML Signature Syntax and Processing, 2nd edn. W3C Recommendation (2008),
    http://www.w3.org/TR/2008/REC-xmldsig-core-20080610/
15. RFC2807. XML Signature Requirements. IETF (July 2000),
    http://www.ietf.org/rfc/rfc2807.txt
16. RFC3986. Uniform Resource Identifier (URI): Generic Syntax. IETF (January 2005),
    http://www.ietf.org/rfc/rfc3986.txt
17. Smetters, D.K., Durfee, G.: Domain-Based Administration of Identity-Based Cryptosystems for Secure Email and IPsec. In: Proceedings of 12th USENIX Security Symposium, pp. 215–229 (2003)

Self-organized Anonymous Authentication in Mobile Ad Hoc Networks

Julien Freudiger, Maxim Raya, and Jean-Pierre Hubaux

LCA1, EPFL, Switzerland
firstname.lastname@epfl.ch

Abstract. Pervasive communications bring along new privacy challenges, fueled by the capability of mobile devices to communicate with,
and thus "sniff on", each other directly. We design a new mechanism
that aims at achieving location privacy in these forthcoming mobile networks, whereby mobile nodes collect the pseudonyms of the nodes they
encounter to generate their own privacy cloaks. Thus, privacy emerges
from the mobile network and users gain control over the disclosure of
their locations. We call this new paradigm *self-organized location privacy*. In this work, we focus on the problem of self-organized anonymous
authentication that is a necessary prerequisite for location privacy. We
investigate, using graph theory, the optimality of different cloak constructions and evaluate with simulations the achievable anonymity in various
network topologies. We show that peer-to-peer wireless communications
and mobility help in the establishment of self-organized anonymous authentication in mobile networks.

1 Introduction

The current model of wireless communication relies heavily on infrastructure:
Two mobile phones have to go through cellular base stations to exchange calls
and data for which users pay, even if they are only a few meters apart. But as
more mobile devices become equipped with ad hoc (peer-to-peer) communication technologies, such as WiFi and Bluetooth, the coexistence of both peer-to-
peer and infrastructure-based communications is inevitable. Moreover, the recent
surge in mobile social networks [1,2] reinforces the need for mobile devices, such
as phones, to be able to talk to each other without going through the infrastructure. These peer-to-peer communications enable *context-based* applications, such
as dating [3], gaming [4], as well as distributed location-based services [43]. But
these communications also make possible the continuous tracking of the location
of these devices. Thus, whereas the standard privacy threat model focuses on
protecting users with respect to the infrastructure (be it the cellular network
or the Internet), pervasive communications will expand it to the whole set of
mobile devices.

The promised ad hoc sharing of information might turn into a pervasive night-
mare if undesired communications cannot be filtered out: For example, if mobile
nodes cannot verify the source of information, they are susceptible to mobile

Y. Chen et al. (Eds.): SecureComm 2009, LNICST 19, pp. 350–372, 2009.

spam. To thwart rogue devices from polluting the network, nodes should authenticate each other: The existence of an authentication feature (and the implied procedure to obtain the appropriate credentials) makes it more difficult for attackers to join the network in the first place and thus increases the cost of misbehavior. Hence, by verifying the authenticity of their interlocutor before exchanging information, mobile nodes reduce the amount of undesired data. For example, users of context-based applications would obtain authentication credentials by subscribing to the service. They could subsequently verify that received messages were sent by other subscribers to the service. But if this is done without appropriate precautions, the authentication mechanism would then reveal the identity of the nodes, thus rendering the privacy problem particularly challenging.

The location privacy of mobile devices is guaranteed if and only if devices are anonymous and untraceable. Hence, our quest for location privacy in the upcoming generation of mobile computing becomes an attempt to devise mechanisms for *untraceable anonymous authentication*. More specifically, mobile devices must authenticate themselves directly to other devices without revealing privacy-sensitive information. In this paper, we assume that the authentication mechanism should not rely on the constant presence of a central authority because of the scalability and accessibility problems that this would cause. We also leave the privacy threat of authentication towards the infrastructure out of the scope of the paper. We show that the seeming disadvantage, privacy-wise, of peer-to-peer communications can actually be turned into an advantage, thus allowing each node to create its own privacy cloak without the need for a central privacy coordination service. We coin this new paradigm *self-organized location privacy*.

In this work, we focus on the analysis of anonymity as it is a prerequisite for untraceability: If nodes cannot be anonymous, they cannot be untraceable. The key enablers of our solution are the groups of users themselves and a cryptographic construction called *ring signatures* [34] that allows a node to authenticate itself to other nodes by using a ring of pseudonyms, instead of its pseudonym alone. This ring constitutes the *anonymity set* of the node and can be constructed out of the pseudonyms of the node's past and present encounters without any interactive protocols. Hence, our mechanism provides *self-organized anonymous authentication*. The advantage of this approach is that each user only owns a *single authenticated pseudonym*. But we show that rings alone are insufficient to protect user privacy: By analyzing the different pseudonyms used in rings, an eavesdropper can link - with a sufficiently high probability - some rings to users. As described in this paper, the problem gets worse if the network of nodes grows. Hence, it is crucial to construct rings using mechanisms that maximize user anonymity. We develop a graph-theoretic model to evaluate different *ring construction strategies* and derive the optimal (in terms of achieved anonymity) ring constructions. Leveraging on each node's local knowledge and history of encounters, we devise self-organizing methods to achieve, in practice, near-optimal anonymity. We show with simulations that mobility and peer-to-peer communications are beneficial for the emergence of self-organized location privacy.

The paper is organized as follows: In Section 2, we review the state of the art. In Section 3, we present the system and threat models assumed throughout the paper. After introducing our proposed solution in Section 4, we analyze in Section 5 the achievable anonymity using a graph-theoretic model and evaluate the solution in Section 6. Finally, in Section 7 we discuss the cost of our approach and present remaining challenges. In that section, we also provide preliminary results addressing the untraceability requirement before concluding in Section 8.

2   Related Work

There are several techniques available to achieve anonymous authentication.

A large body of work focuses on the use of multiple pseudonyms [15] and, in particular, in mobile scenarios [7,24]. Instead of using a single pseudonym, mobile devices are preloaded with a set of pseudonyms and change over time the pseudonym used for sending messages. To impede an adversary from linking old and new pseudonyms, the change of pseudonyms must be spatially and temporally coordinated among mobile nodes in regions called *mix zones* [8]. The analysis in [7,20,21] shows that the achieved location privacy depends on the node density and on the unpredictability of node movements in mix zones. The main drawbacks of mix zones is that they are inefficient when the node density in the mix zone is low and can be costly in terms of pseudonym management. A related technique uses frequently changing pseudonyms, silent periods, and power control to hide privacy-sensitive information [26]. As we will see, our approach alleviates the problem of low densities in mix zones by relying on the history of encounters of mobile nodes, instead of strictly using their current neighbors. In addition, we alleviate the problem of pseudonym management by allowing a single pseudonym per device.

Another solution relies on *group signatures* [16] that allow a group member to sign on behalf of a group without revealing the identity of the signer. Nowadays, highly efficient group signatures schemes exist with constant size signatures and efficient signing and verification even when nodes are revoked [10,13,31]. But group signatures require a group manager to add and revoke group members, thus making the flexibility of groups dependent on the availability and computational capacity of the group manager. In contrast, with ring signatures, nodes can change the members of their rings without central coordination.

Anonymous credential systems (e.g., Idemix [11]) allow mobile nodes to anonymously authenticate to third parties with the help of an online credential issuer. The online availability of a credential issuer is often not possible in wireless networks. To circumvent the issue, techniques based on unclonable identifiers, such as e-tokens [12], allow nodes to anonymously authenticate themselves a given number of times per period. However, such techniques lack flexibility, in particular in the case of a prolonged unavailability of the credential issuer.

To the best of our knowledge, we are the first to investigate the potential of ring signatures to achieve anonymity and untraceability in mobile networks. Until now, most of the work focused on proving properties of ring signatures [41] or on the anonymity and unlinkability of the signature generation process [28].

Recently, ring signatures were proposed in [29] as a building block for anonymous routing in MANET but without investigation of the ring creation process.

3 Preliminaries

3.1 System Model

We assume a mobile network with n mobile nodes and a single offline Certification Authority (CA) run by an independent trusted third party. We focus on scenarios where the mobile nodes are autonomous entities equipped with WiFi or Bluetooth-enabled devices that communicate with each other upon coming in range. In other words, we describe a pervasive communication system in which mobile nodes automatically exchange information upon meeting.

Prior to entering the network, each mobile node registers with the CA that preloads a single *public/private key* pair (K_i, K_i^{-1}) and a digital certificate in the nodes' device. The CA verifies the identity of each user upon registration. The public key K_i serves as the identifier of node i and is referred to as its *pseudonym* P_i. The private key K_i^{-1} permits mobile node i to digitally sign messages, while the digital certificate validates the authenticity of the signature.

We assume that mobile nodes automatically exchange information as soon as they are in communication range. To do so, mobile nodes advertise their presence by periodically broadcasting proximity beacons containing the node's authenticating information (i.e., the sender attaches its pseudonym to signed messages). When a node receives a beacon, it verifies the authenticity of the sender before reading the message.

3.2 Threat Model

We assume that a *passive* adversary $\mathcal{A}$ aims to track the location of mobile nodes. In practice, the adversary can be a rogue individual, a set of malicious mobile nodes or may even deploy its own infrastructure (e.g., by placing eavesdropping devices in the network). In the worst case, $\mathcal{A}$ obtains a complete coverage and tracks nodes throughout the entire network. We characterize this type of adversary as *global*.

$\mathcal{A}$ collects identifying information (i.e., pseudonyms) from the entire network and attempts to break the anonymity provided by ring signatures in order to track the location of mobile nodes. If the adversary is successful, it can implicitly obtain the true identity of the owner of a mobile node from the analysis of its mobility [27]. Hence, the *location privacy* of mobile nodes cannot be taken for granted.

Finally, we assume that the key-pair generation process cannot be altered or controlled by the adversary.

3.3 Problem Statement

The location of mobile nodes can be tracked based on the information leaked from authentication messages. To thwart this threat, we define the following design goals:

- **Anonymous authentication:** The nodes should be able to authenticate to each other without being identifiable. Anonymous authentication permits mobile nodes to verify the origin of received messages without revealing their identity (neither to the receiver, nor to an eavesdropper).
- **Self-Organization:** The anonymity of nodes should not depend on the constant presence of a central authority because of the scalability and accessibility problems this would cause (the CA distributes pseudonyms to nodes prior to their entrance in the network but is not always accessible). With self-organization, the cost of anonymity management is distributed among all the nodes.

It should be noted that we do not consider *accountability* as a design goal. Indeed, like many Internet applications, the peer-to-peer wireless scenarios we study do not require it.

4   Self-organized Anonymous Authentication

In this section, we describe the techniques that permit the emergence of self-organized anonymous authentication.

4.1   Overview

With standard asymmetric cryptography, nodes authenticate themselves to others by signing their messages with their private key and providing the public key for signature verification, thereby revealing their identity. Instead, self-organized anonymous authentication, explained in detail in the next section, allows a node to select a set of pseudonyms called a *ring* and then sign its messages with a *Ring Signature* (RS) [34]. A RS preserves the *cryptographic anonymity* of the signer because it cannot be distinguished among the members of the ring.[1] Besides, rings are setup-free: The knowledge of the pseudonyms of the other nodes is sufficient to create a ring without any interaction. Hence, unlike group signatures, RSs have no group managers and do not require any coordination among ring members. Finally, two signatures generated by the same signer with the same ring are *cryptographically unlinkable*. Of course, to be able to generate a RS, each node must always use its own pseudonym in its ring, thus guaranteeing the authentication requirement.

The pseudonyms used for constructing rings can be collected by downloading sets of rings from online databases, much like PGP keyrings, or, in the case of the mobile network considered here, by recording the pseudonyms of neighboring nodes in a *history* S_i. Each node constructs a ring of pseudonyms by selecting a subset of pseudonyms from its history of encounters. This allows nodes to have an anonymity set without any central coordination: Rings are *dynamically* and

---

[1] In this paper, anonymity and untraceability are evaluated with respect to the pseudonyms used in rings and not with respect to the signature generation process, thus the distinction "cryptographic".

independently created by mobile nodes. A node i can thus authenticate itself to other nodes at time t by sending a message m with a ring signature $RS_{i,t}(m)$ created with ring $R_{i,t}$. It is worth making a clear distinction between the notions of "node" i, "pseudonym" P_i and "ring" $R_{i,t}$. Mobile nodes are indexed by a counter i (that does not refer to any ordering of the nodes). A mobile node i is represented in the network by its pseudonym P_i. In order to avoid being tracked by its pseudonym, i actually uses a set of pseudonyms of other nodes together with its own pseudonym P_i, to create its ring $R_{i,t}$ at time t.

Authenticating the source of information is a crucial primitive in pervasive communication systems to limit the spread of undesired data. By signing a message with a ring of pseudonyms, a signer proves its membership to a club of nodes (e.g., a mobile social network). The verifier can then be sure that a message originates from a member of the club. Of course, all members of a ring should have the appropriate credentials - the pseudonyms have been certified by the CA - and belong to the club; otherwise their presence in the ring would invalidate its authenticity. For simplicity of presentation, we consider in the rest of the paper that there is a single club of members encompassing all the legitimate nodes of the network.

4.2   Anonymous Authentication with Ring Signatures

Ring signatures were formalized by Rivest, Shamir and Tauman in [34] as an anonymous signature scheme. A ring signature allows a member of an ad hoc collection of users, i.e., the *ring*, to prove to any verifier that a message was sent by a member of the ring. An authenticated message does not leak the identity of its signer. Every node i has a ring $R_{i,t}$ at time t that is composed of a finite subset of the collection $\mathcal{P}$ of all pseudonyms in the network: $R_{i,t} = \{..., P_i, P_j, ...\}$. Let $\mathcal{R}_t = \{R_{1,t}, R_{2,t}, ...\}$ be the set of rings in the network at time t. Based on the pseudonyms in their local histories, mobile nodes decide which pseudonyms to use in their rings. We call this the *ring construction strategy*.

Ring signatures can be constructed upon any type of public key cryptographic primitive [6]. What is common to these schemes is that ring signatures are based on combining functions:

$$\mathcal{C}_{\mathcal{H},v}(T_0, T_1, ..., T_{r_i-1}) = v \tag{1}$$

where $\mathcal{H}$ is a secure cryptographic hash function, v is a random glue value, r_i is the size of the ring (constant over time) and T_k, $k = 0, ..., r_i - 1$, are randomly generated values except for one that requires the knowledge of a secret key to solve (1).

For efficiency reasons, we consider the ring signature scheme presented in [41] in which the combining function $\mathcal{C}$ relies on bilinear pairings and the public key cryptosystem is identity-based (i.e., ID-based cryptography [36]). In ID-based cryptography, the knowledge of the identifier (i.e., pseudonym) of a node is sufficient to validate the authenticity of its signature. This reduces the communication overhead because it avoids the use of certificates accompanying signatures generated by traditional cryptosystems such as RSA and ECC. The Achilles' heel of ID-based cryptosystems has always been their slower speed compared to other cryptosystems. But the recent introduction of efficient algorithms for computing pairings starts showing its feasibility on mobile devices [40]. We will elaborate more on the corresponding costs in Section 7. In ID-based cryptography the CA must be replaced by a Private Key Generator (PKG). A common critique of ID-based cryptography is that the PKG must be trusted to generate/protect private keys, and can forge signatures on behalf of the nodes (i.e., the key escrow problem). But for the applications considered here, we assume that the PKG (i.e., the CA) is trusted.

Let $\mathcal{G}$ be a Gap Diffie Hellman (GDH) group of prime order q. When a mobile node i wants to send a message m at time t, it first constructs a ring $R_{i,t}$ by selecting r_i pseudonyms (including its own pseudonym) out of its history. The ring signature is an $r_i + 1$ tuple of random values $T_k \in \mathcal{G}$ for $k = 0, 1, ..., r_i - 1$ and of $c_0 \in \mathcal{G}$:

$$(c_0, T_0, T_1, ..., T_{r_i-1}) \tag{2}$$

where c_0 is an initialization value for the ring creation; it contains the hash of the message m. T_k are randomly generated values except for one (only known to user i) that solves (1) with $v = 0$ and requires the knowledge of the secret key K_i^{-1}. We denote $RS_{i,t}(m) = (c_0, T_0, T_1, ..., T_{r_i-1})$ the ring signature on a message m sent by node i. To avoid replay attacks, the message m also contains a timestamp. The entire packet sent over the air looks as follows:

$$m, R_{i,t}, RS_{i,t}(m) \tag{3}$$

4.3 Anonymous Communications

Upon receiving a message, a node validates its signature before reading it. The receiver can reply to the message to initiate a communication session. To do so, two nodes establish a security association through an authenticated key exchange, e.g., ring signatures can be used in conjunction with the Diffie-Hellman protocol [29].

However, in order to allow for bidirectional communications, mobile nodes must be identifiable in the short term. Much to the detriment of privacy, mobile nodes already make use of long term identifiers, such as MAC (Medium Access Control) addresses, to communicate on the data link. For example, in IEEE 802.11, the MAC addresses are 48-bit values included in frames to identify the source or destination of a frame. Hence, whereas rings can provide an appropriate layer of anonymity at the application layer, the MAC addresses have to be anonymized to serve uniquely for short term communications. One approach consists in changing the MAC address [24] every time the ring changes, to preserve the anonymity created by the ring while still being able to identify nodes in the short term. The MAC address can be generated randomly, taking into account that collisions must be avoided. In [23], the authors suggest another approach based on an identifier-free link layer protocol. Basically, their solution increases the difficulty of profiling users from the link layer by obscuring long term explicit identifiers.

Finally, it must be noted that, at the physical layer, the wireless transceiver has a wireless *fingerprint* that can identify mobile devices in the long term [33]. However, this requires a costly installation for the adversary and stringent conditions on the wireless medium. A more generic approach consists in the analysis of the signal power of mobile devices to track their locations. It is still an open problem to determine how much identifying information a sophisticated adversary can extract from the physical layer.

5   Anonymity Analysis

In this section, we evaluate the anonymity provided by rings, considering a passive adversary. We show how to optimally construct rings to maximize the achievable anonymity.

5.1   Attack Description

A global and passive adversary observes the rings used by the nodes to authenticate each other over time (Fig. 1). Based on this information, it attempts to de-anonymize rings signatures.

Given a ring alone, an adversary is unable to determine the identity of the ring owner because of the cryptographic anonymity of ring signatures. However, if an adversary obtains all the rings used at time t in the network, it can infer the most probable owner of each ring by analyzing the ring members. For example, node i constructs a ring $R_{i,t}$ of size r_i. It uses its pseudonym P_i and selects $r_i - 1$ pseudonyms out of its history. If no other ring in the network uses pseudonym P_i, the adversary can conclude that ring R_i corresponds to pseudonym P_i (e.g., node u_4 in Fig. 2 (a)). A methodic analysis of ring members can thus reverse the *anonymity* provided by rings. Repeating this attack for each t, the adversary can track the locations of mobile nodes. In this section, we focus on the analysis of anonymity, which, as explained above, is a prerequisite to untraceability. The adversary will thus analyze snapshots of rings (columns in Fig. 1). Without loss

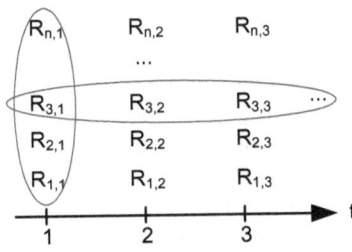

Fig. 1. Rings over time. An adversary will observe sets of rings changing over time and try to track the locations of mobile nodes. $R_{i,t}$ is the ring of node i at time t.

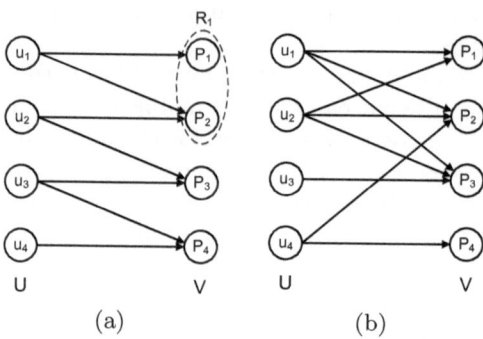

Fig. 2. Two examples of bipartite graphs G. (a) Rings are $R_1 = \{P_1, P_2\}$, $R_2 = \{P_2, P_3\}$, $R_3 = \{P_3, P_4\}$, and $R_4 = \{P_4\}$. (b) Rings are $R_1 = \{P_1, P_2, P_3\}$, $R_2 = \{P_1, P_2, P_3\}$, $R_3 = \{P_3\}$, and $R_4 = \{P_2, P_4\}$.

of generality, we write in the following $R_{i,t} = R_i$ and $RS_{i,t}(m) = RS_i(m)$. The adversary can also try to defeat the untraceability of rings by linking resembling rings over time to the same ring owner (rows of Fig. 1). Section 7 will give preliminary results on the untraceability analysis.

5.2   Graph-Theoretic Model

A set of rings can be modeled with a bipartite graph $G = (U \cup V, E)$, where $U = \{u_i\}_{i=1}^n$ is the set of nodes, $V = \{P_j\}_{j=1}^n$ is the set of pseudonyms and $E \subseteq U \times V$ is the set of edges. A graph is *bipartite* if its vertices can be partitioned into two sets such that no edge connects vertices in the same set. If a pseudonym P_j is in R_i, then we say that the node u_i using R_i is connected to P_j and we create an edge $(u_i, P_j) \in E$. We consider a balanced graph, that is, there are $|U| = |V| = n$ nodes in the system. There are $|E| = e$ edges directed from U to V. We denote d^{in} the in-degree of a node in V, i.e., the number of edges directed towards the node. Similarly, d^{out} denotes the out-degree of a node in U, i.e., the number of edges directed away from the node (the size of the ring). Two possible bipartite graphs are illustrated in Fig. 2. Graphs are *simple* if there are no multiple edges between two nodes.

After modeling rings with a graph G, $\mathcal{A}$ aims to discover which among the pseudonym $P_j \in R_i$ corresponds to the node u_i. To do so, $\mathcal{A}$ must find the most likely mapping of pseudonyms in V onto nodes in U. In graph-theoretic terms, $\mathcal{A}$ is looking for an *assignment* of nodes in V to nodes in U in the bipartite graph G. An assignment is a *matching* if no two edges share a common vertex. A *perfect matching* is a matching that covers all vertices of the graph. $\mathcal{A}$ must thus find the most probable perfect matching.

To do so, $\mathcal{A}$ assigns probabilities to all edges of the graph: $p_{j|i}$ is the probability that pseudonym P_j in V corresponds to node u_i in U. Hence, the graph G is weighted with probabilities computed by the adversary. Finally, $\mathcal{A}$ can find the most probable perfect matching by computing the maximum-weight perfect matching over the weighted bipartite graph G.

Measuring Anonymity. The individual anonymity of a node u_i (i.e., the uncertainty of an adversary about the identity of node u_i [35]) can be measured by:

$$H_i = -\sum_{j=1}^{r_i} p_{j|i} \log_2(p_{j|i}) \tag{4}$$

which is the entropy of the random variable $p_{j|i}$ and where $r_i = |R_i|$ is the size of the ring of node u_i.

A priori, the adversary will choose probability $p_{j|i}$ equal to $1/d_i^{out}$ as each outgoing edge is equally likely to be chosen. In Fig. 6(a), the entropy yields with this approach a non-zero anonymity for all but the last node. However, by doing so, the adversary focuses on the anonymity of individual nodes and overlooks some important properties of the system as a whole [19]. A clever adversary would eliminate many possible assignments by working backwards from vertices with a degree of one and the entropy would then yield a zero anonymity for all nodes.

$\mathcal{A}$ can thus first consider all assignments m_k of the elements of V onto U before computing $p_{j|i}$ a posteriori [7,38]. The probability of an assignment m_k is given by:

$$p(m_k) = \prod_{l \in m_k} w_l$$

where w_l is the weight of edge l in G. The weight is the a priori probability $p_{j|i}$ and we write $w_l = 1/d_i^{out}$ where node u_i is the origin of edge l. Because all the weights of the edges leaving a node are equal, all perfect matchings are equally likely and we have: $p(m_k) = p(m)$.

Hence, the probability of a perfect matching, i.e., the probability that an assignment is perfect knowing the set of all perfect matchings M, is:

$$p(m_k|M) = \frac{p(m_k, M)}{p(M)} = \frac{p(m_k)}{p(M)} = \frac{p(m_k)}{\sum_{k=1}^{|M|} p(m_k)} = \frac{1}{|M|}$$

where $m_k \in M$ for $k \in [0, |M|]$, and $p(M)$ is the sum of probabilities of all perfect matchings. The a posteriori probability $p_{j|i}$ is finally computed by considering all perfect matchings containing the pair (u_i, P_j).

$$p_{j|i} = \sum_{m_k \in M | (u_i, P_j) \in m_k} p(m_k|M) \tag{5}$$

In words, the number of perfect matchings going over an edge determines the weight of an edge. Hence, the anonymity of a node not only depends on its out-degree but also on the distribution of perfect matchings, i.e., the structure of the bipartite graph. Considering again the example in Fig. 2 (a), there is a single perfect matching in the graph, and consequently the anonymity of each node

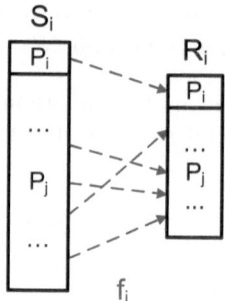

Fig. 3. Ring construction. To construct its ring R_i, node u_i uses its pseudonym P_i and selects, according to its strategy f_i, $r_i - 1$ pseudonyms out of its history S_i.

is null. To further illustrate the result, consider the example in Fig. 2 (b) and observe node 4. To compute $p_{4|4}$, we consider all the perfect matchings with the pair $(4, 4)$. In fact, every perfect matching in the graph contains that pair because $d_4^{in} = 1$, hence $p_{4|4} = 1$. The same analysis is true for node 3 as $d_3^{out} = 1$. Still, nodes 1 and 2 have a non-zero anonymity. In other words, both the in-degrees of nodes in V and out-degrees of nodes in U affect the distribution of perfect matchings and determine the probability $p_{j|i}$.

Complexity. The analysis presented above is difficult to carry out in practice because of its complexity: All perfect matchings must be found. Itai *et al.* introduce in [25] a polynomial time algorithm to find all perfect matchings in a bipartite graph. The algorithm starts from a perfect matching to iteratively produce them all in $\mathcal{O}(e \cdot (\sqrt{n} + |M|))$ time. The algorithm remains hard to use, as the number of nodes n can be extremely large and the number of matchings $|M|$ increases exponentially with the number of nodes. The adversary could thus focus on small sets of rings using, for example, the divide and conquer approach presented in [22].

5.3   Ring Construction Problem

Fig. 3 illustrates the ring construction process: Each node u_i obtains a ring R_i of size r_i by using its pseudonym P_i and choosing $r_i - 1$ pseudonyms from its history S_i. Rings must be carefully created to obtain high anonymity. The *ring construction strategy* of node u_i gives the criteria to include a pseudonym from u_i's history in its ring. We define it as a function $f_i : 2^{\mathcal{P}} \to 2^{\mathcal{P}}$. The selected pseudonyms must belong to u_i's history: $f_i(X) \subseteq X$ where $X \subseteq \mathcal{P}$ is a set of pseudonyms. The number of selected pseudonyms must not exceed the ring size: $|f_i(X)| \leq r_i$. In this paper, we consider that all nodes use the same ring construction strategy: $f_i = f$.

The *ring construction problem* consists in finding the ring construction that maximizes anonymity. To do so, we must obtain the optimal graph that maximizes the achievable anonymity for all nodes:

$$\max_{G}(H_i) \ \forall u_i \in U \tag{6}$$

subject to:

$$1 \le d_i^{out} \le d^{\max} \tag{7}$$

$$(u_i, P_j) \in E \Leftrightarrow \Gamma_j \in f(S_i) \tag{8}$$

Equation (7) confines the out-degrees d_i^{out} to a maximum $d^{\max}$. The graph construction is constrained by (8): The resulting graph G depends on the information collected by the nodes (their knowledge of pseudonyms). In other words, we seek to obtain a graph that maximizes the level of anonymity of every node constrained by the maximum out-degree and using a distributed construction function f (i.e., self-organization).

Optimal Graph G. As illustrated in Fig. 2, the distribution of the node in-/out-degrees affects the anonymity of each node. Let us introduce the following notation: $G_{\setminus(u_i, P_j)}$ corresponds to graph G without the edge (u_i, P_j).

The following Theorem identifies the graphs that provide the maximum anonymity. The proof is provided in the Appendix.

Theorem 1. *Anonymity is maximal if and only if every vertex in the bipartite graph G has the same degree $d^{\max}$ and for each i, the subgraphs $G_{\setminus(u_i, P_j)} \subseteq G$ for all $P_j \in R_i$ are isomorphic to each other. The anonymity of each node is then $\log_2(d^{\max})$.*

Theorem 1 characterizes the optimal graph G that maximizes the achievable anonymity. Basically, all subgraphs of G obtained by removing an edge starting at one node must be isomorphic. The isomorphism property captures the notion of *similarity* between subgraphs: If subgraphs are similar (i.e., have the same structure), it is more difficult to distinguish nodes in G. A large body of work has studied the existence of graph isomorphism and shown that the problem is NP: It belongs to its own complexity class, neither known to be solvable in polynomial time nor NP-complete [9,17]. In other words, theory says that it might be hard to determine whether two graphs are isomorphic. However, in practice, the graph isomorphism problem is easy to solve in polynomial time with heuristics [5]. It is thus possible in principle to determine whether subgraphs are isomorphic and, as a consequence, whether a graph G provides maximum anonymity.

In the following, we will compare the anonymity provided by different graph constructions and will see that regular graphs perform best. In fact, the local structure of the graph (i.e., the way each node is connected) determines whether subgraph isomorphisms can exist. In particular, the regularity of graphs is a necessary condition in our scenario for subgraphs to be isomorphic to each other (see Appendix).

6   Ring Construction Strategies

In the previous section, we examined the achievable anonymity with rings and derived necessary conditions (i.e., regular and isomorphic) to maximize it. In this section, we evaluate the performance of different graph constructions by means of simulations. The simulations are carried out in C++ using the LEDA library [30] to manipulate graphs. First, we assume that the nodes know the entire network and show the superiority of regular constructions over random graphs. The results are averaged over 20 runs with a running example of 10 nodes, which is sufficient to evaluate the effect of ring construction strategies with a reasonable simulation complexity. Then, we approximate the achievable anonymity on geometric graphs with 100 mobile nodes that only know a portion of the entire network.

6.1   Random Graphs

Let us assume that the nodes are aware of all the pseudonyms in the network (i.e., $S_i = \mathcal{P}, \forall u_i$). With a random graph construction f^{rand}, mobile nodes choose pseudonyms randomly: We consider a bipartite Erdos-Renyi random graph $G(n, p)$ where n is the number of nodes and each edge is included in the graph with probability p independently of others. With such graphs, the in-/out-degree distribution is binomial $Pr(d_i = k) = \binom{n}{k} p^k (1-p)^{(n-k)}$ with average $E[d_i] = np$ and variance $var[d_i] = np(1-p)$. Fig. 4 (a) shows the average distribution of the achieved entropy. We observe that the average anonymity increases with the edge density p, whereas the average variance decreases when p approaches 0 or 1. In other words, with a low or high density of edges, the achievable anonymity has a narrow distribution. As $p \to 1$, the graph becomes complete (i.e., all nodes are connected) and thus optimal in terms of anonymity.

We compare the performance of random and r-regular graphs in Fig. 4 (b) by computing their minimum and mean anonymity. We observe that regular graphs have a near-optimal behavior as they approach the maximum achievable anony-

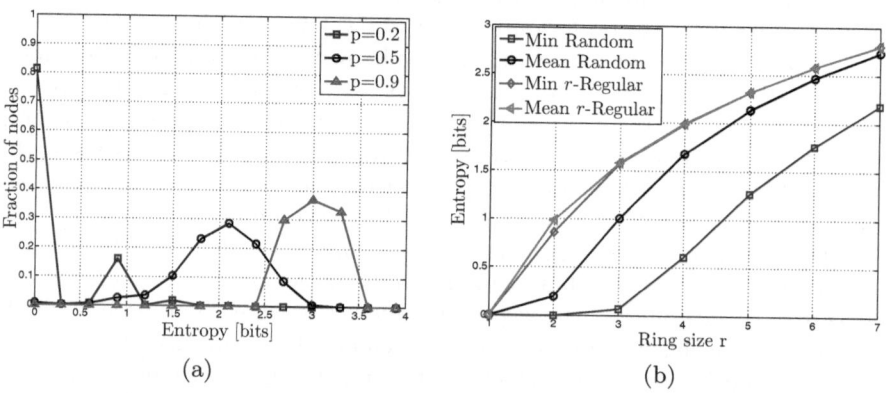

(a)   (b)

Fig. 4. Comparison of random and regular graphs. (a) Entropy distribution of random graphs with an increasing edge density p. The x-axis is divided into bins of size 0.3 and the y-axis represents the fraction of nodes in each bin. (b) Minimum and mean entropy levels of random and r-regular graph constructions.

mity $\log_2(r)$. Random graphs perform poorly, illustrating the importance of regular degree distributions: Nodes with a low in-/out-degree have lower anonymity, which is hardly compensated by nodes with a higher degree as the anonymity is logarithmic. The node degree variance induces a larger anonymity variance. Thus, to guarantee a minimal level of anonymity, the mean degree must be even larger.

As $n \to \infty$, the node degree distribution is approximated by a Poisson distribution with parameter $\lambda = np$. As the variance of the degree distribution equals the mean, it will be large and reduce the average anonymity. Bollobas in [9] notably investigates the asymptotic distribution of the degree sequence of graphs and proves that random graph constructions do not permit to guarantee predictable minimal and maximal degrees: the minimum and maximum degrees are *essentially determined* (by a function whose exact value is unknown). Bollobas further demonstrates that for some values of p (Theorem 3.5, [9]), there is a minimal degree $d^{min} \geq 2$. However, in this case, the maximum degree is not finite. In other words, as the graph grows larger, the degree sequence of random graphs is unpredictable and the performance gap with regular graphs increases.

6.2 K^{out} Graphs

We evaluate whether introducing a structure in the graph construction increases the achievable anonymity. We impose the same fixed out-degree $d^{out} = K$ to every node to obtain a K^{out} graph [9]. We consider various ring construction strategies with and without the help of a central entity.

Centralized Algorithm. The central entity is a network coordinator that knows the in-/out-degrees of each node and generates regular graphs from K^{out} graphs. Bollobas in [9] suggests a pairing model (i.e., a ring construction f^{reg}) to construct regular graphs with a centralized algorithm: Every vertex of the graph is connected to K nodes uniformly at random forming Kn pairs. If there are no multiple edges between two same nodes, the resulting graph is a random regular graph.

Distributed Algorithm. In the absence of a central entity, the nodes must decide individually with whom to connect. Each vertex $u_i \in U$ uses its pseudonym P_i and randomly selects $K - 1$ vertices from V. As all $\binom{n-1}{K-1}$ choices are equiprobable, the probability that a pseudonym P_j is chosen by another node is the ratio of assignments containing P_j over all possible assignments:

$$p = Pr(\text{"Node } u_i \text{ picks } P_j \text{ after } K - 1 \text{ tries"})$$

$$= \frac{\binom{n-2}{K-2}}{\binom{n-1}{K-1}} = \frac{K - 1}{n - 1} \tag{9}$$

The node in-degree distribution is then $Pr(d_i^{in} = k) = \binom{n}{k}p^k(1 - p)^{n-k}$. Thus, the average in-degree distribution is $E[d_i^{in}] = \frac{n}{n-1}(K - 1)$ and the variance is

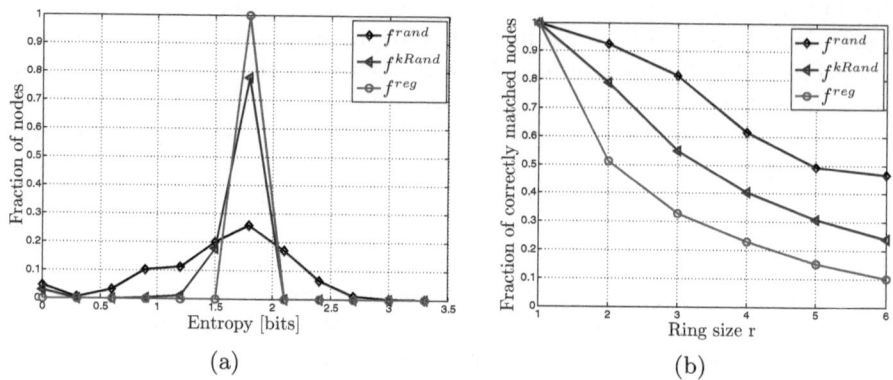

Fig. 5. Comparison of various ring constructions f. (a) Entropy distributions. The x-axis is divided into bins of size 0.3 and the y-axis represents the fraction of nodes in each bin. (b) Fraction of correctly matched pseudonyms to rings.

$var[d_i^{in}] = \frac{n}{n-1}(K-1)(1-\frac{K-1}{n-1})$. Consequently, with the distributed algorithm, the ring construction strategy f^{kRand} heavily depends on the degree K.

Fig. 5 (a) compares the entropy distribution for various ring constructions. We consider equivalent graphs constructions with $n = 10$ nodes, $p = 4/10$, $K = 4$ and $d = 4$ for d-regular graphs. We observe that f^{kRand} obtains a narrower distribution of entropies than f^{rand}, thus illustrating the importance of a regular out-degree. If random graphs obtain the maximum entropy among all constructions, they also have a smaller minimum entropy and lower average entropy. f^{reg} obtains very good results, close to the maximum achievable entropy $\log_2(d)$. This is due to the regularity of the in-degree distribution. With f^{reg} and $d = 4$, the majority of the nodes are indistinguishable as their entropy is $2^2 = 4$ equal to d. Like with random graphs, as $n \to \infty$, the in-degree distribution of K^{out} graphs is Poissonian: The mean and variance approach $K - 1$. In other words, the difference between regular and K^{out} graphs will increase as K becomes large.

In Fig. 5 (b), we observe that the proportion of successful matchings of pseudonyms onto nodes (i.e., the adversary success ratio) varies significantly among graphs. In the worst case, the adversary cannot infer information statistically and thus makes random attempts. The probability of success of the adversary is then equal to $1/r$. For regular graphs, $\mathcal{A}$'s success is limited and approaches its worst case strategy. In other words, $\mathcal{A}$ would do better by randomly matching rings to pseudonyms. With random constructions however, the adversary can infer significant information: Even with rings composed of 6 nodes, 5 out of 10 nodes in the example are correctly matched.

6.3 Geometric Graphs

As discussed above, the introduction of a structure in the ring construction dramatically increases the achievable anonymity. Still, the nodes were aware of all the pseudonyms in the network. In practice, mobile nodes will only have access to information gathered from the network, i.e., the rings of their encounters. In this section, we evaluate how the topology affects the achievable anonymity. In particular, as regular graph provide high anonymity, we study several ring construction strategies to obtain a regular graph G. To take the network topology into account, we consider a geometric graph G_g in which each vertex is associated with a physical device. Two vertices are connected (i.e., learn each other's rings) if and only if they are within distance $D(u, v) \leq \Gamma$ of each other, where Γ is a fixed radius (i.e., the unit disk graph model). The geometric graph G_g models the connection between nodes. We assume that the nodes are homogeneous (i.e., identical devices) and equipped with omnidirectional antennas. We consider both static and mobile scenarios.

Static Scenario. In static scenarios, nodes learn the pseudonyms in the rings of their direct neighbors. Indirectly, they also learn the pseudonyms of the neighbors of their neighbors as they are passed along. Given a history S_i and a ring size $r_i = r$, $\forall u_i$, the probability that node u_i chooses pseudonym P_j from its history in its ring is: $Pr(P_j \in R_i) = \min(1, r/|S_i|)$. For first-hop neighbors of u_i, the probability of learning P_j corresponds to the probability that u_i uses P_j, i.e., $Pr(P_j \in R_i)$. In other words, the ring size determines the propagation rate of pseudonyms in the network. For a x-hop neighbor u_l of u_i where x is larger than 1, the probability that pseudonym P_j is used by all rings on a path $\Delta_{i,l}$ from node u_i to node u_l is: $Pr(P_j \in R_k, \forall k \in \Delta_{i,l}) = \prod_{\forall k \in \Delta_{i,l}} \min(1, r/|S_k|)$. However, nodes belonging to disconnected sets of the graph G_g are isolated from each other and have zero probability of learning each other's pseudonyms. Hence, the propagation of pseudonyms is limited by the graph connectivity as well, reducing the potential size of anonymity sets. The topology of the network thus critically affects the achievable privacy. With the ring construction f^{static}, we consider that mobile nodes randomly choose pseudonyms from their local history to construct their rings.

Mobile Scenario. We examine how mobility can lessen the negative impact of topology: As nodes move in the network, they discover a larger portion of the set of pseudonyms $\mathcal{P}$. We consider the *restricted random waypoint model* introduced in [14]. In the *random waypoint model*, a mobile node moves on a continuous plane from its current position to a new position by randomly choosing its destination coordinates, its speed and the amount of time it will pause when it reaches the destination. After its pause, a node chooses a new destination and speed. This is repeated for each node until the end of the simulation time. In the restricted model, the choice of destination points is restricted with some probability ϕ to a set Ψ of fixed points on a plane. With probability ϕ a node randomly chooses a point from Ψ, and with probability $1 - \phi$, a node will choose a random point on the plane. This model is close to reality as users do not choose their destinations randomly, but instead meet at cafés, bus stops, etc.

In this mobile environment, we evaluate various ring construction strategies that aim at obtaining the most regular graph. These strategies capitalize on the frequency and freshness of the appearance of pseudonyms.

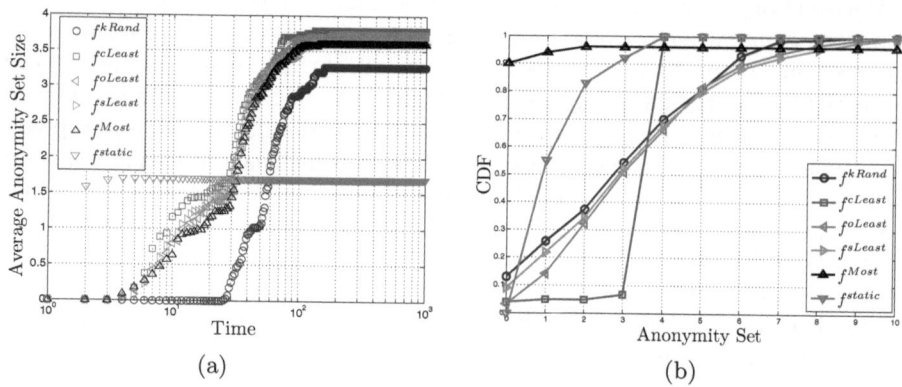

Fig. 6. Average anonymity set size for several ring construction strategies f in a mobile scenario with $\phi = 0.5$ and $r = 4$. (a) Average anonymity set size over time. (b) Cumulative distribution function of the anonymity set.

Least Popular Strategy. With the *least popular ring construction strategy*, each node maintains a counter for each pseudonym and selects the pseudonyms with the lowest counter value (i.e., the least popular). We consider three variations of the strategy: In f^{cLeast}, a central server informs the nodes of the in-degree of members of their histories. In f^{sLeast}, nodes choose in their histories the pseudonyms that were used the least often. In f^{oLeast}, nodes choose in their histories the pseudonyms that were used the least often in the rings of others.

Most Popular Strategy. We consider a *most popular ring construction strategy* f^{Most} in which the most popular nodes are chosen with the help of a central server.

Random Strategy. With the *random construction strategy* f^{kRand}, nodes choose their ring members randomly from their local history.

We ran 20 simulations on a $500 \times 500 \ m^2$ torus with $n = 100$ nodes, transmission range $= 25m$, pause $= 20s$, history $|S| = 10$, ring size $r = 4$, $|\Psi| = 5$ and $\phi = 0.5$. For simplicity and clarity, instead of computing the entropy, we compute the anonymity sets of mobile nodes, which corresponds to the in-degree distribution of graph G_g.

Fig. 6 (a) shows the evolution over time of the average anonymity set size of mobile nodes. We observe that the achieved anonymity set in mobile scenarios surpasses by far the static scenario but takes longer to converge. In general, we observe a percolation region ($[10, 10^2]$ seconds) where the anonymity set of the nodes increases quickly, and then a region of convergence ($[10^2, 10^3]$ seconds). f^{static} reaches a small anonymity set and is topped by all mobile strategies. Comparing mobility scenarios, we observe that f^{cLeast}, f^{oLeast} and f^{sLeast} improve the average size of the anonymity set with respect to the f^{kRand} (10% to 20% improvement). We notice that f^{oLeast} performs slightly better than f^{sLeast} as it takes better advantage of mobility (i.e., nodes have a better global knowledge of rings) and approaches the performance of the centralized algorithm f^{cLeast}. The f^{Most} approach seems to perform as well as the least popular approaches.

Fig. 6 (b) gives the cumulative distribution function (CDF) of the anonymity set showing the fraction of nodes per anonymity set size. The spread of the curve represents the variance across mobile nodes in the anonymity set sizes. We observe that f^{cLeast} performs well: It has a small variance as the majority of the nodes has an anonymity set size equal to 4 (i.e., the ring size). f^{oLeast} and f^{sLeast} have a smaller variance than f^{kRand}. Notably, with the f^{oLeast} approach, fewer nodes (20% less) have a small anonymity set. Finally, we observe that the f^{Most} approach actually performs worse than all other strategies: As a small number of nodes become extremely popular, the majority of nodes (90%) has a small anonymity set. Hence, although the average anonymity set size is large (Fig. 6 (a)), only a few nodes are actually anonymous, while others are easily identifiable. As social networks (usually modeled with scale-free graphs) tend to have this form, social networks based ring constructions would perform poorly.

In conclusion, the knowledge of the *popularity* of a pseudonym helps to achieve high anonymity (i.e., the least popular strategy). The nodes can thus independently aggregate information about their encounters and achieve anonymity in a self-organized way (without harming the anonymity of other nodes). Hence, peer-to-peer communications between mobile nodes enable privacy to emerge in ad hoc wireless networks.

7   Discussion

In this section, we present preliminary results on the untraceability of rings, explain their resilience to Sybil attacks, detail how revocation works and finally discuss the cost of ring signatures.

7.1   Untraceability

Untraceability of rings is also required in order to achieve self-organized location privacy. Similar to mix zones [7], mobile nodes can change their rings simultaneously upon meeting in the network. An external adversary will have to infer the most probable matching of old and new rings. Mobile nodes are untraceable if the adversary is unlikely to successfully match rings. Unlike the multiple pseudonym approach, in self-organized location privacy rings are correlated over time. Hence, by analyzing the similarity of ring members over time, an adversary could statistically estimate the matching of rings and track mobile nodes in the network. For example, if ring members remain constant, an adversary trivially tracks the whereabouts of mobile nodes. Ring members must thus vary: Except for the pseudonym P_i of the ring creator u_i, a ring $R_{i,t+1}$ can be entirely different from the previous ring $R_{i,t}$. Still, if all but one pseudonym are systematically updated, an adversary tracks mobile nodes by identifying persistent ring members.

In order to defeat an attack on untraceability by an external adversary $\mathcal{A}$, ring members must evolve with time depending on both past ring members and new encounters. Therefore, on top of the self-organization involved to achieve anonymity, mobile nodes must coordinate the *evolution* of their ring members to obtain untraceability. One possible way to coordinate the evolution of rings is to *cluster* ring members. The clustering coefficient of a vertex is used to quantify how close the vertex and its neighbors are to being a complete graph [39]. In our case, the clustering coefficient of a node measures the number of common ring members it shares with nearby nodes. The clustering coefficient of ring members results in an overlap of rings, which hardens the attack by $\mathcal{A}$. Mobile nodes can cluster their rings in a self-organized way by favoring pseudonyms recently observed: Newly acquired pseudonyms have a higher probability of being chosen in a ring. Preliminary results have demonstrated the success of this approach. We leave the formal investigation of this method for future work.

7.2   Sybil Attacks and Revocation

If a single node can present multiple identities, it can control a substantial fraction of the system and thereby undermine its security. These Sybil attacks [18] are not possible if there is a central entity to vouch for a one-to-one correspondence between entity and identity. In our model, the offline CA attributes a single pseudonym to every node after proper identification and rings are only used for authentication purposes. Hence, rings are unaffected by Sybil attacks. Actually, as privacy is generated by the nodes, RSs can be viewed as a *Sybil defense* that exploits the redundancy of mobile networks to generate a self-cloak.

Typical misbehavior remains possible in peer-to-peer wireless networks: For example, a mobile node can engage in denial of service attacks. However, the CA can exclude misbehaving nodes by revoking their keying material (as a signer must own a private key to generate a ring signature). Thus, keys can be blacklisted using certificate revocation lists (CRLs) like traditional revocation algorithms [42].

7.3   Cost

As ring sizes affect the anonymity level, users will tend to create the largest possible rings. But as ring signatures incur a communication and computation overhead, ring sizes will be bounded by the acceptable performance overhead.

Computation Overhead. RS *computational requirements* depend on the underlying trapdoor permutation, i.e., with ID-based ring signatures, one bilinear pairing computation is required for each member of the ring. In other words, for a node u_i, the *signature cost* C_{sign} is:

$$C_{sign} \approx r_i \cdot C_{BP} \tag{10}$$

where C_{BP} captures the cost of a Bilinear Pairing. The verification of a message has the same complexity. Using FPGA hardware accelerators for bilinear pairings [37], one bilinear pairing takes $61\mu s$. In total, for a ring of size $r_i = 10$, $C_{sign} = 610\mu s$. Without hardware accelerators, the efficiency of ring signatures in mobile phones depends on software optimizations: Currently, one bilinear pairing takes 478ms on a 225MHz ARM9 processor [40]. If this is not usable, the computation cost will eventually decrease as mobiles' hardware improves.

Transmission Overhead. To sign messages with ring signatures, only the first authenticated message between two nodes must contain the ring. Subsequent messages will thus have a smaller overhead. A RS is an $r_i + 1$ tuple: $(c_0, T_1, ..., T_{r_i-1})$. Each of those tuples is taken out of the group $\mathcal{G}$ of prime order q. Hence, the total size of the signature is $(r_i + 1) \cdot \mathcal{M}$ bits where $\mathcal{M} = \log_2(q)$. On top of the signature, a ring R_i is composed of r_i pseudonyms of $\mathcal{M}$ bits. Hence, the size of the signature grows linearly with the size of the ring. The *transmission cost* C_{trans} is:

$$C_{trans} \approx (r_i + 1) \cdot \mathcal{M} + r_i \cdot \mathcal{M} = (2r_i + 1)\mathcal{M} \tag{11}$$

For example, assume that a node u_i with pseudonym P_i creates a ring of size $r_i = 10$. For 128-bit security, NIST [32] recommends $\mathcal{M} = 283$ with elliptic curves defined over a *binary* underlying finite field of characteristic two ($\mathcal{F}_{2^\mathcal{M}}$). The communication overhead of the first message of each node is then $21 \cdot \mathcal{M}$ bits and $11 \cdot \mathcal{M}$ for subsequent messages.

8   Conclusion

We introduced the *self-organized location privacy* paradigm to solve the problem of location privacy in wireless mobile networks. With this approach, the network protects the location privacy of its nodes in a self-organized manner relying on Ring Signatures. Using graph theory, we theoretically measured the efficiency of the approach to provide anonymous authentication and derived its optimum. We examined numerically different ring construction strategies at the mobile nodes and showed that regular constructions achieve near-optimal anonymity. Then, we demonstrated that enabling nodes to communicate with each other increases their respective privacy levels by means of simulations. Despite their lack of knowledge of the entire network, mobile nodes achieve a high anonymity level by relying, for example, on the popularity of pseudonyms. In particular, choosing to connect to the *least popular pseudonyms* tops the achievable anonymity. Another particularly interesting result is that mobility helps in establishing self-organized anonymous authentication by improving the network awareness of every node without compromising their anonymity.

Future Work. We will investigate the effect of stronger adversary models on the achievable anonymity, such as an adversary that compromises members of the network. We will also extend the study of the effect of social networks on the construction of rings. In particular, social networks could provide information (e.g., the social graph) to improve the efficiency of ring constructions. Finally, we intend to complete our preliminary study on the untraceability of rings.

Acknowledgments

We would like to thank Levente Buttyan, Rafik Chaabouni, Mario Cagalj, Marcin Poturalski, and Serge Vaudenay for their insights and suggestions on earlier versions of this work, and the anonymous reviewers for their helpful feedback.

References

1. http://www.techcrunch.com/2007/09/11/the-holy-grail-for-mobile-socialnetworks
2. http://www.aka-aki.com/
3. http://en.wikipedia.org/wiki/Bluedating/
4. http://www.gamemobile.co.uk/bluetoothmobilegames/
5. http://cs.anu.edu.au/~bdm/nauty/
6. Abe, M., Ohkubo, M., Suzuki, K.: 1-out-of-n signatures from a variety of keys. In: Zheng, Y. (ed.) ASIACRYPT 2002. LNCS, vol. 2501, pp. 415–432. Springer, Heidelberg (2002)
7. Beresford, A.R.: Location privacy in ubiquitous computing. Ph.D. thesis, University of Cambridge (2005)
8. Beresford, A.R., Stajano, F.: Mix zones: User privacy in location-aware services. In: PerSec (2004)
9. Bollobas, B.: Random Graphs. Cambridge University Press, Cambridge (2004)
10. Boneh, D., Boyen, X., Shacham, H.: Short group signatures. In: Franklin, M. (ed.) CRYPTO 2004. LNCS, vol. 3152, pp. 41–55. Springer, Heidelberg (2004)
11. Camenisch, J., Van Herreweghen, E.: Design and implementation of the Idemix anonymous credential system. In: CCS (2002)
12. Camenisch, J., Hohenberger, S., Kohlweiss, M., Lysyanskaya, A., Meyerovich, M.: How to win the clone wars: efficient periodic n-times anonymous authentication. In: CCS (2006)
13. Camenisch, J.L., Lysyanskaya, A.: Dynamic accumulators and application to efficient revocation of anonymous credentials. In: Yung, M. (ed.) CRYPTO 2002. LNCS, vol. 2442, p. 61. Springer, Heidelberg (2002)
14. Capkun, S., Hubaux, J.-P., Buttyan, L.: Mobility helps peer-to-peer security. IEEE Transactions on Mobile Computing (2006)
15. Chaum, D.: Untraceable electronic mail, return addresses, and digital pseudonyms. Communications of the ACM 24(2) (1981)
16. Chaum, D., van Heyst, E.: Group signatures. In: Davies, D.W. (ed.) EUROCRYPT 1991. LNCS, vol. 547, pp. 257–265. Springer, Heidelberg (1991)
17. Corneil, D.G., Gotlieb, C.C.: An efficient algorithm for graph isomorphism. J. ACM 17(1), 51–64 (1970)
18. Douceur, J.R., Donath, J.S.: The sybil attack. In: Druschel, P., Kaashoek, M.F., Rowstron, A. (eds.) IPTPS 2002. LNCS, vol. 2429, p. 251. Springer, Heidelberg (2002)
19. Edman, M., Sivrikaya, F., Yener, B.: A combinatorial approach to measuring anonymity. Intelligence and Security Informatics (2007)
20. Freudiger, J., Raya, M., Felegyhazi, M., Papadimitratos, P., Hubaux, J.-P.: Mix zones for location privacy in vehicular networks. In: WiN-ITS (2007)
21. Freudiger, J., Shokri, R., Hubaux, J.-P.: On the optimal placement of mix zones. In: PETS (2009)

22. Gierlichs, B., Troncoso, C., Diaz, C., Preneel, B., Verbauwhede, I.: Revisiting a combinatorial approach toward measuring anonymity. In: WPES (2008)

23. Greenstein, B., McCoy, D., Pang, J., Kohno, T., Seshan, S., Wetherall, D.: Improving wireless privacy with an identifier-free link layer protocol. In: MobiSys (2008)

24. Gruteser, M., Grunwald, D.: Enhancing location privacy in wireless LAN through disposable interface identifiers: a quantitative analysis. Mob. Netw. Appl. (2005)

25. Itai, A., Rodeh, M., Tanimoto, S.: Some matching problems for bipartite graphs. Journal of the Association for Computing Machinery (1978)

26. Jiang, T., Wang, H.J., Hu, Y.-C.: Preserving location privacy in wireless LANs. In: MobiSys (2007)

27. Krumm, J.: Inference attacks on location tracks. In: LaMarca, A., Langheinrich, M., Truong, K.N. (eds.) Pervasive 2007. LNCS, vol. 4480, pp. 127–143. Springer, Heidelberg (2007)

28. Lin, H.-C., Yen, S.-M., Chen, H.-S.: Protection of mobile agent data collection by using ring signature. In: International Conference on Networking, Sensing and Control (2004)

29. Lin, X., Lu, R., Zhu, H., Ho, P., Shen, X., Cao, Z.: ASRPAKE: An anonymous secure routing protocol with authenticated key exchange for wireless ad hoc networks. In: ICC (2007)

30. Mehlhorn, K., Naher, St.: The LEDA Platform of Combinatorial and Geometric Computing. Cambridge University Press, Cambridge (1999)

31. Nakanishi, T., Fujii, H., Hira, Y., Funabiki, N.: Revocable group signature schemes with constant costs for signing and verifying. In: PKC (2009)

32. NIST. Recommended elliptic curves for government use. White Paper (1999)

33. Rasmussen, B., Capkun, S.: Implications of radio fingerprinting on the security of sensor networks. In: SecureComm (2007)

34. Rivest, R.L., Shamir, A., Tauman, Y.: How to leak a secret. In: Boyd, C. (ed.) ASIACRYPT 2001. LNCS, vol. 2248, p. 552. Springer, Heidelberg (2001)

35. Serjantov, A., Danezis, G.: Towards an information theoretic metric for anonymity. In: Dingledine, R., Syverson, P.F. (eds.) PET 2002. LNCS, vol. 2482, pp. 41–53. Springer, Heidelberg (2003)

36. Shamir, A.: Identity-based cryptosystems and signature schemes. In: Blakely, G.R., Chaum, D. (eds.) CRYPTO 1984. LNCS, vol. 196, pp. 47–53. Springer, Heidelberg (1985)

37. Shu, C., Kwon, S., Gaj, K.: FPGA accelerated Tate pairing based cryptosystem over binary fields. In: FPT (2006)

38. Tóth, G., Hornák, Z.: Measuring anonymity in a non-adaptive, real-time system. In: Martin, D., Serjantov, A. (eds.) PET 2004. LNCS, vol. 3424, pp. 226–241. Springer, Heidelberg (2005)

39. Watts, D.J., Strogatz, S.: Collective dynamics of small-world networks. Nature (1998)

40. Yoshitomi, M., Takagi, T., Kiyomoto, S., Tanaka, T.: Efficient implementation of the pairing on mobile phones using BREW. IEICE Transactions on Information and Systems (2008)

41. Zhang, F., Kim, K.: ID-based blind signature and ring signature from pairings. In: Zheng, Y. (ed.) ASIACRYPT 2002. LNCS, vol. 2501, pp. 533–547. Springer, Heidelberg (2002)

42. Zheng, P.: Tradeoffs in certificate revocation schemes. SIGCOMM Comput. Commun. Rev. (2003)

43. Zhong, G., Goldberg, I., Hengartner, U.: Louis, lester and pierre: Three protocols for location privacy. In: Borisov, N., Golle, P. (eds.) PET 2007. LNCS, vol. 4776, pp. 62–76. Springer, Heidelberg (2007)

A   Proof of Theorem 1

Proof. We first show that each node u_i must have an out-degree $d_i^{out} = d^{\max}$ and then obtain the condition for achieving maximum anonymity. Assume a bipartite graph G' where at least one node $u_i \in U$ has $d_i^{out} < d^{\max}$. We add new edges to G' such that $d_i^{out} = d^{\max} \; \forall u_i$, and obtain the graph G. Because no edges were removed, G will contain at least the same number of perfect matchings as G'. Adding new edges might actually increase the number of existing perfect matchings and consequently increase the anonymity of the nodes. In other words, to maximize their anonymity, each node must choose $d_i^{out} = d^{\max}$.

To maximize the entropy of each node, the random variable $p_{j|i}$ must have a uniform distribution. Given a node u_i, $p_{j|i}$ is uniform if and only if the number of perfect matchings over (u_i, P_j) is the same for all $P_j \in R_i$. A simple way to verify this consists in comparing whether the subgraphs obtained by removing any pair $G_{\backslash (u_i, P_j)} \subseteq G \; \forall P_j \in R_i$ yield the same number of perfect matchings. The number of perfect matchings without (u_i, P_j) will be the same for any pair (u_i, P_j) with $P_j \in R_i$, if and only if all subgraphs $G_{\backslash (u_i, P_j)}$ have the same number of perfect matchings. This will be true if all subgraphs are isomorphic to each other (i.e., belong to the same equivalence class). Consider two subgraphs $G_{\backslash (u_i, P_1)}$ and $G_{\backslash (u_i, P_2)}$. An isomorphism of graphs $G_{\backslash (u_i, P_1)}$ and $G_{\backslash (u_i, P_2)}$ is defined as $\mathcal{I} : \mathcal{V}(G_{\backslash (u_i, P_1)}) \rightleftarrows \mathcal{V}(G_{\backslash (u_i, P_2)})$ where $\mathcal{V}(G_{\backslash (u_i, P_1)})$ is the vertex set of graph $G_{\backslash (u_i, P_1)}$. $\mathcal{I}$ defines an assignment of the nodes of $G_{\backslash (u_i, P_1)}$ onto the nodes of $G_{\backslash (u_i, P_2)}$ such that $\forall (u_i, P_j) \in G_{\backslash (u_i, P_1)}$, there is $(\mathcal{I}(u_i), \mathcal{I}(P_j)) \in G_{\backslash (u_i, P_2)}$. A necessary (but not sufficient) condition for the graph isomorphism to exist in this case is that the graph is *d-regular*: Each vertex has the same degree d. Indeed, if the degrees of vertices of two subgraphs cannot be matched (e.g., a subgraph has a node of degree 5 while the other does not), then it is impossible for the subgraphs to be isomorphic. Hence, we know that the graph will be $d^{\max}$-regular and that the entropy of each node will be $\log_2(d^{\max})$.

An Active Global Attack Model for Sensor Source Location Privacy: Analysis and Countermeasures

Yi Yang, Sencun Zhu, Guohong Cao, and Thomas LaPorta

Department of Computer Science and Engineering,
Pennsylvania State University, University Park, PA 16802, USA
{yy5,szhu,gcao,tlp}@cse.psu.edu

Abstract. Source locations of events are sensitive contextual information that needs to be protected in sensor networks. Previous work focuses on either an active local attacker that traces back to a real source in a hop-by-hop fashion, or a passive global attacker that eavesdrops/analyzes all network traffic to discover real sources. An *active global* attack model, which is more realistic and powerful than current ones, has not been studied yet. In this paper, we not only formalize this strong attack model, but also propose countermeasures against it.

As case studies, we first apply such an attack model to two previous schemes, with results indicating that even these theoretically sound constructions are vulnerable. We then propose a lightweight dynamic source anonymity scheme that seamlessly switches from a statistically strong source anonymity scheme to a k-anonymity scheme on demand. Moreover, we enhance the traditional k-anonymity scheme with a spatial l-diversity capability by cautiously placing fake sources, to thwart attacker's on-site examinations. Simulation results demonstrate that the attacker's gain in our scheme is greatly reduced when compared to the k-anonymity scheme.

Keywords: Active Global Attacker, Source Location Privacy, Wireless Sensor Network, L-diversity, K-anonymity.

1 Introduction

Source location privacy is an important privacy issue in both civilian and military applications of sensor networks, because the exposure of source location information may result in catastrophic damages. In an asset monitoring network [1,2], when an endangered animal (e.g., panda) appears in the network, an event notification message will be delivered to the base station (BS). A nonconforming hunter may identify the source location and capture the animal by monitoring network traffic. In a battlefield scenario, the communication between soldiers and their surrounding sensors could reveal the positions of the soldiers, putting them in great danger as the opposing force may locate and accurately attack them.

Prior work on sensor source location privacy has explored two different adversarial models. In an *active local* attack model[1] [1,2,3], an attacker's hearing range is assumed

---

[1] Note that our differentiation of "active" and "passive" attackers is based on whether the attacker actively takes actions to visit suspicious spots or not. This is different from the traditional one based on whether the attacker actively manipulates packets or not.

Y. Chen et al. (Eds.): SecureComm 2009, LNICST 19, pp. 373–393, 2009.
© Institute for Computer Science, Social-Informatics and Telecommunications Engineering 2009

to be comparable to that of regular sensors. The attacker tries to trace back to the real source in a hop-by-hop fashion, given that the real event source emits packets continuously for a period of time. Countermeasures in this category [1,2] focus on confusing or misleading the attacker by introducing random or additional paths. Although such solutions have been shown to be effective, the local adversarial model is relatively weak. An attacker, with a hearing range more than three times of individual sensors, may locate the real source with a chance as high as 97% [1].

Recently, a *passive global* attack model has been studied [4,5,6,7], where the attacker is assumed to be capable of monitoring all the network traffic by either deploying simple sensors covering the network or employing powerful site surveillance devices with hearing range no less than the network radius. With the collected network-wide traffic, the attacker can conduct traffic analysis to identify the potentially real sources. Under such a strong attack model, the corresponding countermeasures focus on making all sensors [4,5,6,7] or k sensors [4] transmit (dummy) messages at the same or similar pattern to disguise the real source location. In general, such approaches are more robust to traffic analysis, at the cost of higher message overhead. This passive global attack model, however, is not realistic because it assumes that *an attacker merely monitors the traffic without taking any action.* Thus, although it is theoretically interesting, its real-world application is unclear. We believe in a real attack, an attacker will try to locate the real source by all means, as in the local attack model.

In this work, we focus on an *active global* attack model, in which the attacker is not only a global eavesdropper but also a realistic tracker that devises an optimal route to traverse suspicious spots one by one to find real events, under certain constraints, such as time, resource, and event duration. Compared with previous attack models, this is a more practical and powerful attack model. We formalize such a strong attack model, analyze it, and propose countermeasures against it.

In particular, we devise a dynamic programming algorithm and a greedy algorithm, based on which the attacker can derive the optimal traversal route to identify real events. To demonstrate both the procedure and the effectiveness of this attack model, we apply it to two existing schemes: a statistically strong source anonymity scheme [5] (referred to as SSSA scheme hereinafter) and a k-anonymity scheme [4]. We show that although the SSSA scheme provides strong source location privacy with statistical testing, under our attack model an attacker can gain some information about the locations of real sources when the message rate of a real event becomes high. The second scheme cannot provide actual k-anonymity because on average the attacker needs to check $k/2$ sources to find out the real one. Indeed, no schemes are perfectly secure under our attack model, because there is always some chance for the attacker to find out the real sources through his investigation, even if the attacker just randomly picks up places to check. For example, the constant-rate based schemes [4,7] are just the special case of the k-anonymity scheme where k equals to n, the total number of cells in the network.

Our research is not only to demonstrate the power of this attack, but also to propose viable solutions to defend against such an attack. Specifically, as no schemes can completely prevent the real sources from being identified, our goal is to devise efficient mechanisms which will *minimize the location information disclosure,* under certain resource constraints of the attacker. We notice that while the SSSA scheme has the advantage of greatly reducing the transmission latency for real event messages compared to the constant-rate schemes, it also introduces *continuous, network-wide* dummy messages. For high message rate applications, the transmission overhead could be prohibitively high. As a tradeoff between privacy and performance overhead, the k-anonymity scheme could largely reduce the message overhead, not only because only k data sources are involved, but also because dummy messages are triggered by real events and stop once the events complete.

To leverage the advantages from both the worlds, we propose a lightweight dynamic source anonymity scheme that seamlessly switches from a low-rate SSSA scheme to a k-anonymity scheme on demand. Moreover, we enhance the k-anonymity scheme with the property of *spatial l-diversity* to maximize the attacker's cost. Our simulation results show that with our defense the attacker's gain can be much reduced while his cost is increased compared to the k-anonymity scheme.

The main contributions of this work are summarized below.

- First, we formalize a new attack model, where an active global attacker designs an optimal route to check suspicious spots in the whole network;
- Second, we apply this attack model to the existing source anonymity schemes and demonstrate their limitations;
- Third, to thwart the attack, we propose a new dynamic scheme that seamlessly transits from a low-rate SSSA scheme to a spatial l-diversity enhanced k-anonymity scheme.

The rest of this paper is organized as follows. The active global adversary model is built up in Section 2. Case studies on existing source anonymity schemes are addressed in Section 3. Then, the dynamic source anonymity scheme is discussed in Section 4. Finally, after describing the related work in Section 5, we conclude this paper in Section 6.

2   An Active Global Adversary Model

In this section, we formalize the active global attack model and discuss details of the attacker's investigation. The attacker may employ a dynamic programming algorithm or a greedy algorithm to devise an optimal route for the investigation. We compare the results of these two algorithms through simulation and make clear the application scenario for each algorithm.

2.1   Modeling of Network

We consider a cell-based (or grid-based) network model. Deployment area of the network is partitioned into cells, which is the smallest unit of event detection: $\mathcal{N} = \{c_1, c_2, \cdots, c_n\}$, where n is the total number of cells. Every cell has a unique id $i (1 \leq i \leq n)$ and multiple sensors may reside in one cell. Each pair of sensors in neighboring cells could directly communicate with each other. A cell head, which is elected and rotated among all sensors in the cell, coordinates all the operations inside the cell. A base station (BS), connecting to the outside infrastructure such as the Internet, collects data from the network and reports them to a remote commander.

2.2 Modeling of Events

We assume that the total *information quantity* of a real event is $y_0(> 0)$ and a real event will last for time $t_0(> 0)$ once it happens. We model the information quantity of a real event at any time t as a function $f(t)$. In general, the choice of $f(t)$ is subject to the characteristics of the application. To be concrete, here we select a linear decrease function. If the attacker checks a real event after time t, the remaining information quantity could be modeled by:

$$f(t) = \begin{cases} y_0 - \frac{y_0}{t_0}t, & 0 \leq t \leq t_0 \\ 0, & t > t_0. \end{cases} \tag{1}$$

This means that if the attacker reaches the spot at the very first beginning of the real event then the attacker can get the maximum information y_0. The quantity of information that the attacker may obtain decreases after that. If the attacker reaches the real event spot after t_0, then he cannot obtain any information.

2.3 Investigation of Attacker

Although an attacker may have resources to check all the cells one by one, this is not an intelligent choice because real events often last only for a short time period. If the attacker spends too much time on fake sources and thus reaches the real source too late, real events may have already disappeared. Hence, the attacker faces the following two challenges:

- First, how many suspicious cells to check?
- Second, what is their visiting sequence to maximize the attacker's gain in information quantity?

Next, we discuss how these challenges can be addressed. After observing and collecting network traffic for some time, the attacker first determines a suspicion level for each cell through traffic analysis (a specific way to determine suspicion levels of cells is discussed later in Section 3.1). Then, the attacker decides a threshold. If a cell's suspicion level is higher than this threshold, this cell will be marked as a suspicious cell. The determination of this threshold value depends on many factors, e.g., the balance between the attacker's gain and the cost involved to achieve that gain. Note that even in the worst case for the attacker: every source have the same communication pattern, the attacker can still randomly select places for investigation and there is a certain chance for the attacker to find out the real sources.

Given the positions of all suspicious cells, the attacker optimizes the checking route. Clearly, the attacker's ultimate goal is to maximize the overall gain. Therefore, suspicious cells with higher suspicion levels should be checked at higher priorities. As shown in Figure 1, we assume that the attacker always starts from the center of the deployment area, traversing along a predetermined checking path under a specific velocity v. The main constraints of the attacker are time and resources. For each round of the real event investigation, there is a time limit τ, by which the attacker shall return to the starting point to start the next-round investigation based on newly collected data. τ could be the same as t_0 or other values determined by the attacker's resources.

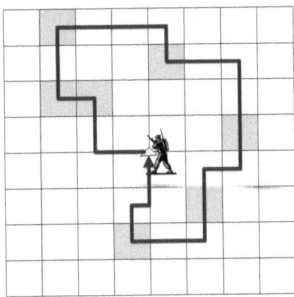

Fig. 1. The attacker traverses suspicious cells (highlighted as gray squares). We consider a network with $n(= \sqrt{n} \times \sqrt{n})$ cells that cover a rectangle deployment area. Each cell has a unique id, ranging from 1 to n.

Here we define *weighted gain*, which equals to the information gained from the suspicious cell if this cell is a real source *times* the probability of the cell being a real source. Assume there are totally s suspicious cells. The weighted quantity of information that the attacker could obtain from the jth ($1 \leq j \leq s$) suspicious cell is:

$$\psi(j) = f(t_j) \cdot \xi_j, \tag{2}$$

where t_j is the time to reach the jth suspicious cell and ξ_j is the suspicion level of the jth cell. Note that $t_j = \sum_{k=1}^{j} \tau_k$, in which $\tau_k = \frac{distance(c_{k-1}, c_k)}{v}$ is the time to travel from the $(k-1)$th cell to the kth cell. Therefore, the total information quantity that the attacker could obtain is

$$info_{total} = \sum_{j=1}^{s} \psi(j). \tag{3}$$

Given all the suspicious cells, intuitively, we can have a brute force method to design an optimal route for the attacker's investigation. In this brute force method, the attacker permutes all the possible traverse sequences and finds one with the maximum gain. For s suspicious cells, there are $s!$ permutations. For each permutation, the total number of summations is s. Hence, the time complexity of a brute force solution is $O(s * s!)$. Since the factorial time complexity in brute force is too high, in the following we discuss other two more efficient algorithms.

A Greedy Algorithm. The attacker prefers to checking the most suspicious cells while trying to maximize the total number of cells that he can check in a limited time τ. In a greedy algorithm (Algorithm 1), at his current location, each time when the attacker selects the next suspicious cell, he chooses the one with the maximum *ratio* of suspicious level to its distance from the current location. The greedy algorithm is efficient since it finishes in polynomial time. However, because every time a local optimum is chosen, there is no guarantee that a global optimum will be output from the greedy algorithm finally.

A Dynamic Programming Algorithm. We also propose a dynamic programming [8] based solution for path selection. This algorithm could output a global optimal result in a relatively efficient way.

Algorithm 1. Attacker's Greedy Traversal Algorithm

Input: a set of suspicious cells: $S = \{1, 2, \cdots, s\}$; each cell i's corresponding suspicion level $\xi_i (1 \leq i \leq s)$; starting point $\mathcal{SP}$;

Output: whether there are real events in suspect cells;

Procedure:

1: current point is $\mathcal{SP}$;
2: time = 0;
3: **repeat**
4:     select a cell c from S with maximum ratio of ξ_c/distance(current point, c);
5:     go to cell c to check;
6:     time += distance(current point, c)/v;
7:     output whether there is real event in cell c;
8:     current point is cell c;
9:     $S = S - \{c\}$;
10: **until** $S = \emptyset$ or time$\approx \tau$

The basic idea is as follows. Let S be a subset of $\{1, 2, \cdots, s\}$ and t_S is $\sum_{j \in S} \tau_j$. We denote $C(S)$ as the maximum gain that could be obtained by traveling all cells in S until time t_S. For $l \in S$, let $S - l$ denote a set obtained by removing element l from set S. Then, the following recurrence could be derived. When the size of S is 1, for any $l \in S$, we have

$$C(\{l\}) = \psi(l); \tag{4}$$

when the size of S is larger than 1,

$$C(S) = max_{l \in S}[C(S - l) + \psi_S(l)], \tag{5}$$

where the subscript S in $\psi_S(l)$ denotes the weighted information quantity obtained from cell l influenced by traversed cells in subset S prior to cell l in the sequence. That is, in an optimal traversal sequence for the cells with indices in S, a certain cell with index l must be the last one to visit and the remaining cells must be traversed in an optimal order in the time interval $[0, t_{S-l}]$. Then the overall maximum gain by such an ordering will be $C(S - l) + \psi_S(l)$. Taking the maximum over all the choices of l, we can derive the above equation (5). Due to page limit, we do not show the details of this algorithm here. We analyze the time complexity of this algorithm, which is $O(s2^s)$. Although it is still exponential, it is better than the factorial time complexity of the brute force algorithm.

Simulation Results. We use simulations to compare the results of greedy algorithm (Algorithm 1) and dynamic programming algorithm. In the simulation setting[2], the total number of cells $num_{cell} = 100(10 \times 10)$. The number of suspicious cells that are checked is $s = 9$ and $t_0 = 50$, $y_0 = 50$.

The results from our dynamic programming algorithm match those from the brute force method very well, which means dynamic programming algorithm can output the optimal traversal sequence accurately. When s is relatively small, e.g., less than 10, this

---

[2] This is a default setting in the following simulations.

algorithm could output results within ten seconds in a 2.0GHz processor PC. On the other hand, we find the greedy algorithm can generate sequences close to those from the brute force method. This algorithm can generate data in one second even when s is relatively large (e.g., in tens).

In conclusion, the dynamic programming algorithm is more accurate but slower, whereas the greedy algorithm quickly generates solutions close to the optimal ones. Which algorithm the attacker should choose depends on the relative criticality of accuracy and time consumption to the attacker. In the following sections, to achieve accuracy, we use the dynamic programming algorithm to find an optimal route for attacker's investigation.

3   Case Studies

To examine the impact of the proposed attack model, we apply it to two existing schemes, an SSSA scheme [5] and a k-anonymity scheme [4]), as case studies. Both schemes exhibit limitations under our attack model.

3.1   The SSSA Scheme

We first apply the active global attack model to the SSSA scheme [5]. We briefly introduce this scheme, followed by simulations to illustrate the attacker's investigation process as well as results.

Scheme Overview. To hide real messages that report the occurrence of real events, a straightforward solution is to employ network-wide constant-rate traffic. Since every cell has the same transmission pattern, an attacker would not be able to distinguish the real sources from the fake ones. To reduce the network-wide message overhead, message transmission rate should be as low as possible. In this case, however, the transmission of real messages will need to be delayed more until the next transmission point. Therefore, there is an intrinsic difficulty to determine the message transmission rate because of the necessary tradeoff among privacy level, message overhead and real message latency.

In [5], a statistically strong source anonymity scheme (SSSA) is proposed to trade privacy level for reduced real message latency. In this scheme, real messages at the sources are transmitted as early as possible while the overall transmission pattern of a cell remains the same, in the sense that some existing well-known statistical testing methods [9,10,11] would not be able to detect the changes. As an instance, if the normal inter-message time intervals follow a predetermined exponential distribution, then after the changes (perturbation) made for real messages, the overall distribution looks the same under statistical tests. As such, the real message transmission latency is reduced and meanwhile a statistically strong source anonymity property for sensor networks is achieved.

Attacker's Detection. We discuss the attacker's operation in two steps: Step I identifying the suspicious cells and Step II investigating suspicious cells.

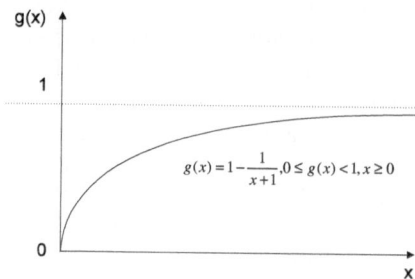

Fig. 2. Suspicion function to evaluate a cell's suspicious level

Step I: Identifying Suspicious Cells Intuitively, when real messages are relatively rare, it is unlikely for an attacker to notice the perturbation of the distribution due to the random nature of the variables forming the distribution. However, when a real event lasts for some time $t_0 (t_0 > 0)$ or when an event report has to be divided into multiple packets (in Mica motes, the typical packet size is only 36 bytes), a real source will need to transmit multiple real messages continuously. Since the SSSA scheme tries to reduce the waiting times of multiple real messages, our intuition is that the the actual distribution will become farther and farther away from the ideal one. In [5], Anderson-Darling test [9] and Kolmogorov-Smirnov Test [10] are employed for goodness of fit test and Sequential Probability Ratio Test (SPRT) [11] is proposed for mean test. Although under these test models the SSSA scheme has been shown to be statistically strong, an attacker may apply some other testing methods. To identify suspicious cells, the attacker may check continuous small message time intervals from each cell and quantify its suspicion level.

Small message time intervals are defined as those smaller than the mean. If the number of continuous small message time intervals is larger than a threshold, say three, then it is called a *cluster* and the number of continuous small time intervals is called *cluster size* n_s. Denote the number of clusters as n_c. An attacker may infer the suspicious level of a cell based on $x = \sum n_s * n_c$. More formally, to normalize the suspicion level of a cell into the $[0, 1)$ range, the attacker may construct a suspicion function $g(x)$ (Figure 2) with x as input parameter:

$$g(x) = 1 - \frac{1}{x+1}, 0 \le g(x) < 1, x \ge 0. \tag{6}$$

For example, if message time intervals from a cell has two clusters of size five, then the suspicion level of this cell is quantified as 90.9%; suppose message time intervals from another cell has three clusters of size six, then the suspicion level of this cell will be quantified to 94.7%.

We use simulations to check the distribution of cells' suspicious levels under high and low real message rates, with results shown in Figure 3 and Figure 4, respectively. In the simulation, the dummy message rate is 0.05. The probability for a cell to become the real source is 0.1. The attacker keeps the most recent 50 message time intervals from each cell. When real message rate is high (i.e., $rate_{real} = 0.2$), in most cases the suspicion levels of cells range from 94% to 97%, because the suspicion levels of

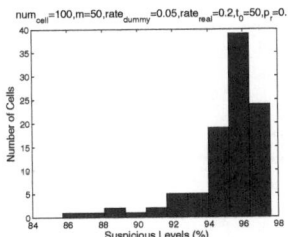

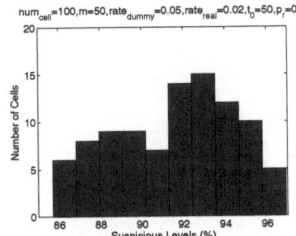

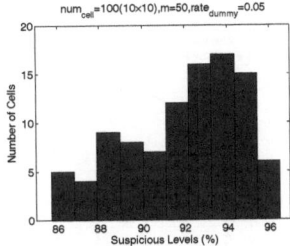

Fig. 3. Distribution of cells' suspicion levels when real message rate is high

Fig. 4. Distribution of cells' suspicion levels when real message rate is low

Fig. 5. Distribution of cells' suspicion levels when there are no real messages

real sources are normally high. On the other hand, when real message rate is low (i.e., $rate_{real} = 0.02$), The suspicion levels of cells are more uniformly distributed over the whole range. The attacker is able to determine a threshold accordingly. Considering the cost in checking suspicious cells, the attacker may want to control the number of suspicious cells to be checked as a relatively low value, e.g., about 10 out of 100 cells. In this case, 96% may be a good choice for the threshold.

In addition, we compare the distribution of suspicion levels under low real message rate with that when there are no real messages (i.e., when no events happen and all messages are dummy). Our purpose is to show the false positive of the attacker's detection. From Figure 4 and Figure 5, we can observe that there are no big difference between these two figures. Thus, the attacker will not gain more from the low-rate SSSA scheme than a perfect secure scheme. Therefore, in our following dynamic source anonymity scheme, we switch from a low-rate SSSA scheme to a k-anonymity scheme.

Step II: Investigating Suspicious Cells After identifying the suspicious levels of cells in Step I, the attacker will arrange an optimal route to check these suspicious cells in Step II. In this section, we use simulations to check the attacker's gain in the SSSA scheme.

We first evaluate the attacker's gain and cost as a function of s, the number of suspicious cells to be checked. As shown in Figure 6, when the number of suspicious cells checked is increased from 3 to 9, the attacker's maximum gain is increased from 124.9 to 342.2. This is because if the attacker checks more suspicious cells a real source is more likely to be discovered. On the other hand, we observe that the attacker's traveling distance also increases from 15.1 to 29.5. This indicates that if the attacker wants to increase the maximum gain by checking more suspicious cells his cost will also increase. In practice, the attacker can decide a maximum traveling distance according to the maximum cost the attacker is willing to pay.

Next, we show how the attacker's gain varies with the real message rate in Figure 7. In the simulation, the probability of a cell being a real source is 0.1, and the average message rate of each cell is 0.05. We observe that when under a fixed t_0, the attacker's gain increases with the real message rate. This is because of two factors. First, with more real messages, more clusters will be observed in a real source and hence the suspicion levels of the real sources will increase. Second, s, the number of suspicious

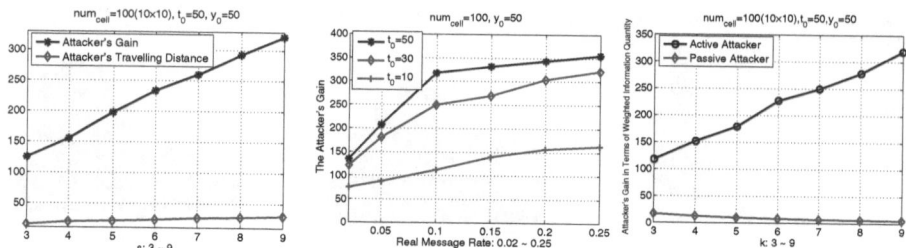

Fig. 6. The attacker's gain and cost as a function of s

Fig. 7. The attacker's gain increases with the real message rate

Fig. 8. Comparing the gains of passive and active attackers in k-anonymity scheme

cells that are actually checked by the attacker, could also increase as more cells have a suspicion level over the threshold value. We also observe that the attacker's information gain increases with real event's duration t_0. This is because the remaining information $f(t)$ increases with t_0 at the same t.

From the above simulation results, we can see that the SSSA scheme, a theoretically sound privacy scheme, exhibit some limitations under the active global attacker model: the information quantity that the attacker could obtain increases with the real message rate. To address this problem, the SSSA scheme has to increase the overall message rate (including dummy and real ones). In other words, dummy message rate has to be adjusted to a larger value to cover the real messages, resulting in potentially prohibitively high message overhead for resource constrained sensor networks. Hence, the SSSA scheme is best applicable when real message rate is low.

3.2 The k-Anonymity Scheme

Next we apply the proposed active global attack model to the k-anonymity scheme [4]. We first briefly introduce this scheme, then check the active global attacker's gain in this scheme and compare it with the gain under a passive global attack.

Scheme Overview. k-anonymity used to be employed to improve the privacy of database without influencing data usability. The basic idea is that each individual data record can be released only when there are at least $k - 1$ other distinct individuals whose associated records are indistinguishable from this record with respect to the quasi-identifiers [12,13,14].

In [4], the idea of k-anonymity is adopted to provide source location privacy under global passive attacks. Basically, to disguise a real source, $k - 1$ fake sources are randomly selected. All k sources start to transmit messages at the similar patterns to confuse the attacker. On one hand, to effectively hide a real source, k should be large enough in the k-anonymity scheme. On the other hand, a large k will lead to higher traffic overhead as more dummy messages will be introduced. One extreme case is when $k = n$, the total number of cells in the network. The highest level of privacy is achieved at the cost of the highest message overhead. Even so, we notice that the k-anonymity scheme has the advantage of on-demand traffic, compared to a constant-rate scheme [4].

That is, fake sources are introduced when real events occur and they are dismissed when real events complete. By adjusting the parameter k, we can have the flexibility of on-demand message overhead rather than a constant quantity of high-volume traffic.

Attacker's Detection. To attack the k-anonymity scheme, Step I for the attacker is to identify suspicious cells. This is quite simple, because it is obvious that the real source is one of the k cells that transmit messages to the BS. Also, according to the property of k-anonymity, ideally all these k cells have the same suspicion level. Then, the attacker may design an optimum route to check the k cells with these k cells as input. The only difference from the attack in the SSSA scheme is that the algorithm runs until the real source has been discovered (instead of time limit τ being reached). Since the real source could be any cell in the k sources, on average the algorithm will stop until $k/2$ sources have been traversed.

We use simulation to check the attacker's gain under the active global attack model. We also compare the result to that from the passive global attack model. Based on the property of k-anonymity, a *passive* attacker cannot differentiate the k sources, so the only thing he can do is to randomly select a source and claim it to be the real source. It is equivalent to pick up and check one out of k cells for the attacker. Actually, the probability for this cell to be the real source is only $\xi = 1/k$. Therefore, according to Equation (2), the weighted information quantity that a *passive* attacker could obtain is $\psi = y_0 \cdot \xi = y_0/k$. As shown in Figure 8, the gain of a passive attacker is much less than that of the active attacker. Clearly, we must take some steps to reduce the active attacker's gain.

4 A Dynamic Source Anonymity Scheme

Our previous discussion showed that a low-rate SSSA scheme is robust to the active global attack. It also has low message overhead. However, it does not adapt well to the case of high-rate real messages. With more real messages, the buffer of the real source might be overflowed if the messages are not delivered promptly. Also, the delivery latency of all the real messages will become very high. In a k-anonymity scheme, as long as the transmission pattern of a high-rate real source can be estimated, $k - 1$ fake sources can be dynamically selected. Based on these observations, we devise a dynamic source anonymity scheme, which seamlessly integrates the merits of both the SSSA scheme and the k-anonymity scheme. The basic idea is when the real event rate exceeds a threshold, the network switches to a k-anonymity scheme. The process starts with an event notification message from the real source to the BS. This message contains information such as how many packets are to be sent and the transmission pattern (e.g., constant rate). It is encrypted and looks the same as all the other messages in the network. The BS then selects $k - 1$ fake sources and notify them to start transmissions at the similar patterns.

Although conceptually straightforward, our scheme has to answer the following questions.

- First, what is an appropriate switching point?
- Second, how to securely bootstrap the k-anonymity scheme?

- Third, how to enhance the security of our dynamic scheme against the active global attack?
- Fourth, how to evaluate the privacy level of our scheme?

The answer to the first question has to take into account many factors, including message overhead, latency, and privacy level. As message overhead is normally the biggest energy expenditure for sensor networks, here we will consider message overhead as the premier criterion in determining the switching point. Note that our proposed techniques are independent of the way a switching point is selected. The second question exists because upon the occurrence of a real event, the event notification message is also under the monitoring of the attacker. If the attacker can figure out which message is an event notification message, he will be able to easily identify the real source. Note that in [4] it does not mention how to bootstrap the k-anonymity scheme. The third question arises because the k-anonymity scheme is not very robust to active attacks, as shown previously. Correspondingly, we will reduce the information gain of the attacker as much as possible. The fourth question has to be answered when evaluating our scheme.

To clearly explain our solutions to the questions, we will first need to make some formal definitions (Section 4.1), then describe the solutions in details in the remaining subsections.

4.1   Problem Definitions

Let $\mathcal{N}$ denote the set of all the n cells in the network. In our case, first, we have a definition of temporal k-anonymity as follows.

Definition 1. *(Temporal k-anonymity).* A real source $r \in \mathcal{N}$ is temporal k-anonymous if there exist at least $k - 1$ other cells $c_1, c_2, \cdots, c_{k-1} \in \mathcal{N}$ such that transmission patterns of all the sources $c_r, c_1, c_2, \cdots, c_{k-1}$ are indistinguishable from each other.

Two transmission patterns are indistinguishable from each other, if their message transmission time intervals follow the same distribution with the same parameters. For example, if message time intervals from two cells follow an exponential distribution with the same mean, we can say that transmission patterns of these two cells are indistinguishable.

According to [15], k-anonymity alone is not sufficient to guarantee database privacy. For example, a person with background knowledge is able to figure out sensitive information from a table with k-anonymity property. l-diversity, which means that for each sensitive attribute there are at least l well-represented different values, is thereby presented to improve the diversity and also the robustness of data items against the above attack. Different from l-diversity in database privacy, we propose a definition of spatial l-diversity, which is adapted properly to our case, in order to improve location diversity of fake sources (and to decrease the attacker's gain). To be more specific, suppose the deployment area of the network is divided into $L(L > 0)$ partitions with almost the same size. Let $\mathcal{C}(\mathcal{C} \subseteq \mathcal{N})$ denote a set of cells and $\mathcal{P}(\mathcal{C})$ denote the total number of different partitions that cells in $\mathcal{C}$ are from. We then have the following definition of spatial l-diversity:

Definition 2. *(Spatial l-diversity).* A set of cells $\mathcal{C}(\mathcal{C} \subseteq \mathcal{N})$ has the property of spatial l-diversity if $\mathcal{P}(\mathcal{C}) \geq l(0 < l \leq L)$.

To quantify the level of source location privacy, next, we exploit the metric of normalized entropy proposed in [16]. This metric is defined based on probability: after the observation, the attacker assigns to the ith subject a probability p_i to be the source, with the sum of probabilities for all the subjects in the set of size n to be 1. For a given distribution of probabilities, the concept of entropy in information theory [17] provides a measure of the information contained in that distribution. Let X be the discrete random variable with probability mass function $p_i = Pr(X = i)$, where $i(1 \le i \le n)$ represents each possible value that X may take. The entropy of X is denoted as $H(X) = -\sum_{i=1}^{n} p_i \log_2(p_i)$. The maximum entropy of the scheme $H_M = \log_2(n)$, which could be achieved when all subjects have the same probability $\frac{1}{n}$ to become the source so that the attacker obtains no information about the source after observation. Therefore, the information that could be learned by the attacker is expressed as $H_M - H(X)$. Apparently, we want the entropy of the scheme $H(X)$ to be as large as possible so that the possible information that could be obtained by the attacker is minimized, since the maximum entropy H_M is fixed under a specific n. Accordingly, we have the following definition on privacy level.

Definition 3. *(Privacy level).* The level of source location privacy is defined as $l_p = \frac{H(X)}{H_M}$, where $H(X)$ is the entropy of the scheme and H_M is the maximum entropy of the scheme.

4.2 Scheme Description

Determining the Switching Point. To cover high-rate real messages, suppose the overall message rate (including dummy and real ones) in the SSSA scheme needs to be increased from λ_1 to λ_2. If $\lambda_2 - \lambda_1 > \delta_1$ where δ_1 is a predetermined threshold, the increase in message overhead is considered intolerable. Then the SSSA scheme needs to be switched to the k-anonymity scheme. Given the total number of cells n, we may determine a proper value for k. Assuming there is a system parameter $\delta_2 > 0$. Then, k needs to satisfy the following two constraints: $1 < k < n$ and $n\lambda_1 - k\lambda_2 > \delta_2$. Such a switch can reduce the overall message overhead significantly.

Secure Bootstrapping. The on-demand switching process is bootstrapped securely as follows. When there are no or low-rate real events, cells send dummy messages to the BS at a low rate following the SSSA scheme. After a real event is detected or a switching is needed, the cell detecting this event will send an event notification message to the BS. After the BS receives this message, it selects $k - 1$ fake sources to generate bogus messages (note that fake sources are carefully chosen). Because the event notification message is just one message and it is easily hidden among the dummy traffic in the SSSA scheme. All the following data messages from the real source are covered by the bogus messages from the fake sources, so these k cells are indistinguishable from each other in the attacker's view. Hence, the k-anonymity scheme can be bootstrapped securely.

In more detail, as shown in Figure 9, after cell u detects a real event:

- Cell u notifies the BS that a real event happens (message 1 in Figure 9(a));
- After an appropriate delay ζ (which will be discussed later), the BS sends out notifications to $k - 1$ fake sources as well as the real source, asking them to start transmitting messages (message 2 in Figure 9(a)) at the same rate or same pattern;

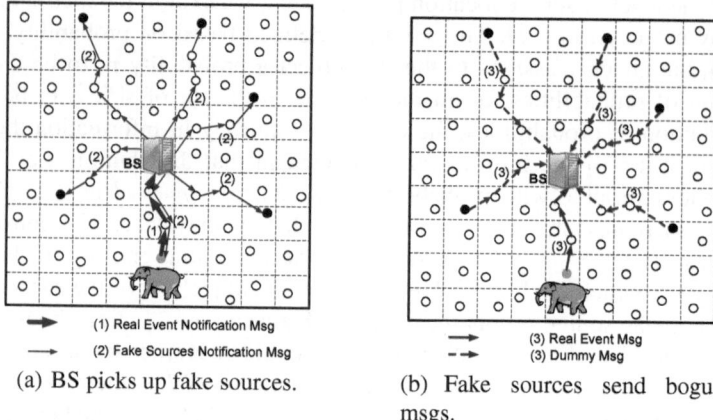

(1) Real Event Notification Msg

(2) Fake Sources Notification Msg

(a) BS picks up fake sources.

(3) Real Event Msg

(3) Dummy Msg

(b) Fake sources send bogus msgs.

Fig. 9. The notification of fake sources in the dynamic source anonymity scheme

- All the cells receiving the notifications start to send messages to the BS (message 3 in Figure 9(b)).

We notice that when the BS receives an event notification message it should not send out k source nomination messages right away. Otherwise, a global observer will easily realize that the message coming to the BS just before BS emits those k messages corresponds to a real source. Therefore, in order to cover the real event notification message, the BS will need to wait for an appropriate time ζ before transmitting k source nomination messages.

To determine ζ, the BS will need to first collect k messages from k different cells including the real source. As can be seen in Figure 10, to keep the k-anonymity property and make the notification message indistinguishable from other $k-1$ normal messages received by the BS, the BS picks k_d, a random number between 0 and k. It will wait until another $k - k_d$ messages are received. These k messages form an anonymity set and their sources are selected as the k sources. After that, it distributes source nomination messages to these cells. Hence, from the attacker's point of view all these k messages are equally likely to be a real notification message and their sources are equally likely to be a real source.

Given a proper k, the average delay ζ before the BS sends out k source nomination messages could be derived as follows. In the SSSA scheme, all the nodes send bogus

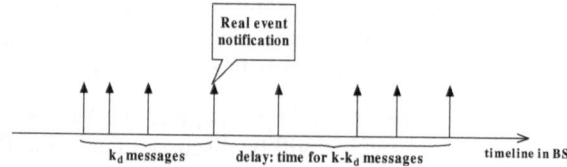

Fig. 10. The introduction of delay in the BS after a real event notification is received

messages with intervals following a mean $1/\lambda_1$. Then, in the BS, the event of incoming messages could be modeled as the sum of n distributions with an overall mean $\frac{1}{n\lambda_1}$. The delay is related to the time for $k - k_d$ messages received by the BS, which is $\frac{k}{2}$ on average since k_d is a random value between 0 and k. Therefore, the average delay in the BS is

$$\zeta = \frac{k}{2n\lambda_1}. \tag{7}$$

Introduction of Spatial l-Diversity. We notice that if it happens that all the k sources are close to each other then the attacker gain a lot within a short time. Therefore, in practice, the BS may selectively choose fake sources that are separated far away from each other. After the BS divides the deployment area into L logical partitions evenly with approximately the same size, we have an algorithm by which the BS achieves spatial l-diversity, as shown in Algorithm 2.

Note that in practice an application may have an upper bound on event notification delay, which is denoted as ω. Hence, the ideal l-diversity may not be attainable because it requires to receive k messages from l partitions (which could take a longer time). So the basic idea of the algorithm is as following. The BS keeps a set $\mathcal{C}$, which is initialized to include the real source and $k - 1$ cells that the first $k - 1$ messages originate from. Each time when a message is received by the BS, the BS tries to swap its source cell with every other cell except the real source in the current set, as long as such a swap could increase the overall distance of the cells in set $\mathcal{C}$. This procedure is repeated until the limit of latency ω is reached and the total number of partitions is larger than l. At this time, the current set $\mathcal{C}$ will be output.

The result of the algorithm is related to the values of ω and l. If they are larger, then k sources may be farther away from each other. Therefore, to reduce the attacker's gain and increase his traversal cost, the BS may wait for a longer time to choose the $k - 1$ fake sources, so that all the k sources are from at least l different partitions.

We use simulation to verify the above statement. First, we check the attacker's gain as a function of the total number of sources k. We compare two options for fake source selection: random selection [4] and spatial l-diversity. In the simulation, the deployment area (10×10) is divided into nine partitions with approximately the same size. $l = k$ under different ks ($3 \leq k \leq 9$). As shown in Figure 11, the attacker's gain increases with k and y_0 (the total information quantity of a real event). Also, the technique of spatial l-diversity could largely reduce the quantity of information that the attacker obtains, compared with random selection.

Second, we check the attacker's traveling distance as a function of k. As shown in Figure 12, the attacker's traveling distance is increased by about 1.5 times because of the spatial l-diversity technique, compared with random selection of fake sources. When $k = 8$, the attacker's traveling distances in random selection of fake sources and spatial l-diversity are 39.7 and 55.9, respectively.

Analysis of Privacy Level. First, we have the following theorem about the initial privacy level of our dynamic scheme (before the attacker's check).

Theorem 1. The k-anonymity scheme with n cells and one real source has an initial privacy level of $l_p = \frac{\log_2 k}{\log_2 n}$, where $k - 1$ is the number of fake sources.

Algorithm 2. The Spatial l-Diversity Algorithm by the BS

Input: a sequence of messages from different cells $\{msg_1, msg_2, \cdots\}$, each message carries such information as which cell and partition it is from;

Output: a set C of size k indicating where sources are, including the real source as a default item and $k - 1$ fake sources;

Procedure:

1: C is initialized to be a set including the real source c_r and cells $c_1, \cdots, c_{k-1}$ where the first $k - 1$ messages are from;
2: calculate $d = sum\_distance(C)$; {function sum_distance() returns the sum of distances of cells in set C, starting from c_r;}
3: P is initialized to be a set including the partition p_r of real source c_r;
4: **for** $j = 1$ to $k - 1$ **do**
5:      p_j is the partition of c_j;
6:      **if** $p_j \neq$ any partition from set P **then**
7:          put p_j into set P;
8:      **end if**
9: **end for**
10: **repeat**
11:     take an incoming message msg as input;
12:     obtain the cell c and partition p of msg;
13:     **for** $i = 1$ to $k - 1$ **do**
14:         $C' = swap(c_i, c)$; {replace c_i with c in set C}
15:         $d' = sum\_distance(C')$;
16:         **if** $d' > d$ **then**
17:             $d = d'$;
18:             $C = C'$; {record set C with maximum distance}
19:             **if** $p \neq$ any partition from set P **then**
20:                 put p into set P;
21:             **end if**
22:         **end if**
23:     **end for**
24: **until** (latency ω is reached)&&(size$(P) \geq l$)
25: return current set C;

Proof: Although there are n cells in the network, the number of active cells transmitting messages at a specific time is only k. Therefore, at any time, the attacker knows the probability for each of the rest $n - k$ cells to be the source is 0. Since the first message sent by the real source is buried in the dummy traffic and the paces of sending messages for all the fake sources as well as the real source are synchronized, the attacker cannot differentiate these k cells. Hence, from the attacker's view the probability for each of the k cells to be the real source is the same. The sum of these probabilities is 1, so every probability equals to $1/k$. Then, the entropy of this scheme

$$H(X) = -\sum_{i=1}^{n} p_i \log_2(p_i) = -\sum_{i=1}^{k} \frac{1}{k} \log_2(\frac{1}{k}) = \log_2(k),$$

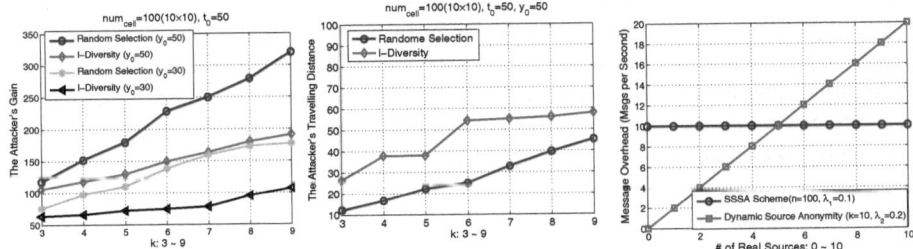

Fig. 11. The attacker's gain in the dynamic source anonymity scheme

Fig. 12. The attacker's traveling distance in the dynamic source anonymity scheme

Fig. 13. Comparison of dynamic schemes in overhead

whereas the maximum entropy of this scheme is $H_M = \log_2(n)$. Therefore, the initial privacy level for this scheme before the attacker's check is $\frac{H(X)}{H_M} = \frac{\log_2(k)}{\log_2(n)}$.

We notice that during the attacker's check the k-anonymity scheme has a dynamic privacy level as follows.

Corollary 1. During the attacker's check, the k-anonymity scheme with n cells and one real source has a dynamic privacy level:

$$l_p = \begin{cases} 0, & \text{if real source;} \\ \frac{\log_2(k')}{\log_2(n)}, & \text{otherwise,} \end{cases} \tag{8}$$

where $k'(k' \leq k)$ is the number of sources that have not been checked by the attacker.

Proof: As presented in Theorem 1, the initial privacy level of the k-anonymity scheme is $\frac{\log_2(k)}{\log_2(n)}$. After the attacker checks one out of k sources, the privacy level of this scheme becomes:

$$l_p = \begin{cases} 0, & \text{if real source;} \\ \frac{\log_2(k-1)}{\log_2(n)}, & \text{otherwise.} \end{cases}$$

In general, when there are $k'(k' \leq k)$ sources that have not been checked by the attacker, all these k' sources have the equal probability $\frac{1}{k'}$ to be the real source. Hence, the entropy of the scheme at this time is

$$H(X) = -\sum_{i=1}^{k'} \frac{1}{k'} \log_2 \frac{1}{k'} = \log_2(k').$$

The privacy level is $\frac{\log_2(k')}{\log_2(n)}$. However, at any time when the attacker discovers the real source, the privacy level of this scheme becomes 0.

Clearly, the selection of k reflects a tradeoff between performance and privacy. A larger k means higher latency and message overhead. Simultaneously, a larger k also leads to higher privacy level based on Theorem 1. In practice, we can decide k according to the application's requirement in latency and overhead. After k is decided, actually the privacy level of the scheme has already been determined. The privacy level of the k-anonymity scheme depends on the ratio of $k (0 < k < n)$ and n. Since the privacy level of the SSSA scheme is close to 100%, normally, the privacy level of the k-anonymity scheme is lower than that of the SSSA scheme.

4.3 Discussions

Mobility of Object. In many cases, an object may go through several cells, which is referred to as a *handoff* problem. After an object moves to another cell, if the BS randomly chooses another $k - 1$ fake sources, the attacker may be able to detect the real source. This is because the locations of the real sources that report the movement of this object actually form a trajectory, whereas the locations of the randomly chosen fake sources do not form a real trajectory. To address this problem, the next fake source should be picked up based on the position of the old fake source, to ensure that positions of these fake sources also form a seemingly real trajectory. This is a hard problem while implementation because building and simulating the object's mobility profile are still open research topics [4]. We may investigate more on this issue, e.g., how to solve the handoff problem in a secure and distributed manner, in our future work.

Multiple Real Sources. Considering the different mobility pattern of different objects, we cannot use the same set of fake sources for different real sources. The starting and ending time for different objects may be different, so using a fake source to serve multiple real sources is not feasible. Therefore, for each real source, the BS needs to assign a group of $k - 1$ fake sources to simulate the real source. The maximum number of real sources that could be serviced at the same time will be $\lfloor n/k \rfloor$. The message overhead of our dynamic scheme increases with the number of real sources. At some point, it may be increased to a value that is more than that of the SSSA scheme (Figure 13). Therefore, the dynamic scheme is best applicable when there are few real sources continuously sending messages at a relatively high rate.

5   Related Work

Protecting location privacy in the context of location-based services has been extensively discussed in the past [18,19,20,21,22]. Location privacy in wireless sensor networks has gained a lot of attention recently. In [23], techniques for hiding the base station (message destination) from an external global adversary are studied. In their schemes, secure multi-path routing to multiple destination base stations is designed to provide intrusion tolerance against isolation of base station and anti-traffic analysis is proposed to disguise the location of base station. [24] proposes a location-privacy routing protocol that provides path diversity combined with fake packet injection to protect receiver-location privacy. Complementary to their work, we are interested in source location privacy.

In [1,2], a random walk based phantom routing scheme is proposed to defend against an external adversary who attempts to trace back to the data source in a sensor network, where sensor nodes report sensing data to a fixed base station for a certain period. A more recent work [3] proposes a two-way random walk algorithm, in which the routing path is obfuscated from both the source and sink. In [25], a path confusion algorithm is presented to increase source location anonymity. Note that these schemes work for a local adversary model. In our scheme, we consider a global attacker who has the view of all the network traffic.

[26] presents pDCS, a privacy-enhanced Data-Centric Sensor networks that offers different levels of data privacy based on different types of cryptographic keys. Under a global attacker model, in [27], two schemes are proposed. The first one is a ConstRate scheme; the second one is a k-anonymity based source-simulation scheme. Analytical results show how much communication overhead is needed to achieve a certain level of privacy. [7] addresses source location privacy against laptop-class attackers by proposing four schemes: naive, global, greedy, and probabilistic. In [6], to provide source event unobservability, schemes like ConstRate or ProbRate are used by the sensors. The focus of this work is to reduce the overall network traffic by proactively dropping dummy messages on their way to the BS.

[5] concentrates on reducing the latency of real messages under a global attacker model, by sending real messages as early as possible, in a way that the disturbance cannot be detected by available statistical tests. [28] also considers anonymous networking with minimum latency. Mixes are used for individual relays. The introduction of a limited number of dummy messages leads to a significant reduction in network latency. Information theoretical measurement is employed to analyze the relationship between anonymity level and latency. [29] provides temporal privacy protection for wireless sensor networks. In our work, we further improve the power of the attacker and consider a more realistic global attacker model in which the attacker can go to suspicious spots and check real events by himself.

6   Conclusion and Future Work

Previous work in sensor source location privacy mainly considers either a local tracker or a global eavesdropper in the attack model. We study a even more powerful and realistic attack model, in which a global attacker goes to suspicious spots and check real events by himself after monitoring all the network traffic. We formalize such a strong attack model and discuss countermeasures against it. An important future direction will be the development of a distributed way to solve the handoff problem under a mobile object in the dynamic source anonymity scheme. Other adversary models such as insider attackers are also of interest to us.

Acknowledgement. We thank the anonymous reviewers for their valuable suggestions. This work was supported in part by NSF 0627382, NSF-0643906 and MURI/ARO W911NF-07-1-0318.

References

1. Ozturk, C., Zhang, Y., Trappe, W.: Source-location privacy in energy-constrained sensor networks routing (SASN 2004) (October 2004)
2. Kamat, P., Zhang, Y., Trappe, W., Ozturk, C.: Enhancing source-location privacy in sensor network routing. In: ICDCS 2005 (June 2005)

3. Xi, Y., Schwiebert, L., Shi, W.: Preserving source location privacy in monitoring-based wireless sensor networks. In: SSN 2006 (2006)
4. Mehta, K., Liu, D., Wright, M.: Location privacy in sensor networks against a global eavesdropper. In: ICNP 2007 (October 2007)
5. Shao, M., Yang, Y., Zhu, S., Cao, G.: Towards statistically strong source anonymity for sensor networks. In: Infocom 2008 (April 2008)
6. Yang, Y., Shao, M., Zhu, S., Urgaonkar, B., Cao, G.: Towards event source unobservability with minimum network traffic in sensor networks. In: WiSec (2008)
7. Ouyang, Y., Le, Z., Liu, D., Ford, J., Makedon, F.: Source location privacy against laptop-class attacks in sensor networks. In: SecureComm (2008)
8. Held, M., Karp, R.M.: A dynamic programming approach to sequencing problems. J. Soc. Indust. Appl. Math. (March 1962)
9. Anderson, T.W., Darling, D.A.: A test of goodness of fit. Journal of the American Statistical Association 49(268) (December 1954)
10. Romeu, J.L.: Kolmogorov-simirnov: A goodness of fit test for small samples. START: Selected Topics in Assurance Related Technologies 10(6) (2003)
11. Wald, A.: Sequential Analysis. J. Wiley & Sons, New York (1947)
12. Samarati, P.: Protecting respondents' identities in microdata release. IEEE Transactions on Knowledge and Data Engineering 13(6), 1010–1027 (2001)
13. Sweeney, L.: k-anonymity: a model for protecting privacy. International Journal on Uncertainty. Fuzziness and Knowledge-based Systems 10(5), 557–570 (2002)
14. Gedik, B., Liu, L.: A customizable k-anonymity model for protecting location privacy. In: ICDCS (2005)
15. Machanavajjhala, A., Gehrkc, J., Kifer, D., Venkitasubramaniam, M.: l-diversity: Privacy beyond k-anonymity. In: ICDE 2006 (2006)
16. Díaz, C., Seys, S., Claessens, J., Preneel, B.: Towards measuring anonymity. In: Dingledine, R., Syverson, P.F. (eds.) PET 2002. LNCS, vol. 2482, pp. 54–68. Springer, Heidelberg (2003)
17. Cover, T.M., Thomas, J.A.: Elements of Information Theory. John Wiley & Sons, Inc., Chichester (1991)
18. Gruteser, M., Grunwald, D.: Anonymous usage of location-based services through spatial and temporal cloaking. In: Proceedings of the 1st international conference on Mobile systems, applications and services (2003)
19. Myles, G., Friday, A., Davies, N.: Preserving privacy in environments with location-based applications. IEEE Pervasive Computing 2(1) (2003)
20. Kido, H., Yanagisawa, Y., Satoh, T.: An anonymous communication technique using dummies for location-based services. In: PICPS (2005)
21. Bettini, C., Wang, X.S., Jajodia, S.: Protecting privacy against location-based personal identification. In: Jonker, W., Petković, M. (eds.) SDM 2005. LNCS, vol. 3674, pp. 185–199. Springer, Heidelberg (2005)
22. Gunter, C.A., May, M.J., Stubblebine, S.G.: A formal privacy system and its application to location based services. In: Martin, D., Serjantov, A. (eds.) PET 2004. LNCS, vol. 3424, pp. 256–282. Springer, Heidelberg (2005)
23. Deng, J., Han, R., Mishra, S.: Intrusion tolerance and anti-traffic analysis strategies for wireless sensor networks. In: DSN 2004 (2004)
24. Jian, Y., Chen, S., Zhang, Z., Zhang, L.: Protecting receiver-location privacy in wireless sensor networks. In: INFOCOM (2007)
25. Hoh, B., Gruteser, M.: Protecting location privacy through path confusion. In: Securecomm, pp. 194–205 (2005)
26. Shao, M., Zhu, S., Zhang, W., Cao, G.: pdcs: Security and privacy support for data-centric sensor networks. In: INFOCOM (2007)

27. Mehta, K., Liu, D., Wright, M.: Location privacy in sensor networks against a global eaves-dropper. In: ICNP (2007)
28. Venkitasubramaniam, P., Tong, L.: Anonymous networking with minimum latency in multi-hop networks. IEEE Security and Privacy (2008)
29. Kamat, P., Xu, W., Trappe, W., Zhang, Y.: Temporal privacy in wireless sensor networks. In: ICDCS 2007 (2007)

Rogue Access Point Detection
Using Innate Characteristics of the 802.11 MAC

Aravind Venkataraman and Raheem Beyah

Cigital Inc. and Georgia State University
avenkataraman@cigital.com, rbeyah@cs.gsu.edu

Abstract. Attacks on wireless networks can be classified into two categories: external wireless and internal wired. In external wireless attacks, an attacker uses a wireless device to target the access point (AP), other wireless nodes or the communications on the network. In internal wired attacks, an attacker or authorized insider inserts an unauthorized (or rogue) AP into the wired backbone for malicious activity or misfeasance. This paper addresses detecting the internal wired attack of inserting rogue APs (RAPs) in a network by monitoring on the wired-side for characteristics of wireless traffic. We focus on two 802.11 medium access control (MAC) layer features as a means of fingerprinting wireless traffic in a wired network. In particular, we study the effect of the Distributed Coordination Function (DCF) and rate adaptation specifications on wireless traffic by observing their influence on arrival delays. By focusing on fundamental traits of wireless communications, unlike existing techniques, we demonstrate that it is possible to extract wireless components from a flow without having to train our system with network-specific wired and wireless traces. Unlike some existing anomaly based detection schemes, our approach is generic as it does not assume that the wired network is inherently faster than the wireless network, is effective for networks that do not have sample wireless traffic, and is independent of network speed/type/protocol. We evaluate our approach using experiments and simulations. Using a Bayesian classifier we show that we can correctly identify wireless traffic on a wired link with 86-90% accuracy. This coupled with an appropriate switch port policy allows the identification of RAPs.

Keywords: Rogue Access Point Detection, 802.11 MAC Protocol, Rate Adaptation, Distributed Coordination Function.

1 Introduction

A dangerous insider attack is one where cheaply available APs are illicitly plugged into the network with the motivation of extending connectivity. Like other insider attacks, the AP stays invisible to a firewall as it is actually behind it, thus making it difficult to detect. Hence, the AP creates a back door for attackers, obviating the need to go through the firewall. This paper presents a practical solution for this attack which can happen in one of two scenarios - wired networks *with* or *without* existing legitimate APs.

The core of our detection scheme is an agent sitting atop a switch, or a separate monitoring device that is connected to the mirror port of a switch, that passively sniffs

Y. Chen et al. (Eds.): SecureComm 2009, LNICST 19, pp. 394–416, 2009.
© Springer-Verlag Berlin Heidelberg 2009

passing traffic streams on the wired-side. Using inherent differences in wireless characteristics as compared to wired traffic, this agent is able to deem the originating link as being wired or wireless. This inference is then followed up with a switch port AP authorization policy to differentiate between rogue and legitimate APs.

Though some of the existing methods work with proven efficacy, they do not try to exploit the underlying facets of the wireless MAC protocol to detect RAPs, but instead attempt to classify wireless traffic based on the greater delay observed in network statistics (e.g., round-trip-time (RTT), inter-packet arrival time (IAT)). This is based on an assumption that the wireless link capacity will never reach that of wired. A more general solution is needed as this may not always be the case. Also, since many of the previous algorithms need to be trained on both wired and wireless traffic for a given network, they cannot be used in networks without existing APs as there would be no prior wireless trace available.

As in other wired-side detection approaches from academia[1], in our method we study the arrival pattern of upstream traffic towards the gateway router (and possibly the Internet) for traces of the 802.11 MAC protocol. Though downstream TCP flows are likely to occupy a significant portion of traffic on the link, our approach is not limited in scope. This is because, as will be shown in Section 4.3, our classifier can work with a minimal input trace. It works with an accuracy ranging from 87% to 91% for upstream data inputs of size ranging from 250 packets to 1000 packets respectively[2]. At any given time, there may be various activities that the RAP is used for in a corporate network, such as, web browsing, email, document uploads to file servers, etc. Web browsing contributes varying levels of upstream data - mostly in the form of HTTP requests - depending on the content and the load of requests. When a web site is crawled by visiting, say five URLs recursively, the amount of upstream data generated varies from 75 data packets (for primarily text based web pages like *www.craigslist.com*) up to 400 packets (for relatively graphic intensive web pages like *www.facebook.com*). Further, it takes about 500 packets to deliver an email of size 750Kb and about 1000 packets to upload a file of 1.5Mb (e.g., saving a file to the company file server). Thus, upstream monitoring is a viable option.

Our first approach exploits the collision avoidance process of the DCF in the 802.11 MAC. To avoid collisions while transmitting, a wireless node has to sense the channel prior to an attempt at sending. Once the channel is clear, the node will wait for a random time period (chosen from 0 time units to a fixed upper bound) before attempting to transmit. If the node senses that the channel is occupied, or in case of a collision, the node has to back-off exponentially before retransmitting (i.e., the fixed upper bound increases exponentially, increasing the probability of choosing a higher back-off value). This procedure, carrier sense multiple access with collision avoidance (CSMA/CA), of the DCF has both fixed components and bounded random components that can be artificially produced and used as a signature for wireless traffic.

The second approach exploits the process of rate adaptation in the 802.11 MAC. Rate adaptation algorithms allow wireless hosts to alter their encoding scheme (transmission rate) to account for channel interference during transmission. When interference is detected, the node adapts its rate and transmits at a slower rate in an

---

[1] The wired-side approaches will be discussed in Section 2.
[2] Refer to Figure 8, the details of which will be discussed in Section 4.4.

attempt at reducing packet loss. As the rate adjusts (lower or higher), there are noticeable and unique 'jumps' in the packet IAT. These 'jumps' can be artificially produced and used as a signature for wireless traffic.

For both of the above techniques, we show that the signature created stays intact and can be detected on the wired-side allowing us to deem specific traffic as originating from a wireless node.

Each of the two approaches work best in specific cases. The first approach works best when there is little interference and the transmission rate essentially stays constant. Intuitively, the second approach works best when the network is more volatile as more 'jumps' are produced during that period. Since network stability is unpredictable, we combine the two schemes and present a solution that accounts for realistic, unpredictable network conditions.

The remainder of this paper is organized as follows. Section 2 outlines previous work in RAP detection. In Section 3 we briefly illustrate why magnitude-based approaches are not optimal. An introduction to the 802.11 MAC protocol collision avoidance mechanism and a breakdown of the delay induced by it on wireless traffic is presented in Section 4. A mathematical representation is derived from its inherent mechanism following which we validate the model using a Bayesian classifier. A similar pattern of presentation is taken in Section 5 as in Section 4, where we perform an analysis of the manner in which rate adaptation occurs, followed by accuracy measures of our model. In Section 6, we perform a comparative study of the two techniques in an attempt to come up with a bridged solution. We present the scalability of our techniques in Section 7 and conclude in Section 8.

2   Related Work

Current work on RAP detection can be classified into three categories. The first two categories contain techniques that use the magnitude of statistics (mean, median, entropy, etc.) of IATs and RTTs as the primary metrics for classification respectively. The third category contains industry work that primarily make use of radio frequency scanning to discover wireless activity within a network.

References [1-6] fall in the first category. Beyah R., et al., [1] were among the earliest to suggest the possibility of using temporal characteristics, such as IATs, for RAP detection. They used the IATs of data packets and TCP ACK packets to identify the type of traffic flow. The authors in [2] take a similar approach as that taken in [1] but extend the work by creating an automated classifier. Wei W., et al., in [3-4] present two similar proposals that examine IATs of TCP ACK pairs to identify the type of traffic flow. However, the use of ACK pairs limits this technique to TCP traffic. A noteworthy effort in the area of traffic classification is [5] which attempts to categorize different types of access links using median and entropy of packet IATs. The approach is however not applicable to detecting RAPs because it is active (requires probing) and requires cooperation (probe responses) from malicious nodes. In [6], the authors create a spectral profile for WLANs based on the entropy of IATs. They assume link quality and unpredictability of the wireless medium as the cause for greater wireless 'uncertainty' and do not study the effect of the DCF.

In the second category, [7-9] make use of RTT as a metric for classification. Since these methods rely on RTT, they cannot accommodate traffic streams other than TCP. Though [7] briefly mentions the effect of the DCF, it does not go into detail to study its mechanics. Reference [8] uses a distinctive approach for segregating network types, complete with traffic conditioning to eliminate noise. However, it demarcates wired and wireless traffic with the help of mean and deviation of the RTT dataset which is not advisable as these parameters differ with varying types, speeds, and congestion levels of networks. Their approach is claimed to be non-intrusive. However, since it involves conditioning of traffic it is still, at minimum, pseudo-active. In [9], although for a disparate motive and in a dissimilar context, Cheng L., et al., were among the first to work on identifying wireless traffic for the purpose of access link type recognition. However, their model employs a probing process to gain information about nodes in the network and thus not likely to be of assistance in the RAP problem space for the same reason that [5], as mentioned above, falls short.

The third category includes several industry solutions [10-17], many of which exhibit non-scalability and limited effectiveness because of the use of either radio frequency (RF) scanning and/or MAC address based authentication. The use of RF scanning is not practical as the malicious user can use directional antennas, can adjust the power of the AP as to not be detected, and in large networks it becomes analogous to finding a needle in a haystack. The use of the MAC address as a parameter for authentication is not appropriate because of the ease of spoofing.

Outside of the three categories, [18-20] propose hybrid frameworks consolidating the above mentioned wired and wireless-side detection models and inherit the flaws from each type.

As previous schemes primarily compare the *relative* behavior of traffic on each link, they require traces of each class of network traffic for their scheme to be effective. This approach is limiting, as a network without existing legitimate APs (e.g., government labs) would not be able to easily provide a wireless trace. Further, because many use threshold-based separation metrics, another limiting assumption made is that wireless networks will always be slower than their wired counterparts. As will be shown in subsequent sections, our method is free of each of the aforementioned assumptions.

3 Problem with Magnitude-Based Classification

As mentioned in the previous section, many of the existing works focus on the difference, in some form, of the magnitude of the IAT or RTT to differentiate wireless from wired traffic. In this section, we illustrate, via simulation, the challenge with these approaches as wireless speeds begin to approach that of wired traffic.

Simulations were performed using *ns2* [24]. The cumulative distribution functions (CDFs) of the IAT and RTT values are shown in Figures (1a, 1b) and (2a, 2b) respectively. Figures 1a and 2a illustrate why the magnitude-based approaches work when the assumption is that WLANs are slower than LANs ($\{IAT_{wl}, RTT_{wl}\} > \{IAT_{wd}, RTT_{wd}\}$). However, as shown in Figures 1b and 2b, these schemes will breakdown if the WLAN speed reaches that of the LAN ($\{IAT_{wl}, RTT_{wl}\} \simeq \{IAT_{wd}, RTT_{wd}\}$).

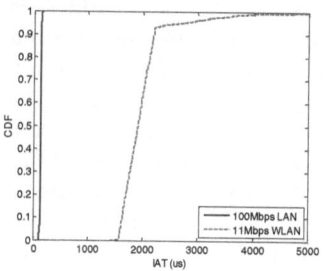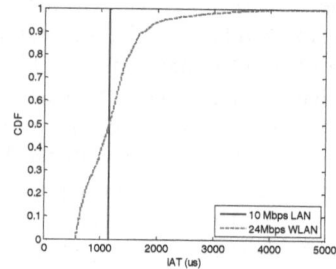

Fig. 1. IAT distribution for (a) slower WLAN, (b) faster WLAN

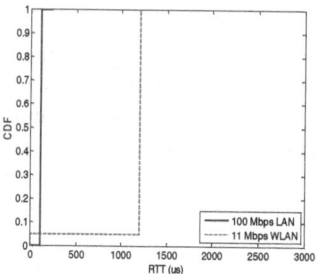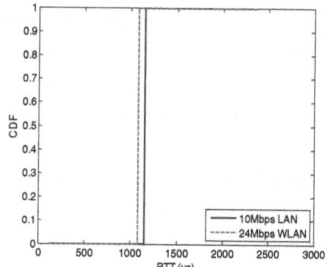

Fig. 2. RTT distribution for (a) slower WLAN, (b) faster WLAN

The results shown in Figures 1 and 2 were obtained from within single trials each of 1000 packets of upstream data for various network type/speeds (LAN - 10Mbps, 100Mbps; WLAN - 11Mbps, 24Mbps). Ethernet and wireless senders were made to send FTP data to a server one hop away on the wired-side. Simulations were performed on a setup similar to the experimental setup that will be described in Section 4.3.

Partially motivated by this argument against threshold based detection, we propose an adaptable solution that makes no assumption about the link speed. In the next section, we introduce our first scheme beginning with an introductory analysis.

4   Scheme I – DCF Based Detection

A wireless node's packet transmission mechanism is regulated by the specifications of the 802.11 MAC layer protocol, the Distributed Coordination Function (DCF). The DCF employs a CSMA/CA distributed algorithm for collision avoidance. In this method, a node that wants to transmit data on a wireless link has to wait for a fixed duration, namely Distributed Inter Frame Space (DIFS) and a bounded random amount of time, called *back-off* (σ), before using the channel. Upon receiving the data, the node at the other end waits for a fixed period, called the Short Inter Frame Space (SIFS), before answering with a MAC-level acknowledgment (MAC-ACK), and the cycle follows thereon. Further, if the channel is sensed busy or if a collision is detected the originating node backs-off before trying again. The bounded random delay - Contention window (*CW*) has an exponentially increasing upper bound to reduce the chances of collisions.

Accordingly, the DCF has both fixed components and bounded random components that can be artificially produced and used as a signature for wireless traffic. The process employed for transmission in a wireless medium and the delay between packet arrivals (IAT_{wl}) as observed at the receiver are shown in Figure 3.

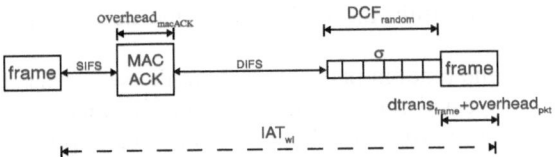

Fig. 3. Illustration of the DCF in 802.11 networks

Drawing from the DCF's basic mode of operation, we deduce a pattern unique to wireless streams that allows one to anticipate packet arrivals at known intervals. This property enables us to artificially construct packet arrival time series that represent wireless traffic.

4.1 Analysis

First, in order to demonstrate the effect of the DCF on the delay, we arrive at representations for the IATs of wired and wireless networks (IAT_{wd} and IAT_{wl}).

In Equations 1 and 2, *dtrans, dprop* and *dqueue* are the transmission, propagation and queuing delays for a network respectively. Since *dtrans* $>>$ *dprop*, the propagation delay is neglected in our analysis. The queuing delay *dqueue* plays an important part in determining the efficacy of wired-side detection. This will be discussed with experimental results in Section 4.3.

$$IAT_{wd} = dtrans_{wd} + dprop_{wd} + dqueue_{wd} \tag{1}$$

$$IAT_{wl} = dtrans_{wl} + dprop_{wl} + dqueue_{wl} \tag{2}$$

$$dtrans_{wd} = dtrans_{frame} + dtrans_{overhead_{wd}} \tag{3}$$

$$dtrans_{wl} = dtrans_{frame} + dtrans_{overhead_{wl}} + DCF_{constant} + DCF_{random} \tag{4}$$

In Equations 3 and 4, $dtrans_{frame}$ is the transmission time per frame; $dtrans_{overhead}$ is the overhead incurred in transmitting the packet header in the wired case, and transmitting the packet header and MAC-ACK in the wireless case.

$$dtrans_{overhead_{wl}} = overhead_{pkt} + overhead_{MAC\text{-}ACK} \tag{5}$$

Note that $dtrans_{wl}$ additionally comprises of the waiting time incurred because of the DCF, the constituents of which are shown in Equations 6 and 7.

$$DCF_{constant} = DIFS + SIFS \tag{6}$$

$$DCF_{random} = \sigma \tag{7}$$

The fixed delay element within the DCF contributed delay is a combination of the DIFS and SIFS periods.

The back-off (σ) is the random period for which the sender has to wait in addition to the DIFS. This is repeated for each unsuccessful transmission attempt. The back-off for the ith retransmission (σ_i) is randomly chosen from within the CW_i which is an increasing function of the number of retransmission attempts and the number of times the channel was sensed as busy by the sender. The DCF uses an exponential algorithm, where for each retry, the CW size is doubled starting at a lower bound (CW_{min}) until a maximum value (CW_{max}) is reached.

$$\sigma_i \in (0, CW_i) \tag{8}$$

$$CW_i = min[2CW_{i-1}, CW_{max}] = min[2^i CW_{min}, CW_{max}] \tag{9}$$

$$\sigma_i \propto CW_i \propto CW_{min} \tag{10}$$

Hence, arrival times can be predicted as a function of CW_{min} in the form of a finite random variable. This is an important result which shows that the DCF provides us with an increasing trend for wireless links, one whose base frequency (θ) is given in Equation 11.

$$\theta = 1/\left(dtrans_{frame} + DCF_{constant} + [0, CW_{min}] \right) \tag{11}$$

Equation 11 forms the basis for our scheme. Specifically, we seek to discover a wireless segment by extracting a basic recurring pattern that exists in all wireless streams. Further, a wireless series can be generated synthetically which spares us from having to train a classifier with real traces.

Since the RAP environment would likely involve a single client node (the malicious intruder), our primary focus is the case where there are minimal collisions as a result of competing traffic in the network, and thus assume that σ varies between 0 and CW_{min}. We plan to address the scenario where multiple users access the RAP in the future.

This wireless time series is not uniform for different traffic types. In light of Equation 4, it is important to consider two transport protocols - TCP and UDP. Figures 4 and 5 show how the IAT distribution would look for the two different classes.

The frame transmission time for each case would differ as shown in Hypothesis 4.1.

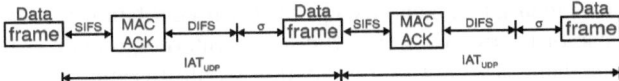

Fig. 4. Packet arrival pattern - UDP

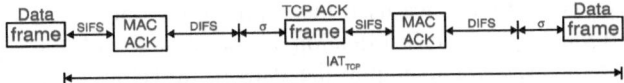

Fig. 5. Packet arrival pattern - TCP

Hypothesis 4.1

1: **if** traffic$_{UDP}$ **then**
2:     $dtrans_{frame} = dtrans_{data}$
3: **else if** traffic$_{TCP}$ **then**
4:     $dtrans_{frame} = dtrans_{data} + dtrans_{tcpACK}$
5: **end if**

Because of the difference in characteristics, considering an 802.11b network as an example, the transmission delay for the two classes would follow from the information in Table 1 (taken from [21]) as shown in Equations 12 and 13.

This difference must be factored in when modeling the traffic behavior.

$$IAT_{wl_{UDP}} \cong dtrans_{wl_{UDP}}$$

$$= DCF_{constant} + DCF_{random} + dtrans_{frame} + dtrans_{overhead_{wl}}$$

$$= DCF_{constant} + DCF_{random} + dtrans_{data} + dtrans_{overhead_{wl}} \qquad (12)$$

$$= 60 + \sigma + 1018 + 215 + 10$$

$$= 1303 + \sigma$$

$$IAT_{wl_{TCP}} \cong dtrans_{wl_{TCP}}$$

$$= 2DCF_{constant} + 2DCF_{random} + dtrans_{frame} + 2dtrans_{overhead_{wl}}$$

$$= 2DCF_{constant} + 2DCF_{random} + dtrans_{data} + dtrans_{tcpACK} \qquad (13)$$

$$+ 2dtrans_{overhead_{wl}}$$

$$= 120 + 2\sigma + 1018 + 30 + 2(215 + 10)$$

$$= 1618 + 2\sigma$$

Note that TCP does not always have to wait for an ACK before transmitting the next packet. In fact, when a node is transmitting TCP traffic with a congestion window size W greater than one (that is, $W>1$), it is likely to exhibit UDP-like behavior (in the form of multiple sequential packets) except for the time when it is waiting for ACKs. In fact, in the case of upstream TCP traffic to the Internet, a node is highly likely to transmit in bursts. Thus, TCP's IAT distribution would resemble that of UDP for the most part. Hence, having taken into account the frequency of packet arrivals for both UDP in Equation 12 and the extreme-case TCP (that is, $W = 1$) in Equation 13, our model is scalable for all traffic types.

As part of our groundwork, we used the expression from Equation 4 - which repeats with the frequency shown in Equation 11, combined with the expected values for each type of WLAN (for example, the data from Table 1 was imported for a 802.11b WLAN) to artificially build a profile set. We used values for $DCF_{constant}$ from the 802.11 standard. Also, we used a pseudo random number generator to emulate DCF_{random}, where random values were generated from within a range equivalent to the initial CW, that is, $(0, CW_{min})$.

Table 1. 802.11b MAC Transmission Overhead

Variable	Parameter	Time (μs)	Formula
$DCF_{constant}$	DIFS	50	2 * slot time + SIFS = 50
	SIFS	10	SIFS
DCF_{random}	Average σ	310	(# of slots * slot time)/2 = (31 * 20)/2 = 310
$dtrans_{frame}$	$dtrans_{data}$	1018	Packet size/data rate = (1400 * 8)/11 = 1018
	$dtrans_{TCP\text{-}ACK}$	30	TCP-ACK/data rate = (40 * 8)/11 = 30
$dtrans_{overhead}$	$overhead_{pkt}$	215	(Preamble + PLCP hdr.)/data rate + MAC hdr./data rate + MAC CRC bits/data rate = (144 + 48)/1 + (30 * 8)/11 + (4 * 8)/11 = 192 + 21 + 2 = 215
	$overhead_{MAC\text{-}ACK}$	10	MAC-ACK/data rate = (14 * 8)/11 = 10

Figures 6a and 6b display the CDF of the IAT of TCP and UDP flows generated via experimentation and simulation, as well as those constructed artificially using Equations 12 and 13. The figures illustrate how closely the experimental and simulated delay distributions follow the ones artificially created.

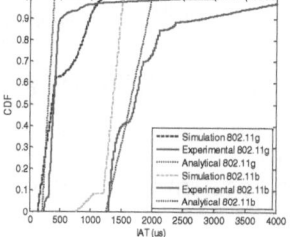 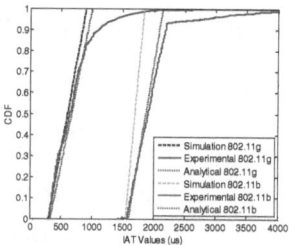

Fig. 6. (a) CDF of IAT for UDP. (b) CDF of IAT for TCP.

Also from Figure 6b, while more than 90% of the sample set follows a uniform random dispersal over the window size, a fraction of the flow tends to deviate out of bounds. We attribute this to the overhead in the network caused by $dprop_{wl}$, possible link-layer retransmissions and packet collisions during transmission.

To arrive at the results in Figure 6, separate TCP and UDP experiments and simulations were performed individually for the 802.11b and 802.11g configurations. For each transport protocol and each WLAN speed setting, 1000 data packets were sent from the wireless client using a socket program to the wired-side server and the IATs were recorded on the wired-side. Correspondingly, the artificial profiles each comprise of 1000 IAT values.

The experiments that produced part of the results in Figure 6 were performed in a lab testbed that will be discussed in Section 4.3. The simulations associated with Figure 6 were performed in a similar setup as the lab testbed using $ns2$.

In this sub-section, we showed that it is possible to independently conjecture how a wireless stream would behave in different types of networks. To demonstrate the accuracy of the technique, a Bayesian classifier is used to compare incoming streams' IAT distributions with the training IAT profile set. The foundation for this classification is presented in the next sub-section.

4.2 Classification Scheme

We use a Naïve Bayes classifier which bins the IAT datasets (the artificial profiles and experimental/simulation traces used for the purpose of testing the system), calculates for each dataset the number of occurrences in each bin, compares the bin frequencies of each profile with those of the trace and predicts the trace as being akin to the profile whose frequency distribution closest resembles that of the trace. The Chi-square Goodness of Fit test is employed to determine the fit between each profile and the unknown trace.

The inputs are binned into 'b' number of bins, where b depends on the bin width and input data size. For both the bin width and input data size, different values are tried with the goal of optimizing 'b' to furnish maximum accuracy. Details about these experiments will be discussed in Section 4.4.

Profiles f_i are compared with an unknown sample f_x based on the frequency of occurrences in each bin. The Bayes theorem is based on the conditional probability model, where the posterior probability is a function of the prior probability and the likelihood.

Because the nature of incoming traffic cannot be predicted, prior probability is unknown and is assumed equally distributed over the n profiles.

$$PriorProbability \quad P(f_i) = 1/n \tag{14}$$

Likelihood (measure of how similar the unknown trace is to a given profile) is calculated for each profile using a two-sample Chi-square test which is run independently on all sample-profile bin frequency pairs.

$$Likelihood \quad P\langle f_x | f_i \rangle \tag{15}$$

Posterior probability (measure of how likely a profile is the closest match for the unknown) is calculated as follows:

$$PosteriorProbability \quad P\langle f_i | f_x \rangle = P\langle f_x | f_i \rangle . P\langle f_i \rangle \tag{16}$$

Since f_x is a random variable $\{x_1, x_2, \ldots x_d\}$,

$$P\langle f_i | f_x \rangle = P\langle (x_1, x_2, \ldots x_d) | f_i \rangle . P\langle f_i \rangle \tag{17}$$

$$P\langle f_i | f_x \rangle = P\langle f_i \rangle \prod_{k=1}^{d} P\langle x_k | f_i \rangle \tag{18}$$

Since the prior probability is constant, the posterior probability essentially depends on the likelihood measures. It is derived by aggregating the Likelihood measures, each of which is calculated using the Pearson's Chi-square test. This test estimates the probability that an unknown distribution fits a Chi-square distribution given a null hypothesis. This null hypothesis is rejected (or accepted) based on the probability of the unknown trace's fit to the Chi-square distribution. This probability is determined as a function of the Chi-square statistic which is obtained as follows:

$$\chi^2 = \sum_{i=1}^{k} \left((X_i - P_i)^2 / P_i \right) \tag{19}$$

X_i and P_i are the bin frequencies of bin i of the two samples to be compared - the unknown and a profile. Note that the profile P is the null hypothesis. In our case, P is the synthetically created wireless profile. The Chi-square statistic is calculated over the bin frequencies of k bins.

4.3 Experimental Setup and Validation of Wired-Side Approach

In this sub-section, we discuss: (i) preliminary experiments that were performed to validate the general idea behind our wired-side approach and (ii) the outline of the experimental test plan we used to evaluate the system's accuracy.

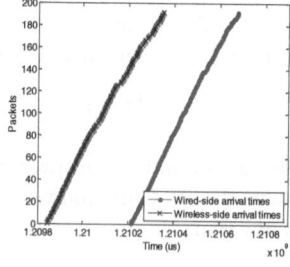

Fig. 7. Packet arrival times on wired and wireless sides

An experimental testbed was built using three Lenovo laptops, three Dell desktops, a Netgear 10/100 Mbps Fast Ethernet switch and a Linksys 2.4Ghz 802.11*b*-*g* AP. The laptops were made to connect to a server on the wired-side through the AP and switch. A desktop was set up as a sink server to receive data from both LAN and WLAN senders. The classifier resides on a desktop connected to the switch immediately linking the AP to the LAN.

To ensure that our technique for wired-side detection is viable, we first determined whether the temporal characteristics of the IAT observed on the wireless link were intact on the wired-side. It is important to check if the DCF induced delay is carried over to the Ethernet backbone with minimal additional overhead delay added to it. In a single hop scenario, the overhead is primarily a function of the router queuing delay and processing delay.

The results shown in Figure 7 are a representative sample of the arrival times of about 200 packets extracted from a trace of a total of 10,000 packets of upstream TCP data sent from a wireless-side sender using a socket program to the wired-side server. The arrival times on the wired-side were recorded at the receiver node. On the wireless side, a laptop acting as a sniffer was used in promiscuous mode to capture traffic from the wireless sender. We observed that the arrival rates were retained albeit with a uniformly witnessed lag (as a result of router queuing) as shown in Figure 7.

Given the simple one-hop path from the WLAN to the classifier on the wired-side, a switch with minimal traffic load exhibits a nearly constant queuing delay (*dqueue*)[3] which is illustrated by the nearly fixed distance between the lines in Figure 7. In Section 7, we discuss how the model scales to networks where the classifier is placed several hops away from the AP.

4.4 Accuracy Measures

Having visually shown why it is likely that the packet IAT from the wireless side is carried over to the wired-side, we evaluate the scheme's accuracy in extracting the DCF imposed delay to determine the packet's originating link.

First, we tuned the bin width and input data size to find the optimal pair - one that maximizes True Positive Rate (TPR) and minimizes False Positive Rate (FPR). This is followed by additional testing with the chosen optimal parameters to obtain the system accuracy.

The classifier was tested on traces from both wired and wireless TCP/UDP data transfers. Sample trials on the LAN were used to measure the FPR and trials on the WLAN to measure the TPR. Trials were performed on the WLAN for both 802.11*b* and 802.11*g* specifications by configuring the AP to operate in the required mode. For each network type (WLAN/LAN) and protocol (TCP/UDP), 50 sets of data were fed into the classifier. The detections from the 50 trials were used in determining TPR/FPR measures for the classifier. This process was repeated for different bin width and input data size combinations. The results shown in Figure 8 are an average of the results from the TCP and UDP trials.

An optimal bin width of $500\mu s$ and input size of 1000 packets were chosen, as the pair gives the minimum FPR of 2% and maximum TPR of 91%. On testing the system

---

[3] Refer to Equations 1and 2 for the definition and Section IV.A for a discussion on *dqueue*.

with the chosen parameters (Bin width = 500*us*, Input size = 1000 packets, and FPR = 2%) for a total of 10 additional trials, it was observed that the technique is accurate in detection approximately 92% of the time for UDP and 89% of the time for TCP traffic, as can be seen from Figure 9.

In the RAP attack scenario, the attacker would likely often hop on the connection for short bursts of time to avoid detection. Given the attacker's short-lived stay online, it is important that the classifier be able to work on a minimum amount of data. Also, considering the relatively negligible portion of WLAN traffic occupied by upstream data (in comparison with TCP downstream data), the classifier might not have much to work with and hence, must be trained accordingly.

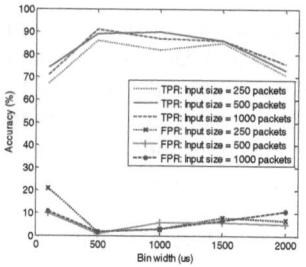

Fig. 8. Parameter tuning

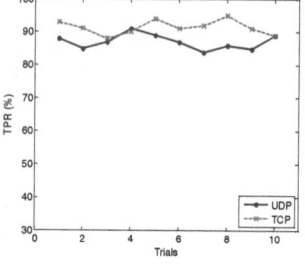

Fig. 9. TPR for chosen bin width and FPR

Each input data size may correspond to different application data on the RAP because each application (e.g., web browsing, email, file upload) contributes different amounts of upstream traffic to the classifier. Accordingly, the results shown in Figure 8 provide a sample of the system's accuracy for different classes of applications - each pertaining to a different input size. For a bin width of 500*us*, the system exhibits maximum accuracy that ranges from 87% to 91% for input sizes varying from 250 packets to 1000 packets. As a result, the attacker is slightly more likely to be detected if he were uploading a file of 1Mb than if he were reading the news at say, *www.cnn.com*, because the former would contribute the sufficient amount of data faster than the latter.

In this section, we discussed our first scheme of detection. In this method, our classifier is trained artificially on IAT signatures individually for different network speeds and different transport protocols for both the LAN and WLAN. We also showed the accuracy measures from lab experiments. This scheme is optimal when there is no interference on the channel and the link is stable. As will be shown in Section 6, its performance degrades as rate adaptation occurs in response to poor link quality. Therefore, in the next section we present a scheme that thrives during rate adaptation.

5   Scheme II – Rate Adaptation Based Detection

The 802.11 MAC protocol provides wireless entities with the ability to change their encoding scheme (data transmission rate) when the need arises. Using automatic rate fallback (ARF), when a node reaches a threshold of not receiving MAC-layer ACKs, it reduces its rate to one that corresponds to a stronger encoding algorithm in order to ensure more robust transmission.

As shown in [21], rate adaptation occurs regularly in WLANs because signal and link-layer interference are common phenomena. Given that rate adaptation occurs regularly, we seek to exploit this property that is specific to wireless streams to distinguish them from their wired counterparts. Particularly, the switching of the physical-layer data rate creates a variation in throughput and packet delay in a wireless transmission that is rarely found in wired traffic. We exploit the unique behavioral characteristics at the time of rate switching to identify wireless traffic.

In this section, we first examine the behavior of the IAT during switches in data rate. Based on this, we derive an artificial profile for the IATs during such shifts in data rate. The artificial profiles are incorporated into a classifier, which is then evaluated for accuracy.

5.1 Analysis

In this sub-section, we illustrate the effect of rate adaptation on a series of packets. Specifically, we show that there exists an IAT pattern that is exhibited only during rate adaptation and not when a pair of successively transmitted packets is sent at a constant physical layer data rate.

First, we visually illustrate how rate adaptation alters the arrival periods of packets transmitted at different data rates. Figure 10 is an example representation of the expected packet arrival sequence for a sample wireless transfer. In Figure 10, note that the IATs vary for each rate R_i because slower rates trigger greater packet delays.

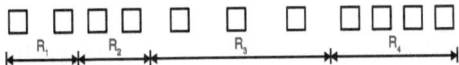

Fig. 10. Packet arrivals during rate adaptation

The probability P_i of the event R_i occurring depends on what we call the *channel interference index* (Ω) which has a range $\{0 \leftrightarrow 1\}$.

$$IAT_{wl} = \sum_i IAT_i P_i \qquad (20)$$

$$If\left(i < k \wedge \Omega \to 0\right)P_i \leq P_k$$
$$If\left(i < k \wedge \Omega \to 1\right)P_i \geq P_k \qquad (21)$$

In other words, the probability of occurrence of a lower transmission rate (in Equation 21, rate i is lower than rate k) is inversely proportional to signal interference and collisions. Our model safely assumes that the measure of interference Ω is not known prior and hence P_i is unknown.

This being the case, unlike Scheme I which assumes minimal to no rate adaptation, we choose to focus not on sets of IAT_i (that is, the IAT of two packets transferred at same rate) but instead on IAT_j (that is, the IAT of two packets transferred at different rates).

Having abstractly shown the influence of rate adaptation on the IATs and having settled on the idea that the inference model should be based on the IAT behavior during the transition in data rate (IAT_j), we proceed to study IAT_j.

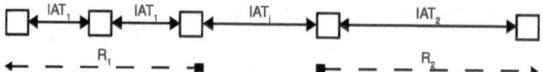

Fig. 11. IAT pattern during a rate switch

As shown in Figure 11, IAT_j is the delay during the *'jump'* from one rate to the next. Note that in Figure 11, IAT_j is of a different magnitude than IAT_i, where IAT_i is the IAT during rate R_i and $i = \{1,2\}$. Accordingly, in our classifier, we associate IAT_{wl} with IAT_j. To determine which link type the test data (that exhibits IAT_x) belongs to, we use the basic premise given in Hypothesis 5.1.

To illustrate the behavior of the jumps, an initial set of experiments were performed on an 802.11*b* WLAN; IATs for packet pairs transmitted at the same rate as well as different rates were extracted. In the absence of notable real channel interference, to stimulate rate adaptation in a simple lab testbed, the experiments were performed in the presence of a running microwave. A laptop was used as a sniffer on the wireless-side to collect the data rates corresponding to the packets within a transmission.

A sample of the IATs from a two minute upstream data transfer is shown in Figure 12. Although the interference resulting from the microwave usage was strong enough to invoke rate switches down to 2Mbps and sometimes 1Mbps, for the purpose of the current argument, the aggregated IATs of packets transmitted at 11Mbps, 5Mbps and

Hypothesis 5.1	**Hypothesis 5.2**
1: **if** $IAT_x \approx IAT_j$ **then**	1: **if** $R_i < R_{i+1}$ **then**
2: Report *Wireless*	2: $IAT_i > IAT_j > IAT_{i+1}$
3: **else**	3: **else**
4: Report *Wired*	4: $IAT_i < IAT_j < IAT_{i+1}$
5: **end if**	5: **end if**

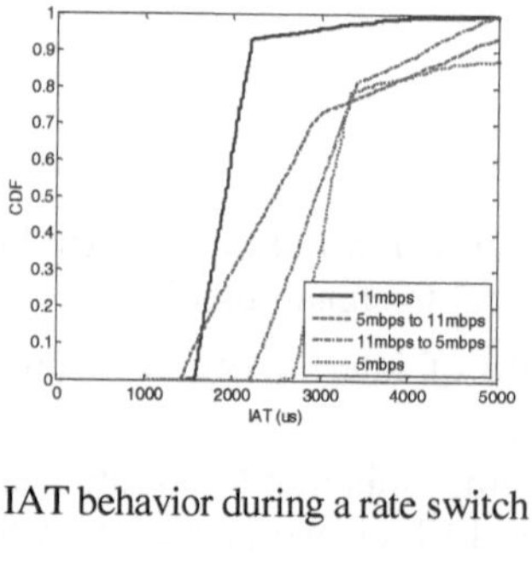

Fig. 12. IAT behavior during a rate switch – TCP

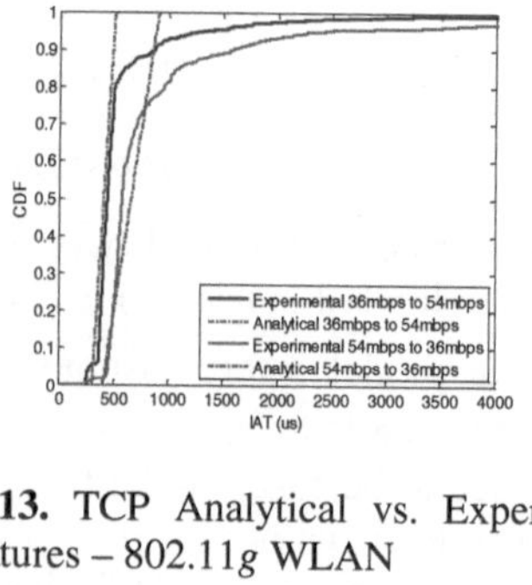

Fig. 13. TCP Analytical vs. Experimental Signatures – 802.11*g* WLAN

packets transmitted immediately after changes in data rate both ways are the only IATs shown in Figure 12.

It can be seen in Figure 12 that the IAT distributions of the jumps fall in between those of the stable rate phases before and after. This leaves us with Hypothesis 5.2.

The rationale behind this (as shown in Figure 14) is that during the transition from R_1 to R_2, the MAC-level ACK is transmitted at R_1 and the subsequent data frame at R_2. That is, a node which decides to reduce its data rate transmits the next data packet at the new rate but the MAC ACK for the previous data packet would still be sent from the AP at the old rate. Also, as can be seen from Figure 12, because of the difference in frame and MAC ACK sizes, the IAT distribution during the jump (IAT_j) is biased towards that corresponding to the rate following the jump. That is, since the frame size >> MAC ACK size and because the data frame is sent at the new rate, IAT_j is closer to the IAT associated with the new rate.

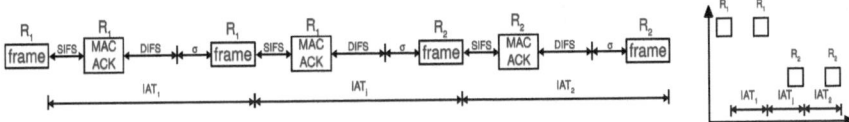

Fig. 14. DCF behavior during a rate switch

This difference in behavior during a rate switch can be exploited by studying how it reflects on individual delay components of the corresponding IATs, as shown below:

$$
dtrans_{wl\ (1,2)} = dtrans_{frame\ (1,2)} + DCF_{constant} + DCF_{random} \\
+ overhead_{MAC-ACK\ (1,2)} + overhead_{pkt\ (1,2)}
$$
(22)

$$
dtrans_{wl\ j} = dtrans_{frame\ 2} + DCF_{constant} + DCF_{random} \\
+ overhead_{MAC-ACK\ 1} + overhead_{pkt\ 2}
$$
(23)

Using Equation 23 as the base for our synthetic profiles, substituting jump-specific $dtrans_{frame}$ and $dtrans_{overhead}$ values, our classifier can be trained as shown in Figure 13. Similar to the synthetic IAT profiles shown for the (36Mbps, 54Mbps) pair in Figure 13, multiple such *jump* signatures were constructed for different data rate pairs as training sets for the classifier. Additionally, the training sets included IAT signatures for 10Mbps and 100Mbps LANs.

5.2 Classification Scheme

The classifier used for this method is similar to the one explained in Section 4.2, with appropriate changes made to incorporate the fact that only the IAT values during *jumps* in rates are considered for training and testing as opposed to the values during a stable rate period. In the Bayesian classifier, instead of comparing the entire trace of IAT readings with the profiles, individual values are inspected for possible *jumps*. That is, a comparison of two datasets (training and testing sets) is not required;

instead, it is sufficient to check individual incoming IAT values to see which IAT jump signatures they are closest to.

5.3 Experimental Setup and Validation of Wired-Side Approach

The experimental setup used to validate the scheme is similar to that used for Scheme I discussed in Section 4.3. As in [21], we use a synthetic means (microwave interference) to force rate switching to investigate Scheme II. One of the laptops is used as a sniffer on the wireless side, while another laptop is used to transfer data to the wired-side desktop sink server.

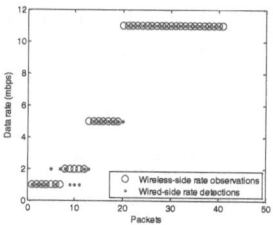

Fig. 15. Rate detection on wired and wireless sides

To determine whether a node is switching rates when capturing packets on the wireless side is simple, as its physical layer header contains the actual transmission rate. However, the rate in the wireless frame is not carried over to the wired-side. Accordingly, on the wired-side, we have to infer the rate by observing the packets' IAT pattern. We verified that this approach is viable by capturing traffic both on the wireless-side and the wired-side, and comparing the data rate observations made on the wireless-side with the data rate predictions made by the classifier on the wired-side. We observed packets that switch rates on the wireless side with a laptop acting as a sniffer capturing promiscuously (by looking at the *radiotap* header in the wireless frame) and concurrently on the wired-side by feeding captured IATs of the same packets into the classifier. From this, we were able to determine that specific IAT values on the wired-side correlated to confirmed rate adaptations on the wireless side.

Figure 15 gives a representative sample of the rates of the packets extracted on the wireless side and the rates inferred by the classifier on the wired-side, illustrating the correlation of rates of the same packets observed at both points. A total of 6000 upstream TCP data packets were transmitted with 81% of the rates predicted correctly. It is important to note that though the accuracy of classification of the data rates on the wired-side was 81%, the classifier is accurate in access link type classification up to an average TPR of 97% for UDP and 91% for TCP (refer to Figure 18). This is because even the IATs corresponding to the incorrectly inferred rates are closer to the synthetic *jump* IAT profiles that the classifier was trained on as opposed to the Ethernet IAT signatures.

The accuracy measures of the classifier used to test Scheme II are shown in the next sub-section.

5.4 Accuracy Measures

As in Scheme I, to validate the system, the bin width used in the Bayesian binning approach was first tuned to determine an optimum value for the classifier. Note that Scheme II operates independent of the input size as it does not compare the dataset as a whole with the profiles and instead studies the input trace a packet at a time.

For each bin width, ten trials were performed, in each of which the classifier was tested on TCP/UDP data packet pairs of upstream Ethernet and WLAN traffic. TPR/FPR were generated as a function of the fraction of the input trace accurately classified each time (Figure 16).

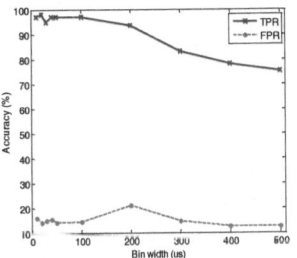
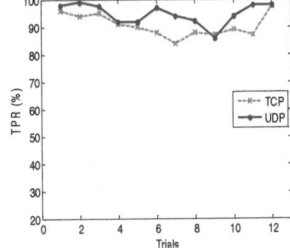

Fig. 16. Bin width tuning   **Fig. 17.** TPR for chosen bin width and FPR

In order to optimize the effectiveness of this technique, we calculate what we call the Effective Accuracy and find the optimum value that maximizes this difference between TPR and FPR in an attempt to make a balanced trade-off between the two metrics. For the chosen parameters (Bin width = $20\mu s$ and FPR = 14%), 12 additional trials are run to observe the TPR distribution (Figure 17).

Note that the accuracy measures shown in Figure 17 hold for WLANs with considerable rate adaptation. As will be shown in the next section, the accuracy of this scheme increases as a function of the amount of interference on the network and thus the method is not suitable for networks with minimal rate adaptation. In the next section, we propose a technique that bridges the strengths of the two schemes discussed so far in an effort to arrive at a comprehensive solution for normal networks (i.e., networks with varying levels of interference).

6 Consolidated Model

While Scheme I compares input sample traces as a whole with each of the profiles, Scheme II checks individual packet pairs within a trace for a switch in data rate. This implies that since the input sample trace to be compared may encompass several rates, Scheme I's accuracy is likely to subside with increased rate adaptation. Conversely, Scheme II will not accurately classify wireless traffic in the absence of a minimum degree of rate adaptation.

6.1 Analysis

In an effort to present a general solution that works both when the link is stable as well as when rate adaptation occurs, we revisit the *channel interference index* (Ω)[4] defining it as follows:

$$\Omega \propto \frac{Accuracy_{SchemeII}}{Accuracy_{SchemeI}} \tag{24}$$

Equation 24 essentially captures the inverse relationship between Schemes I and II. Scheme I works better when there is little to no interference, while Scheme II works better during interference. Thus, it is important to consolidate the pros of the two approaches in a way that the resulting system is effective regardless of the link stability.

6.2 Classification Scheme

To combine the two schemes, we partition the input data set into blocks of a constant size with the expectation that each block will be comprised of data at a specific rate. Of course this need not be the case. So, in addition to this, we exploit the fact that Scheme I detects the access link types of stable rate periods well and Scheme II detects the *jumps* well. For the combined solution, the input trace is fed into the classifier one block at a time. Scheme I contributes the network type/speed observation for each of the partitions and Scheme II points out where two stable rate periods intersect (that is, the *jumps* in data rate), the aggregation of which gives us the temporal distribution of rates for a series of packet pairs. This technique is illustrated in Figure 18, where x and y are the inferred data rates. Based on the inferred rates, the combined scheme determines the access link type of the individual partitions. The final access link type classification decision for the whole block of data is made as a function of the WLAN-to-Ethernet classification ratio of individual partitions. That is, the classifier decides between WLAN and Ethernet based on which link type is classified in majority of the partitions. The general idea behind this unified model is that if one of the two schemes fail, a healthy net effect is maintained as the other scheme chips in.

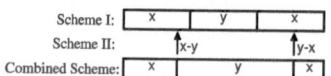

Fig. 18. Depiction of combined scheme

6.3 Experimental Setup

The experimental setup used to test the first two schemes is employed to validate the combined scheme. A block size of 250 packets is chosen. Accordingly, in our experiments, each input trace of 1000 packets is partitioned into four blocks of 250 packets each.

---

[4] Note that this metric was previously introduced in Section V.A.

In the next section, we evaluate the accuracy of the combined scheme (in comparison with that of the first two schemes) as a function of Ω by testing against data sets that differ in the number of times rate adaptation is invoked.

6.4 Accuracy Measures

The accuracy measures of the consolidated system (in comparison with those of the other two schemes) are shown in Figures 19.

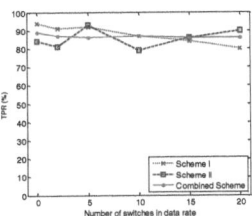

Fig. 19. Scheme accuracy comparison

A total of 14 trials were performed to assess how the TPR varies with an increase in the degree of rate adaptation. This testing set comprised of 2 trials each for the 7 different degrees of rate adaptation. The degree of rate adaptation is devised as a function of the number of switches in data rate invoked within the 1000 packet input data set. Figure 19 shows the results of such experiments performed individually for each of the three schemes. Results shown in Figure 19 are an average of the outcomes from separate TCP and UDP trials. Note that the combined scheme's accuracy is not as high as that of Scheme I. However, this technique is nonetheless effective and unlike the initial two schemes, the combined technique is realistic as it makes no assumption about the link quality.

7 Measure of Robustness and Scalability

In this section, we discuss how the system's performance scales to larger, more realistic networks. We evaluate the system's scalability in two scenarios - (i) a network where the classifier is placed multiple hops away from the AP via simulation, and (ii) a real network (as opposed to a lab testbed).

First, to test the combined scheme's scalability as a function of the classifier's distance from the AP, simulations were performed where detection takes place several hops upstream instead of the switch immediately connecting the AP to the LAN. This is important because the AP to be detected may not always be one hop away from the classifier node. We consider the effect of different fixed access-link and bottleneck delays at each hop, including the best-case (1*ms*, 10*ms*, and 50*ms*) as well the worst-case (300*ms* and 500*ms*) delays. The measurements observed indicate that despite a decrease in accuracy with an increase in the distance, the system averages a worst-case accuracy of above 60%, average-case accuracy of above 75% and best-case accuracy of above 85% (Figure 20a). The results shown in Figure 20a were obtained from simulations done using *ns2* and varying the number of hops between the AP and the classifier node. The TPR measurements shown in the figure are an average of results from 10 trials - each of 10,000 upstream data packets - performed separately for each delay value and tested individually for a given number of hops. The 10 trials comprised of 5 TCP and 5 UDP trials. The trace of 10,000 packets in each trial was fed into the classifier 1000 packets at a time.

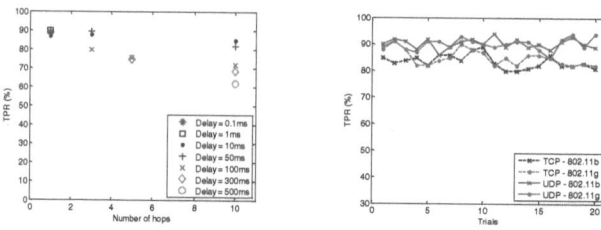

Fig. 20. Multi-hop accuracy: (a) Simulation, (b) Experiment

Next, we conducted experiments on a real network to arrive at the accuracy measures of the classifier when tested with traces from a real environment. Trials were performed on a multi-hop fiber-optic university backbone. A wireless node was made to connect via an AP from a classroom building to the wired-side server three blocks away in the Computer Science Department. The accuracy of the combined scheme was measured over a total of 20 trials performed individually for TCP/UDP data transfers and for 802.11*b/g* network configurations. In each trial, the classifier was tested on a 10 minute long trace for TPR measures. As shown in Figure 20b, the classifier is accurate up to approximately 90% of the time for UDP and 85% of the time for TCP.

8   Conclusion and Future Work

The proposed method detects RAPs by extracting characteristics unique to a wireless stream from network traffic. It makes use of two 802.11 MAC specifications to fingerprint wireless attributes from the wired-side making the process simple and scalable.

In this paper, we have studied the working and validated the accuracy of our detection techniques in several environments. This method is immediately deployable and is shown to scale well to realistic scenarios outside of a lab testbed. In the future, we will continue in this direction and further test the system for robustness to other use cases.

We plan to extend this work by scaling it to networks of greater traffic density by taking into consideration the effect of collisions in the network as a result of multiple users on the RAP. To this end we will study various error models and incorporate the traffic behavior during each of these into our design. Further, we intend to study the effect of link delay on the accuracy of the system in an attempt to derive a metric that the classifier shall be tuned for when placed multiple hops away from the AP. Also, we will test the system's robustness using different real network traces from publicly available archived sources (e.g., CRAWDAD).

Further, looking ahead in RAP detection, we must assume that the misfeasor could be tech savvy and aware of RAP defenses. To this end, we will analyze possible options that an attacker has to evade detection by cleverly altering his transmission pattern. Threat strategies that an attacker may employ include reducing or increasing his packet delay and interleaving his wireless transmissions with other types of traffic to bypass the classifier's signatures. Note that the DCF parameters can be manipulated in open source 802.11 drivers.

References

1. Beyah, R., Kangude, S., Yu, G., Strickland, B., Copeland, J.: Rogue access point detection using temporal traffic characteristics. In: IEEE GLOBECOM (2004)
2. Shetty, S., Song, M., Ma, L.: Rogue Access Point Detection by Analyzing Network Traffic Characteristics. In: MILCOM (2007)
3. Wei, W., Suh, K., Gu, Y., Wang, B., Kurose, J.: Passive online rogue access point detection using sequential hypothesis testing with tcp ack-pairs. In: IMC (2007)
4. Wei, W., Jaiswal, S., Kurose, J., Towsley, D.: Identifying 802.11 Traffic from Passive Measurements Using Iterative Bayesian Inference. In: IEEE INFOCOM (2006)
5. Wei, W., Wang, B., Zhg, C., Kurose, J., Towsley, D.: Classification of access network types: Ethernet, Wireless LAN, ADSL, Cable Modem or Dialup? In: IEEE INFOCOM (2005)
6. Baiamonte, V., Papagiannaki, K., Iannaccone, G.: Detecting 802.11 wireless hosts from remote passive observations. IFIP/TC6 Networking (2007)
7. Beyah, R., Watkins, L., Corbett, C.: A Passive Approach to Rogue Access Point Detection. In: GLOBECOM (2007)
8. Mano, C., Blaich, A., Liao, Q., Jiang, Y., Cieslak, D., Salyers, D., Striegel, A.: RIPPS: Rogue Identifying Packet Payload Slicer Detecting Unauthorized Wireless Hosts Through Network Traffic Conditioning. ACM TISSEC 11(2) (2007)
9. Cheng, L., Marsic, I.: Fuzzy reasoning for wireless awareness. International Journal of Wireless Information Networks 8(1) (2001)
10. Bahl, P., Padhye, J., Ravindranath, L.: Enhancing the Security of Corporate WI-FI Networks Using DAIR. In: ACM MobiSys (2006)
11. http://www.netstumbler.com
12. http://www.wimetrics.com/Products/WAPD.htm
13. http://www.proxim.com/learn/library/whitepapers/Rogue_Access_Point_Detection.pdf
14. http://www.airdefense.net
15. http://www.airmagnet.com
16. http://www.airwave.com
17. http://www.cisco.com/en/US/products/sw/cscowork/ps3915
18. Chirumamilla, M.K., Ramamurthy, B.: Agent based intrusion detection and response system for wireless LANs. In: ICC (2003)
19. Ma, L., Cheng, X.: A Hybrid Rogue Access Point Protection Framework for Commodity Wi-Fi Networks. In: IEEE INFOCOM (2008)
20. Songrit, S., Kitti, W., Anan, P.: Integrated Wireless Rogue Access Point Detection and Counterattack System. In: ISA (2008)

21. Beyah, R., Corbett, C., Copeland, J.: A Passive Approach to Wireless NIC Identification. In: ICC (2006)
22. Bianchi, G.: Performance analysis of the IEEE 802.11 distributed coordination function. Journal on Selected Areas of Communications 18(3) (2000)
23. Bing, B.: Measured Performance of the IEEE 802.11 Wireless LAN. LCN (1999)
24. Chatzimisios, P., Vitsas, V., Boucouvalas, A.C.: Throughput and Delay analysis of IEEE 802.11 protocol. In: 5th IWNA (2002)
25. http://www.isi.edu/nsnam/ns

A Novel Architecture for Secure and Scalable Multicast over IP Network

Yawen Wei, Zhen Yu, and Yong Guan

Department of Electrical and Computer Engineering,
Iowa State University, Ames IA 50011, USA
{weiyawen,yuzhen,yguan}@iastate.edu

Abstract. Currently, multicast services can be implemented at the IP layer or the application layer. While IP multicast violates the stateless paradigm of Internet and incurs great difficulties to congestion and flow control, application-layer multicast is lack of scalability due to the unreliability and resource constraints of end-hosts. Moreover, security is a main weakness in Internet-wide group communications. We propose in this paper a novel architecture for secure and scalable multicast in the Internet. In our architecture, a *Multicast Agent* in each Autonomous System (AS) is responsible for delivering multicast packets at the AS-level, relaying packets to end-hosts, and generating and updating keys to secure group communications. The proposed membership management protocol enables no-delay to membership updating; the proposed inter-domain routing protocol reduces the worst-case link stress by one magnitude compared to state-of-the-art protocols, and bounds the extra bandwidth cost within one percent compared to traditional IP multicast.

Keywords: IP Multicast, Routing protocol, Security, Inter-domain, Source-encoding.

1 Introduction

Multicast is an important and efficient mechanism to support many applications such as multimedia teleconferencing, news distribution, software updates and network games. Previous research efforts have been devoted to implementing multicast service either at the IP-layer or at the application-layer, however, the protocols implemented at both layers have drawbacks and have never been Internet-widely deployed.

In application-layer multicast protocols [5,6,7,8,11,13,14,16,20,22,27], end-hosts are organized into tree-based or mesh-based overlays to forward packets. Since end-hosts are limited in bandwidth and often experience abrupt crashes or failures, the overlay they form always suffers from large end-to-end latency and low data delivery rate. Meanwhile, since end-hosts need to periodically measure link quality to add/drop certain overlay links to improve tree or mesh topology, expensive operation overhead will be incurred, especially for large groups. Therefore, application-layer protocols cannot become a practical solution for large-scale group communications in the Internet.

Y. Chen et al. (Eds.): SecureComm 2009, LNICST 19, pp. 417–436, 2009.

We thus pass the hope on network-layer protocols [3,4,9,10,12,15,17,21,24]. However, IP multicast also has some limitations and has not been deployed through the Internet either. First, IP multicast requires routers to maintain per-group state and violates the stateless paradigm of the Internet. Second, it raises great difficulties in providing reliability, flow and congestion control at higher layers. Finally, IP multicast lacks a strategic business model and a security architecture. In the current open usage model, any host may send packets to an existing multicast group. Besides, flooding and DoS attacks will render multicast service unreliable or unavailable, and make the accounting for providing multicast service infeasible.

In this paper, we propose a secure and scalable architecture for Internet-wide multicast applications. In the proposed architecture, a border router called *Multicast Agent (MA)* exists in each Autonomous System (AS). These MAes are in charge of delivering multicast packets at the inter-domain level, relaying multicast packets to end-hosts at the intra-domain level, and generating and updating keys to secure group communications. The security are achieved in a hierarchical manner: the packets delivered between MAes are encrypted by a *global key*, and the packets delivered between end-hosts in a local domain is encrypted using a *local key*.

In the proposed architecture, we first design a membership and key management protocol. In our protocol, the membership information is explicitly distributed using *augmented-packets*, rather than using the traditional way that membership information are periodically exchanged between neighboring domains. By our design, not only bandwidth is saved from the exchanging traffic but also the propagation latency is reduced.

To achieve efficient inter-domain routing, we design an inter-domain routing protocol using the source-encoding technique. The MA at the source domain constructs and encodes dissemination tree information into each multicast packet. The benefits of such source-encoding are as follows: (1) The source domain knows all in-group domains, thus service fee can be properly charge by ISPs based on the scalability of the multicast group. (2) The privilege of receiving/sending packets is restricted only to in-group members, so a more secure usage model can be enforced. (3) No state information needs to be maintained at intermediate routers, i.e., the stateless nature of Internet is maintained. (4) Since the source can specify the targeted recipients of each packet, hence, subgroup communications can be conveniently supported. The last feature is especially beneficial in some applications where the participating members have heterogeneous interests. For example, in the Commercial Mobile Alert System (CMAS) [1], the text alerts related to disaster, immanent and child abductions are required to send to geographically targeted subgroups of people's cell phones.

In our inter-domain routing protocol, instead of requiring all tree information to be encoded into the packet header, we decompose in-tree nodes into two hierarchical levels, and *shim header* and *shim payload* of a packet encodes the two levels respectively. Such hierarchical design can effectively mitigate the packet duplication problem. It is proved by simulations that our protocol can achieve good scaling, e.g., the number of duplicated packets is around twenty on the most stressful link.

The rest of this paper is organized as follows. Section 2 provides an overview of our proposed multicast architecture. Section 3 describes the group membership management protocol. Section 4 describes the key management issues. Section 5 proposes an inter-domain multicast routing protocol. Section 6 evaluates the proposed multicast architecture through simulations and compares it with state-of-the-art multicast protocols. We discuss the related work in Section 7 and conclude the paper in Section 8.

2 Overview

In our proposed multicast architecture, we assume a border router called *Multicast Agent (MA)* exists in each AS. The MAes are in charge of delivering multicast packets at the AS-level, relaying multicast packets to end-hosts that are interested in sending/receiving these packets, and generating and updating keys to secure group communications. All multicast traffic in and out of an AS will be handled by the responsible MA (we will discuss multiple MAes within a local domain in the Discussion Section 3.3).

At the source domain, when the MA receives a multicast packet from an end-host, it first constructs an AS-level dissemination tree then inserts the tree information into the multicast packet. The encoded tree information can lead downstream MAs to correctly forward the packet. When a border router receives the packet, it forwards it to the local MA. The MA checks if any end-hosts in its domain are interested about this packet. If yes, it multicasts the packet within the local domain; otherwise, it does not perform the intra-domain multicasting. Then, the MA decodes the tree information in the packet and forwards the packet to border routers in other domains. In our architecture, the security are achieved in a hierarchical manner: the packets delivered at the inter-domain level are encrypted using a *global key* shared by MAes, while the packets within a local domain are encrypted using a *local key* shared by end-hosts and the local MA.

In the following, we will describe the group membership management protocol in Section 3. We then discuss the key management issues in Section 4. In Section 5, we propose the inter-domain routing protocol which is used to construct, decompose and encode/decode AS-level dissemination trees.

3 Group Membership Management

3.1 Intra-domain Management

To multicast a packet to end-hosts in a local domain, the MA should know which end-hosts are group members. We do not specify any intra-domain multicast protocols used in a local AS, because most existing protocols can scale well at the domain level. Specifically, if PIM-SM or CBT protocol is used, then the MA constructs a unicast tunnel to the local Rendezvous Point (RP) from which it knows the membership information; it also relays multicast packets through the tunnel to/from the RP. If PIM-DM or MOSPF protocol is used, then the MA can participate as an active member and perform like a normal end-host to multicast to the group.

3.2 Inter-domain Management

The inter-domain level group membership is managed by MAes. Since any in-group MA may become the source of a multicast packet and need to know destination domains to construct a dissemination tree, the membership information should be available to every in-group MA. A simple approach to achieve such group-wide awareness is maintaining membership information at a central server. However, this will lead to large query traffic towards the server and cause it overloaded or even out-of-service. Therefore, we suggest every in-group MA keeps a copy of the member list and collaboratively updates the list when membership changes.

In this subsection, by *membership* we refer to the membership at the domain-level. In other words, only when the number of end-hosts in an AS domain rises above zero or decreases to zero, the MA of this domain joins or leaves the multicast group and becomes an in-group or out-group MA.

Augmented Packet. We first introduce the *Augmented Packet* which is a basic technique used in our inter-domain membership management. The format of an augmented packet is shown in Fig. 1 where a normal packet is augmented with a *membership payload*. In this membership payload, a header contains 32 bits. The first 8 bits (denoted by n_j) indicate the number of newly-joined MAes; the second 8 bits (denoted by n_l) indicate the number of leaving MAes; the last 16 bits are the checksum computed over the entire membership payload to ensure its integrity. Followed the header are the IDs of joining MAes then the IDs of leaving MAes.

The reason for introducing such augmented packet is to avoid the extensive traffic caused by sending a separate updating message for every member join or leave, especially if we consider the fact that millions of multicast groups may exist in the Internet simultaneously.

Now the question is which MA should be in charge of appending the membership payload to its multicast packet? A reasonable answer is that the first MA

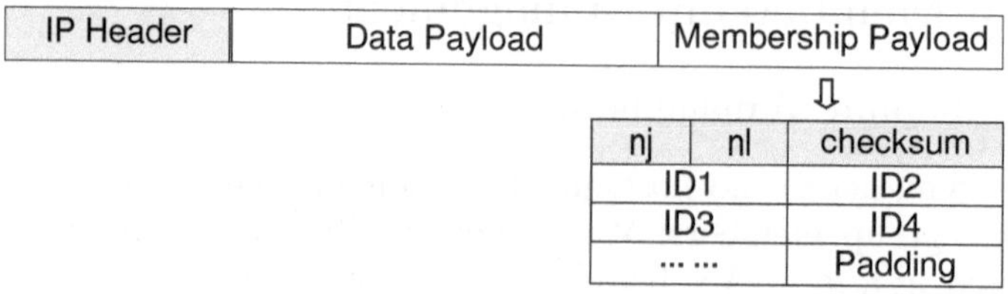

IP Header	Data Payload	Membership Payload

⇩

nj	nl	checksum
ID1		ID2
ID3		ID4
... ...		Padding

Fig. 1. Augmented packet format

that multicasts packet after membership changes should take the responsibility, because the delay in membership updating can be minimized. However, the next source MA is not directly available except for some application (for example, in IPTV, all data are originated from a source domain where the TV station resides). In our approach, we suggest MAes use self-learning algorithms to make predictions based on historical records on previous source domains. They can use those algorithms that are used to handle page replacements in virtual memory management, or they can predict the next source domain as the one that most recently or most frequently sends packets, or they can consider both frequency and recentness and apply aging algorithm for the prediction.

Membership Updates upon Member Joins. The detailed process for MA to join a multicast group is described in the following steps (Fig. 2).

- Step 0: If a MA (say, MA-1) wants to join a group, it should first contact the group registry server and get bootstrapped with a list of in-group MAes. (Here, we consider close usage model in which every member explicitly goes through registration process to obtain the privilege of sending/receiving packets from this group. Close usage model provides many benefits such as better control, traffic engineering and accounting.) The list obtained in bootstrap does not need to be complete, i.e., it may just contain several in-group MAes.
- Step 1: The new member MA-1 randomly selects a MA (say, MA-2) from its bootstrap list, and sends to it a join message.
- Step 2: MA-2 predicts the next source MA based on the record of source domains during a past period of time, and transmits the AS number of this MA (say, MA-3) together with a full list of all in-group MAes back to MA-1.
- Step 3: If MA-1 happens to have a multicast packet originated from its domain at this moment, it directly multicasts an augmented packet with its ID included in the membership payload, then goes to step 6. Otherwise, MA-1 will solicit help from MA-3 by sending to it a request, asking MA-3 to send an augmented-packet with MA-1's membership information.

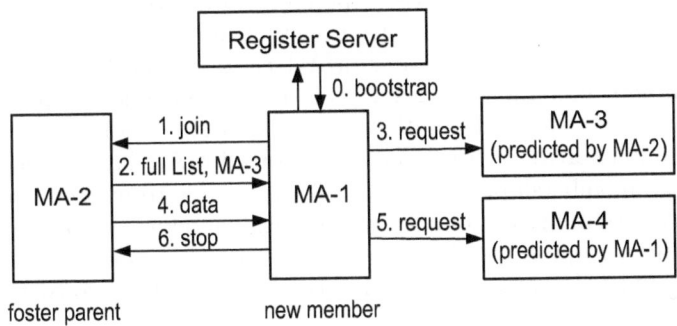

Fig. 2. A new MA joins a multicast group

- Step 4: Notice the next source prediction may not be accurate, i.e., MA-3 may not become the source during a period of time in the future. In this case, MA-1 will not receive any data for this group during this time since other in-group MAes are not aware of the existence of the new member. To mitigate this problem, we designate MA-2 as the *foster-parent* of MA-1. That is, when MA-2 receives any packet for this multicast group, it should relay the packet to MA-1 through a unicast tunnel between them.
- Step 5: If after a certain period of time t after joining, MA-1 has not received any packet containing its ID in the membership payload, it knows MA-3 has not sent out any augmented-packet yet. Then MA-1 will predict another MA (say, MA-4) based on its own historical record of the source domains, and send to it a request message. MA-1 repeats this step by periodically contacting different MAes until its membership can be group-widely notified. However, if more than a certain number of predications fail, MA-1 will multicast a separate message by itself to announce its membership.
- Step 6: In the end, MA-1 informs its foster-parent to remove it from the foster-children list.

Membership Updates upon Member Leaves. If an existing MA wants to leave a multicast group, it either multicasts an augmented-packet itself, or predicts the next source MA and informs it about its leaving. We notice that before a leaving MA is finally removed from the member list, the MA may continue receiving multicast packets, so it should perform as an in-group MA for a period of time to guarantee correct data delivery.

3.3 Discussion

In this subsection, we would like to discuss the consistency issue and multiple MAes issue associated with our inter-domain group membership management protocol.

Consistency. We notice that there may be some timing and delay in communicating MA joins/leaves at the inter-domain level. Consider a scenario where a users is channel surfing. The corresponding MA, MA-1, will first announce join then announce leave. Assume MA-1 contacts a predicted source MA, MA-2, and sends a join request to it. Later, MA-1 contacts another predicted source MA, MA-3, and announces its leave. Due to the possible timing/delay, some MAes may receive leave information by MA-3 prior to join information from MA-2, which may create a situation where not all MAes have a consistent view about MA-1's membership. Our fix to this problem is to use sequence numbers for each MA in sending out join/leave updates. Therefore, when in-group MAes receive conflicting group membership information, they can ignore the obsolete information and only keep the latest information to achieve consistency.

Multiple MAes. In our architecture, we assume one border router plays the role as the local MA. The selection can be based on which border router is the exit point to most other ASes, or which has the smallest IP address in the local domain. However, the vulnerability and unreliability associated with a single MA is present. First, when the current MA crashes or reboots, another border router has to take over the responsibility and the handoff process will introduce some delay. Second, since all group traffic in/out of the domain is through a single MA, the traffic concentration problem is obvious given the inter-domain traffic is high in the backbone. We are considering to allow multiple MAes to share the workload of handling group traffic in a local domain. Some design challenges exist to guarantee a consistent view among multiple MAes of both the intra- and inter-domain membership, and achieve the smooth cooperations between them, and these will be our interested research subjects in the future.

4   Group Key Management

Given that security is one of the main weaknesses of IP multicast, the need to secure multicast packets is particularly apparent and crucial. Secure group communication systems mostly rely on a group key, which is a secret only known to group members and used to encrypt multicast messages. When group membership changes, a new group key should be established to guarantee forward security and backward security, that is, members who have left the group cannot decrypt messages in later sessions, and new members cannot decrypt messages in previous sessions. The challenging problem is to design key management schemes that can scale to large groups or groups with highly dynamic memberships. Previous key management schemes, including the key graph approach [25,26] and its extensions, require at least O(logN) computation and communication per rekeying operation, where N is the number of group members. In many Internet multicast applications such as the massive multiplayer games, the value of N, i.e. the number of participating players, can be several millions, which will make rekeying overhead particularly huge.

Instead of requiring all end-hosts use a single group key to secure their communications, in our architecture, we suggest to organize the end-hosts into a hierarchy that is consistent to the Internet topology. Namely, the end-hosts within an AS form a subgroup, and the domain's MA is the subgroup head. Within a local domain, multicast packets are encrypted using a *local key* shared by all in-group end-hosts and the local MA; at the inter-domain level, packets are encrypted using a *global key* shared by all in-group MAes. Therefore, if an end-host joins or leaves a group, only the local key needs to be changed, but the global key is maintained the same, which can greatly mitigate the scalability problem.

When a multicast packet enters/leaves a domain, decryption and re-encryption should be performed. Precisely, when a MA relays a multicast packet into its domain, it needs to decrypt the packet using the global key then re-encrypt it using the local key. To enhance the efficiency of the cryptographical operations, we can adopt techniques suggested by previous works such as [18]. Such that the local/global keys are not used to encrypt and decrypt a packet directly, instead, they are used to encrypt and decrypt a random *session key*, and the session key is the real key that encrypts the packet. In this way, decrypting and re-encrypting a packet is reduced to decrypting and re-encrypting the session key.

In the following subsections, we discuss the key management at the intra- and inter-domain level in more detail.

4.1   Local Key Management

The MA serves as *key server* within each domain. The MA is responsible to distribute a local key for each group, and update the key whenever membership changes. We emphasize that when an end-host leaves or a new end-host joins the group, only the local key has to be updated, while the global key will remain the same. The MA shares a pair-wise secret key with every end-host. (1) When a new host joins, the MA generates a new local key, encrypts it using the shared key with the new host and unicasts to the host. Meanwhile, it encrypts the new local key using the old local key, and multicasts to previous hosts. (2) When an existing host leaves, the MA also generates a new local key, encrypts it using each of the shared keys with the remaining members, and multicasts one message containing all the encrypted keys to them. The computation overhead for member join and leave is $O(1)$ and $O(n)$ respectively, where n is the local group size. Although scalability may not be a severe issue within a local domain (compared to Internet-wide groups), we do not limit ourselves to adopt any key management algorithms that have less computation, communication or storage overhead.

4.2   Global Key Management

We assume each MA shares a pair-wise key with every other MA in the Internet. Since it is not reasonable to assume a single Internet-wide key server for all multicast groups, we require the global key for each group is maintained by in-group MAes. Precisely, for a group, one in-group MA is selected as the key server, and is responsible of distributing a new global key to the current in-group MAes, whenever a new MA joins or an existing MA leaves the group. We can adopt any re-keying algorithm for the key server to update the global key.

If the key server itself wishes to leave, it has the authority to designate another in-group MA to be the key server. The designation can be based on reliability, bandwidth, membership length and local group size. For instance, the MA with the largest number of participating end-hosts is less possible than other MAes to quit the group, thus it can be selected as the next key server. For a newly-joined MA whose membership have not been notified to other in-group MAes (thus not known to the key server either), the multicast packets relayed from its foster-parent to the new member can be encrypted and decrypted using their pair-wise key.

5   Inter-domain Multicast Protocol

In this section, we propose an efficient inter-domain multicast protocol. We first discuss how a dissemination tree is constructed, decomposed and encoded at

AS Number	AS Path	Next Router	Interface
4515	34225 41692 3491 4515	193.138.164.1	m0
6356	34225 1299 6830 22773 22318 6356	193.138.164.1	m1
... ...	... ...	... ...	... ...

Fig. 3. Routing table at MA34225

the source MA, and then we describe how the tree information is decoded and updated at downstream MAes.

5.1 Preliminary Work

Before we introduce our inter-domain routing protocol, we first take a look at a previous work related to our protocol, i.e., the Free Riding Multicast (FRM) [21] protocol. In FRM, the border router of a source AS computes a dissemination tree from the union of unicast paths, then puts the tree information into the fixed-size *shim header* of each multicast packet. If the tree is very large, then multiple packets have to be transmitted to carry the encoded tree information. The packet duplication problem can be very severe, e.g., it is reported that when the dissemination tree spans on all AS domains in the Internet, the worst-case physical link has to transmit about 150 duplicated packets.

Although two approximation methods were suggested to mitigate packet duplication problem associated with FRM, there are some practical issues with the suggested methods. (1) The first method is to omit customer ASes at tree leaves when encoding the dissemination tree. However, this requires the border router at the source AS to know customer-provider relationships between other ASes. Although some techniques [23] can help guess AS relationships by exploring the AS graph, the guess cannot be validated because the customer information is proprietary information of an ISP. Therefore, the guesses would be wrong and packets would not be able to be delivered to some valid in-group members at the leaves. (2) The second approximation method is to replace all the tree links connecting a node by one *aggregated link*, if the number of tree edges from the node is a large fraction of its total edges. It is claimed that an AS domain A forwards packets to its neighbor B only when B-X (X is a neighbor of B) lies on the path from A to some destination domain. However, this cannot ensure A-B is an in-tree link because the path containing B-X do not necessarily pass domain A. The consequence is that some packets will be sent on non-tree links.

In the following, we propose a novel inter-domain multicast protocol that can effectively mitigate packet duplications and achieve efficient data delivery.

5.2 Construction of Hierarchical Dissemination Tree

For a multicast group, the source MA can learn all in-group MAes from the membership management protocol we have described in Section 3. To construct a dissemination tree, the source MA looks up its *MA routing table* to find out

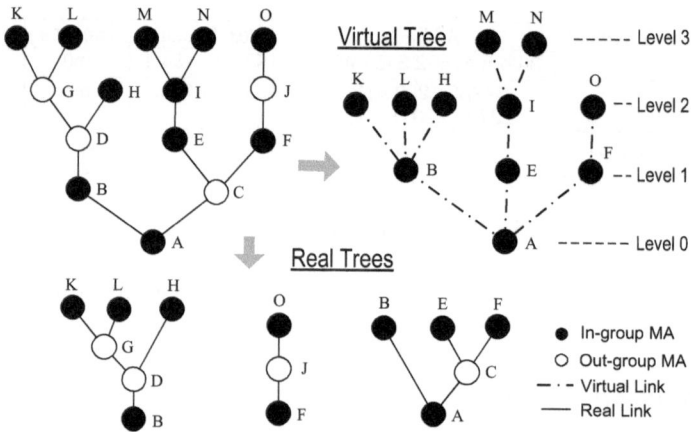

Fig. 4. Decomposition of the flat tree

the unicast path leading to each in-group domain (an example is shown in Fig. 3). The routing table is indexed by AS numbers and each entry provides the AS paths, next-hop, interface, and other path attribute information. This table can be easily constructed using BGP RIB and updated using BGP routing updates. If multiple policy-permitted paths exist leading to a same AS, then the best-quality path can be selected. For instance, the path with the smallest AS-hops or the shortest geographical distance [19] can be selected.

After consulting this table, the source MA can construct a flat tree by the union of unicast pathes leading to all destination domains. Then, it decomposes the flat tree into one *virtual tree* and multiple *real trees*. Fig. 4 presents an example of the decomposition. The virtual tree consists of *virtual links* connecting in-group MAes, and the real trees consist of *real links* connecting out-group MAes. The source MA puts the information about the virtual tree and the real tree rooted at itself into the packet. The real tree information can guide out-group MAes to forward the packet properly until it reaches in-group MAes at level 1 in the virtual tree. Then, each level-1 MA replaces the real tree in the packet with the one rooted at itself, which leads the packet to level-2 MAes, and so forth. Finally, the packet will traverse through the whole tree and visit every in-group MA.

5.3   Shim Header and Shim Payload

Now the problem is how the source MA encodes and attaches tree information to each multicast packet. Since real-tree information is used by out-group MAes and virtual-tree information is used by in-group MAes, we can encode them into *shim header* and *shim payload* of a packet, respectively (Fig. 5).

We adopt the technique of bloom filter to encode tree information into shim header and shim payload. Bloom filter is a space-efficient probabilistic data structure that can support membership queries. It uses k independent hash functions to map every member to k different positions in a m-bit vector. When using

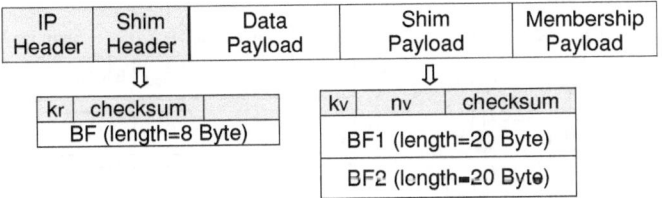

Fig. 5. Shim header and shim payload formats

bloom filter, false negative is guaranteed to be zero, but false positives is nonzero and will increase with the number of members hashed into the filter. Given the highest acceptable false positive, the maximum number of members one bloom filter can afford can be determined. We assume all MAes in the Internet share a same set of hash functions, therefore, an MA can use the set of functions to decode the links from filters constructed by other MAes.

Fig. 5 shows the format of shim header and shim payload. Shim header consists of 32 control bits and a 8-byte bloom filter to contain real tree links. The control bits include 4 bits (denoted by k_r) that indicate the number of hash functions, 16 bits checksum that are computed over the entire shim header to ensure the integrity, and the remaining 12 bits for future use. Shim payload consists of 32 control bits and multiple fixed-length bloom filters (20 bytes in our design) . The control bits include 4 bits (denoted by k_v) that indicate the number of hash functions, 12 bits (denoted by n_v) that indicate the number of bloom filters, and 16 bits of checksum that are computed over the entire shim payload to ensure its integrity.

5.4   Tree Encoding on Source MA

In this subsection, we give an example that illustrates the tree encoding process at a source MA. As we can see in Fig. 6, source node A sends packet P1 to an in-group node (node B) and packet P2 to an out-group node (node C). In packet P1, the shim header contains link A:B, which is both a real link and a virtual link, and the shim payload contains virtual links B:K, B:L and B:H. In packet P2, the shim header contains links A:C, C:E and C:F, and the shim payload contains two bloom filters corresponding to two subtrees, one with virtual link F:O and the other with virtual links E:I, I:M and I:N. In this example, the whole dissemination tree is first decomposed into one virtual tree and three real trees, then the virtual tree is decomposed into three sub virtual trees. Given the maximum number a bloom filter can afford, the decomposition of the virtual tree is to ensure no bloom filter contain more links than the threshold.

5.5   Tree Decoding and Updating on Transit MAes

We now discuss the checking and updating process at downstream MAes. If the MA is an out-group MA, it simply checks all its AS-neighbor-links to find out

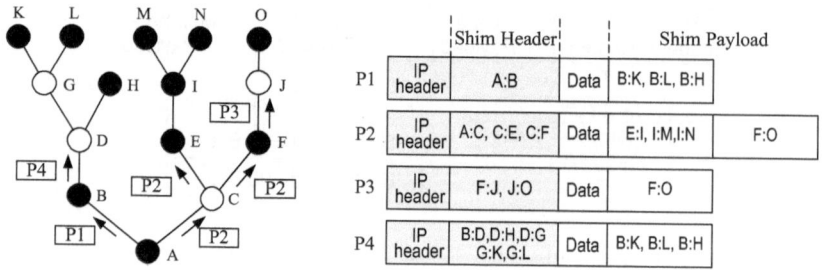

Fig. 6. Example of tree encoding, decoding and updating

the present ones in the shim header, then forwards the packet accordingly. No updating on the shim header or payload is needed. If the MA is an in-group MA, it first checks the shim payload and decodes the present virtual links, then it rewrites the shim head to encode the links of the real tree rooted at itself, and removes some bloom filters in the shim payload. As the example shown in Fig. 6, a transit MA node, node F, receives packet P2 and sends packet P3. It rewrites the shim head to contain links F:J and J:O, then it removes the second bloom filter in the shim payload, because none of the virtual links E:I, I:M and I:N in that filter will be of future use as the packet traverses deeper into the tree through node F.

When an in-group MA checks the bloom filters in the shim payload, it does not need to check the virtual links connecting itself and every other in-group MA, instead, it only needs to check those connecting itself and its children MAes in the virtual tree rooted at itself. We can prove the correctness of such checking by the following proposition.

Preposition 1. In a multicast group G, node i's child in a virtual tree rooted at node j $(j \neq i)$ must be i's child in the virtual tree rooted at itself.

Proof. We prove by contradiction. We denote the virtual tree constructed for multicast group G and rooted at node t by T_t^G. Assume in the virtual tree T_j^G, there exists a child k of node i, such that k is not i's child in the virtual tree T_i^G, which means another node s must lie on the shortest path from node i to node k in T_i^G, denoted by L_{ik}. Since L_{ik} must be a part of L_{jk} in T_j^G (otherwise L_{jk} will not be the shortest), node s will also lie between node i and k in the path L_{jk} in T_j^G, which implies that node k is not a child of i in T_j^G. Contradicted.

5.6   Discussions

We now discuss some properties of our proposed inter-domain routing protocol.

IP Fragments. During the packet delivery process, IP fragmentation may take place at an intermediate router. Our protocol deals with such situations by doing the followings: first, both the IP header and the shim header from the original IP datagram should be copied to new datagrams. Therefore, the shim header is per-packet based, which enables out-group MAes to solely look at a packet's shim header to forward it correctly. Second, the shim payload should be fragmented and inserted into multiple packets. Since in-group MAes are the destinations of all the fragmented packets, IP reassembly will be performed at every in-group MA and the shim payload will be recovered.

Bandwidth Consumption. The bandwidth consumption is minimized in our protocol by adopting two important techniques.

First, the hierarchical decomposition effectively alleviates packet duplications. In the basic source-encoding approaches, the entire flat tree is encoded into the shim header of a packet. Since the shim header has very limited length and cannot accommodate too many links, multiple shim headers have to be constructed, which directly causes duplicated packets. In our protocol, the virtual tree is inserted into a packet's payload which can contain up to 65KB data. Only a small real tree needs to be inserted into the shim header. Fig. 6 showcases the benefits. The shim header of packet P2 only contains three links A:C, C:E and C:F in our protocol; without tree decomposition, the shim header would have to contain eight links A:C, C:E, C:F, E:I, I:M, I:N, F:J and J:O.

Second, we encode virtual links into multiple bloom filters instead of a single large one, which further reduce bandwidth consumptions. As a packet traverse deeper into the dissemination tree, in-group MAes can continuously remove some "useless" filters that contain virtual links not present in its subtrees. Therefore, packet size can be reduced and bandwidth can be saved. This advantage can be seen clearly in Fig. 6, where node F removes the second bloom filter in the shim payload, since none of the virtual links E:I, I:M and I:N will be of future use.

Processing Delay. In our protocol, there are three procedures where delay may be introduced.

First, the processing delay can be introduced at the source MA by constructing shim header and shim payload. However, we notice a MA can pre-construct the dissemination tree and cache the shim header and payload information. For example, a MA will cache for groups for which it has a large number of end-host users, because it is very probable for it to become the source MA in the future.

Second, since an out-group/in-group MA should check the shim header/payload for present real/virtual links, there is a delay associated with the look-ups in bloom filters. Fortunately, bloom filters can be implemented using very efficient hardware like TCAM [28], such that the to-be-checked links can be hashed into multiple rows and accessed in parallel to achieve high efficiency.

Third, delay may be incurred at an in-group MA by the rewriting operation of the shim header of a packet. To mitigate this delay, a MA can pre-compute and cache the bloom filter containing the real links associated with each virtual link, then it can rapidly construct the shim header by simply XORing these filters. As an example, node B in Fig. 6 can cache three bloom filters: filter-1 contains B:D and D:H corresponding to the virtual link B:H, filter-2 contains B:D, D:G and

G:K corresponding to virtual link B:K, and filter-3 contains B:D, D:G and G:L corresponding to the virtual link B:L. After receiving packet P1, node B checks the presence of virtual link B:H, B:K and B:L. So it XORs filter-1, filter-2 and filter-3 and inserts the result into the shim header of packet P4. Currently, less than 30,000 AS domains exist in the Internet, which means the memory cost for caching the bloom filters associated with virtual links will be no more than 240KB at all MAes.

6   Simulation Result

In this section, we conduct simulations to evaluate the performance of the proposed multicast architecture. We mainly focus on the network cost of our inter-domain multicast protocol and compare it with other state-of-the-art protocols. The protocols we use to compare with our protocol include the followings. (1) IP multicast: the dissemination tree is composed of shortest reverse paths from the source AS to destination ASes. (2) Per-AS unicast: the source AS sends a separate unicast packet to each destination AS. (3) FRM: the dissemination tree is constructed by the union of unicast paths from the source AS to destination ASes, and the whole tree is encoded into shim header of every packet. (4) AS-level overlay: the dissemination tree is constructed using our proposed protocol, but the packets are unicast between different MAes.

We use the following metrics to measure the network costs associated with different multicast protocols. (1) *Link stress* is defined as the number of *duplicate* packets transmitted on a physical link. By duplicate packets we mean the packets that have identical *application* payload, though they may have different protocol-related headers or payloads. Obviously, the stress on all physical links is one in IP multicast. (2) *Protocol overhead* is defined as the extra bandwidth consumed by the protocol-related data in the packets. In IP multicast and per-AS unicast, the protocol overhead is zero; in FRM, AS-level overlay and our protocol, the protocol overhead is not zero because the tree information is present in the packets' headers/payloads. (3) *Bandwidth cost*: This metric evaluates the total bandwidth consumption to multicast one packet to all receivers. Essentially, this metric reflects the combined impacts of the link stresses and protocol overheads.

Our simulations are conducted using real BGP data from RIS [2]. RIS is a RIPE NCC project that collects and stores routing data from the Internet. We download one day's files of BGP data collected by the Remote Route Collectors (RRCs) in MRT format. After removing the incomplete measurements and IPv6 paths, we select twenty IPv4 full tables as the basis of our experiments. The results are averaged over 200 runs using the 20 BGP tables with 10 runs per table. In our further work, we will implement our protocol on real border routers for better understanding of the protocol's behavior in dynamic real-world environments.

6.1   Link Stress

Fig. 7(a) compares the CDF of link stresses of our protocol and other multicast protocols in a typical run when the multicast group consists of 10,000 in-group domains. The per-AS unicast and AS-level overlay have very high link stresses, with the worst-case link stress reaches four and three orders of magnitude respectively. In FRM, about 99.6% links see one transmission, but the worst link stress is over one hundred. Since the worst case always happens on links between the root and its children domains in the dissemination tree, the congestion of these links will impact many downstream members. Our protocol effectively reduces the stresses on all physical links: more than 99.9% links see exactly one transmission, and the worst-case stress can be reduced to only 14.

Fig. 7(b) plots the worst link stress for our protocol and other multicast protocols for different group sizes. We select in-group AS domains randomly and increase group size from 10 domains to 20,000 domains. In per-AS unicast protocol, the worst stress increases linearly with the group size. In AS-level overlay, the worst link stress first increases then decreases to a few hundreds. This is because the duplicate transmissions are incurred by the unicast between overlay nodes. As the group size increases, an overlay node will have more children, resulting in more stresses on physical links leading to these children domains. However, when the group gets even larger, in-group ASes get closer and the unicast paths between them become shorter, thus, less links are shared between the unicast paths and the link stresses drop accordingly. In FRM, the worst link stress gradually increases with the group size, with the highest one around 100. In our protocol, the max worst link stress is only 21 and happens when group size is around 5,000. Since the duplicate packets are caused by multiple shim headers that encode real tree links, hence, the fact that the size of real trees first increases then decreases with the group sizes directly causes the same tendency on link stresses.

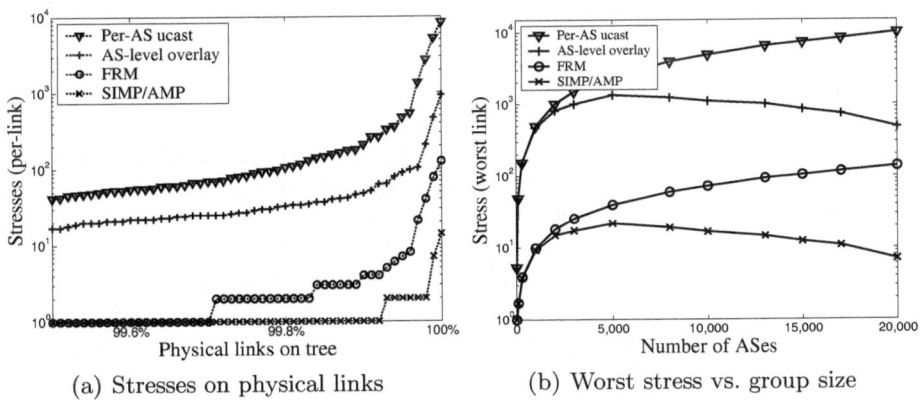

(a) Stresses on physical links   (b) Worst stress vs. group size

Fig. 7. Link stress distribution and worst-case Links stress

6.2   Protocol Overhead

Fig. 8 plots the total protocol overhead involved in our protocol and FRM for different group sizes. Since this metric measures overhead due to shim headers and payloads, the overhead increases with the to-be-encoded tree size for both schemes. However, the growing speed is quite different. In FRM, the protocol overhead grows almost linearly with the group size, which is expected because the shim header encodes all tree links. In our protocol, the protocol overhead grows much more slowly, and when in-group domains is 20,000, a total of 0.5MB bandwidth is consumed. We attribute the conservation on protocol overhead when using our protocol by two major reasons: first, the tree decomposition hides the real tree links from virtual tree, resulting in less information to be encoded in shim headers; second, the shim payload updating at in-group domains enables further reducing of shim payloads.

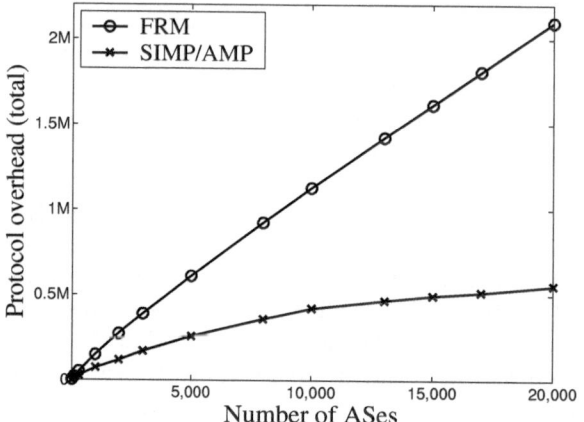

Fig. 8. Protocol overhead vs. group size

6.3   Bandwidth Cost

Fig. 9 shows the bandwidth cost against group sizes. We use 1 KB as the size of the data packet, and normalize the bandwidth costs of different multicast protocols with respect to IP multicast. We have repeated this study with different packet sizes and observed similar ratios for all protocols, which implies that the bandwidth consumption is largely due to packet duplications rather than the protocol overheads. This also explains the similar shapes of the curves in this figure compared to the curves in Fig. 7(b). The two upper curves correspond to per-AS unicast and AS-level overlay, which show that they introduce at least 100% and 10% more bandwidth cost respectively. The other two curves correspond to FRM and our protocol. We see our protocol performs the best again, and incurs extra bandwidth cost no more than 1% in the worst case compared to traditional IP multicast.

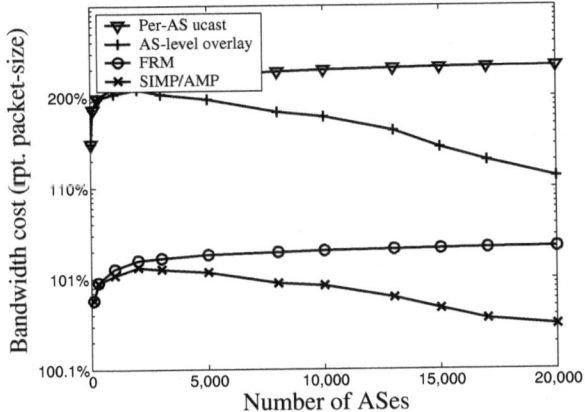

Fig. 9. Bandwidth cost vs. group size

7   Related Work

Numerous protocols have been proposed to provide multicast service at the network layer. Intra-domain multicast protocols include DVMRP [24], PIM-DM [3] and MOSPF [17], etc. Inter-domain multicast protocols include CBT [4], PIM-SM [10], BGMP [15], PIM-SSM [12], etc. In DVMRP/PIM-DM [24,3], the multicast data is first broadcast to all routers, then every router that receives unwanted multicast data sends a pruning packet to its parent. MOSPF [17] protocol is an extension to OSPF protocol. Every router refers to the link state database and the group membership knowledge, and constructs shortest-path tree from any source to all receivers. CBT [4] and PIM-SM [10] constructs a shared tree rooted at a group-specific Rendezvous Point (RP). BGMP [15] constructs bidirectional shared tree that is rooted at the home domain whose address allocation includes the group's address. However, all of above protocols cannot become efficient solutions for Internet-wide multicast services because of their scalability limitations. PIM-SSM [12] protocol bypasses the discovery of RP, and constructs shortest path trees rooted at the single source domain. This protocol can be used only for single-source multicast model.

FRM [21] uses the source-encoding forwarding technique, where the source router forms the dissemination tree and inserts the tree information into the header of multicast packet. Their protocol has severe packet duplications especially when the group size is large. While our protocol utilizes hierarchical decomposition and shim payload (instead of shim header) to accommodate large trees, we can effectively reduces the number of duplicated packets.

Many application-layer protocols have been proposed in recent years. They can be classified as tree-based [5,11,13,20,27] or mesh-based [7,8,16] protocols, depending on whether the dissemination tree is maintained directly, or a mesh is maintained and the tree is constructed over the mesh on demand. The proxy based overlays [6,14,22] can provide more reliable and efficient multicast service, where the proxies are application servers deployed throughout the Internet. They self-organize into overlays to disseminate multicasting packets, and relay packets to attached end-hosts.

Our protocol can be viewed as a variant of proxy based overlay where the MA in each domain plays the role of multicast proxy. However, both the membership management and the routing method in our protocol are completely different from those in traditional proxy overlays. (1) To manage membership information, proxy overlays propagate updating information to all members through periodically neighboring exchanging. We do not adopt such techniques in our protocol because of the following reasons: first, MA servers are much more reliable than end-hosts, it is not worthwhile to detect the rare abrupt failures of members using periodic refreshing messages at the expense of high bandwidth consumptions. Second, the constraints on the exchange frequency between border routers introduce latency to the propagations of membership information. In our protocol, we use foster-parent technique to reduce the join-delay, and use augmented-packets to minimize the communication overhead. (2) For the routing mechanisms, proxy overlays construct mesh and measure link qualities periodically, our protocol explores the knowledge about the domain-level unicast paths available at BGP border routers, and constructs source-optimal dissemination trees directly.

8   Conclusion

In this paper, we proposed a secure and scalable multicast architecture over IP Network. In our architecture, the AS-level group membership are explicitly maintained by in-group Multicast Agents (MA), the inter domain routing is based on source-encoded information in multicast packet, and the multicast packets are encrypted using two-level keys: a global key at the inter-domain level and a local key at the intra-domain level. Our future work involves implementing the proposed multicast architecture in real Internet environments.

Acknowledgments

This work was partially supported by NSF under grants No. CNS-0644238, CNS-0626822, and CNS-0831470. We appreciate anonymous reviewers for their valuable suggestions and comments.

References

1. Rules for Delivery of CM Alerts to the Public During Emergencies (April 2008), http://hraunfoss.fcc.gov/edocs_public/attachmatch/FCC-08-99A1.pdf
2. Routing Information Service (October 2007), http://www.ripe.net/projects/ris/index.html
3. Adams, A., Nicholas, J., Siadak, W.: Protocol Independent Multicast - Dense Mode (PIM-DM) Protocol specification (Revised). Internet Draft (October 2003)

4. Ballardie, T., Francis, P., Crowcroft, J.: Core based trees (CBT) an architecture for scalable inter-domain multicast routing. Technical report, San Francisco, CA (September 1993)
5. Banerjee, S., Bhattacharjee, B., Kommareddy, C.: Scalable application layer multicast. In: Proceedings of ACM SIGCOMM (September 2002)
6. Banerjee, S., Kommareddy, C., Kar, K., Bhattacharjee, B., Khuller, S.: Construction of an efficient overlay multicast infrastructure for real-time applications. In: Proceedings of IEEE INFOCOM (April 2003)
7. Chawathe, Y.: Scattercast: An Architecture for Internet Broadcast Distribution as an Infrastructure Service, Ph.D. Thesis, University of California, Berkeley (December 2000)
8. Chu, Y., Rao, S.G., Zhang, H.: A case for end system multicast. In: Proceedings of ACM SIGMETRICS (June 2000)
9. Deering, S., Cheriton, D.: Multicast routing in datagram internetworks and extended LANs. ACM Transactions on Computer Systems 8(2), 85–110 (1990)
10. Fenner, B., Handley, M., Holbrook, H., Kouvelas, I.: Protocol Independent Multicast sparse mode (PIM-SM): Protocol specification (October 2003); Internet Draft
11. Francis, P.: Yoid: your own internet distribution (March 2001), http://www.isi.edu/div7/yoid/
12. Fenner, B., Handley, M., Holbrook, H., Kouvelas, I.: Protocol Independent Multicast - Sparse Mode (PIM-SM): Protocol Specification (Revised). Internet Draft (March 2001)
13. Helder, D.A., Jamin, S.: End-host multicast communication using switch-tree protocols. In: Proceedings of the Workshop on Global and PeertoPeer Computing on Large Scale Distributed Systems (GP2PC) (May 2002)
14. Jannotti, J., Gifford, D., Johnson, K., Kaashoek, M., OToole, J.: Overcast: reliable multicasting with an overlay network. In: Proceedings of the Symposium on Operating Systems Design and Implementation (October 2000)
15. Kumar, K., Radolavov, P., Thaler, D., Alaettinoglu, D., Estrin, D., Handley, M.: The MASC/BGMP architecture for inter-domain multicast routing. In: Proceedings of SIGCOMM, Vancouver, Canada (September 1998)
16. Liebeherr, J., Beam, T.: HyperCast: a protocol for maintaining multicast group members in a logical hypercube topology. Networked Group Communication, 72–89 (1999)
17. Moy, J.: RFC 1585: MOSPF. Analisys and Experience. Proteon Inc. (March 1994)
18. Mittra, S.: Iolus: A framework for scalable secure multicasting. In: ACM SIGCOMM, pp. 277–288 (1997)
19. Oliveira, R., Lad, M., Zhang, B., Zhang, L.: Geographically Informed Inter-domain Routing. In: Proceeding of IEEE International Conference on Network Protocols (ICNP) (October 2007)
20. Pendarakis, D., Shi, S., Verma, D., Waldvogel, M.: ALMI: An Application Level Multicast Infrastructure. In: Proceedings of 3rd Usenix Symposium on Internet Technologies & Systems (USITS) (March 2001)
21. Ratnasamy, S., Ermolinskiy, A., Shenker, S.: Revisiting IP Multicast. In: Proceeding of SIGCOMM 2006, Pisa, Italy, September 2006, pp. 11–15 (2006)
22. Shi, S., Turner, J.: Routing in overlay multicast networks. In: Proceedings of IEEE INFOCOM (June 2002)
23. Subramanian, L., Agarwal, S., Rexford, J., Katz, R.H.: Characterizing the Internet Hierarchy from Multiple Vantage Points. In: Proceedings of IEEE INFOCOM (June 2002)

24. Waitzman, D., Partridge, C., Deering, S.: Distance Vector Multicast Routing Proto-
    col. ARPANETWorking Group Requests for Comment, DDN Network Information
    Center (November 1988); RFC-1075
25. Wallner, D., Harder, E., Agee, R.: Key management for multicast: Issues and ar-
    chitectures. IETF Request For Comments, RFC 2627 (June 1999)
26. Wong, C.K., Gouda, M.G., Lam, S.S.: Secure group communications using key
    graphs. In: ACM SIGCOMM, pp. 68-79 (1998)
27. Zhang, B., Jamin, S., Zhang, L.: Universal IP multicast delivery. In: Proceedings of
    the International Workshop on Networked Group Communication (NGC) (October
    2002)
28. Content Addressable Memory Cypress Semiconductor, http://www.cypress.com

Reliable Resource Searching in P2P Networks[*]

Michael T. Goodrich[1], Jonathan Z. Sun[2],
Roberto Tamassia[3], and Nikos Triandopoulos[3,4]

[1] Dept. of Computer Science, U. California, Irvine, USA
[2] School of Computing, Univ. of Southern Mississippi, USA
[3] Dept. of Computer Science, Brown University, USA
[4] Dept. of Computer Science, Boston University, USA

Abstract. We study the problem of securely searching for resources in p2p networks where a constant fraction of the peers may act maliciously. We present two novel hashing-based schemes that can be employed to reliably support *resource location* and *content retrieval* queries, limiting the ability of adversarial nodes to carry out attacks. Our schemes achieve scalability and load balancing and have small authentication overhead. In particular, for a network with n peers, resources are securely located with $O(\log^2 n)$ messages and content from a collection of m data items is securely retrieved with $O(\log n \log m)$ messages.

Keywords: peer-to-peer, overlay networks, distributed hash tables, one-way hash functions, digital signatures.

1 Introduction

An overlay peer-to-peer (p2p) network is a distributed structure imposed on a set of machines, called nodes or peers, for sharing data and computing resources. A p2p network can achieve load balancing and scalability by allowing peers to efficiently join and leave the network and users to efficiently store and retrieve data content. Data storage is typically supported by realizing a *distributed hash table* (DHT) that exports a basic put/get API. A data item can be inserted into the DHT with a put operation under a key and can be retrieved from the DHT with a get operation given its key. At a lower level, any resource is mapped to some peer that is responsible for this resource and can be efficiently located.

In this paper, we study the problem of *verifying* the resource searching functionality in a p2p network in the presence of faulty or malicious nodes. While faulty nodes are trouble enough, adversarial p2p nodes—a considerable threat since most p2p systems do not impose any restrictions on membership in the network—can be especially troublesome. For example, a coalition of adversarial nodes may wish to degrade the network performance by falsifying responses to redirection queries during resource location. Alternatively, nodes responsible for

---

[*] Work supported in part by NSF grants 0713046, 0713403, and 0724806, the RISCS Center at Boston University and the Center for Geometric Computing at Brown University.

some stored data may respond with content that appears to be a file of interest, but is in fact of degraded quality, virus infected, or outdated. Even more insidiously, adversaries may collude to systematically misdirect queries to a "parallel" p2p network that has invalid content. Finally, a group of nodes may mimic normal behavior and only aim at taking control over a small set of target items, by maliciously subverting resource locations or falsifying content retrievals.

To defend against such attacks, we are interested in designing techniques that protect the integrity of *resource location* at the network level and of *content retrieval* at the application level. That is, we wish to authenticate the p2p routing paths (followed by the distributed location process) as well as the p2p content (returned through a get operation over a DHT). With respect to routing protection, we aim at defending against *shunting* attacks, where adversaries misdirect queries and updates to malicious nodes. With respect to content protection, we would like to detect *content forgery*, where invalid data is returned, and *replay attacks*, where out-of-date content is retrieved.

We consider an adversary that inserts corrupted machines into the system and controls their behavior, stored content and routing information. However, we assume that at all times the adversary controls a constant fraction of the participating peers—$\frac{1}{4}$ in our case, which is reasonable for any large-scale p2p network. Thus, we do not consider denial-of-service attacks. Our goal is to design *crypto-enhanced* schemes that employ lightweight cryptographic primitives, such as collision-resistant hashing. When malicious behavior is sporadic or selective, we additionally wish that if no attack is in place, our verification mechanisms asymptotically incur no extra overhead. To the best of our knowledge no existing work has this *mode-adaptability* property for crypto-based secure routing. Applying standard techniques for content authentication results in solutions that do not achieve either scalability or load balancing. For instance, storing signed items results in a linear-size overhead as all items must be resigned to prevent reply attacks. Also, hash-tree based schemes introduce "hot-spots" in the system, as nodes storing hash values close to the root are more heavily accessed.

We assume the existence of a PKI where the public keys of the users (not necessarily peers of the p2p network) publishing data into the DHT are known to all parties. In addition, for an owner of a data collection of size m we assume the availability of a public and reliable storage of size $O(\log m)$ (e.g., a web page managed by the owner). Thus far, existing works have addressed the two problems in isolation, solely providing either route or content verification.

Related work. Early work on *secure routing* [3] in p2p networks tolerates certain attacks by assigning verifiable identifiers to network nodes. Numerous DHTs (e.g., [14,17]) have been shown to tolerate random failures. Other schemes (e.g., [10,16]) have been designed to deter misbehavior in adversarial models, and in particular some schemes (e.g., [2,5,6]) use *quorum-based* approaches, where regular network nodes correspond to large random blocks of machines and faulty behavior is prevented through majority-voting. Schemes using *redundancy* in searching (e.g., [9,11]) have also been proposed to tolerate random Byzantine behaviors. Most p2p storage systems (e.g., [4,15]) support content authentication using *"sign-all"* techniques where data items are each individually signed. Signature amortization (i.e., signing a single digest) is used in some systems by storing the so-called *self-certified* data [7], but only for large individual items in static data sets. A *distributed Merkle tree (DMT)* is presented in [18]; realizing a p2p extension of Merkle's hash tree, this scheme lacks load-balancing.

Our contributions. We present two new authentication schemes for efficiently and securely verifying resource location and content retrieval operations, respectively. Our schemes are based on corresponding novel *hashing schemes*, which constitute extensions of hash trees to general directed acyclic graphs (DAGs).

Our first authentication scheme, called *skip-DHT*, is based on *skip graphs* [1]. Its hashing scheme embeds a set of DAGs in a skip graph so that source-to-sink paths in each DAG correspond to search paths in the skip graph. Our construction efficiently authenticates all possible search paths that can be used for resource location. By combining this scheme with *quorum-based* techniques (e.g., [2]), an n-node skip-DHT supports resource locations that are cryptographically verifiable with $O(\log n)$ messages in $O(\log n)$ time, has near-optimal query complexity in the absence of faulty nodes and is attack-resistant in the presence of a constant fraction $\leq \frac{1}{4}$ of adversarial nodes using $O(\log^2 n)$ messages.

Our second authentication scheme is a middleware component that can operate on top of *any* DHT to efficiently verify put/get operations on a data set of size m owned by a given data source. We define a hashing scheme of high expansion over the data items and we store this structure in the DHT so that retrieved content can be associated with many equivalent verification hash paths, linking each item to one of $O(\log m)$ publicly available digests that are signed by the data source. As paths can be retrieved with uniform workload, we obtain a distributed implementation of Merkle's hash tree where load balancing is preserved. Using certain algorithmic techniques, item insertions and delayed item deletions can be supported with $O(\log n \log m)$ amortized time and communication overheads.

Combined together, our two schemes yield a new distributed hash table with certain unique features: (*i*) it is the first DHT to provide cryptographic security guarantees for both resource location and content retrieval; (*ii*) it has near optimal searching performance in the absence of adversarial nodes; (*iii*) it achieves

Table 1. Qualitative comparison of our authentication schemes with other approaches

path verification	secure routing	quorum-based	redundancy	scheme 1
attack-resistant	•	•	•	•
crypto-enhanced	–	–	–	•
mode-adaptable	•	–	–	•

content verification	self-certified	sign-all	DMT	scheme 2
dynamic	–	•	•	•
replay-safe	N/A	–	•	•
load-balanced	–	•	–	•

both scalability and load balancing. Table 1 compares our work with previous approaches for path and content verification.

2   Resource Location Authentication

In this section, we describe a new attack-resistant DHT that is based on the structure induced by skip graphs [1]. To authenticate the search paths in the DHT, we design a hashing scheme that is defined over its graph structure.

A skip graph on a set K of keys is a distributed structure that supports operation $\mathsf{suc}(k)$, returning the smallest key $k' \in K$ such that $k' \geq k$. Although designed for the purpose of supporting order-based queries, skip graphs provide a natural method for support $\mathsf{put}/\mathsf{get}$ operations. The structure of a skip graph can be viewed as a distributed extension of skip lists [13]. Both skip lists and skip graphs consist of a set of increasingly sparse, sorted, doubly-linked lists ordered by levels starting at level 0, where membership of a particular key $k \in K$ in a list at level i is determined by the first i bits of a potentially infinite sequence of random bits associated with k, referred to as the *membership vector of k*, and denoted by $m(k)$. We denote the first i bits of $m(k)$ by $m(k)|i$. In the case of skip lists, level i has only one list, for each i, which contains all keys k such that $m(k)|i = 1^i$, i.e., all keys whose first i coin flips all came up heads. As this leads to a bottleneck at the single node present in the uppermost list, skip *graphs* have 2^i lists at level i, which we will index from 0 to $2^i - 1$. Key k belongs to the jth list of level i if and only if $m(k)|i$ corresponds to the binary representation of j. Hence, each key is present in one list of every level until it eventually becomes the only member of a singleton list. The set of all lists to which a particular key k belongs meets the definition of a skip list, with membership in level i determined by comparison to $m(k)|i$ rather than to 1^i. We refer to such a skip list as the skip list defined by $m(k)$ and denote it by $SL(m(k))$ or $SL(k)$.

To search for the successor of k' we begin from the sparsest list at, say, level x and traverse the list by pointers to the right as far as possible without moving to a node whose key is greater than the key sought; we proceed downward to the list at level $x - 1$ in the same way, until level 0 is reached, where we move rightward to get the result. In a distributed setting, we map each graph node at level i in to a network node, so that each node stores only four links: pointers L and R for the network addresses of the machines assigned to the nodes of the list immediately before and after the given node, and pointers U and D for the machines responsible for the same key at levels $i + 1$ and $i - 1$. To search for the successor of key k', a machine mapped to key k performs a search in the skip list $SL(k)$ defined by $m(k)$. If $k < k'$, a rightward search is performed beginning at the top-most list of $SL(k)$, which we refer to as the root of k and denote by $r(k)$. This root list must contain key k and hence can be reached by following the U pointers of the lower level nodes of k. Based on the analysis in [13] it can be shown that queries in a skip graph take $O(\log n)$ time and messages.

Search-path hashing scheme. We present a hashing scheme consisting of a collection of DAGs embedded in the skip graph so that the authenticity of any root-to-leaf search path may be verified by the querier. Overall, our hashing scheme is an extension of the one used in [8]. Let h be a cryptographic collision-resistant hash function, and let $h(a, b) \triangleq h(h(a)\|h(b))$. Let v be a level-i node in a skip list $SL(k)$, with neighboring nodes $w = R(v)$ and $u = D(v)$, and denote by $d(v)$ the digest of v. Node w is called a *plateau* node if its key does not appear at level $i+1$ in $SL(k)$ or a *tower* node otherwise. Then skip list $SL(k)$ is hashed as follows: If v is at level 0, then $d(v) = h(ID(v), ID(w))$ if w is a tower node or $d(v) = h(ID(v), d(w))$ if w is a plateau node; if v is not at level 0, then $d(v) = d(u)$ if w is a tower node or $d(v) = h(d(u), d(w))$ if w is a plateau node. All these digests can be computed efficiently.

Lemma 1. *If L is a distributed skip list with n nodes with a longest search path of size H and where each node maps to a machine, then the digests of L can be computed by respective nodes in H rounds using at most three messages per node (sent plus received) and $n-1$ messages in total.*

Given the true digest of a root node digest in the above hashing scheme, a querier is able to verify the value returned by the node at the end of the search path, since every search in a skip graph is also a search in a skip list. Thus, if the digests of every skip list in the skip graph were computed and stored at those nodes, the searches could be verified given the true value of the root node's digest. On the surface, this seems an unsatisfactory solution, as nodes are present in as many as $\frac{n}{\log n}$ different skip lists, and hence would seem to need to store an equally large number of digests. However, consider a level-i list L in a skip graph corresponding to membership prefix s. Suppose the above hashing scheme is applied to two skip lists, which have membership prefix sb, $b \in 0, 1$. Then the digests of all nodes in L are identical between the two skip lists. Therefore, it turns out that each node takes on only two distinct digest values, one for those skip lists in which the node is a plateau and one for those in which it is not.

Lemma 2. *A skip graph can be hashed to authenticate the search path of each membership query, with each node maintaining two digests. The two digest values at each node can be computed using $O(1)$ messages per node and $O(n)$ messages for all nodes. With high probability, this process takes $O(\log n)$ rounds.*

Quorum-based extension. To make this path authentication scheme resilient to shunting attacks, we need to satisfy the following requirements: 1) the digests are computed correctly at each node and passed correctly to neighboring nodes, and 2) the querier machine knows the true value of the root node's digest. We observe that the only machines that begin a query from a particular root node v are those machines assigned to v or some node directly below v. As such, the hashing algorithm should pass each root digest down to the nodes directly beneath it. We therefore need a message-passing scheme that is resistant to adversarial tampering in order to satisfy these both requirements.

We consider the quorum-based extension (e.g., as in [2]) of our skip-graph, where each skip-graph node corresponds to a *supernode* consisting of $\Theta(\log n)$ machines. Members of a supernode are completely connected, forming a clique, edges between supernodes correspond to complete sets of edges between their members, and data mapped to a supernode is stored by all of its members. Doing so increases the degree of each machine and the number of stored keys by a factor of $O(\log n)$. For a constant fraction of adversarial nodes, if it is guaranteed that each supernode contains a random subset of machines, then, with high probability, every supernode will consist of a majority of honest machines.

In this redundant DHT, reliable search and update operations can be performed using a voting scheme in which each step in the traversal of the skip graph is verified by requesting the correct local answer from every machine in the current supernode. Also, when a search reaches the supernode responsible for the key sought, all of its members are polled to determine what data, if any, matches the search key. Our scheme employs this polling-based search whenever the path-verification protocol indicates an error in the resource location execution. Furthermore, we use voting-based computations to ensure that the digests are correctly reported to neighboring supernodes and the root digests are correctly reported to the supernodes beneath the root. This increases the asymptotic message costs by a factor of $\log^2 n$, as each node-to-node communication because hash updates now uses $O(\log^2 n)$ messages.

Updates. As described, our authentication scheme supports secure resource locations in a static collection of keys, since the digests of all root nodes need to change when a key is added or removed from the skip graph. We regain efficient support for updates as follows. We assume that the fraction of bad nodes is less than $\frac{1}{4}$ and use the construction of [2] to assign node identifiers from the interval $[0, 1)$ to the machines in such a way that w.h.p. every interval of length $\frac{c \log n}{n}$ contains a $\frac{3}{4}$-majority-good supernode of $\Theta(\log n)$ machines. We construct a skip graph whose keys consist of the node identifiers of the smallest member of each supernode. Data items are stored to supernodes through a pseudo-random hash function mapping arbitrary strings to $[0, 1)$ and then to the closest supernode identifier. Thus, exact searches for data items are supported by searching for the hash of the desired key. We refer to this structure as a *skip-DHT*.

We use the skip-graph hashing scheme to certify query results in the skip-DHT in a way that avoids recomputing the digests after every key or machine update. We use the machine identifiers in a supernode as the data that will be hashed as the digest of leaf nodes. Therefore, data updates no longer must yield an update in the root digests, since supernode membership is verified instead—we do not need to know the current list of machines in a supernode to have confidence in the query results. Instead, we rely upon knowing that the majority of the remaining original machines can be trusted. This is true as long as less than $\frac{1}{2}$ of the original nodes have left the network.

Overall, when a resource location query is executed, a *non-redundant* search is carried out given the query resource key—that is, a pointer to a single, arbitrary member of the next node in the search path is requested from a (single, arbitrary) member of the current node. When a destination machine is reached that claims to be a member of the supernode nearest the search key, it must provide a list of the identifiers of the original members of the supernode to which it belongs. The querier then computes the hash of this list and checks it against the verification path. This verifies the successor supernode in the skip list, thus the correct resource location. Each of these steps requires $O(\log n)$ messages.

Theorem 1. *An n-node skip-DHT satisfies the following properties: (1) In the absence of faulty nodes, verifiable exact-match queries are executed with $O(\log n)$ messages in $O(\log n)$ time; (2) In the presence of a constant fraction of adversarial nodes, queries are answered and securely verified with $O(\log^2 n)$ messages in $O(\log n)$ time; (3) The hashing scheme adds only a constant number of messages to amortized bandwidth usage for adding and removing machines.*

3   Content Retrieval Authentication

In this section, we study the problem of authenticating content at the application level through the put/get core functionality of any DHT. Our goal is to design a distributed scheme that verifies that data items claimed to have been added by a data source were really put in the DHT by this entity and have not been modified by malicious nodes. We wish this scheme to achieve *load balancing*, that is, to evenly distribute the workload related to authentication across the network nodes. We consider a standard query model where an underlying DHT stores key-value pairs of the type (k, x), each added through operation $put(k, x)$, where keys are unique identifiers and values are associated with keys. We assume that the DHT supports operation $get(k)$, which returns the value associated with key k, with $O(\log n)$ expected time and message costs.

For simplicity, we assume that a single data source is storing items in the system; for more data sources, we make use multiple invocations of our scheme. We assume that the public key of each data source storing data in the DHT is known to any entity querying the DHT. Also, we assume the availability of some *public reliable storage* that is associated to a given data source and that can be easily accessed and updated independently of the underlying DHT. The size of this information is only *logarithmic* in the number of data items published by the source. In practice, this assumption is easily implementable through a web service that posts to a web-site a small amount of data regarding a data source.

Load-balanced hashing scheme. Our data structure achieves signature amortization by applying a hashing scheme over the data items stored in the DHT. The main idea in our construction is to use a hashing scheme G of high expansion rate, namely with a structure that resembles the FFT computation graph or a butterfly network, such that for any data item, there exist many equivalent verification paths. We distribute DAG G to the network nodes of the underlying DHT by appropriately indexing the digests and storing them as special data items. We preserve the structure of the hashing scheme G in the DHT as follows: the network node storing the digest of node v in G also stores the keys under which the digests of the immediate successors and predecessors of v in G are stored in the DHT. We then *randomize* the generation of the verification paths to achieve a uniform workload over the visited network nodes.

We describe our hashing scheme G for m data items and its embedding into an n-node DHT. For simplicity and without loss of generality, we assume that $m = 2^k$. The nodes of G are partitioned into $k+1$ levels, each having m nodes. The nodes at level 0 are source DAG nodes, each associated with a data item. Each of the nodes at one of the remaining levels has two predecessors nodes at the previous level. The edges in G are defined so that the nodes at level k are the roots of m perfect binary trees over the data set. More formally, let us number the nodes on each level and denote with $v_{i,j}$ the j-th node of G on level i, $i = 0, \ldots, k$, $j = 0, \ldots, m-1$. For $i > 0$, node $v_{i,j}$ has two incoming edges from nodes $v_{i-1,j}$ and $v_{i-1,j+\delta(i,j)}$, where $\delta(i,j) = (-1)^{\lfloor j/2^{i-1} \rfloor} 2^{i-1}$. Let h be a cryptographic collision-resistant hash function. For $i = 0$, we set $d(v_{i,j}) = h(k\|x)$, where (k,x) is the data item associated with $v_{i,j}$. For $i > 0$, we set $d(v_{i,j}) = h(d(v_{i-1,j}) \| d(v_{i-1,j+\delta(i,j)}))$. By symmetry, the nodes of G at level i store 2^{k-i} distinct digests. The data source signs the single digest stored at nodes of level k and makes it available as public information. Then, each DAG node $v_{i,j}$ is indexed by a unique identifier $id_{i,j}$, where in particular node $v_{0,j}$ that is associated with data item (k,x) is indexed by k, and is inserted in the DHT as a special data item, using $id_{i,j}$ as the key and the digest and identifiers of its predecessors and successors in G ($O(1)$ information) as the value.

Query and verification. We now describe how **get** operations are handled. We begin by performing a query according to the underlying DHT structure (e.g., as discussed in the previous section). Given that data item (k,x) stored at network node W is located by the DHT, node W initiates a randomized process for generating a verification path for (k,x). Namely, W flips a coin to determine which of its two parents at level 1 (next node in the path) to contact next (through a resource location operation, first). In general, a network node V at level j randomly chooses the next network node (to be contacted while forming the verification path) independently and with probability $\frac{1}{2}$. Thus, any query results in a verification path of length $O(\log m)$, using $O(\log m)$ location operations, with $O(\log m \log n)$ computation and communication costs. Through the randomized search process, every verification path for a fixed data item is actually an independent and identically distributed random variable and no hot-spots are created while accessing the authentication structure. The verification path is returned by the DHT and given this, one can authenticate the answer of operation **get** by processing the digests contained in the path, verifying the publicly available signed digest and checking their consistency. The total storage required is $O(m \log m)$; that is, assuming perfect mapping functions from keys to network nodes (usually through a cryptographic hash function), the storage is logarithmic in n per network node, when $m = O(n)$—i.e., still optimal, since most DHTs use routing tables of logarithmic size. Using a caching technique as in [18], we can further improve the creation of the verification paths.

Updates. To support updates, we modify the scheme described above using a dynamization technique due to Overmars [12], which allows to transform a static data structure into a corresponding dynamic structure. The idea is to partition a data set of size m into sequence of $O(\log m)$ *blocks*, where the size of each block is twice the size of the previous block, and to completely rebuilt blocks after updates, as necessary. We apply this technique to support insertions of data items with new keys. Let D be a data set of size m and let $(b_k, b_{k-1}, \ldots, b_1, b_0)_2$ be number m written in binary, with $b_k = 1$. Note that items in D are not assumed to be sorted. We partition D into $\lfloor \log m \rfloor + 1$ blocks $B_0, B_1, \ldots, B_k$, each a subset of D, according to the weights of the bits of m, i.e., $|B_i| = b_i \cdot 2^i$. Let then $G(i)$, $0 \leq i \leq k$, denote the hashing DAG described in previous section that is built for the items of block B_i. DAG $G(i)$ has $b_i \cdot 2^i \cdot i$ nodes. DAGs $G(0), G(1), \ldots, G(k)$ are used separately as authentication structures: that is, for $i = 0, \ldots, k$, if $b_i = 1$, the source signs the top-level digest h_i of DAG $G(i)$ and each $G(i)$ is distributed over the network nodes as before. For any queried data item in block B_i, the corresponding verification path in $G(i)$ is retrieved using $O(i)$ location operations. Thus, $O(\log m)$ signed time-stamped digests (one for each block) are made available as public information.

We perform insertions of data items through operations put as follows. Let i be the smallest i such that $b_i = 0$ or $i = k+1$ if no such i exists. To insert an item x into D, we merge DAGs $G(0), G(1), \ldots, G(i-1)$ to create DAG $G(i)$ for the new block $B_i = B_0 \cup \ldots \cup B_{i-1} \cup x$. Note that $|B_i| = 1 + \sum_{j=0}^{i-1} 2^j = 2^i$. The insertion of a data item into a set of size m stored into a DHT of size n takes $O(\log m \log n)$ expected amortized time. Accordingly, we update the public information: the data source creates new fresh time-stamps and re-signs the publicly available digests. This occurs for all blocks after every update of a block, independently of whether or not the corresponding block structure has been altered in the most recent update. Thus, at any point in time, we maintain $O(\log m)$ fresh signed digests as public information. At asymptotically no additional cost and using similar ideas with the verification of queries, the data source can verify the correctness of an operation put performed by the DHT: any change in the hashing scheme is checked for consistency with the $O(\log m)$ signed digests.

We can also support *delayed deletions*, defined in our context as item removals that do not actually occur on-line, but instead occur at some future time and during the insertion of new items. Asymptotically, these deletions incur no additional communication or computational cost. In particular, we schedule the deletion of an item in block B_i during the construction phase of a new DAG $G(j)$, $j > i$, where j depends on the exact state of the authentication structure. This deletion procedure requires minor modifications to the above insertion algorithm. Replay attacks are eliminated by having the data source S performing controlled delayed deletions of items before they are replaced by new items in the system. Moreover, using delayed deletions, our structure supports data item expiration and content revocation: we remove expired or revoked items during the construction of some particular new DAG $G(j)$. In this case, our structure has the following important *self-correction* property that limits the window of opportunity for replay attacks:

any expired or revoked item is automatically removed from the structure the first time that the corresponding block containing the item is restructured (rebuilt). Thus, the system supports item expiration/revocation in the sense that no old item can stay forever in the system; in particular, no item can be more than $m/2$ steps old, where m is the current number of items, and depending in the exact application, items can be scheduled to leave the storage system such that no replay-attacks can be launched by the DHT.

Theorem 2. *Given an n-node DHT where resource location has $O(\log n)$ expected time and message cost, there exists a distributed authentication scheme for verifying content from an m-item data set such that: (1) The scheme uses $O(\log m)$ public reliable storage and $O(m \log m)$ distributed storage; (2) Retrieved content is verified in $O(\log m)$ time with one signature and proofs of $O(\log m)$ size computed with $O(\log n \log m)$ expected time and message cost and with uniform workload over the DHT nodes; (3) Data-item insertions have $O(\log n \log m)$ expected amortized time and message cost; (4) The scheme is resilient to content forgery and replay attacks and supports delayed data-item deletions.*

References

1. Aspnes, J., Shah, G.: Skip graphs. In: SODA, pp. 384–393. ACM, New York (2003)
2. Awerbuch, B., Scheideler, C.: Towards a scalable and robust DHT. In: SPAA, pp. 318–327. ACM, New York (2006)
3. Castro, M., Druschel, P., Ganesh, A., Rowstron, A., Wallach, D.S.: Secure routing for structured P2P overlay networks. In: OSDI, pp. 299–314. ACM, New York (2002)
4. Dabek, F., Kaashoek, M.F., Karger, D., Morris, R., Stoica, I.: Wide-area cooperative storage with CFS. In: SOSP, pp. 202–215. ACM, New York (2001)
5. Fiat, A., Saia, J.: Censorship resistant peer-to-peer content addressable networks. In: SODA, pp. 94–103. ACM, New York (2002)
6. Fiat, A., Saia, J., Young, M.: Making Chord robust to Byzantine attacks. In: Brodal, G.S., Leonardi, S. (eds.) ESA 2005. LNCS, vol. 3669, pp. 803–814. Springer, Heidelberg (2005)
7. Fu, K., Kaashoek, M.F., Mazieres, D.: Fast and secure distributed read-only file system. Transactions on Computer Systems 20(1), 1–24 (2002)
8. Goodrich, M.T., Tamassia, R., Schwerin, A.: Implementation of an authenticated dictionary with skip lists and commutative hashing. In: DISCEX 2002, p. 1068. IEEE, Los Alamitos (2001)
9. Kapadia, A., Triandopoulos, N.: Halo: High assurance locate for distributed hash tables. In: NDSS, pp. 61–79 (2008); Internet Society
10. Kothapalli, K., Scheideler, C.: Supervised peer-to-peer systems. In: I-SPAN, pp. 188–193. IEEE, Los Alamitos (2005)
11. Nambiar, A., Wright, M.: Salsa: a structured approach to large-scale anonymity. In: CCS, pp. 17–26. ACM, New York (2006)
12. Overmars, M.H.: The Design of Dynamic Data Structures, vol. 156. Springer, Heidelberg (1983)
13. Pugh, W.: Skip lists: a probabilistic alternative to balanced trees. Communications of the ACM 33(6), 668–676 (1990)

14. Ratnasamy, S., Francis, P., Handley, M., Karp, R., Shenker, S.: A scalable content-addressable network. In: SIGCOMM, pp. 161–172. ACM, New York (2001)
15. Rhea, S., Godfrey, B., Karp, B., Kubiatowicz, J., Ratnasamy, S., Shenker, S., Stoica, I., Yu, H.: OpenDHT: A public DHT service and its uses. In: SIGCOMM, pp. 73–84. ACM, New York (2005)
16. Saia, J., Fiat, A., Gribble, S.D., Karlin, A.R., Saroiu, S.: Dynamically fault-tolerant content addressable networks. In: IPTPS, pp. 270–279. Springer, Heidelberg (2002)
17. Stoica, I., Morris, R., Karger, D., Kaashoek, F., Balakrishnan, H.: Chord: A scalable P2P lookup service for Internet applications. In: SIGCOMM, pp. 149–160 (2001)
18. Tamassia, R., Triandopoulos, N.: Efficient content authentication in peer-to-peer networks. In: Katz, J., Yung, M. (eds.) ACNS 2007. LNCS, vol. 4521, pp. 354–372. Springer, Heidelberg (2007)

The Frog-Boiling Attack:
Limitations of Anomaly Detection for Secure Network
Coordinate Systems

Eric Chan-Tin, Daniel Feldman, Nicholas Hopper, and Yongdae Kim

University of Minnesota
{dchantin,feldman,hopper,kyd}@cs.umn.edu

Abstract. A network coordinate system assigns Euclidean "virtual" coordinates to every node in a network to allow easy estimation of network latency between pairs of nodes that have never contacted each other. These systems have been implemented in a variety of applications, most notably the popular Azureus/Vuze BitTorrent client. Zage and Nita-Rotaru (CCS 2007) and independently, Kaafar et al. (SIGCOMM 2007), demonstrated that several widely-cited network coordinate systems are prone to simple attacks, and proposed mechanisms to defeat these attacks using outlier detection to filter out adversarial inputs. We propose a new attack, Frog-Boiling, that defeats anomaly-detection based defenses in the context of network coordinate systems, and demonstrate empirically that Frog-Boiling is more disruptive than the previously known attacks. Our results suggest that a new approach is needed to solve this problem: outlier detection alone cannot be used to secure network coordinate systems.

Keywords: Vivaldi, Anomaly Detection, Network Coordinate Systems.

1 Introduction

Network coordinate systems assign virtual coordinates to every node in a network. These coordinates allow efficient estimation of the latency between any pair of nodes in the network: instead of directly measuring the $O(n^2)$ pairwise latencies, each of the n nodes computes its coordinates based on the round-trip time to a few other nodes and their coordinates, greatly reducing the communication costs. Several possible uses of network coordinate systems include choosing peers to download from in a filesharing network [1], choosing peers for routing in a DHT [2], or finding the closest node in a content-distribution network. A popular BitTorrent client, Azureus (now called Vuze [3]), is currently using a network coordinate system to prioritize lookups based on network distance and to find closer nodes [4].

There have been several network coordinate systems proposed in the literature; these schemes can be categorized into centralized or "landmark"-based systems [1,5,6] that depend on a small set of "trusted" nodes, and decentralized systems [7,8]. A widely-implemented and studied example of decentralized coordinate systems is Vivaldi [7], which has been shown to produce accurate estimations and converge quickly under various network conditions. Although it is decentralized, Vivaldi can be easily disrupted

Y. Chen et al. (Eds.): SecureComm 2009, LNICST 19, pp. 448–458, 2009.

by spurious or malicious nodes, rendering the network coordinate system useless and impractical since the nodes never reach a stable coordinate. Zage and Nita-Rotaru [9] proposed a mechanism, based on real-time statistical analysis of nodes' coordinates, to detect and discard adversarial inputs. A similar mechanism was proposed and evaluated by Kaafar *et al.* [10]. Both methods rely on outlier detection using statistical models – respectively, the Mahalanobis distance and Kalman filters – of coordinate evolution.

In this paper, we demonstrate the inherent challenge in designing a secure network coordinate system using outlier detection. We propose the *Frog-Boiling attack*, where an adversary disrupts the network while consistently operating within the threshold of outlier detection. This is analogous to the popular account that a frog put in hot water will quickly jump out but a frog placed in cold water that is gradually brought to a boil will not notice the change and boil to death. The adversary sends "small-step" fake updates (fake RTTs or self-reported error or coordinate)[1] to nodes in the network. The "step" is small enough that it does not trigger the anomaly detection but the nodes attacked are still affected. Thus, the coordinates of the nodes in the attacked network quickly become very different from the coordinates of the same nodes in the original network. The effectiveness of the attack can also be significantly increased when conducted in conjunction with a *Sybil* attack.

We implement, and empirically evaluate, three variants of the Frog-Boiling attack to demonstrate its effectiveness against outlier-detection based defenses. All three attacks rely on a simple concept: lying can be harmful but telling consistent, believable lies is even more harmful. Our evaluation on a PlanetLab deployment of Vivaldi shows that even the basic frog-boiling attack is *more disruptive* against the security mechanism proposed in [9] than the attacks they defend against. In particular, with only 5% of attackers in the network, Frog-boiling causes a median relative error of 0.28 after two hours and 0.57 after 14. The same network with no attackers has a median relative error of 0.11, and under Zage and Nita-Rotaru's "random" attack, the insecure coordinate scheme has a maximum median relative error of 0.22, even when the fraction of attackers is above 10%. Thus the outlier detection mechanism is completely ineffective against frog-boiling. We note that while the step size of the attack is small – nodes are pushed "little by little" – the result of the attack is neither slow nor small, resulting in similar errors just as quickly as previously known attacks but causing greater damage over time. See Section 5.3 for more details.

While similar attacks on outlier detection mechanisms appear in the literature (including [12, 13]), to our knowledge we are the first to demonstrate the effectiveness of frog-boiling in the context of network coordinate systems. Furthermore, we demonstrate that the attacks are *more disruptive* than previous work and are completely unmitigated by the existing approaches to securing network coordinate systems. These results suggest that new approaches and/or stronger assumptions are needed to construct secure network coordinate systems.

The remainder of the paper is organized as follows. We give a brief background on network coordinate systems, existing attacks, and the outlier detection mechanisms in Section 2. A detailed description of the attacks outlined above is given in Section 3.

---

[1] This is possible since updates are usually done via the application level, and an adversary can easily delay or hasten [11] replies.

The evaluations of our experiments on a wide area network are shown in Section 4 and Section 5. Finally, we conclude in Section 6.

2 Background

2.1 Network Coordinate Systems

The first network coordinate systems developed were centralized – trusted infrastructure nodes compute coordinates for all other nodes. Centralized systems typically require a significant fraction of all network nodes to act as trusted servers, which is not possible for large networks. Centralized network coordinate systems include IDMaps [6], GNP [1] and NPS [5].

To improve the ease of deployment of network coordinate systems, decentralized network coordinate systems were introduced. A decentralized network coordinate system has no infrastructure nodes. Instead, normal nodes pick peers out of the set of all nodes, and compute their own coordinates with respect to those peers only. Finding potential peers is delegated to the underlying network. Decentralized network coordinate systems are attractive for P2P applications, since they can be deployed alongside the client software. Moreover, decentralized network coordinate systems are scalable as there are no centralized servers which could become overloaded.

Vivaldi. Vivaldi [7] is a decentralized network coordinate system. It is based on a spring model. Its behavior is analogous to a physical model made of springs and balls, in which each ball represents a network node and the spring connecting any two balls is longer when the latency between those nodes is larger. Over time, such a model reaches a stable equilibrium. A Vivaldi node begins by selecting an arbitrary set of peers, and sets its initial coordinate to the origin. It then begins an iterative algorithm that pulls it closer to peers with lower latencies, and pushes it away from peers with higher latencies. After many iterations, the coordinate system reaches an equilibrium, and subsequent changes are due only to the changing latency between nodes. Each node will pick 64 other nodes in its reference set – 32 nodes are "close" and 32 nodes are "far". On each iteration, a Vivaldi node sends a probe packet (which could be piggybacked on top of application-level messages) to each of its peers. It receives a response to each probe packet containing the peer's current coordinate and self-reported error estimate (can also be piggybacked on top of application-level messages), and learns its latency to that peer from the RTT of the transaction. It then computes a new position that is closer to the peer if the estimated latency is too large, and farther from the peer if the estimated latency is too small. Vivaldi's coordinate system is n-dimensional. It was shown in [7] that 2 dimensions plus height work well for most cases. Moreover, Vivaldi boasts a low convergence time, a low reported error, and an accurate mapping of the virtual coordinate network. Vivaldi also deals well with *churn* – the constant change in membership of a P2P network due to its public nature – because of its low convergence time. However, Vivaldi was not designed for an adversarial environment and it is simple for an attacker to disrupt the whole network.

Pyxida. Pyxida [14] implements a virtual coordinate network. It is being used in both academia and commercially – to track the coordinates of all the PlanetLab [15] nodes; in the Azureus [3] BitTorrent client; and to study selfish neighbor selection in P2P networks [16]. It is designed to work on a P2P network and implements the Vivaldi algorithm. Pyxida coordinates use 4 dimensions plus height. Moreover, it is open-source, enabling easy modification to implement the countermeasures and attacks. We used Pyxida in our experiments since it implements the Vivaldi algorithm, provides a stable network coordinate system, and has been used in a large-scale deployment [17]. A detailed description of Pyxida is given in [18].

2.2 Existing Attacks

Several attacks have been proposed [19, 10, 9]. They are the *Disorder attack*, *Repulsion attack*, *Colluding Isolation attack*, *Inflation/Deflation attack*, and the *Oscillation attack*. The Repulsion and Colluding Isolation attacker sends the same coordinates each time in an attempt to move the victim nodes to some coordinate space. The other attacks consist of the attacker reporting random coordinates and a low error. The reader is referred to those papers for a more detailed description of the attacks.

2.3 Countermeasures

Several mechanisms, based on outlier detection, have recently been proposed to secure network coordinate systems.

Kalman Filter. Kaafar *et al.* [10] propose to implement a Kalman filter [20] to detect outlier hosts in the network, that is, hosts that are lying or behaving strangely. The Kalman filter works by comparing the previous trajectory of a node's coordinates with its coordinates after an update. If the distance between the expected coordinates and the update is larger than the threshold for the Kalman filter, then the update is rejected. The authors estimate that in order to resist the disorder attack, about 10% of the network must be trusted "surveyor" nodes.

Mahalanobis Distance. Zage *et al.* [9] proposed a countermeasure that uses two statistical filters to ignore peers that report unusually large or rapidly changing coordinates. The first filter is called the *spatial filter*, while the second is called the *temporal filter*. Each node applies both filters to incoming data from its peers, and discards data that do not pass both filters. The Mahalanobis outlier detection function used by the spatial filter determines if the new spatial vector falls inside an ellipsoid defined by previously-seen vectors. The temporal filter looks at the change in the last iteration. Since the data set is much larger, a constant-time and constant-space but slightly less accurate variant of the Mahalanobis function is used for this filter. Since the cost of a false positive is small, nodes can afford to set their thresholds very low. However, if the thresholds are too low, nodes will only accept data points that fit into a small range, leading to inaccurate coordinates. To our knowledge, the correct choice of thresholds to maximize security vs correctness has not been studied. When a peer's data fails either the spatial or temporal filter, there are two consequences. First, that peer's data is not used to update the node's current coordinate. Second, that peer's data is not used as history for the filters in the next iteration. However, there is no permanent blacklist of nodes which failed the filters. For a more detailed description, see [9].

In this paper, we attack Pyxida with Mahalanobis distance-based outlier detection. However, because the Kalman filter approach also features a threshold region in which updates will be accepted (and incorporated into the filter) we do not expect the Kalman filter to offer any significant defense against frog-boiling.

3   Proposed Attacks

Recall that the ellipsoid used to determine whether a new data point falls within acceptable bounds has axes with lengths that are multiples of the variances of the variables used in each filter. New data points are accepted if they fall inside this ellipsoid, and rejected otherwise. This mechanism correctly identifies a small number of spurious nodes that return random coordinates with low error. Since correctly operating nodes are unlikely to change coordinates much faster than average while still reporting low error, nodes that do so must be spurious.

However, an intelligent adversary can send "random" data points that still fall inside the Mahalanobis ellipsoid. Thus, the data points will be accepted although they are "wrong". We call this approach the *Frog-Boiling attack*. If the adversary lies too much, its peers won't accept its updates. If it lies too little, the attack won't succeed in disrupting the network. The Frog-Boiling attack can be used to disrupt the whole network by continuously lying to all the nodes.

As a simple example, assume there are only two nodes A and B in the network and they have converged to stable coordinates. An attacker node C is introduced and obtains its coordinates from both A and B. However, each time C receives a request (say from A), it replies with $Coord_C = Coord_C + \delta$, where δ is a small offset. For example, if its coordinates in 2-dimensions (Pyxida uses 4-dimensions with height) are $(120, 100)$, the reported coordinate will be $(120.5, 100.5)$. Since the coordinate reported is not outside of the Mahalanobis thresholds, A will accept the coordinate and update its own coordinate accordingly. Then whenever B queries A, the response will be a coordinate that is slightly higher than what the "real" coordinate should have been. Thus, B's coordinate changes slightly as well. This process continues with the attacker continuously lying in small increments about its own coordinate. This whole process might just shift the coordinates, but not affect the estimated distance between any two nodes. Thus a targeted attack can be performed and as we show in Section 5, our attack effectively renders the network coordinate ineffective.

The targeted frog-boiling attack works as follows. The attacker attempts to move some victim nodes (a fraction of the whole network) to some arbitrary network coordinates. The targeted location in this case is far from the rest of the network. Although those nodes can still communicate with the rest of the network, they will not be able to calculate a correct coordinate for themselves and will report a "false" coordinate and error to the rest of the network. The Mahalanobis distance will flag those nodes as outliers and will not accept their updates. This effectively isolates the victim nodes from the rest of the network.

One way of performing this attack is for the attacker to consistently report its coordinates to the victim nodes so that the latter end up to coordinate space A. Note that the attacker will not be able to pull the victim nodes all the way to A, but the victims will be closer to A than the rest of the network. This is because, although the rest of the network might not accept updates from the victim nodes, the latter will still accept updates from the rest of the network. Thus, the victims are pushed to A by the attacker but also pulled back to the rest of the network. The success of the attack is for the attacker nodes to exert a greater force on the victim nodes than the rest of the network.

In this paper we evaluate three variants of this attack against Zage and Nita-Rotaru's secure network coordinate system. All three attacks rely on the same concept of consistently and progressively lying:

- The **Basic-Targeted** attack is as described above.
- The **Network-Partition** attack is an extension of the previous attack, where the whole network is partitioned into two subnetworks or clusters.
- The **Closest-Node** attacker tries to become the closest node (in terms of coordinate space) to the victim nodes. Becoming the closest node might not be important by itself. However, if the network coordinate system is used with an application such as in Azureus, then the closest node could be used to initiate file transfer. If the attacker becomes the closest node to a victim node, it will then be the first node that the victim contacts for a file. This can have various implications such as preventing any node in a file-sharing network from being able to download a file. This attack is performed in a similar way to the targeted attack. Instead of pulling the victim node to a certain coordinate space, the attacker pushes itself close to the victim node. One way of doing this is for the attacker (after learning the victim's coordinate) to report its network coordinates as being very close to that of the victim's.

4 Experimental Setup

To evaluate the impact of our attacks on a secure network coordinate system, we deployed a standalone Pyxida service (see Section 2) on PlanetLab [15]. Since the original Pyxida code implements the basic Vivaldi coordinate system, the Mahalanobis distance outlier detection mechanism proposed in [9] was added to the Pyxida code using a third-party library [21].

We made some small modifications to Pyxida before deploying it. The neighbor list was modified to contain a maximum of 32 nodes (due to an estimated PlanetLab network size of 400). We used 50 nodes as the common "bootstrap" nodes, that is, all the Pyxida nodes contact those nodes when they first start. We wait until the network stabilizes before introducing any adversaries in the network.

The metric we used is the median relative error (henceforth just called error). It is calculated as $\frac{|RTT_{estimated}-RTT_{actual}|}{RTT_{actual}}$, where RTT_{actual} is the actual RTT between two nodes and $RTT_{estimated}$ is the RTT obtained by taking the difference in the coordinates of the two nodes. The lower this number is, the more accurate the network coordinate system is (each node believes it has the right coordinate). This is the same metric used in various other papers [9, 17, 18].

We use both a spatial and temporal threshold of 5 for our experiments. The network starts to stabilize after only 2 hours, indicating a low convergence time. The median relative error was 0.1. The attackers join the network at time 2 hours. The experiments for determining the best thresholds, as well as the other metrics used (such as relative rank loss [22]), will be described in the full version of this paper. We note that most of the experiments were also performed using a simulated network to verify implementation correctness. The results of these simulations are consistent with experimental results and are thus omitted due to space constraints.

5   Attack Evaluations

5.1   Previous Attacks

To establish a baseline for comparison with the effectiveness of our attacks, we implemented the previously proposed "coordinate oscillation" attack [9] (in which attacker nodes report completely random coordinates with low relative error) and measure the performance of the attack against our Pyxida deployment (without the Mahalanobis distance filter). The progress over time of the median relative error with 11% attacker nodes is shown below.

Time (mins)	100	250	500	750	1000
Relative Error	0.23	0.21	0.23	0.22	0.2

5.2   Basic-Targeted Attack

The Basic-Targeted attacker targets a victim node and attempts to change the victim's coordinate in small steps. We attempt to change the coordinate of the victim nodes to be $Loc_T = (2000, 2000, 2000, 2000)$ with height 2000. Initially, for each victim node (say coordinate C), the attacker node will report its coordinate to be $C' = C + \delta$. For each subsequent time that victim node contacts our attacker node, the latter reports its coordinate as $C'' = C' + \delta$, until $C'' = Loc_T$. Thus, the victim's coordinate is moved in small steps to the target coordinate.

Recall from Section 2 that a Pyxida node only updates its coordinate when it has sent a "ping" request. Thus, the victim nodes have to contact the attacker nodes for the attack to work. With 10% of attackers, the victim will contact one attacker node 10% of the time. Once an attacker node becomes a neighbor of the victim, it will stay in the neighbor's list for at least the next 32 iterations, which is long enough for another attacker to be contacted and added to the list. The probability of an attacker node being part of the neighbor list after 32 iterations is $1 - 0.9^{32} = 96.5\%$. Thus, there is a very high probability that a victim node will have at least one attacker node in its neighbor list. Recall that the neighbor list is used every 10 seconds in Pyxida to calculate the current force. Since the attacker is updating its coordinate to be closer to the target coordinate at each time step, the victim will thus go closer to the target coordinate progressively. The Mahalanobis distance does not work in this case because the attacker is within the thresholds (since δ is small). The attacker only attacks the victim nodes and does not respond to other nodes in the network. Since there is no gossiping in Pyxida, this does not affect the attack.

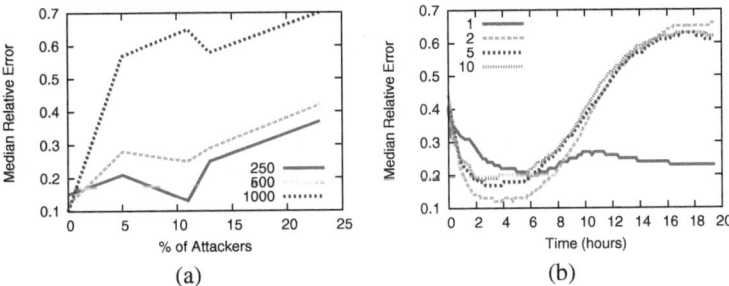

Fig. 1. The average median relative error for (a) varying % of attackers at different timestamps, (b) the targeted nodes with 11% of attackers over time and with different values of δ

Figure 1(a) shows the error with varying percentage of attackers. (We note that 20% of attacker nodes may seem high, but many of the applications that implement network coordinate systems are vulnerable to Sybil attacks that make it trivial to control a large fraction of the nodes) The different lines show the error at different times (250 minutes, 500 minutes, and 1000 minutes). Adding more adversaries significantly increases the error (by more than 100% with only 11% of attackers). The error is increased from 0.12 with no attackers to 0.25 with 11% of attackers, an increase of 108%. After 1000 minutes (a little over 16 hours), it can be seen that the network coordinate is unusable even with only 5% of the network being malicious – the error is greater than 0.5.

The frog-boiling attack on the secure network coordinate system is as effective as a random attack on the original network coordinate system. At time 500 minutes, the error for the random attack is 0.23 while the error for the frog-boiling attack is 0.25 with 11% of attackers. This means that the Mahalanobis distance does not provide any extra protection to a network coordinate system. This reinforces our belief that an outlier detection system is not suitable to secure a network coordinate system.

5.3   Aggressive Frog-Boiling

Our attack works by moving the victims in small steps to some coordinate. In the previous section, the step size δ was $2ms$. In this section, we varied the value of δ to test the effect of a more aggressive attack, which will produce an impact on the network earlier – in other terms, we show how quickly our attack can have an impact on the network. Figure 1(b) shows the error with 11% of attackers in the network. The different lines show the different δ values used – 1, 2, 5, and 10. With δ equal to 1 and 2, the error stays the same until time 6 hours, so it take 4 hours for the attack to start having an effect. On the other hand, with δ equal to 5 or 10, the relative error starts to increase at time 4 hours – after only 2 hours, the victim's network coordinates start to be disrupted. Thus, out attack is fast and efficient.

5.4   Network-Partition Attack

The Network-Partition attack is similar to the Basic-Targeted attack. Instead of just moving the victim nodes (*Network1*) to some far-away coordinate, the rest of the network

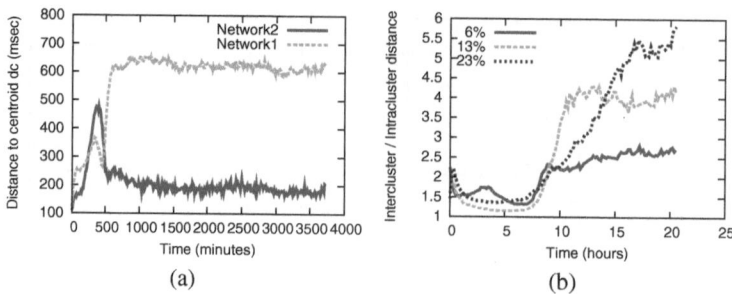

Fig. 2. (a) The coordinate distance to the centroid and (b) the intercluster / intracluster ratio for the Network-Partition attack

(*Network2*) is also moved to some other location. This effectively partitions the network into two subnetworks. The targeted coordinate for Network1 was set to $P_1 = (1000, 1000, 1000, 1000)$ with height 1000 and the targeted coordinate for Network2 was set to $P_2 = (-1000, -1000, -1000, -1000)$ with height -1000.

In our experiment, 6% of the nodes were adversaries, 37% of the network was assigned to Network1 and 57% of the network was assigned to Network2. Figure 2(a) shows the distance to the origin of the network for Network1 and Network2. At the beginning, the two clusters are close together. At time 500 minutes, which is how long it takes for the attack to have an effect, the two networks start to diverge. Network1 is pushed toward P_1 while Network2 is pushed toward P_2. Since the two clusters continue to exert some pull on each other, the intended coordinates are not reached, but the network is still effectively partitioned.

Figure 2(b) shows the ratio of the intercluster distance to the intracluster distance. The intercluster distance is the average of the distance from Network1 to the centroid of Network2 and the distance from Network2 to the centroid of Network1. The intracluster distance is the average of all the nodes in a cluster to the centroid of that cluster. The ratio shows how far apart the two clusters are moving from each other. The figure shows that over time, the two networks are getting pulled further apart from each other. The different lines show different fractions of attackers. This shows that our attack effectively partitions the whole network into two smaller networks far apart from each other. We note that this attack could easily be extended to support partitioning into an arbitrary (constant) number of clusters with arbitrary membership ratios.

5.5   Closest-Node Attack

An adversary tries to become the closest node (in terms of coordinate space) to a victim in the Closest-Node attack. The attacker node queries the victim nodes constantly to obtain their coordinates. When a victim node queries the attacker node, it will reply back with that victim node's coordinate $+\delta$. The attacker node does not reply to other nodes in the network. We took a snapshot at 500 minutes and determine how many times one of the attacker nodes was reported as being the closest neighbor of a victim node (this reporting is done every 10 minutes). With only 11% of attackers, we find that an attacker is able to become the closest neighbor to a victim node 41% of the time.

6   Conclusion

A stable, decentralized network coordinate system could potentially provide a beneficial service for many Internet applications. However, existing systems provide no protection against malicious participants: even a single adversary can cause the entire coordinate system to fail. The apparent solution to such a dilemma is to add an anomaly detection mechanism to the coordinate system. Previous studies have shown that such a mechanism can prevent adversaries from disrupting the network. However, protection against more complicated adversaries is fraught with difficulty.

Consider a node in a network coordinate system that has some outlier detection mechanism. In order for the node to determine its coordinates, it must learn about the coordinates of its peers – it must accept some updates. The range of updates it accepts must be based on recent history, since network topologies and conditions vary widely. However, under these two assumptions an adversary can slowly expand the range of data accepted by the node by influencing the node's recent history. We call this attack the *Frog-Boiling* attack. In this paper we have introduced three variants of the frog-boiling attack and empirically demonstrated that the attack effectively disrupts the Vivaldi network coordinate system to a greater extent than previous attacks, and that the attack is completely unmitigated by Mahalanobis distance-based outlier detection. There is no reason to believe that Frog-Boiling would not be equally effective against Kalman filter-based outlier detection; we leave the evaluation of this claim for future work.

The task of securing a distributed network coordinate system against adversaries seems very challenging. The problem is that the current distributed network coordinate system mechanisms (secure or not) rely only on a node's local view of the network. Because of this, it is a challenge for a node to know whether a reported coordinate and RTT is correct or faked. Thus, a secure network coordinate system will need to provide some mechanism to verify a node's reported coordinates and/or RTTs. The success of the Frog-Boiling attack demonstrates that outlier detection is not a secure mechanism to provide this service. Recent work based on reputation or trust mechanisms [23, 24] may provide an alternative approach, but the difficulty of constructing secure reputation systems suggests that these schemes will also require careful evaluation.

Acknowledgments. We thank Jonathan Ledlie and Peter Pietzuch for their help with Pyxida, and Eugene Vasserman for pointing out the analogy to "boiling a frog." This work was supported by the NSF under grant CNS-0716025. No frogs were harmed in the writing of this paper.

References

1. Ng, T.S.E., Zhang, H.: Predicting Internet Network Distance with Coordinates-Based Approaches. In: Proceedings of IEEE, INFOCOM (2002)
2. Dabek, F., Li, J., Sit, E., Robertson, J., Kaashoek, M.F., Morris, R.: Designing a DHT for low latency and high throughput. In: Proceedings of the 1st USENIX Symposium on Networked Systems Design and Implementation, NSDI (2004)

3. Azureus, `http://azureus.sourceforge.net`
4. Vuze Forums, `http://forum.vuze.com/thread.jspa?threadID=80764`
5. Ng, T.S.E., Zhang, H.: A network positioning system for the internet. In: Proceedings of the USENIX annual technical conference (2004)
6. Francis, P., Jamin, S., Jin, C., Jin, Y., Raz, D., Shavitt, Y., Zhang, L.: IDMaps: A Global Internet Host Distance Estimation Service. IEEE/ACM Trans. Netw. 9(5), 525–540 (2001)
7. Dabek, F., Cox, R., Kaashoek, F., Morris, R.: Vivaldi: A Decentralized Network Coordinate System. In: Proceedings of ACM SIGCOMM (2004)
8. Costa, M., Castro, M., Rowstron, A., Key, P.: PIC: Practical Internet Coordinates for Distance Estimation. In: Proceedings of the IEEE International Conference on Distributed Computing Systems (ICDCS) (2004)
9. Zage, D.J., Nita-Rotaru, C.: On the accuracy of decentralized virtual coordinate systems in adversarial networks. In: Proceedings of the 14th ACM conference on Computer and communications security, CCS (2007)
10. Kaafar, M.A., Mathy, L., Barakat, C., Salamatian, K., Turletti, T., Dabbous, W.: Securing Internet Coordinate Embedding Systems. In: Proceedings of ACM SIGCOMM (2007)
11. Su, A.J., Choffnes, D.R., Kuzmanovic, A., Bustamante, F.E.: Drafting Behind Akamai (Travelocity-Based Detouring). In: Proceedings of ACM SIGCOMM (2006)
12. Denning, D.E.: An Intrusion-Detection Model. IEEE Transactions on Software Engineering SE-13(2) (1987)
13. Barreno, M., Nelson, B., Sears, R., Joseph, A.D., Tygar, J.D.: Can Machine Learning Be Secure? In: Proceedings of the ACM Symposium on InformAtion, Computer and Communications Security, ASIACCS (2006)
14. Pyxida, `http://pyxida.sourceforge.net`
15. PlanetLab, `http://planet-lab.org`
16. Selfish Neighbor Selection, `http://csr.bu.edu/sns`
17. Ledlie, J., Pietzuch, P., Seltzer, M.: Network coordinates in the wild. In: Proceedings of the USENIX Symposium on Networked Systems Design and Implementation, NSDI (2007)
18. Ledlie, J., Pietzuch, P., Seltzer, M.: Stable and accurate network coordinates. In: Proceedings of the IEEE International Conference on Distributed Computing Systems (ICDCS) (2006)
19. Kaafar, M.A., Mathy, L., Turletti, T., Dabbous, W.: Real attacks on virtual networks: Vivaldi out of tune. In: Proceedings of the SIGCOMM workshop on Large-scale Attack Defense (2006)
20. Kalman, R.E.: A new approach to linear filtering and prediction problems. Transactions of the ASME–Journal of Basic Engineering 82(Series D), 35–45 (1960)
21. CommonSense,
`http://www.kimvdlinde.com/professional/programming/`
`statistics/commonSense/body.html`
22. Lua, E.K., Griffin, T., Pias, M., Zheng, H., Crowcroft, J.: On the Accuracy of Embeddings for Internet Coordinate Systems. In: Proceedings of ACM SIGCOMM-Usenix Internet Measurement Conference, IMC (2005)
23. Sherr, M., Blaze, M., Loo, B.T.: Veracity: Practical Secure Network Coordinates via Vote-based Agreements. In: USENIX Annual Technical Conference (2009)
24. Zhao, X., Lua, E.K., Chen, Y., Song, X., Deng, B., Li, X.: Sniper: Social-link Defense for Network Coordinate Systems. IEEE Conference on Computer Communications (INFOCOM) (2009); poster

Author Index